Lecture Notes in Computer Sc

Edited by G. Goos and J. Hartmanis

R. S. Bird C. C. Morgan
J. C. P. Woodcock (Eds.)

Mathematics of Program Construction

Second International Conference,
Oxford, U.K., June 29 - July 3, 1992
Proceedings

Springer-Verlag
Berlin Heidelberg New York
London Paris Tokyo
Hong Kong Barcelona
Budapest

Series Editors

Gerhard Goos
Universität Karlsruhe
Postfach 69 80
Vincenz-Priessnitz-Straße 1
W-7500 Karlsruhe, FRG

Juris Hartmanis
Cornell University
Department of Computer Science
4130 Upson Hall
Ithaca, NY 14853, USA

Volume Editors

Richard S. Bird
C. Carroll Morgan
James C. P. Woodcock
Oxford University Computing Laboratory, Programming Research Group
11 Keble Road, Oxford OX1 3QD, U.K.

CR Subject Classification (1991): D.1-2, F.2-4, G.2

ISBN 3-540-56625-2 Springer-Verlag Berlin Heidelberg New York
ISBN 0-387-56625-2 Springer-Verlag New York Berlin Heidelberg

Printed in Germany

Typesetting: Camera ready by author/editor
Printing and binding: Druckhaus Beltz, Hemsbach/Bergstr.
45/3140-543210 - Printed on acid-free paper

Preface

Not very long ago, the uninhibited use of mathematics in the development of software was regarded as something academics should do amongst themselves in private. Today, there is more and more interest from industry in formal methods based on mathematics. This interest has come from the success of a number of experiments on real industrial applications (see, for example, *LNCS*, Vol. 551). Thus, there is not only a belief, but also evidence, that the study of computer programs as mathematical objects leads to more efficient methods for constructing them. However, if we are to be of service to those actually creating computing systems in industry, we must extend and improve our work.

The papers in this volume were presented at the Second International Conference on the *Mathematics of Program Construction*, held at St Catherine's College, Oxford, during the week of 29 June – 3 July, 1992. The conference was organised by Oxford University Programming Research Group, and continued the theme set by the first—the use of crisp, clear mathematics in the discovery and design of algorithms. In this second conference, we see evidence of the ever-widening impact of precise mathematical methods in program development. There are papers applying mathematics not only to sequential programs, but also to parallel and oncurrent applications, real-time and reactive systems, and to designs realised directly in hardware.

The scientific programme for the conference consisted of five invited lectures delivered by distinguished researchers, a further 17 papers selected by the programme committee, and six *ad hoc* contributions presented on the final day. These were as follows:

A Short Problem
J.L.A. van de Snepscheut

Compiler Verification
Greg Nelson

An Alternative Derivation of a Binary Heap Construction Function
Lex Augusteijn

A Derivation of Huffman's Algorithm
Rob R. Hoogerwoord

Galois Connexions
Roland Backhouse

An Elegant Solution
J.L.A. van de Snepscheut

A record of Augusteijn's and Hoogerwoord's contributions may be found at the end of this volume.

Acknowledgments

The conference received sponsorship from BP Research and Prentice Hall International. The administration was provided by the Continuing Professional Development Centre of the Department for Continuing Education, Oxford University.

I am most grateful to Miss Frances Page for her expert assistance.

Oxford, February 1993 J.C.P. Woodcock

Organising Committee

R.S. Bird, C.C. Morgan, J.C.P. Woodcock

Programme Committee

J.-R. Abrial (Paris)
E. Astesiano (Genova)
R.-J.R. Back (Åbo)
R.S. Bird (Oxford)
W.H.J. Feijen (Eindhoven)
E.C.R. Hehner (Toronto)
L.G.L.T. Meertens (CWI)
B. Möller (Augsburg)
C.C. Morgan (Oxford)
J.M. Morris (Glasgow)
B. Nordstrom (Böteborg)
G. Nelson (DEC SRC)
E.-R. Olderog (Oldenburg)
B. Ritchie (RAL)
D. Sannella (Edinburgh)
M. Sheeran (Glasgow)
J.L.A. van de Snepscheut (CalTech)
W.M. Turski (Warsaw)
J.C.P. Woodcock (Oxford)

Table of Contents

Invited Lectures

Contributed Lectures

Additional Contributions

On the Economy of doing Mathematics

prof. dr. Edsger W. Dijkstra

Department of Computer Sciences
The University of Texas at Austin, U.S.A.

Every honest scientist regularly does some soul searching; he then tries to analyse how well he is doing and if his progress is still towards his original goal. Two years ago, at the previous conference on the Mathematics of Program Construction, I observed some of such soul searching among the scientists there present, and I observed more doubt than faith. The first part of this talk is therefore devoted to my explanation of why I think most of those doubts unjustified.

One general remark about an unexpected similarity between the programming community and the mathematical community first. In the past, art and science of programming have seriously suffered from the concept of "the average programmer" and the widespread feeling that his (severe) intellectual limitations should be respected. Industrial pressure and propaganda promoted the view that, with the proper programming language, programming did not pose a major challenge and did not require mental gifts or education, thus creating an atmosphere in which self-censorship withheld many a scientist from exploring the more serious forms of programming methodology. I would like to point out that, in very much the same way, mathematics has suffered and still suffers from the tyranny of "the average mathematician". Improvements in notation, in language or concepts are rejected because such deviations from the hallowed tradition would only confuse the average mathematician, and today's mathematical educators who think that they are doing something great by turning a piece of mathematics into a video game are as misguided as earlier recruiters for the programming profession. The similarity is striking, the only difference between the average mathematician possibly being that the latter tends to believe that he is brighter than everybody else. And this concludes my general introductory remark.

* * *

Let us now have a look at various supposed reasons for losing faith in mathematical construction of programs.

1. "Formal program derivation is no good because industry rejects it." Well, what else do you expect? For four decades, the computing industry has been guided by businessmen, bean-counters, and an occasional electronic engineer. They have carefully seen to it that computing science had no influence whatsoever on their product line, and now the computer industry is in problems, computing science gets the blame. (Ironically, many CS departments now feel "guilty" and obliged to turn out more "practical" graduates, i.e. people with the attitudes that got the computer industry into trouble in the first place.) Having put the blame on our doorstep, they have "justified" rejecting our work, and they have to continue to do so so as to save their own image.

Extended Calculus of Constructions as a specification language

Rod Burstall

Department of Computer Science
University of Edinburgh
Edinburgh
Scotland

Abstract

Huet and Coquand's Calculus of Constructions, an implementation of type theory, was extended by Luo with sigma types, a type of pairs where the type of the second component depends on the value of the first one. This calculus has been implemented as 'Lego' by Pollack. The system and documentation is obtainable thus:

```
ftp ftp.dcs.ed.ac.uk
cd export/lego ,
```

after which one should read the file **README**.

The sigma types enable one to give a compact description of abstract mathematical structures such as 'group', to build more concrete structures of them, such as 'the group of integers under addition' and to check that the concrete structure is indeed an instance of the abstract one. The trick is that the concrete structure includes as components proofs that it satisfies the axioms of the abstract structures. So 'group' is a sigma type and 'group of integers under addition' is an n-tuple of types, operators and proofs which is an element of this sigmatype. We can define functions which enrich such structures or forget them to simpler ones.

However the calculus is intentional and it is too restrictive to identify the mathematical notion of 'set' with 'type'. We will discuss how sets and functions between them may be represented and the notational difficulties which arise. We put forward a tentative suggestion as to how these difficulties might be overcome by defining the category of sets and functions in the calculus, then using the internal language of that category to extend the type theory. This would be a particular example of a more general reflection principle.

In short: when industry rejects formal methods, it is not because formal methods are no good, but because industry is no better.

2. "Formal methods are no good because the practitioner rejects them." This may sound convincing to you, but only if you don't know the practitioners. My eyes were opened more than 20 years ago on a lecturing trip to Paris with, on the way back, a performance in Brussels. It was the first time I showed to foreign audiences how to write programs that were intended to be correct by construction. In Paris, my talk was very well received, half the audience getting about as excited as I was; in Brussels, however, the talk fell flat on its face: it was a complete flop. I was addressing the people at a large software house, and in my innocence had expected them to be interested in a way of designing programs such that one could keep them firmly under one's intellectual control. The management of the company, which derived its financial stability from its maintenance contracts, felt the design of flawless software not to be in the company's interest, so they rejected my recommendations for sound commercial reasons. The programmers, to my surprise, rejected them too: for them, programming was tedious, but debugging was fun! It turned out that they derived the major part of their professional excitement from *not* quite understanding what they were doing and from chasing the bugs that should not have been introduced in the first place.
In short: formal techniques are rejected by those practitioners that are hackers instead of professionals.

3. "Formal methods are no good because quite a few colleagues in your own CS department are opposed to them." This sounds like a much more serious objection, but fades away as soon as you know the reasons why your colleague objects. The one main reason is that, deep in his heart, your colleague is a hacker, that by advocating formal techniques you are promoting quality standards he cannot meet, and that, consequently, he feels discriminated against. Such a colleague should be pitied instead of taken seriously. The other main reason is political. Once growth of the Department has been accepted as the target, one can no longer afford to present computing as a serious science, the mastery of which is sufficiently demanding to scare away the average student. Such a political colleague has adopted the morals of the best–seller society, and should be despised instead of taken seriously.

4. "Formal methods are no good, for even mathematicians have not accepted them." This sounds really serious, for mathematics embodies an intellectual tradition of thousands of years. Closer inspection shows that that is precisely one of its problems. While the Renaissance sent all of classical "science" to the rubbish heap, much of classical mathematics was accepted as the standard. This is one of the explanations of the fact that, to this very day, the Mathematical Guild is much more medieval than, say, the Physical Guild, and that informality — not to say, handwaving — is the hallmark of the Mathematical Guild Member: frantically trying to distinguish between form and contents, he neglects the form and thus obscures the contents. Never having given formal techniques, i.e. the manipulation of uninterpreted formulae, a fair chance, he does not have the right to say that they are no good.

This may be the place to point out that most programmers and most mathematicians work in linguistically comparable situations. The programmer uses symbols of a "programming language", the mathematician uses symbols of a "reasoning language". For the sake of compatibility, both languages are accepted as *de facto* standards, although both are bad, and totally defective as carriers of a formal system. These people do, indeed, work in an environment in which formal techniques are hard to apply directly, but this is the consequence of well-identified shortcomings of the "languages" they have accepted as standards.

* * *

Suddenly having to run a children's party for kids ranging from 8 to 13 years of age, I gave them pencil and paper and asked them the well-known question whether it is possible to cover with 31 2x1 dominoes the 8x8 square from which 2 unit squares at opposite corners have been removed. They all tried doing it and failed each time. After about 20 minutes, the consensus was that it could not be done, but they agreed that impossibility could not be concluded from a sample of failed efforts. They also agreed that neither the party nor their patience would suffice for an exhaustive test, and their attention was on the verge of turning to something else when I showed to them how the impossibility can be established by means of the well-known counting argument inspired by the colouring of the chessboard. You should have seen the excitement on those young faces! They were absolutely fascinated, and rightly so, for even in this simple setting they had caught a glimpse of the unequalled power of mathematics.

In connection with this problem, I would like to point out two things. Firstly, it is not just brute force versus reasoning, for even the exhaustive "experiment" requires an argument that no cases have overlooked. In other words, the two approaches represent two extremes on a spectrum of arguments: a very clumsy one and a very effective one. Secondly, the problem has effortlessly been generalized to many other sizes than 8x8. The importance of these observations is that mathematics emerges as something very different from what the Oxford Dictionaries give you, *viz.* "the abstract science of space, number, and quantity"; mathematics rather emerges as "the art and science of effective reasoning", regardless of what the reasoning is supposed to be about. The traditional mathematician recognizes and appreciates mathematical elegance when he sees it. I propose to go one step further, and to consider elegance an essential ingredient of mathematics: if it is clumsy, it is not mathematics.

For the improvement of the efficiency of the reasoning process, three main devices have been found effective:

1. adopting a notation that captures what is essential and nothing more (among other things to enable what A.J.M. van Gasteren has called "disentanglement")
2. restricting steps to a well-defined (and modest) repertoire of manipulations
3. adopting a notation that yields formulae that are well-geared to our manipulative needs.

Remark Alfred North Whitehead almost understood this; he was, however, willing to sacrifice ease of manipulation to brevity. A small study of the ergonomics of formula writing shows that the stress on his kind of brevity was misguided. (End of Remark.)

All these devices are closely connected to notation, and it is worth pointing out that the typical mathematician is very ambivalent about the importance of notation: on the one hand he knows that a good notation can make the difference between day and night, on the other hand he considers explicit attention to notation wasted. Quite typical was the reaction of a rather pure mathematician that had attended several days of lectures that W.H.J. Feijen and I gave in the second half of the 70s: he was very impressed by the power of our formal apparatus but regretted our deviations from standard notational conventions, not realising that these deviations had been a *conditio sine qua non*. The probable explanation of the mathematical ambivalence towards notation is that the average mathematician is still a Platonist, for whom formulae primarily have a descriptive rôle and are not meant for uninterpreted manipulation, "in their own right" so to speak; moreover, his unfamiliarity with formal grammars makes many notational issues impossible for him to discuss.

But let me give you three tiny examples of the importance of notation.

1. Among a large number of electronic engineers I have established that, while all are certain that "and" distributes over "or", a significant majority is ignorant of or very uncomfortable with the inverse distribution. This shocking observation has a simple observation: in their crummy notation, the familiar distribution is rendered by the familiar $x(y+z) = xy + xz$, the unfamiliar distribution is rendered by the unfamiliar $x + yz = (x = y)(x + z)$.
2. With $\Sigma, \prod, \forall, \exists, \uparrow$ or $\downarrow$ for the quantifier **Q**, the "1-point rule" is

   $$\langle \mathbf{Q}i : i = n : t.i\rangle = t.n$$

   For Σ and $\prod$, mathematicians know it, probably in the forms

   $$\sum_{i=1}^{n} t(i) = t(n) \text{ and } \prod_{i=1}^{n} t(i) = t(n).$$

   For $\forall$ and $\exists$, they do *not* know the 1-point rule, working, as they do, in these cases without "range" for the dummy; they would have to code them as something like

   $$(\forall i)(i = n \Rightarrow t(i)) = t(n) \text{ and} (\exists i)(i = n \wedge t(i)) = t(n),$$

   forms sufficiently opaque to make these formulae unknown. (And not knowing these 1-point rules really hurts, as they are the main devices for equating a quantified expression to an expression without quantifier.)
3. As a final example: recently I saw for a nonassociative relation between predicates an infix operator, say $\underline{M}$, being introduced. An important property of $\underline{M}$ was rendered by

   $$[\langle \forall i :: p.i\underline{M}q.i\rangle \Rightarrow (\langle \forall i :: p.i\rangle \underline{M} \langle \forall i :: q.i\rangle)]$$

   Had we written $\underline{M}.(x, y)$ for $x\underline{M}y$, we could have used that quantification distributes over pair-forming, i.e.

$$[\langle\forall i :: (p.i, q.i)\rangle \equiv (\langle\forall i :: p.i\rangle, \langle\forall i :: q.i\rangle)],$$

and with the dummy w of type "predicate pair", the above property of $\underline{M}$ could have been written

$$[\langle\forall w :: M.w\rangle \Rightarrow M.\langle\forall w :: w\rangle],$$

an implication familiar from conjunctivity. But it is hard to exploit the fact that quantification distributes over pair-forming if the pair has been pulled apart by an infix operator.

The ambivalence towards notation is reflected in what I am tempted to call "hybrid theories"; these are very traditional and are warmly recommended to this very day. The designer of a hybrid theory has the feeling of creating two things: firstly, the subject matter of the theory, and, secondly, a "language" to discuss the subject matter in. Besides metaphors, jargon and definitions, "language" includes here formalisms as well.

The idea behind the hybrid theories is that arguments and calculations, by definition formulated in the new language, are to be understood, interpreted, and given "meaning" in terms of the new subject matter. Usually the language is a semi-formal system, in which many steps of the argument are "justified" by an appeal to our intuition about the subject matter.

By giving enough care to the design of the language, by making it sufficiently elaborate and unambiguous, the validity of conclusions about the subject matter becomes a linguistic characteristic of the sentences describing the argument. In the ultimate case, reasoning about and in terms of the subject matter is replaced by the manipulation of uninterpreted formulae, by calculations in which the original subject matter has disappeared from the picture. The theory is now no longer hybrid, and according to your taste you may say that theory design has been degraded to or simplified to a linguistical exercise.

I have observed that even in the presence of a simple and effective formalism, mathematicians of a more conservative persuasion are opposed to such calculational manipulation of formulae and advocate their constant interpretation in terms of the constituents of the intended model. They defend this interpretation, also called "appeal to intuition" mainly on two grounds. Their first argument is that the close ties to the intended model serve as a source of inspriation as to what to conjecture and how to prove it. I have my doubts about this but I don't need to elaborate on those because the sudden focus on "the source of inspiration" reveals a lack of separation of concerns: because there is a job to be done we are interested in the structure of arguments and not in the psychological habits and mental addictions of individual mathematicians. Their second argument for constant interpretation is that it is an effective protection against the erroneous derivation of what, when interpreted, is obvious nonsense: in short, constant interpretation for safety's sake. It is my contention that the second argument reflects a **most harmful mistake**, from which we should recover as quickly as possible: adhering to it is ineffective, paralyzing, misleading, and expensive.

- It is ineffective in the sense that it only protects us against those erroneous conclusions for which the model used readily provides an obvious counter example.

It is of the same level as trying to base confidence in the correctness of a program on testing it.

- It is paralyzing. You see, a calculus really helps us by freeing us from the fetters of our minds, by the eminently feasible manipulation of formulae whose interpretation in terms of a model completely defies our powers of imagination. Insistence on constant interpretation would rule out the most effective use of the calculus.
- It is misleading because the model is always overspecific and constant interpretation invites the introduction of steps that are only valid for the model the author has at that moment in mind, but are not valid in general. It is this hybrid reasoning that has given all sorts of theorems a very fuzzy range of validity. People working with automatic theorem provers discovered that in geometry most theorems are wrong in the sense that "the obvious exceptions" are not mentioned in the theorem statements; more alarming is that those exceptions don't surface in the published proofs either.
- It is expensive because the constant "translation" from formulae to interpretation is a considerable burden. (In order to reduce this burden, P. Halmos strongly recommends to avoid formulae with quantifiers!)

The moral is clear: for doing high-quality mathematics effectively verbal and pictoral reasoning have to be replaced by calculational reasoning: logic's rôle is no longer to mimic human reasoning, but to provide a calculational alternative. We have no choice, and, as a result, we are entitled to view the design of a mathematical theory as a linguistic exercise, and to rate linguistical simplifications as methodological improvements.

A major such improvement is the restriction to unambiguous formulae. It is easily implemented and is a great simplification compared to the unwritten rules of disambiguation that are invoked by "but everyone understands what is meant". I remember how I was ridiculed by my colleagues, and accused of mannerism and arrogance, just because I avoided a few traditional ambiguities, but that was a quarter of a century ago and by now the mathematical community could be wiser.

A next step is to recognize that it is not string or symbol manipulation that we are doing, but **formula** manipulation; we replace not just substrings but subexpressions and formula manipulation is therefore simplified by easing the parsing. Context-dependent grammars, for instance, are not recommended. I remember a text in which $x = y \wedge z$ had to be parsed as $x = (y \wedge z)$ for boolean x, y, but as $(x = y) \wedge z$ otherwise, and such a convention is definitely misleading. These are not minor details that can be ignored. (The only calculational errors I am aware of having made in the last decade were caused by parsing errors.)

I would now like to draw your attention briefly to a linguistic simplification of a somewhat different nature, which took place in an area familiar to all of us, *viz.* program semantics. In this area, the 60s closed after the work of Naur and Floyd with Hare's Axiomic Basis, where the rules of the games were couched as "inference rules", expressed in their own formalism. A main reason for publishing in the 70s the predicate tranformer semantics was that the latter eliminated the whole concept of inference rules by subsuming them in the boolean algebra of predicates, which was needed anyhow. The paper was written under the assumption that this simplification would be noticed without being pointed out; in retrospect I think that assumption was a mistake.

Let me now turn to what might turn out to be the most significant simplification that is taking place these decades. If it is as significant as I estimate, it will need a name; my provisional name, descriptive but a bit long, is "domain removal". Let me illustrate it with a very simple but striking example.

Let us consider the transitive relation $\propto$ ("fish"), for all p, q, r

$$p \propto q \wedge q \propto r \Rightarrow p \propto r,$$

and the two ways of correcting the erroneous conclusiòn that $p \propto q$ holds for and p, r. The erroneous argument is "choose a q such that $p \propto q$ and $q \propto r$ both hold; $p \propto r$ then follows from transitivity". The obvious error is that such a q need not exist, i.e. we can only demonstrate

$$\langle \exists q :: p \propto q \wedge q \propto r \rangle \Rightarrow p \propto r.$$

The traditional argument goes as follows. Assume the antecedent $\langle \exists q :: p \propto q \wedge q \propto r \rangle$; this means that we can choose a value that we shall call k such that $p \propto k \wedge k \propto r$; and now we can conclude $p \propto r$ on account of the transitivity of $\propto$.

The alternative argument can be rendered by:

We observe for any p, r

$\langle \exists q :: p \propto q \wedge q \propto r \rangle \Rightarrow p \propto r$
$= \{\text{predicate calculus}\}$
$\langle \forall q :: p \propto q \wedge q \propto r \Rightarrow p \propto r \rangle$
$= \{\propto \text{ is transitive}\}$
true

These two arguments illustrate the difference *in vitro*, so to speak. In the traditional argument, identifiers are used to name **variables** of the appropriate type, identifier k, however, names a **value** of that type. In the alternative argument, the identifiers all name variables, and, consequently, the possible values of the variables have disappeared from the picture. By not mentioning the values, their domain has been removed. In demonstrating an existential quantification by **constructing a witness** and conversely, in using an existential quantification by **naming a witness**, Gentzen's Natural Deduction advocates proofs of the needlessly complicated structure.

Whether identifiers are names of variables or (unique!) names of values turns out to be a traditional unclarity. "Consider points A, B, C and D in the real Euclidean plane, etc.": it is now unclear whether coincidence is excluded or not.

Proofs of uniquness traditionally suffer from this complication. It means that for some given P one has to show

$$P.x \wedge P.y \Rightarrow x = y \text{forall} x, y.$$

Authors that are unclear about the status of their identifiers are uncomfortable about this because, with x and y unique names of values, $x \neq y$ holds by definition. They feel obliged to embed the above, packaging it as a *reductio ad absurdum*:

"Let $P.x$; assume that there is another value y such that $P.y$. But $P.x \wedge P.y$ implies $x = y$, which is a contradiction, hence no such y exists."

Another complication is that value-naming easily induces case analysis. Some time ago the problem of the finite number of couples circulated. You had to show that in each couple husband and wife were of the same age if

(i) the oldests of either sex had the same age, and
(ii) if two couples did a wife swap, the minimum ages in the two new combinations were equal to each other.

The traditional argument is in two steps. It first establishes on account of (i) and (ii) the existence of a couple of which both husband and wife are of maximum age. In the second step it establishes that then on account of (ii) in any other couple husband and wife are of the same age. When formulating this carefully, taking into account that there need not exist a second couple and that more than one man and one woman may enjoy the maximum age, I could not avoid all sorts of case distinctions. The trouble with this argument is of course, that it has to refer to a specific couple. Here is, in contrast, the calculation in which no specific husband, wife, age or couple is named; the variables x, y are of type "couple". We are given

1. $\langle \uparrow y :: m.y \rangle = \langle \uparrow y :: f.y \rangle$
2. $\langle \forall x, y :: m.x \downarrow f.y = f.x \downarrow m.y \rangle$

and observe for any x

$m.x = f.x$
$= \{\uparrow\downarrow$ calculus: law of absorption$\}$
$m.x \downarrow \langle \uparrow y :: m.y \rangle = f.x \downarrow \langle \uparrow y :: f.y \rangle$
$= \{(\text{i})\}$
$m.x \downarrow \langle \uparrow y :: f.y \rangle = f.x \downarrow \langle \uparrow y :: m.y \rangle$
$= \{\uparrow\downarrow$ calculus: $\downarrow$ distributes over $\uparrow\}$
$\langle \uparrow y :: m.x \downarrow f.y \rangle = \langle \uparrow y :: f.x \downarrow m.y \rangle$
$= \{(\text{ii})\}$
true

which concludes the proof without any case analysis. (Some of the simplification is due to the fact that the range in (ii) includes $x = y$, the rest of the simplification is due to the fact that the calculation names no specific couple, only variables of type "couple".)

The fact that in useful calculations like the above the possible values of the variables are totally irrelevant and can therefore profitably be ignored has been seen and explained in 1840 by D.F. Gregory — in "On the Real Nature of Symbolic Algebra" [Trans. Roy. Soc. Edinburgh 14, (1840), 208-216]:

> "The light, then, in which I would consider symbolic algebra, is, that it is the science which treates of the combination of operations defined not by their nature, that is by what they are or what they do, but by the laws of combination to which they are subject."

Gregory knew that the domain of values could be ignored. So did John D. Lipson, in whose book "Elements of Algebra and Algebraic Computing" I found the above quotation. The amazing thing is that *all* texts about algebra I saw, Lipson's book included, introduce algebras as defined on a (named) set of values! Gregory's insight seems uniformly ignored. Sad.

All this has a direct parallel in programming. In order to establish the correctness of a program one can try a number of test cases, but for all other applications of the program one has to rely on an argument as to why they are correct. The purpose of the argument is to reduce the number of test cases needed, and I would like you to appreciate the dramatic change that occurs when the number of test cases needed is reduced to zero. Suddenly we don't need a machine any more, because we are no longer interested in computations. That is, we are no longer interested in individual machine states(because that is what computations are about). Reasoning about programs then becomes reasoning about subsets of machine states as characterised by their characteristic predicated, e.g.

$$\{\ s|X.s\ \}$$

The next step — known as "lifting" — is to rewrite relations between predicates, e.g. rewrite

$$\langle\forall s : s \in S : X.s \equiv Y.s \wedge Z.s\rangle$$

as

$$[X \equiv Y \wedge S],$$

because the most effective way of eliminating machine states from our considerations is by removal of the last variable that ranges over them. From variables of type "machine state" — like "s" — we have gone to variables of type "predicate" — like "$X, Y, \text{and} Z$" —. I cannot stress enough the economy of not having to deal with both types of variables. In other words, besides being ineffective, quality control of programs by means of testing is conceptually expensive.

The characteristic functions of singletons are called "point predicates". My great surprise over the last ten years is the power of what became known as "pointless logic", i.e. predicate calculus without the axiom that postulates the existence of point predicates. The calculus is extremely elegant: it is like purified set theory, for which traditional sets composed of elements are an over-specific model. It is extremely powerful. It admits extension to "pointless relational calculus" which for me had many surprises in store. For notions like transitivity and well-foundedness I grew up with definitions in terms of individual elements, bu the use of those elements is an obfuscation, for the concepts can be beautifully embedded in pointless relational calculus. Of all I have seen, it is the ultimate of removal of the inessential.

Finally, let me add to these pleasant surprises a pleasant observation: the average calculational proof is very short. I can truly do no better than relay to you the valuable advice my dear mother gave me almost half a century ago: "And, remember, when you need more than five lines, you are probably on the wrong track.".

Pretty-printing: An Exercise in Functional Programming

John Hughes

Department of Computing Science,
University of Glasgow
GLASGOW
Scotland
from August '92:
Informationsbehandling,
Chalmers Tekniska Högskola,
S-41296 GÖTEBORG,
Sweden.

1 Introduction

At the heart of functional programming is the treatment of functions as data. Just as arithmetic operators compute numbers from numbers, so higher-order functions can be used to compute new functions from old ones. In the limit, as Backus advocated, the programmer might write no first-order code at all but simply compute the desired program using the operators of an algebra of programs. More commonly, higher-order functions such as *map*, *filter* and *foldr* are used to express commonly occurring idioms, and other parts are programmed in a first-order style.

While these particular higher-order functions, or combinators, are quite general purpose there is also considerable mileage in defining application-specific combinators to capture the idioms of a particular application area. A family of such combinators can be thought of as defining the abstract syntax of a special-purpose programming language, and their implementations as defining its semantics. Language design is difficult: combinator families are correspondingly hard to invent. But the payoff can be great — a well-chosen set of combinators can give a powerful handle on an application area. For example, Wadler's parsing combinators and LCF's tacticals make parsers and proof strategies respectively very easy to write.

A good choice of combinators should not only be expressive, but should also satisfy a collection of algebraic laws. Such laws are useful for reasoning about or optimising programs, and can also help in the design of combinator implementations.

In this paper we illustrate the approach by designing a combinator library for pretty-printing.

1.1 The Pretty-printing Problem

Almost every program operating on symbolic data needs to display that data from time to time. Yet the fixed-format output that most programming languages handle easily is quite unsuited to symbolic data, whose shape and size cannot be known in advance. There is thus a recurring need for pretty-printers, which choose dynamically between alternative layouts depending on the actual data to be printed. Our goal is to design a combinator library to ease the programming of such pretty-printers.

2 Layouts

A pretty-printer's task is to break a string into lines and indent these lines appropriately. We model the output of a pretty-printer as a list of lines (strings) paired with integer indentations. We call such a list a *layout*.

A layout is a representation of a string, obtained by ignoring indentation and concatenating the lines, taking a line break to be equivalent to a space. A layout also defines an *image* — a two-dimensional arrangement of characters.

We introduce four operators on layouts:

- *text* s is a layout representing the string s,
- *nest* x indents the layout x,
- $x \sim y$ combines x and y horizontally,
- $x \wr y$ combines x and y vertically.

In specifying these operators we have two principles in mind:

- The string represented by the result of an operation should depend only on the strings represented by its operands.
- The image of the result of an operation should be a combination of (translations of) the images of its operands.

These principles constrain the design of the operators strongly. After specifying them we go on to study the algebra of the layout operators.

3 Documents

We choose to factor each pretty-printer into a composition of two functions:

- a layout-independent function from the objects to be pretty-printed to a *document*,
- a function which given a document and information such as the page width, selects a pretty layout for that document.

A document is something with many possible layouts: formally it is a set of layouts.

We extend the layout operators to apply to documents as well, and define a new n-ary operator *sep* with two alternative layouts, a horizontal one and a vertical one. This set of operators seems to offer enough flexibility to define a wide variety of pretty-printers, and has similarities with other proposals for language-independent pretty-printing.

We study the algebra of documents, and as a payoff prove that every document constructed with the operators above can be expressed in a canonical form. We design a Haskell datatype to represent (the abstract syntax of) canonical forms, and use the algebraic laws to derive implementations of the document operators as functions from canonical forms to canonical forms. We define what we mean by 'best' layout (there are many possible definitions) and derive a Haskell function to construct the best layout of a canonical form, thus completing an implementation of a pretty-printing library.

4 Documents as Functions

This first implementation is sometimes inefficient, so to improve it we choose an alternative representation of documents as functions from the page width *etc.* to the best layout. The document operators are then implemented as higher-order functions building pretty-printing programs. Once the representation of documents is specified, the implementations of the document operators can be derived using the document algebra.

5 Conclusion

We designed a very small library (around one page) defining five combinators for pretty-printing. Despite, or perhaps because of their simplicity, these combinators have proved sufficiently efficient and expressive for quite extensive practical use. The combinators are simple to specify and enjoy many algebraic properties. The algebra is useful for reasoning about pretty-printers that use the library, and played an important role in deriving its implementations.

True Concurrency: Theory and Practice

Ugo Montanari

Dipartimento di Informatica
University of Pisa
ugo@di.unipi.it

Extended Abstract

The true concurrency approach aims at equipping concurrent languages with process semantics which are more expressive than usual. The classical approach can be exemplified by Milner's CCS, where agents $p = a.nil \mid b.nil$ and $q = a.b.nil + b.a.nil$ have the same semantics. More generally, every CCS agent is equivalent to another agent built only with action prefixing, nil and nondeterministic sum, i.e. in CCS concurrency can always be reduced to nondeterminism, via interleaving. Instead, one would often like to say that agent p has more intrinsic parallelism than q, or that actions a and b, when executed by q are causally related, while they are concurrent when executed by p. Another point of view is that agent p extends on two locations, say l and r, while q has only one location. Thus p performs a and b in two different locations, while q uses only one.

The above considerations suggest to make the testing mechanisms of CCS stronger, e.g. by allowing multisets of actions, or partial orderings, or localities, as observations. This defines what could be called an extensional truly concurrent semantics. For instance, a recent result is reported in [20]. However, there is a more intensional sense (i.e. not related to abstract semantics, but rather to direct operational behavior) in which agents p and q are different. In fact, while agent q can perform two independent two-step computations (the first consisting of the two transitions $a.b.nil + b.a.nil \rightarrow b.nil \rightarrow nil$; the other of the two transitions $a.b.nil + b.a.nil \rightarrow a.nil \rightarrow nil$), one can assert that p has only one possible two-step computation, consisting of the "diamond" $a.nil \mid b.nil \rightarrow nil \mid b.nil \rightarrow nil \mid nil = a.nil \mid b.nil \rightarrow a.nil \mid nil \rightarrow nil \mid nil$. In other words, one can express the fact that certain events are concurrent, by asserting that the computations where these events are executed in all possible different orders are identified, i.e. they are the same abstract computation.

Thus in general the intensional truly concurrent semantics of a language, or rather of a model of computation, consists of defining suitable equivalence classes of computations. This equivalence should of course be a congruence with respect to sequential composition of computations, and with respect to any other operation defined on computations.

There are several reasons for defining a truly concurrent semantics of our favorite model of computation. One reason is that specification languages based on truly concurrent models are more informative, e.g. testing sequences defining partial orderings may carry the same information as an exponentially larger number of interleaving traces. Another reason is that in truly concurrent models the existing fine parallelism of the application is fully specified. It is left to the implementor to take advantage of it by allocating concurrent events to different processors, or to partition events into coarser classes performed by a few concurrent processes. Finally, truly concurrent semantics carries extra information, and thus it is usually straightforward to recovery interleaving semantics from it when needed.

When we consider a large variety of models of computations, it turns out that truly concurrent semantics is actually the most natural, and that a special effort is needed to derive the interleaving semantics instead. The simplest case study refers to Petri nets. In [14], it is noticed that a place-transition net is a just a graph where the nodes form a free commutative monoid. The usual operational net semantics can be recovered by means of a free construction, which builds a monoidal category starting from the graph. In this category, certain obvious axioms (e.g. the functoriality of the monoidal operation) hold on arrows. Interestingly enough, arrows can be shown to be isomorphic to Petri nonsequential processes [6, 7], and thus the axioms above provide a truly concurrent semantics for Petri nets. In this setting, it is possible to avoid the axioms (thus recovering the interleaving semantics of nets, based on firing sequences), only by forgetting first the monoidal structure of nets.

The truly concurrent, intensional semantics of CCS can be reduced to that of Petri nets. In fact, in [9, 10] it is shown that CCS agents and CCS proofs can be inductively mapped on nets. Then the net axioms above, when referred back to CCS computations, identify exactly all the diamonds like $a.nil \mid b.nil$ discussed above. Constructions similar to those for Petri nets hold for phrase structure grammars (in the context free case abstract computations are syntactic trees) and for Horn clauses [1].

The construction yielding the truly concurrent semantics of a structured transition system can be parametrized within category theory [3]. Another (possibly related) general framework has been proposed by Meseguer for concurrent term rewriting systems[13]. Also in this case the model of computation is a suitable category, where the arrows are defined via certain axioms (in addition to functoriality of term constructors, every rewrite rule defines a natural transformation). Interestingly enough, the classical concurrent semantics of lambda-calculus can be recast in this framework as a special case, and thus an axiomatization of its truly concurrent semantics can be derived [12].

The case of graph rewriting systems is of special interest. In fact, graph grammars have been equipped from the beginning with a truly concurrent semantics, defined via exchanging sequentially independent production applications [8]. In addition, graph grammars have the ability of defining, for every production application, a subgraph which is needed for the rewriting but which is not modified by it. This leads to an interesting, general notion of context dependent rewriting systems, of which classical graph grammars are, at our knowledge, the pioneering instance. Their interest is that they generalize the intuitive readers-writers schema, where several readers of the same data can proceed in parallel. Graph grammars have been given also a partial ordering semantics[18], which has been proved consistent with the classical semantics based on exchanging productions. An ongoing research aims at recasting graph grammars in the general framework described above, based on structured transition systems. A simple algebra based on three operations (parallel composition, hiding and renaming) and on several, rather obvious axioms is actually able to directly represent classical graph rewriting [2, 4].

On a more application-oriented side, graph rewriting systems have been used for equipping concurrent constraint programming with a truly concurrent semantics [15, 16, 19]. The concurrent constraint programming paradigm extends both concurrent logic programming and constraint logic programming, thus providing an unifying framework for most logic languages. In addition, it has interesting higher order capabilities, e.g. it can simulate both the lazy and the call by value lambda calculus in much the same was the pi-calculus does it [11]. The truly concurrent semantics of concurrent constraint programming may turn out to be a satisfactory basis for handling both AND and OR parallelism of logic languages.

The context-dependent version of Petri nets has been introduced in [17]. Apparently, most interesting constructions and properties can be extended to contextual nets. As an application, it has been shown that the classical serializability problem of concurrent access control to data bases can be stated as the equivalence (in the sense of corresponding to the same nonsequential process) of firing sequences in a certain contextual net [5]. More generally, there is apparently the possibility of characterizing those computations which yield the same result in a (uninterpreted) model of computation based on shared memory as those with the same truly concurrent semantics. An interesting application to the parallelization of microprograms is proposed in [21].

References

1. Corradini, A. and Montanari, U., An Algebraic Semantics of Logic Programs as Structured Transition Systems, invited talk, Proc. North American Conference on Logic Programming (NACLP) 1990, Austin, Texas, MIT Press, pp.788-812.
2. Corradini, A. and Montanari, U., An Algebra of Graphs and Graph Rewriting, invited paper, Proc. 4th Conference on Category Theory and Computer Science, Spriger LNCS 530, pp. 236-260, 1991.
3. Corradini, A. and Montanari, U., An Algebraic Semantics for Structured Transition Systems and its Application to Logic Programs, to appear in TCS.
4. Corradini, A., Montanari, U. and Rossi, F., CHARM: Concurrency and Hiding in an Abstract Rewriting Machine, Proc.FGCS'92, Vol.2, Ohmsha 1992, pp887-896.
5. De Francesco, N., Montanari, U. and Ristori G., Modeling Serializability via Process Equivalence in Petri Nets, draft, March 1992.
6. Degano, P., Meseguer, J. and Montanari, U., Axiomatizing Net Computations and Processes, Proc. 4th Symp. on Logics in Computer Science, IEEE 1989, pp. 175-185.
7. Degano, P., Meseguer, J. and Montanari, U., Axiomatizing the Algebra of Net Computations and Processes, Technical Report TR-1/91, Dipartimento di Informatica, University of Pisa, January 1991.
8. Ehrig, H., Introduction to the Algebraic Theory of Graph Grammars, Proc. International Workshop on Graph Grammars, Springer LNCS 73, 1978.
9. Gorrieri, R. and Montanari, U., Distributed Implementation of CCS, Proc. 12th International Conference on Applications and Theory of Petri Nets, 1991, pp. 329-348.
10. Gorrieri, R. and Montanari, U., On the Implementation of Concurrent Calculi into Net Calculi: Two Case Studies, draft, December 1991.
11. Laneve, C. and Montanari, U., Mobility in the cc-Paradigm, to appear in Proc. MFCS'92.
12. Laneve, C. and Montanari, U., An Axiomatization of the Concurrent Semantics of the Lambda Calculus, to appear in Proc. ALP'92, Pisa, September 1992.

13. Meseguer, J., Conditional rewriting Logic as a Unified Model of Concurrency, Theoretical Computer Science 96 (1992), pp. 73-155.
14. Meseguer, J. and Montanari, U., Petri Nets are Monoids, Info and Co, Vol. 88, No. 2 (October 1990), 105-155. Also SRI-CSL-88-3, January 1988.
15. Montanari, U. and Rossi, F., True-Concurrency in Concurrent Constraint Programming, in V. Saraswat and K. Ueda, Eds., Proc. International Logic Programming Symposium '91, MIT Press, pp.694-713.
16. Montanari, U., and Rossi, F., Graph Rewriting for a Partial Ordering Semantics of Concurrent Constraint Programming, to appear in TCS.
17. Montanari, U. and Rossi, F. , Contextual Nets, draft, Jan. 1992.
18. Montanari, U. and Rossi, F., Graph Grammars as Context-Dependent Rewriting Systems: A Partial Ordering Semantics, in: J.-C. Raoult, Ed., CAAP'92, Springer LNCS 581, pp. 232-247.
19. Montanari, U. and Rossi, F., An Event Structure Semantics for Concurrent Constraint Programming, Proc. Workshop on Constraint Logic Programming, Tokyo, June 1992.
20. Montanari, U. and Yankelevich, D., Partial Order Localities, to appear in Proc. ICALP'92.
21. Mowbray, M., A Proposal to Look For a New Scheduling Strategy for VLIW Compilers, Hewlett Packard Pisa Science Centre, April 1992.

Programming for Behaviour *

Władysław M. Turski

Institute of Informatics
Warsaw University
Banacha 2, 02-097 Warsaw

Abstract. Traditional programming is for computations. Many contemporary applications suffer when squeezed into programming paradigm. This presentation investigates desirable properties of another paradigm: programming for behaviours and explores some technical consequences of a possible formalism to describe such programs.

1 A view of history

For centuries a computer was a person performing strictly prescribed operations on numbers recorded on a sheet of paper. The object of the computation — as far as the computer was concerned — was to evaluate an algebraic formula, perhaps repeatedly, for a number of data sets. In a computation, the major portion of human effort was consumed by acts of executing arithmetical operations. The evolution of computing machinery concentrated on ever faster means of executing the four arithmetical operations.

With the invention of electronic digital circuits the long established proportions between the time needed to execute arithmetical operations and that consumed by human operator in following the prescribed sequence of operations changed dramatically. Further increases in the speed of executing arithmetical operations would be aimless if the control actions were not performed at approximately the same rate.

The concept of programmed computer was a natural and necessary consequence of the spectacular speed-up of arithmetical operations. The observation that programs can be stored just as easily as data and in the same medium (present already in Babbage's designs, but usually attributed to von Neumann) led to the stored-program computers.

With these machines becoming ever more readily available and faster, the appetite for computations soon met another barrier that could not be breached by improving the hardware technology. The programs controlling the order in which arithmetical operations were performed, the machine language programs, were exceedingly time-consuming to prepare. However, this limitation was not quite universal. Most acute in the environments characterised by the need to perform a variety of ever fresh computations, it was hardly at all felt by those who were satisfied with repeated executions of the same programs.

* The work reported here was supported in part by the Polish Research Council under grant KBN-211999101.

At this point in development of programming two different attitudes, indeed two subcultures emerged. One was concerned primarily with program construction, another with program use.

When the use of a program is the main concern, one tends to consider it as a part of machinery which, once installed, is expected to provide a continued service. In other words, a program is treated as (a part of) a product, purchased for its continued utility in a work environment. Its quality is measured with respect to the services it provides, and therefore is subject to empirical evaluation (testing).

When the main concern is with the program-making, one's chief objective is to reduce the effort of program production. This implies two things: (i) inventing and using program-production tools, and (ii) applying such production methods which provide for quality control through design and its implementation rather than by the a posteriori testing of finished products.

At first glance, no serious controversy follows from such separation of concerns. But appearances — at least in this case — are highly misleading.

The best, most productive program-making tools and the most reliable methods of program design and implementation are based on the assumption of inviolability of program specification. There is no proof that there is no other way, but the weight of circumstantial evidence makes this conjecture highly probable. The program makers tend to swear by the principle: the progam is no better than its specification.

On the other hand, program users tend to regard a program good if its execution on a computer provides a satisfactory service. Since the notion of satisfaction is related to the current needs which, just as computer configurations, tend to vary in a less than predictable way, the program users put a premium on program flexibility.

The problem is that with the traditional view of a program "flexibility" is hard to specify. The controversy about the value of calculational (formal) methods of programming is a symptom of an essential divergence between the actual use of computers and the traditional paradigm of computing. Formal methods of programming are geared to making programs for computations, where fixed specifications are possible and natural, whereas computers are increasingly used for service purposes, for which the very notion of fixed specifications is ill-fitting.

The traditional paradigm of computing did not consider any interaction between the process of computing and its environment. This is not to say that computations were undertaken for no practical purpose, merely that the process of computation and the application of its outcome were physically separate. The necessary link was often established by various tables produced by computers and used in an application domain, such as science and commerce, but also in some control applications, such as navigation and artillery.

To facilitate the field use of tables many ingenious techniques were applied, both analytic and computational. Some of them were so closely associated with a particular problem that they became inseparable from it. Even today, when the availability of the outline computing technology vitiates any need for such techniques, we are following them because we were taught that they constitute a part of the solution or, in extreme cases, of the problem itself. A prime example of such habits, formerly necessary and — despite their present technical irrelevance — faithfully maintained, is the presentation (and measurement) of physical process variables at regular time intervals, rather than at occurrences of significant events (as, e.g., when we cook an egg for three minutes rather than until its white/yolk consistence reaches a desired state).

The traditional physical separation between a computation and the application of its results was, of course, entirely justified in the past, both because the time-scale of computation process was usually totally different from the time-scale of the process to which its results were applied, and because the nature of a data-gathering (as well as results-utilizing) apparatus was totally different from that of computing hardware. Now, however, when all machinery involved uses compatible information media and when it is patently expedient to physically link up control computations with their applications, we discover that the computing paradigm inherited from the bygone days clashes with what is popularly known as the real-time computing.

Three factors from the core of difficulties with the real-time computing:

1. Formal (calculational) logic, which is the foundation of all known sound methods of programming, is impervious to any notion of physical time. At best, we can analyse programs in terms of ticks of an abstract flip-flop, but its "true" (physical) frequency cannot be ascertained by programming means. This is in stark contrast to all other aspects of the computing machinery, such as numerical correctness of results delivered by arithmetical circuits, memory size and retention, all of which can be — at least in principle — verified by executing suitably programmed tests.
2. Inclusion of a trusted clock, which by the above-mentioned reasons cannot be controlled by the program, brings us into the realm of systems with two (or more) independent processors. (This is not surprising since all notions of time, except, perhaps, theological ones, presuppose the presence of at least two independent actors.) Unfortunately, there is no algorithm based on discrete mathematics which can achieve a perfect synchronization of two independent processors, no matter how complex their communication protocol.
3. The traditional view of programming makes absolutely no allowance for a spontaneous outside interference with the process of computing. Of course, an algorithm may include an instruction to consult a table of trigonometric functions, or even to read a thermometer, and branch according to the reading obtained. These interactions with the environment occur, however, when the elaboration of the algorithm reaches corresponding instructions; they are controlled by the program being executed. The technological device to allow the environment to interact spontaneously with the process of computation, i.e. the interrupt, has never been fully accepted in the theory of programming, and by most methods, as well as in the literature, is treated as a second-class citizen.

From the first two points it rather follows that programming for the real-time computations is most unlikely to find as solid a basis as the programming for formula evaluation has in logical calculi, whereas from the third we conclude that the alternative principle for the on-line control computations, the interrupt-, or, more generally, event-driven algorithms, do not fit well the traditional paradigm of computing. On the other hand, it would be very short-sighted indeed not to admit that the on-line control computations constitute an ever more important class of applications of computing technology.

Digital technology made multiple computing systems (such as multiprocessors, array processors, systolic computers and machines with multiple memory banks) not only possible and quite accessible, but also very attractive in terms of the speed with which information-carrying signals can be processed. It is quite apparent that the recently developed human appetite for fast signal- processing is insatiable. It is also pretty obvious that in this race the classical architectures are not capable of providing a good value for money. The multiple architectures, with their practically unlimited potential for ever faster processing, are sure to become a standard rather than an exception. In fact, even with the present-day technology, a highly multiple computer would be a norm, if it were not for the absence of satisfactory programming methods.

Basically, there are two ways in which this problem is being approached. One consists in inventing a method to split programs written for a single processor into as many subprograms as the multiple architecture of a given machine admits. Alternatively, a program written for a sequential machine is modified so that some of its parts can be used to control a parallel machine of a given kind. The second method does not start with an algorithm for a single processor. Instead, individual processors of a structure are programmed in such a way that results of operations they perform combine via established communication links into a meaningful whole. As a rule, in order to provide for a sensible interpretation of the achieved gross results, the actions of individual processors are so controlled that they can be uniquely mapped into an equivalent sequence of actions of a virtual single-processor machine; the meaning of the parallel machinery operations is then derived in a standard way from the serial equivalent of a parallel process.

Neither of the two approaches is entirely satisfactory. The first one, because a general "parallelisation" algorithm does not exist and, consequently, each time we want to effect a split, we must rely on particular circumstances; obviously it is not very efficient first to invent a serial algorithm and then to invent the special trick that would convert it into a parallel one. The second approach falls short of the expectations because it a priori excludes such multiple processes which do not have a serial equivalent, while it is generally hoped that multiple architectures would be capable of an essential extension of what is computable.

The traditional programming paradigm is ill-suited for multiple digital processing architectures.

It appears, therefore, that we should look not only for further technical improvements within the classical view of programming, but also for an extension of this view:

1. It should allow an enlarged view of functional specification, one that would permit the "service approach" to computation in addition to the "formula evaluation" one.
2. It should allow a natural treatment of spontaneous environmental interaction with the process of computing, thus opening the doors to control applications other than by the "real-time" computations.
3. It should accept the multiplicity of the computing machinery not as an added complication but as a primary property; the sequential machines being treated as a special case of $n = 1$.

2 A Proposal

At the risk of appearing utterly naive, we shall start looking for the extended programming paradigm by analysing two entirely informal examples of instructions a mother may give to her little daughter.

Instruction 1. How to make a sandwich.
Take a loaf of bread. Cut two slices from it. Spread butter evenly on one side of each. Put a slice of ham on one buttered slice of bread. Put a wedge of cheese on the other. Add some lettuce or sliced tomatoes. Fold the two pieces together, if too squirmy, fasten with a toothpick.

Instruction 2. What to do in the afternoon.
You may watch TV. If you get hungry make yourself a sandwich. When it gets dark switch on the lights. If you feel cold put on the red jumper.
When your father calls tell him I went out shopping and should be back by supper time.

Let us call the first one a ***process instruction***, and the second, ***behaviour instruction***. Both kinds of instructions are quite common in our lives, but only the process instructions are consistent with the traditional view of programming.

What are the essential differences between the two kinds of instructions?

Any process instruction is intrinsically sequential. Note its sequentiality is not merely a matter of the write-up, it cannot be separated from the "physics" of the process; one cannot butter the slices before they are cut from the loaf, and in a properly made sandwich the ham and cheese are put on after the buttering is completed.

A behaviour instruction has no intrinsic sequentiality in it (apart from "local" sequentiality in executing individual actions, such as making a sandwich).

For a process, there is a well-defined initial state (explicit or implicit, does not matter), a progression of steps, each bringing an overall goal a little closer, and — hopefully — a final step in which the goal is achieved. Actions that constitute the process are expected to be executed in a state established by their predecessors (with the trivial exception of the initial action).

There is no clear initial state for a behaviour, nor an overall goal-like purpose to the collection of actions. No action is expected to happen unless its "guarding" condition is established, but individual action need not be responsible for establishing the conditions propitious for other actions. The individual actions are in fact *reactions* to certain conditions that may have arisen, rather than steps aimed at establishing conditions for possible future actions.

A behaviour is defined in terms of reactions to the "current" environment. The guarding conditions are evaluated in states which are not assumed to be the results of the "past" behaviour. Indeed, the analysis of the "historical trace" hardly ever helps in the analysis of "current" behaviour, the "past" need not influence the "present" at all: the little girl may get hungry regardless of whether she has the red jumper on or not. (On the other hand, the analysis of the collection of past events, i.e. associations of conditions with reactions, may be very helpful; it is the *sequence* of past events that is largely irrelevant and may be misleading).

In terms to which we are accustomed, the meaning of a process instruction depends on the transitivity of action composition. Hence the programmer's concern with issues of termination, hence also the essentially transformational flavour of all formal semantics of programs. A correct program computes the right final output for any acceptable input.

For behaviour instruction the compositionality of actions is of no concern. There is no global transformation to be performed by transitively composing individual actions. Termination (apart, possibly from a simplifying assumption that each individual action itself terminates) is of no concern, either. Correctness of behavioural programs depends on two questions:

- are the condition/action associations appropriate? ("When hungry, switch on TV" would, apparently, provide an example of an inappropriate association!), and
- are all possible contingencies covered by the guarding conditions? (e.g., the little girl was not instructed what to do when she feels thirsty).

Lets us consider now a somewhat more complicated example.

2.1 Example of a simple vending machine

This machine has a finite magazine containing chocolates and a slot through which coins can be pushed in. It also has an externally accessible bay from which a chocolate can be picked up, a box for coins accepted in payment, and a drawer for coins to be returned.

In order to prescribe the behaviour of the vending machine, we introduce several boolean variables (in a more general case we would probably introduce predicates):

$RC \Leftrightarrow$ the machine is ready to accept a coin,
$CS \Leftrightarrow$ there is a coin in the slot,
$CP \Leftrightarrow$ a customer has paid up but no chocolate has been yet put for him into the dispensing bay,
$CA \Leftrightarrow$ the magazine contains at least one chocolate,

and an integer counter, cc, of chocolates in the magazine. We also assume the availability of three special "primitive operations":

accept_coin: defined when there is a coin in the slot; on completion of this operation the coin is no more in the slot (presumably it has been dropped into the box),
reject_coin: defined when there is a coin in the slot; on completion of this operation the coin is no more in the slot (presumably it has been dropped into the return drawer),
dispense_chocolate: defined when the magazine is not empty; on completion of this operation a chocolate from the magazine is put into the externally accessible bay.

The vending machine may be now prescribed as follows:

$$\{RC, CS, CP, CA, cc := \textbf{true}, \textbf{false}, \textbf{false}, N \geq 1, N\}$$
$$CS \wedge RC \rightarrow [\text{accept\_coin}; RC := \textbf{false}; CP := \textbf{true}]$$
$$\| CS \wedge \neg RC \rightarrow [\text{reject\_coin}]$$
$$\| CP \wedge CA \rightarrow [\text{dispense\_chocolate}; cc, CP := cc - 1, \textbf{false};$$
$$\textbf{if } cc = 0 \textbf{then } CA := \textbf{false else } RC := \textbf{true fi}]$$
$$\| CP \wedge \neg CA \rightarrow [\text{reject\_coin}; CP := \textbf{false}]$$
$$\| (CP \vee CS) \rightarrow [\text{skip}]$$

where the multiple assignment in braces denotes the initialization of variables, and N is the number of chocolates put into the magazine. The main body of the behaviour program consists of five *guarded actions* separated by $\|$ symbols. Each guarded action consists of a boolean expression (the guard), followed by $\rightarrow$ symbol, and by an action program in brackets. Action programs are written here in a familiarly vague fashion; it is assumed that they terminate when initiated in states in which the corresponding guards are satisfied. Operational semantics of such a behaviour program are given by requesting that whenever there obtains a state in which a guard is satisfied, the corresponding action program is executed.

Note that no sequence of the vending machine actions is ever mentioned. We never say (or need to) what the machine does first, and what it does next. We do not even say that the machine with no coin in the slot and no paid-up customer *waits* for some external event to change its state. The behaviour program merely prescribes that the vending machine by itself does not change its state in which there is no paid-up customer nor coin in the slot. In any other state — however obtained — the program prescribes a state-changing action.

For instance, whenever there is a coin in the slot and the machine is not ready to accept it, the coin is rejected. In "normal" operation of the vending machine the "not ready to accept a coin" attribute of the state is established by the action associated with guard $CS \wedge RC$, but any environmental interference which sets RC to **false** (such as, e.g., hitting the "suspend trading" button by the the maintenance engineer) also establishes the condition rule under which a coin from the slot will be rejected.

Because the environment can change the state of the vending machine while the latter is performing an action (e.g., coins may be pushed in very rapidly), we must refine the way in which behaviour programs are constructed and interpreted. The guards — as we understood them so far — described situations in which corresponding actions were *allowed* to happen; let us call them *preguards*. In order to handle the state changes occurring during the performance of an action but not caused by it, we introduce two assumptions (introduce two restrictions on admissible semantics):

1. Actions are performed on a *private copy* of the state. This copy is (atomically) taken from the global (public) state in which the preguard is satisfied and the action is chosen for execution.
2. To each action there is given a second guard, known as *postguard*, evaluated on the *public* state at the end of the action. If the postguard holds, the results of action (atomically) update the public state. If the postguard fails, the private copy modified by the action is abandoned, the action has no visible effects.

Of the two assumptions, the second one is less obvious. In particular, one could argue that the postguard should be evaluated not on the public state, but on the private copy, which would take into account the results of the just performed action. The decision to evaluate the postguard on the public state is based on the following reasoning. If — while the action was being performed (on a private copy of the state) — no changes were made to the public state, the postguard is not really needed, assuming, of course, that the action is well-designed for the situation in which it was initiated. Therefore, if there is any doubt as to the appropriateness of the results of the just completed action, such doubts could be only due to these changes in the environment which took place while the action was being performed. These changes could not possibly be due to the action itself, as it was performed on a private copy of the state. Hence, the postguard evaluation on the public state corresponds to asking the question "Are the results of this action still needed in the new situation?" If the answer is positive, the results are made public, otherwise they are simply discarded.

The notation for pre- and postguarded actions could be quite simple:

$$(P, Q) \rightarrow [p]$$

denotes action p with preguard P and postguard Q.

2.2 Examples

1. If we want to prescribe an action p that takes place in situation where P holds, and whose results are acceptable regardless of any environmental changes that may happen while it is being performed, we write:

   $$(P, \text{true}) \rightarrow [p]$$

2. If we want to prescribe an action p having two conditions: P to start, and Q to be acceptable, we may delay checking of Q by making it into the postguard. A particularly interesting case arises when $Q \equiv \neg P$, i.e.

   $$(P, \neg P) \rightarrow [p]$$

   which prescribes an action p which has a public effect only if during its execution the environment changes the public state to "opposite" of what it was at the start of this action.

3. This principle can be applied to the design of a simple event counter. Assume the event of interest consists in a public variable x changing its value from 0 to 1, or from 1 to 0. Then

   $$(x = 0, x = 1) \rightarrow [n := n + 1]$$

   $$\|(x = 1, x = 0) \rightarrow [n := n + 1]$$

   prescribes the behaviour of the desired counter, n.
   An objection could be raised that if the event is more rapid than execution of $n := n + 1$ then the counter may miss some events. This objection is, unfortunately, well-founded. On the other hand, the same criticism applies to ***any*** design of a discrete counter which ***does not control*** the event, as only "counters" interfering with the event (via some synchronizing mechanism), can be certain not to miss it. Incidentally, this form of Heisenberg's principle is often overlooked in designs of process control software. (End of examples.)

The use of behaviour instructions rather than process instructions obviously satisfies the postulate on spontaneous environmental interaction. It appears that it also meets the other two.

Indeed, since behaviour specifications do not rely on any particular textural ordering of individual guarded instructions, they will not be unduly disturbed if an extra instruction is added, or an existing one is removed, or its "working part" is modified. Similarly, should a condition under which an action was deemed appropriate turn out to be too coarse; it is an easy matter to remove the corresponding instruction and substitute two or more, the union of whose preguards would be equivalent to the original one. The need for such modifications would be dictated by the changing view of the services required under prevailing conditions; the proposed style merely facilitates the technical implementation. At the heart of the gained flexibility is the loosening of mutual dependencies between the units of which the whole "program" is composed. Since they do not transitively compose into a single computation, the individual units can be modified without endangering others.

Naturally, a loosened set of constraints can be abused, the consistency of the specified behaviour may suffer. Fortunately, this can now be checked by purely logical computations, viz. by the analysis of the set of guards and by verifying the appropriateness of the guards/actions associations.

The looseness of the links between constituents of a behaviour program may also be exploited as a basis for a design method, in which actions will be specified and implemented as the specific conditions for their desirability and admissibility become gradually known.

Somewhat less obvious is the manner in which the proposed extension caters for the multiple architecture postulate.

With a single processor and no spontaneous environmental interference with the process of computation, both guards of an action could be collapsed into a single preguard: $(P, Q) \longrightarrow [p]$ would be entirely equivalent to $P \wedge Q \longrightarrow [p]$. Thus, $P, \neg P \longrightarrow [p]$ is an example of a guarded action that cannot be executed by a single-processor machine (without a spontaneous outside interference). Therefore, in the proposed extension we have simple means of prescribing behaviours which cannot be serialized.

The postguards have been introduced ostensibly in order to prevent the delivery of action results in case they are not needed in the state of environment which obtains by the time of action complete. However, for a particular processor assigned to execute an action, the collective accepted results of whatever actions were performed in the meantime by other processors constitute but a change of the public state (environment); from the "point of view of a processor" there is not much difference between spontaneously interfering environment and concurrently running processors. Thus, a properly designed behaviour program may be executed by as many processors as there may be available. The notion of a processor (or processors) becomes superfluous: a behaviour program specifies a *field of actions*, the system becomes a *point probe* in the field.

Hence the behaviour programming contains two main ingredients for an essentially novel approach to multiprocessor computing: a means to prescribe non-serializable computations and a logical immunity with respect to the actual number of processors used.

3 Some technical points

Consider a system whose states are fully characterised by values of a finite number of state-variables $z^0, \ldots, z^M$. We shall employ one-letter designations for the tuples, marking individual states, when necessary, by subscripts. Thus, e.g., z_b, z_m and z_e represent three states of the system.

Total, boolean-valued functions of system state are known as ***predicates***. When P is a predicate, $P.z$ denotes the value of P in state z (in order to simplify the notation, throughout this section we use the "dot convention" for function application).

Let a be a program with ***input variables*** x, and ***output variables*** y, where both x and y are tuples of state-variables drawn from $z^0, \ldots, z^M$. Denote by dom.a the set of states starting from which the evaluation of program a terminates, and assume that terminating programs establish a functional dependence of output variables on input ones. Thus $z \in \text{dom}.a \Rightarrow y = a.x$.

Consider now a guarded action

$$(P, Q) \rightarrow [x|a|y]$$

where a is a program with input variables x and output variables y (x and y constitute the *frame* of a), P and Q are predicates over z and, in particular, it is assumed $P.z \Rightarrow z \in \text{dom}.a$.

Let $P.z_b$ hold and execution of a start in z_b, i.e. with input variables x_b. The execution process does not influence values of the state-variables, it being carried out over local variables of the action. The set of local variables includes x_{loc} and y_{loc}, local correspondents of input and output state-variables of a. At the onset of execution $x_{loc} := x_b$ is executed. Other than that, no state variable is accessible to program a execution process.

At some future instant the execution of a terminates with local variables $y_{loc} = a.x_b$. At this same instant the system is in state z_m, possibly different from z_b.

If z_m does not satisfy Q, the completed execution of the action has no effect on the state of the system. If, however, $Q.z_m$ holds, the system is instantaneously transferred to state z_e defined as follows:
for state-variables which are output variables of a:

$$y_e = y_{loc} = a.x_b$$

for the remaining state-variables:

$$z_e^i = Z_m^i. \text{ for all } i, \text{ except when } z^i \text{ is an output variable .}$$

Note 1. Often it is possible to determine the input and output variables of a program by purely syntactic means, for instance, by identifying all variables occurring in the right-hand side expressions of assignments within a as input variables, and all variables occurring in the left-hand sides of such assignments as output variables. In such cases one can use the simplified notation for guarded actions $(P, Q) \rightarrow [a]$, omitting the explicit listing of the frame. Observe, however, that the removal of a seemingly pointless assignment $u := u$ from the body of a changes the meaning of the guarded action. Indeed, with this assignment in place, variable u is included in the list of output variables and, therefore, when $Q.z_m$ holds, $u_e = u_b$; with the assignment removed, under identical circumstances we obtain $u_e = u_m$, a possibly different value of u and, consequently, a possibly different end state of the system. (End of Note 1.)

Thus the semantics of a completed guarded action are quite complex. In relational terms, for an accepted action it is a ternary relation, the three states z_b, z_m, z_e satisfy:

$$P.z_b \wedge Q.z_m \wedge ((\mathsf{sp}.(y := a.x_b)).Q).z_e$$

where sp is Dijkstra's strongest postcondition; if the completed action is not accepted, it is a binary relation, the two states z_b and z_m satisfy

$$P.z_b \wedge \neg Q.z_m$$

These relations are nearly useless for computing the trace of a system governed by a number of guarded actions because the "middle" state z_m depends on the duration of action execution. Fortunately, in the behavioural programming, we are not overly concerned with system traces.

Directly from the definition of guarded actions one can obtain some useful rules to manipulate "similar" programs.

Indeed, consider a behavioural program in which there occurs the pair of guarded actions:

$$\begin{array}{l}(P_1, Q_1) \rightarrow [x \mid a \mid y] \\ \|(P_2, Q_2) \rightarrow [x \mid a \mid y]\end{array} \qquad (PA)$$

1. If $P_1 \equiv P_2$ then (PA) is equivalent to a single guarded action

$$(P_1, Q_1 \vee Q_2) \rightarrow [x \mid a \mid y]$$

2. If $Q_1 \equiv Q_2$ then (PA) is equivalent to a single guarded action

$$(P_1 \vee P_2, Q_1) \rightarrow [x \mid a \mid y]$$

3. If $P_1 \Rightarrow P_2$ then (PA) is equivalent to the pair of guarded actions

$$\begin{array}{l}(P_1, Q_1 \vee Q_2) \rightarrow [x \mid a \mid y] \\ \|(P_1 \wedge P_2, Q_2) \rightarrow [x \mid a \mid y]\end{array}$$

4. If $Q_1 \Rightarrow Q_2$ then (PA) is equivalent to the pair of guarded actions

$$\begin{array}{l}(P_1 \wedge P_2, Q_1) \rightarrow [x \mid a \mid y] \\ \|(P_2, \neg Q_1 \wedge Q_2) \rightarrow [x \mid a \mid y]\end{array}$$

Rules 3 and 4 are particularly useful in conjunction with 5 or 6.

5. Guarded action

$$P, \mathbf{false}) \rightarrow [x \mid a \mid y]$$

may be eliminated from any program.

6. Guarded action

$$(\mathbf{false}, Q) \rightarrow [x \mid a \mid y]$$

may be eliminated from any program.

Selection (and, therefore, useful nondeterminism) may be traded between action bodies and behavioural program structure:

7. The pair of guarded actions

$$(P_1, Q) \rightarrow [x \mid a \mid y]$$
$$\| \ (P_2, Q) \rightarrow [x \mid b \mid y]$$

is equivalent to a single guarded action

$$(P_1 \vee P_2, Q) \rightarrow [x \mid \mathbf{if}\ P_1\ \mathbf{then}\ a\ \mathbf{else}\ b\ \mathbf{fi} \mid y]$$

or, admitting guarded statements[2] in bodies of actions, to

$$(P_1 \vee P_2, Q) \rightarrow [x \mid \mathbf{if}\ P_1 \Longrightarrow a\ []\ P_2 \Longrightarrow b\ \mathbf{fi} \mid y]$$

Rule 7 can be extended to actions with different lists of input variables. Indeed, let $x_1 + x_2$ stand for the set-theoretic sum of x_1 and x_2, then

7' The pair of guarded actions

$$(P_1, Q) \rightarrow [x_1 \mid a \mid y]$$
$$\|(P_2, Q) \rightarrow [x_2 \mid b \mid y]$$

is equivalent to the single guarded action

$$(P_1 \vee P_2, Q) \rightarrow [x_1 + x_2 \mid \mathbf{if}\ P_1 \Longrightarrow a\ []\ P_2 \Longrightarrow b\ \mathbf{fi} \mid y]$$

Note 2. For reasons explained in Note 1, a similar trick does not always work for output variables. (End of Note 2.)

An important methodological principle in reasoning about behavioural programs is the realization that it is the *middle* state, z_m, which should be in the focus of attention. We shall illustrate this principle by establishing sufficient conditions for a deadlock (termination) and progress in behaviourally programmed systems.

Consider a behavioural program

$$\|(P_i, Q_i) \rightarrow [x_i \mid a_i \mid y_i]\ i \in [0, \cdots, N] = \alpha \qquad (BP)$$

in which no P_i nor Q_i is equivalent to a boolean constant.

A state z_m such that $\forall i \in \alpha : \neg Q_i.z_m$ is called a *non-receptive* state.

---

[2] In order to differentiate between guarded actions and Dijkstra's guarded statements, in the latter we employ double arrows, $\Longrightarrow$.

Proposition 1. A system governed by (BP) which reaches (or is placed) in a non-receptive state remains in it forever or until is moved from it by an external action (i.e. action other than those of (BP)).

Proof.

By definition of non-receptive state no action of (BP) can register its results in such a state. □

Note that actions of (BP) may be ***initiated*** in a non-receptive state z_m when $\exists i \in \alpha : P_i.z_m$. Note also that when the system reaches z_m there may be actions of (BP) still being executed ("pending" actions initiated at some past instants). Thus the non-receptiveness of a state does not necessarily imply quiescence.

Insofar as "termination" and "deadlock" are meaningful terms for behaviourally programmed systems, corresponding states need not be non-receptive. Indeed, precise "location" of termination and deadlock states may depend on the system history. In this sense, the non-receptiveness, being a history-independent property, is more stable than just termination or deadlock.

Consider now a receptive state z_m and let $\beta \subset \alpha$ be a set of indices such that $\forall i \in \beta : Q_i.z_m$.

Proposition 2. If

$$\exists i \in \beta : a_i \neq \mathsf{skip} \wedge P_i.z_m$$

(or, alternatively, $\exists i \in \beta : y_i \neq \emptyset \wedge P_i.z_m$) then the system cannot remain forever in z_m.

Proof.

Indeed, assume a system with a limited number of processors is in z_m and all processors are busy executing some actions. Each of the pending actions can fall into one of the three categories:

(i) it is unacceptable at z_m,

(ii) it is acceptable at z_m but, being a skip, it does not move the system to another state,

(iii) it is acceptable at z_m and, not being a skip, it moves the system to another state.

Even if none of the actions being executed actually belongs to category (iii), as the processors are released one by one at completion of actions from the first two categories, they are assigned to actions whose preguards are satisfied at z_m. Sooner or later a β-action will be assigned a processor and, assuming the system will not have moved yet, upon completion of this action the state z_m will be changed.

Of course, when there is an unlimited number of processors a β-action will move the system from z_m straight away. □

Note 3. The above proof for a system with a limited number of processors depends on an underlying assumption of fairness: if processors become continually available, no action with a satisfied preguard may be forever denied a processor. (End of Note 3.)

Corollary. For a behaviourally programmed system with no constant guards and without skip actions, a sufficient condition to ensure progress from any receptive state is $\forall i \in \alpha : Q_i \rightarrow P_i$.

Examples. {Because of the simplicity of both examples, the lists of input and output variables are omitted. All state-variables are integer-valued.}

$$\begin{array}{ll} (y \leq 0, y \leq -\text{abs}.x - 1) \rightarrow [x := x - \text{sign}.x; y := y + 1] & \\ \|(y \geq 0, y \geq \quad \text{abs}.x + 1) \rightarrow [x := x - \text{sign}.x; y := y - 1] & \qquad (E1) \end{array}$$

The set of non-receptive states of (E1) is determined by

$$\neg(y \leq -\text{abs}.x - 1) \wedge \neg(y \geq \text{abs}.x + 1)$$

i.e.

$$-\text{abs}.x - 1 < y < \text{abs}.x + 1$$

Since (E1) contains no skip actions and conditions of the Corollary are met, the system placed in any receptive state will progress until it reaches a non-receptive state. In fact, it can easily be seen that (E1) describes a terminating behaviour, leading to (non-receptive) state (0,0).

$$\begin{array}{ll} (0 < y \leq 2, \;\; 0 \leq x \leq y \leq 2) \rightarrow [y := y - 1] & \\ (-2 \leq x < 0, \; -2 \leq x \leq y \leq 0) \rightarrow [x := x + 1] & \qquad (E2) \end{array}$$

(E2) is an example of a behavioural program with a receptive state in which progress is not guaranteed. Indeed, in (0,0) both postguards are satisfied, hence it is a receptive state. Since, however, no preguard is satisfied in (0,0) progress cannot be guaranteed, although it cannot be asserted that the system placed in (0,0) will remain there forever (remember pending actions!). (End of examples.)

4 Bibliographic note

The programming paradigm extension proposed here and its background have been described in several articles published in various journals, cf.

BIT, **28**(1988), 473-486;
Acta Informatica, **27**(1990), 685-696;
Information Processing Letters, **37**(1991), 171-174;
Theoretical Computer Science, **90**(1991), 119-125;
Structured Programming, **13**(1992), 1-9.

Calculating a Path Algorithm

Roland C. Backhouse and A.J.M. van Gasteren

Department of Mathematics and Computing Science,
Eindhoven University of Technology,
P.O. Box 513, 5600 MB Eindhoven,
The Netherlands.

Abstract. A calculational derivation is given of an abstract path algorithm, one instance of the algorithm being Dijkstra's shortest-path algorithm, another being breadth-first/depth-first search of a directed graph. The basis for the derivation is the algebra of regular languages.

1 Problem Statement

Given is a (non-empty) set N and a $|N| \times |N|$ matrix A, the rows and columns of which are indexed by elements of N. It is assumed that the matrix elements are drawn from a regular algebra[1] $(S, +, \cdot, *, 0, 1)$ having the two additional properties that

(1) the ordering $\leq$ induced[2] by $+$ is a total ordering on the elements of S, and

(2) 1 is the largest element in the ordering.

In addition to N and A, one is given a $1 \times |N|$ matrix b. Hereafter $1 \times |N|$ matrices will be called "vectors", 1×1 matrices will be called "elements" and $|N| \times |N|$ matrices will be called "matrices".

The problem is to derive an algorithm to compute the vector $b \cdot A*$ whereby the primitive terms in the algorithm do not involve the "$*$" operator.

2 Interpretations

The relevance and interest of the stated problem is that it is an abstraction from several path problems on labelled, directed graphs. Let $G = (N, E)$ be a directed graph with node set N and labelled-edge set E, the labels being drawn from some regular algebra $(S, +, \cdot, *, 0, 1)$. Then (as we deem known, see for instance [3] and [7],) there is a correspondence between edge sets E and $|N| \times |N|$ matrices A whereby the (i,j)th element of A equals the label of the edge from node i to node j in the graph, if present, and 0 otherwise. (For graphs with multiple edges, the correspondence will be adjusted later; note that absent edges and edges with label 0 are not distinguished.)

---

[1] Sometimes known as the algebra of regular languages. See [2] for the axiomatisation of regular algebra assumed here.

[2] $x \leq y \equiv x + y = y$

Since S forms a regular algebra $(S,+,\cdot,*,0,1)$, matrix multiplication can be defined in the usual way with the usual properties. In fact one can prove that the set of square matrices of a fixed size with entries drawn from S itself forms a regular algebra, with the usual definitions of the zero and identity matrices, matrix addition and matrix multiplication. The derivation to be presented makes extensive use of this fact. (For details see [3] .)

If the label of a path is defined to be the product of its constituent edge labels (taken in the order defined by the path), the (i,j)th entry of A^k is the sum of the labels of the paths of length k from node i to node j. Note that this still holds true if we redefine the (i,j)th entry of A to be the *sum* of the labels of all edges from i to j, thus admitting multiple edges. Moreover, if b is a vector that differs from the zero vector only in its ith entry, for some node i, then the jth entry of $b \cdot A*$ is the sum of all labels of paths beginning at node i and ending at node j.

Three interpretations of a regular algebra satisfying (1) and (2) are given in the table below. (Note that property (2) implies that $x* = 1$ for all $x \in S$. For this reason the interpretation of $*$ has been omitted.)

Table 1. Regular algebras

	S	+	·	0	1	≤
Shortest paths	nonnegative reals	$\downarrow$	$+$	∞	0	$\geq$
Reachability	booleans	$\vee$	$\wedge$	false	true	$\Rightarrow$
Bottlenecks	reals	$\uparrow$	$\downarrow$	$-\infty$	∞	$\leq$

The first interpretation is that appropriate to finding shortest paths; the length (i.e label) of a path is the (arithmetic) sum of its constituent edge lengths (labels) and for each node x the minimum (denoted in the table by "$\downarrow$") such length is sought to node x from a given, fixed start node. In this case, the algorithm we derive is known as Dijkstra's shortest path algorithm [5] (the carrier set, S, being restricted to *nonnegative* reals in order to comply with requirement (2)).

The second interpretation is that appropriate to solving reachability problems; the (i,j)th entry of matrix A is **true** if and only if there is an edge in E from node i to j, and the "length" of a path is always **true**. The algorithm we derive forms the basis of the so-called depth-first and breadth-first search methods for determining for each node in the graph whether or not there is a path in the graph to that node from the given start node. (The choice of "breadth-first" or "depth-first" search depends on further refinement steps that we do not discuss in detail.)

The third interpretation is appropriate to bottleneck problems; the edge labels may be construed as, say, bridge widths and the "length" of a path is the minimum width of bridge on that path. Sought is, for each node in the graph, the minimum bridge-width on a route to the node from a given start node that maximises that minimum. In this case the algorithm does not appear to have any specific name.

For more discussion of these interpretations see [3].

Another, somewhat simpler, application of the calculational techniques discussed here is a derivation of the Floyd/Warshall all-paths algorithm. For details see [1].

3 Selectors

Although the matrices we consider form a regular algebra, they will obviously never satisfy the requirements (1) and (2), even if their elements do. Given that we wish to appeal to these properties from time to time, it is important to keep track of which terms in our calculations denote elements, which denote vectors and which denote matrices. The rules for doing so are simple and (hopefully) familiar: the product of an $m \times n$ and an $n \times p$ matrix is an $m \times p$ matrix, and addition and $*$ preserve the dimension of their arguments. It remains, therefore, to adopt a systematic naming convention for the variables that we use. This, and a primitive mechanism for forming vectors, is the topic of this section.

During the course of the development the following naming conventions will be used.

$\mathbf{0}$	$1\times\mid N\mid$ vector that is everywhere 0.
k, j	nodes of the graph (i.e. elements of N)
L, M, P	subsets of N
V, W, X, Y, Z	$\mid N\mid \times \mid N\mid$ matrices
u, v, w, x	$1\times\mid N\mid$ vectors

For each node k we denote by $k\bullet$ the $1\times\mid N\mid$ vector that differs from $\mathbf{0}$ only in its kth component which is 1. Such a vector is called a *primitive selector vector*. The transpose of $k\bullet$ (thus a $\mid N\mid \times 1$ vector) is denoted by $\bullet k$. We define the *primitive selector matrix* $\underline{k}$ by the equation

(3) $\quad \underline{k} \;=\; \bullet k \cdot k\bullet$

Any sum (including the empty sum, of course) of primitive selector vectors (respectively matrices) is called a *selector vector* (respectively *matrix*). In fact, in most cases, instead of using one of these four terms we shall just use the term *selector*, it being clear from the context whether the designated selector is primitive or not, and a vector or a matrix.

This terminology is motivated by the interpretation of the product of a matrix and a selector. Specifically, $j\bullet \cdot Y$ is a vector consisting of a copy of the jth row of matrix Y, and $j\bullet \cdot Y \cdot \bullet k$ is the (j,k)th element of Y. Furthermore, there is a (1-1) correspondence between subsets of N and selector matrices given by the function mapping M to $\underline{M}$ where, by definition,

(4) $\quad \underline{M} \;=\; \Sigma(k : k \in M : \underline{k})$

(Note that

(5) $\quad \underline{\{k\}} \;=\; \underline{k}$

This silent "lifting" of a function from elements to sets is not uncommon and very convenient; it should not be a cause for confusion since we do not mix the two forms, preferring always to use the shorter form.)

For all vectors x and all matrices Y, $x \cdot \underline{M}$ is a copy of x except that all elements of x with index outwith M are zero, and $Y \cdot \underline{M}$ (respectively, $\underline{M} \cdot Y$) is a copy of matrix Y except that all columns (respectively, rows) of Y with index outwith M are zero.

The derivation that follows is not dependent on knowing these interpretations of the selectors; rather, we make use of a small number of characteristic algebraic properties. The first is that matrix product is associative. This is the most important property and its exploitation is the reason for introducing the primitive selectors. However, we shall nowhere explicitly mention the use of associativity, in line with the doctrine that the application of the most important properties should be invisible. The second property is that $\underline{\phi}$ (where ϕ denotes the empty set of nodes) is the zero element in the algebra of $|N| \times |N|$ matrices. In particular, for all vectors u,

(6) $\quad u \cdot \underline{\phi} \; = \; 0$

and, for all matrices X,

(7) $\quad X \cdot \underline{\phi} \; = \; \underline{\phi} \cdot X \; = \; \underline{\phi}$

and

(8) $\quad X + \underline{\phi} \; = \; X$

(These properties are immediate consequences of the definition of the "underlining" function.) Two other properties are that, for all nodes k,

(9) $\quad k_\bullet \cdot {}_\bullet k \; = \; 1$

and, for all distinct nodes j and k,

(10) $\quad j_\bullet \cdot {}_\bullet k \; = \; 0$

It follows, by straightforward calculation, that, for all sets of nodes M and all nodes k,

(11) $\quad k \in M \;\Rightarrow\; (\underline{M} \cdot {}_\bullet k = {}_\bullet k \;\wedge\; \underline{M} \cdot \underline{k} = \underline{k})$

and, for all sets of nodes L and M,

(12) $\quad \underline{L} \cdot \underline{M} \; = \; \underline{L \cap M}$

In particular,

(13) $\quad \underline{M} \cdot \underline{M} \; = \; \underline{M}$

and

(14) $\quad L \cap M = \phi \;\Rightarrow\; \underline{L} \cdot \underline{M} \; = \; \underline{\phi}$

The final property is that all matrices and all vectors are indexed by the given node set N. This we render by the equations:

(15) $\quad X \cdot \underline{N} \; = \; X \; = \; \underline{N} \cdot X$

and

(16) $\quad x \cdot \underline{N} \; = \; x$

for all matrices X and all vectors x.

4 The Key Theorem

A key insight in deriving an algorithm is that, because of properties (1) and (2), for any vector x and any matrix Y, at least one element of $x \cdot Y*$ is easy to compute. Specifically, choose k such that

(17) $\quad \forall(j : j \in N : x \cdot \bullet j \leq x \cdot \bullet k)$

Thus the kth element of x is the largest. (Such a choice can always be made because the ordering on elements is total.) Then we claim that

(18) $\quad x \cdot Y * \cdot \bullet k \; = \; x \cdot \bullet k$

In other words, no computation whatsoever is required to compute this one element.

Inevitably, the property we need in the development is more general; for instance, we shall have to comply with an additional condition $k \in M$, for some given M. As a consequence, we have to weaken (17) by strengthening its range to $j \in M$. We take the liberty of immediately stating and establishing the required generalisation, because its proof is virtually the same.

Theorem 19 For all $M \subseteq N$, $k \in N$, matrices Z and vectors u,

$$k \in M \;\wedge\; \forall(j : j \in M : u \cdot \bullet j \leq u \cdot \bullet k) \;\wedge\; Z \geq \underline{N}$$
$$\Rightarrow \quad u \cdot \underline{M} \cdot Z \cdot \bullet k \; = \; u \cdot \bullet k$$

Proof The proof is by mutual inclusion. Assume the antecedent of the implication. Then,

$$
\begin{array}{ll}
 & u \cdot \underline{M} \cdot Z \cdot \bullet k \\
= & \quad \{ \text{ definition of } \underline{M} \text{ } \} \\
 & u \cdot \Sigma(j : j \in M : \bullet j \cdot j\bullet) \cdot Z \cdot \bullet k \\
= & \quad \{ \text{ distributivity } \} \\
 & \Sigma(j : j \in M : u \cdot \bullet j \cdot j\bullet \cdot Z \cdot \bullet k) \\
\leq & \quad \{ \text{ properties of } k \text{, monotonicity } \} \\
 & \Sigma(j : j \in M : u \cdot \bullet k \cdot j\bullet \cdot Z \cdot \bullet k) \\
\leq & \quad \{ \; j\bullet \cdot Z \cdot \bullet k \text{ is an element. Hence,} \\
 & \quad\;\; j\bullet \cdot Z \cdot \bullet k \leq \{ (2) \} \; 1 \text{, monotonicity } \} \\
 & \Sigma(j : j \in M : u \cdot \bullet k \cdot 1) \\
\leq & \quad \{ \text{ calculus } \} \\
 & u \cdot \bullet k
\end{array}
$$

and,

$$
\begin{array}{ll}
 & u \cdot \underline{M} \cdot Z \cdot \bullet k \\
\geq & \quad \{ \text{ assumption: } Z \geq \underline{N} \text{ } \} \\
 & u \cdot \underline{M} \cdot \underline{N} \cdot \bullet k \\
= & \quad \{ \text{ (15) } \} \\
 & u \cdot \underline{M} \cdot \bullet k \\
= & \quad \{ \text{ assumption: } k \in M, \text{ (11) } \} \\
 & u \cdot \bullet k
\end{array}
$$

□

The reader is invited to check that property (18) does indeed follow from (17) by application of the theorem: with u instantiated to x, M instantiated to N and Z to $Y*$ the condition $k \in M$ is automatically satisfied and $Z \geq \underline{N}$ is satisfied by virtue of the fact that $Y* \geq \underline{N}$ for all $|N| \times |N|$ matrices Y. (The latter property follows since, by definition, $Y* = \underline{N} + Y*\cdot Y$.)

5 A Skeleton Algorithm

The algorithm we develop is based on an iterative process in which at each iteration theorem 19 is used to "eliminate" one node from the calculation. At some intermediate stage some set of nodes has been eliminated and some set M of nodes remains to be considered. These heuristics are captured formally by postulating as loop invariant

$$(20) \qquad b\cdot A* \;=\; x\cdot \underline{M}\cdot Y* \;+\; w$$

and as postcondition

$$(21) \qquad b\cdot A* \;=\; w$$

where x and w are vectors and Y is a matrix. (In the course of our calculations we shall be obliged to strengthen this invariant.)

Using elementary properties of regular algebra together with the property (15) of the selector matrix $\underline{N}$ it is easy to see that the invariant is established by the assignment

$$(22) \qquad x,Y,M,w \;:=\; b,A,N,0$$

For the guard of the loop there is an abundance of choice. We shall therefore "hedge our bets" for the moment by choosing the weakest possible guard, namely the negation of the postcondition (21). Note, however, that the postcondition is implied by (although not equivalent to) the conjunction of $M = \phi$ and the loop invariant. Since N is a finite set, progress towards $M = \phi$ (and thus towards termination) is guaranteed if at each iteration one node is removed from M. The key theorem discussed in the last section provides a mechanism for choosing such a node. This is the main reason why invariant (20) is preferable to an otherwise equally reasonable invariant like

$$b\cdot A* \;=\; x\cdot Y* \;+\; w$$

Thus, the main design decision for the loop body is to choose k, $k \in M$, in such a way that

$$(23) \qquad x\cdot \underline{M}\cdot Y*\cdot \underline{k} \;=\; x\cdot \underline{k}$$

and to try to "transfer" $x\cdot \underline{k}$ from $x\cdot \underline{M}\cdot Y*$ to w. The key theorem, indeed, enables this choice since, by multiplying both sides of its consequent by $k\bullet$ and simplifying $\bullet k\cdot k\bullet$ to $\underline{k}$, we obtain the lemma: for all vectors u and matrices Z,

$$(24) \qquad Z \geq \underline{N} \;\Rightarrow\; u\cdot \underline{M}\cdot Z\cdot \underline{k} \;=\; u\cdot \underline{k}$$

This leads to the following skeleton algorithm. (For greater clarity the different components of the simultaneous assignment to M, w, x and Y have been separated by "||" symbols.)

$$
\begin{array}{lll}
 & x, Y, M, w \ := \ b, A, N, 0 & \\
; & \{ \text{ Invariant: } b \cdot A* \ = \ x \cdot \underline{M} \cdot Y * \ + \ w & \\
 & \qquad\qquad \text{Variant: } |M|\} & \\
\textsf{do} & b \cdot A* \ \neq \ w \ \longrightarrow & \\
 & \qquad \textsf{choose } k \in M \text{ s.t. } \forall(j : j \in M : x \cdot \bullet j \leq x \cdot \bullet k) & \\
; & \qquad\quad M := M - \{k\} & \\
 & \qquad \| \ w := w' & \\
 & \qquad \| \ x, Y := x', Y' & \\
\textsf{od} & & \\
\{ b \cdot A* \ = \ w\} & &
\end{array}
$$

For convenience we let P denote $M - \{k\}$; i.e.

(25) $\quad \underline{M} \ = \ \underline{P} + \underline{k}$

Then, by a standard use of the assignment axiom, we obtain a requirement on the three unknowns, namely:

(26) $\quad x \cdot \underline{M} \cdot Y * \ + \ w \ = \ x' \cdot \underline{P} \cdot (Y')* \ + \ w'$

Before embarking on the calculation dealing with (26) it may be helpful to summarise the properties of regular algebra that are useful for dealing with expressions $Z*$. For all matrices Z the defining equation for $Z*$ is

(27) $\quad Z* \ = \ \underline{N} + Z* \cdot Z$

and, hence,

(28) $\quad Z* \ \geq \ \underline{N}$

Finally, two general properties of a regular algebra are the rule we call "the leapfrog rule": for all X and Y,

(29) $\quad V \cdot (W \cdot V)* \ = \ (V \cdot W)* \cdot V$

and the rule.we call "star decomposition":

(30) $\quad (V{+}W)* \ = \ (V* \cdot W)* \cdot V*$

6 The Assignment to w

We begin our calculation by considering the transfer of $x \cdot \underline{k}$ from $x \cdot \underline{M} \cdot Y*$ to w. That is to say, inspired by (23) and (25), we propose choosing $w' \ = \ x \cdot \underline{k} \ + \ w$ and seek a suitable term Z such that

(31) $\quad x \cdot \underline{M} \cdot Y* \ = \ Z + x \cdot \underline{k}$

We have:

$$
\begin{array}{ll}
 & x \cdot \underline{M} \cdot Y* \\
= & \quad \{ \ \underline{M} = \underline{M} \cdot \underline{M} \\
 & \quad \bullet \ \text{assume } Y = Y \cdot \underline{M} \\
 & \quad \text{in preparation for the use of the leapfrog rule } \} \\
 & x \cdot \underline{M} \cdot \underline{M} \cdot (Y \cdot \underline{M})* \\
= & \quad \{ \ \text{leapfrog rule } \} \\
 & x \cdot \underline{M} \cdot (\underline{M} \cdot Y)* \cdot \underline{M} \\
= & \quad \{ \ \underline{M} = \underline{P} + \underline{k} \ \} \\
 & x \cdot \underline{M} \cdot (\underline{M} \cdot Y)* \cdot (\underline{P} + \underline{k}) \\
= & \quad \{ \ \text{distributivity } \} \\
 & x \cdot \underline{M} \cdot (\underline{M} \cdot Y)* \cdot \underline{P} + x \cdot \underline{M} \cdot (\underline{M} \cdot Y)* \cdot \underline{k} \\
= & \quad \{ \ \text{choice of } k, \text{(28) and (24) } \} \\
 & x \cdot \underline{M} \cdot (\underline{M} \cdot Y)* \cdot \underline{P} + x \cdot \underline{k}
\end{array}
$$

In this calculation the first step is the most crucial. The goal is to reap as much advantage as possible from the key property (24). Knowing that "$\underline{k}$" can always be introduced into our calculations by the identity $\underline{M} = \underline{P} + \underline{k}$, the strategy is to retain "$x \cdot \underline{M}$" whilst simultaneously introducing a second occurrence of "$\underline{M}$" on the right of the expression being manipulated. (Quotation marks are used in this discussion because the considerations are entirely syntactic.) To do this an extra property of the matrix Y is required — as indicated by the bullet in the first hint. In order to accommodate this extra requirement we strengthen the invariant to:

(32) $\quad b \cdot A* \ = \ x \cdot \underline{M} \cdot (Y \cdot \underline{M})* \ + \ w$

whereby Y has been replaced by $Y \cdot \underline{M}$. Incorporating the assignment to w and the new invariant we arrive at the following refinement of our initial skeleton algorithm.

```
  x,Y,M,w := b,A,N,0
;   { Invariant:  b·A*  =  x·M·(Y·M)* + w
                  Variant:  |M|}
  do    b · A*   ≠  w   —→
                  choose  k ∈ M s.t. ∀(j : j ∈ M : x·•j ≤ x·•k)
    ;               M := P where P = M − {k}
                  || w := x·k + w
                  || x,Y := x',Y'
  od
{ b·A*  =  w}
```

Strengthening the invariant demands that we check that the initialisation is still correct — which it is on account of (15) — and revise the requirement on x' and Y'. Applying the assignment axiom once again, maintenance of the invariant is guaranteed by the identity

(33) $\quad x \cdot \underline{M} \cdot (Y \cdot \underline{M})* \ + w \ = \ x' \cdot \underline{P} \cdot (Y' \cdot \underline{P})* \ + \ x \cdot \underline{k} \ + \ w$

However, by the calculation above,

$$
\begin{array}{ll}
 & x \cdot \underline{M} \cdot (Y \cdot \underline{M})* \;+\; w \\
= & \\
 & x \cdot \underline{M} \cdot (\underline{M} \cdot Y)* \cdot \underline{P} + x \cdot \underline{k} + w
\end{array}
$$

Hence, it suffices to choose x' and Y' so that

$$(34) \qquad x \cdot \underline{M} \cdot (\underline{M} \cdot Y) * \cdot \underline{P} \;=\; x' \cdot \underline{P} \cdot (Y' \cdot \underline{P})*$$

7 Using Star Decomposition

One half of our calculation is now complete. In the remaining half we deal with the assignments to x and Y. Our first tactic is to get into a position to invoke the star-decomposition rule. Considering the left side of (34) but omitting the final term in the product, we have:

$$
\begin{array}{ll}
 & x \cdot \underline{M} \cdot (\underline{M} \cdot Y)* \\
= & \quad \{ \underline{M} = \underline{P} + \underline{k}, \text{ distributivity } \} \\
 & x \cdot \underline{M} \cdot (\underline{P} \cdot Y + \underline{k} \cdot Y)*
\end{array}
$$

The desired position has been reached. We now aim to use the choice of k — more precisely, (24) — with Z instantiated to as large a value as possible. This involves exposing "$\underline{k}$" outwith the "$*$" operator. For brevity and clarity introduce the temporary abbreviations V for $\underline{P} \cdot Y$ and W for $\underline{k} \cdot Y$. Then we proceed as follows.

$$
\begin{array}{ll}
 & x \cdot \underline{M} \cdot (\underline{P} \cdot Y + \underline{k} \cdot Y)* \\
= & \quad \{ \text{ definitions of } V \text{ and } W \ \} \\
 & x \cdot \underline{M} \cdot (V + W)*
\end{array}
$$

= { consider the term $(V+W)*$. Recalling that $W = \underline{k} \cdot Y$, we try to derive an equal expression in which W is exposed as far to the right as possible.

$$
\begin{array}{ll}
 & (V + W)* \\
= & \quad \{ \text{ star decomposition } \} \\
 & (V* \cdot W)* \cdot V* \\
= & \quad \{ \ (27) \ \} \\
 & (\underline{N} + (V* \cdot W)* \cdot V* \cdot W) \cdot V* \\
= & \quad \{ \text{ star decomposition } \} \\
 & (\underline{N} + (V + W)* \cdot W) \cdot V*
\end{array}
$$

}

$$
\begin{array}{ll}
 & x \cdot \underline{M} \cdot (\underline{N} + (V + W)* \cdot W) \cdot V* \\
= & \quad \{ \text{ distributivity, definition of } W \ \} \\
 & (x \cdot \underline{M} \cdot \underline{N} + x \cdot \underline{M} \cdot (V + W)* \cdot \underline{k} \cdot Y) \cdot V* \\
= & \quad \{ \ (15), \text{ and } (24) \text{ using } (V{+}W)* \geq \{ (28) \} \ \underline{N} \ \} \\
 & (x \cdot \underline{M} + x \cdot \underline{k} \cdot Y) \cdot V* \\
= & \quad \{ \text{ definition of } V \ \} \\
 & (x \cdot \underline{M} + x \cdot \underline{k} \cdot Y) \cdot (\underline{P} \cdot Y)*
\end{array}
$$

Multiplying both sides of the obtained equality by the omitted term $\underline{P}$ and continuing we get:

$$
\begin{array}{ll}
 & x \cdot \underline{M} \cdot (\underline{M} \cdot Y)* \cdot \underline{P} \\
= & \qquad \{ \text{ above calculation } \} \\
 & (x \cdot \underline{M} + x \cdot \underline{k} \cdot Y) \cdot (\underline{P} \cdot Y)* \cdot \underline{P} \\
= & \qquad \{ \text{ leapfrog rule } \} \\
 & (x \cdot \underline{M} + x \cdot \underline{k} \cdot Y) \cdot \underline{P} \cdot (Y \cdot \underline{P})* \\
= & \qquad \{ \text{ distributivity; and } P \subseteq M, \\
 & \qquad \quad \text{thus } x \cdot \underline{M} \cdot \underline{P} \; = \; \{ (12) \} \; x \cdot \underline{P} \\
 & \qquad \} \\
 & (x + x \cdot \underline{k} \cdot Y) \cdot \underline{P} \cdot (Y \cdot \underline{P})*
\end{array}
$$

Our calculation is now complete. Comparing the last line of the above calculation with the stated requirements on x' and Y' (see equation (34)), one sees that $Y = Y'$ and $x' \cdot \underline{P} = (x + x \cdot \underline{k} \cdot Y) \cdot \underline{P}$ suffice. Hence possible choices for x' are

$$x + x \cdot \underline{k} \cdot Y$$

and, since $\underline{P} \cdot \underline{P} = \underline{P}$,

$$(x + x \cdot \underline{k} \cdot Y) \cdot \underline{P}$$

We choose the latter for reasons discussed in the next section, the main reason being that this choice, by construction, yields

$$x \cdot \underline{M} = x$$

as an additional invariant. (It holds initially, i.e. for $M = N$.) Thus, apart from simplification of the termination condition, the complete algorithm is as follows:

$$
\begin{array}{lll}
 & x, Y, M, w \; := \; b, A, N, 0 & \\
; & \{ \text{ Invariant: } & b \cdot A* \; = \; x \cdot \underline{M} \cdot (Y \cdot \underline{M})* \; + \; w \\
 & & \text{Variant: } \; |M|\} \\
\textsf{do} & b \cdot A* & \neq \; w \; \longrightarrow \\
 & & \texttt{choose} \;\; k \in M \text{ s.t. } \forall(j : j \in M : x \cdot \bullet j \leq x \cdot \bullet k) \\
 ; & & \quad M \; := \; P \\
 & & \| \;\; w \; := \; x \cdot \underline{k} + w \\
 & & \| \;\; x \; := \; (x + x \cdot \underline{k} \cdot Y) \cdot \underline{P} \\
 & & \quad \texttt{where } P = M - \{k\} \\
\textsf{od} & & \\
\{ \; b \cdot A* & = \; w\} &
\end{array}
$$

8 Elementwise Implementation

In this final section we make one last significant modification to the invariant, we indicate how to reexpress the vector assignments in terms of elementwise operations and we at long last make a decision on the termination condition. Some remarks on the relationship between the algorithm presented here and conventional descriptions of Dijkstra's shortest-path algorithm and of traversal algorithms are also included.

A simple, but nevertheless significant, observation is that, with L defined by

(35) $\quad L \;=\; N - M$

the assignments to w establish and subsequently maintain invariant the property:

(36) $\quad w \;=\; w \cdot \underline{L}$

(The set L would conventionally be called the "black" nodes; see, for example, Dijkstra and Feijen's account [6] of Dijkstra's shortest-path algorithm.) Exploitation of this property was anticipated when we expressed our preferred choice of assignment to x. For, as stated in the previous section, the assignments to x establish and maintain invariant the property

(37) $\quad x \;=\; x \cdot \underline{M}$

Now, suppose we introduce the vector u where, by definition,

(38) $\quad u \;=\; x + w$

Then, since $L \cap M \;=\; \phi$, we have by (13) and (14)

(39) $\quad x \;=\; u \cdot \underline{M}$

and

(40) $\quad w \;=\; u \cdot \underline{L}$

In other words, there is a (1-1) correspondence between the vector u and the pair of vectors (x, w). We are thus free to replace this pair in the algorithm by u. This we do as follows.

In the invariant we use (37), (39) and (40) to remove occurrences of x and w; we obtain:

(41) $\quad b \cdot A* \;=\; u \cdot \underline{M} \cdot (Y \cdot \underline{M})* \;+\; u \cdot \underline{L}$

In the statement 'choose k ...', in the body of the loop, we replace occurrences of x by u; this is allowed since by (39) and (11) we have, for $j \in M$,

$$x \cdot \bullet j \;=\; u \cdot \bullet j$$

For the assignments to u we simply add the right sides of the assignments to x and w (cf. (38)) and again use (39) and (40) to remove all traces of these variables. We have

$$
\begin{array}{ll}
 & (x + x \cdot \underline{k} \cdot Y) \cdot \underline{P} + x \cdot \underline{k} + w \\
= & \quad \{ \text{ distributivity } \} \\
 & x \cdot \underline{P} + x \cdot \underline{k} \cdot Y \cdot \underline{P} + x \cdot \underline{k} + w \\
= & \quad \{ \text{ commutativity of } + \ \} \\
 & x \cdot \underline{P} + x \cdot \underline{k} + w + x \cdot \underline{k} \cdot Y \cdot \underline{P} \\
= & \quad \{ \text{ distributivity, } \underline{M} = \underline{P} + \underline{k} \ \}
\end{array}
$$

$$
\begin{array}{ll}
& x \cdot \underline{M} + w + x \cdot \underline{k} \cdot Y \cdot \underline{P} \\
= & \quad \{ \ (37), (38) \text{ and } (39) \ \} \\
& u + u \cdot \underline{M} \cdot \underline{k} \cdot Y \cdot \underline{P} \\
= & \quad \{ \ k \in M, (11) \ \} \\
& u + u \cdot \underline{k} \cdot Y \cdot \underline{P}
\end{array}
$$

Thus, in the body of the loop the assignment to u is

(42) $\quad u \ := \ u + u \cdot \underline{k} \cdot Y \cdot \underline{P}$

What this entails at element level can be ascertained by postmultiplying by $\bullet j$, for each node j. For $j \in P$, we calculate as follows:

$$
\begin{array}{ll}
& (u + u \cdot \underline{k} \cdot Y \cdot \underline{P}) \cdot \bullet j \\
= & \quad \{ \ \text{distributivity} \ \} \\
& u \cdot \bullet j + u \cdot \underline{k} \cdot Y \cdot \underline{P} \cdot \bullet j \\
= & \quad \{ \ \text{definition of } \underline{k}, (5); \text{ assumption } j \in P \text{ and } (11) \ \} \\
& (u \cdot \bullet j) + (u \cdot \bullet k) \cdot (k\bullet \cdot Y \cdot \bullet j)
\end{array}
$$

Parentheses have been included in the last line in order to indicate which subexpressions are elements. For $j \notin P$ the second summand —in the penultimate formula of the calculation— reduces to zero, since $\underline{P} \cdot \bullet j \ = \ 0$. I.e.

(43) $\quad (u + u \cdot \underline{k} \cdot Y \cdot \underline{P}) \cdot \bullet j \ = \ u \cdot \bullet j$

The final decision we have to make is on the termination condition. For our own objectives here we make do with the simplest possible termination condition, namely $M \ = \ \phi$, thus obtaining the following algorithm:

$u, Y, M \ := \ b, A, N$
$\quad ; \quad \{$ Invariant: $b \cdot A* \ = \ u \cdot \underline{M} \cdot (Y \cdot \underline{M})* \ + \ u \cdot \underline{N - M}$
$\qquad$ Variant: $|M|\}$
do $M \neq \phi \ \longrightarrow$
$\qquad$ **choose** $k \in M$ s.t. $\forall (j : \ j \in M : \ u \cdot \bullet j \leq u \cdot \bullet k)$
$\qquad ; \quad M \ := \ P$
$\qquad \| \ $ **parfor** $j \in P$ **do**
$\qquad\qquad u \cdot \bullet j \ := \ u \cdot \bullet j \ + \ (u \cdot \bullet k) \cdot (k\bullet \cdot Y \cdot \bullet j)$
$\qquad$ **where** $P \ = \ M - \{k\}$
od
$\{ \ b \cdot A* \ = \ u \}$

Note that, since $k \notin P$, the parallel **for** statement can be replaced by a sequential **for** statement.

As for other termination conditions, we remarked earlier that the set L corresponds to the so-called "black" nodes in Dijkstra and Feijen's description of Dijkstra's shortest-path algorithm. They also distinguish "white" and "grey" nodes. The "grey" nodes are just nodes $j \in M$ for which $u \cdot \bullet j \neq 0$, and the "white" nodes are the remaining nodes in M; in some circumstances (e.g. if the graph is not connected) advantage can be made from the special multiplicative and additive properties of the zero element of the algebra. In this case, a suitable choice for the termination condition would be $u \cdot \underline{M} = 0$: then $u \cdot \underline{N - M} = u$ and, hence, $b \cdot A* \;=\; 0 \cdot (Y \cdot \underline{M})* + u \;=\; 0 + u \;=\; u$. (In conventional terminology, this termination condition reads "the set of grey nodes is empty"; conventionally, this is, indeed, the choice made.)

A termination condition $u \cdot \underline{M} = 0$ requires the algorithm to keep track of the nodes $j \in M$ for which $u \cdot \bullet j \neq 0$. We won't go into the precise details of such an addition to the algorithm, but we finish by noting that, in the interpretation pertaining to reachability problems (see the table in section 2), keeping track of the "grey" nodes in a queue leads to a breadth-first traversal algorithm; using a stack instead of a queue leads to depth-first traversal.

9 Commentary and Credits

The goal of this report has been to show how a class of standard path algorithms can be derived by algebraic calculation. This is, of course, not the first and nor (we hope) will it be the last such derivation. The algebraic basis for the calculation given here was laid in [2], and some of its details were influenced by Carré 's derivation [3] of the same algorithm. A great many other authors have described and applied related algebraic systems to a variety of programming problems; Tarjan's paper [7] includes many references. The main distinguishing feature of the development presented here, however, is its reliance on calculations with matrices rather than with matrix elements, resulting (in our view) in a pleasingly compact presentation.

References

1. R.C. Backhouse. Calculating the Floyd/Warshall path algorithm. Eindhoven University of Technology, Department of Computing Science, 1992.
2. R.C. Backhouse and B.A. Carré. Regular algebra applied to path-finding problems. *Journal of the Institute of Mathematics and its Applications*, 15:161–186, 1975.
3. B.A. Carré. *Graphs and Networks.* Oxford University Press, 1979.
4. P. Chisholm. Calculation by computer. In *Third International Workshop Software Engineering and its Applications*, pages 713–728, Toulouse, France, December 3-7 1990. EC2.
5. E.W. Dijkstra. A note on two problems in connexion with graphs. *Numerische Mathematik*, 1:269–271, 1959.
6. E.W. Dijkstra and W.H.J. Feijen. *Een Methode van Programmeren.* Academic Service, Den Haag, 1984. Also available as *A Method of Programming*, Addison-Wesley, Reading, Mass., 1988.
7. R.E. Tarjan. A unified approach to path problems. *Journal of the Association for Computing Machinery*, 28:577–593, 1981.

Acknowledgements

Preparation of this report was expedited by the use of the proof editor developed by Paul Chisholm [4].

Solving Optimisation Problems with Catamorphisms

Richard S. Bird and Oege de Moor*

Oxford University Programming Research Group
11 Keble Road, Oxford OX1 3QD

Abstract. This paper contributes to an ongoing effort to construct a calculus for deriving programs for optimisation problems. The calculus is built around the notion of initial data types and *catamorphisms* which are homomorphisms on initial data types. It is shown how certain optimisation problems, which are specified in terms of a relational catamorphism, can be solved by means of a functional catamorphism. The result is illustrated with a derivation of Kruskal's algorithm for finding a minimum spanning tree in a connected graph.

1 Introduction

Efficient algorithms for solving optimisation problems can sometimes be expressed as homomorphisms on initial data types. Such homomorphisms, which correspond to the familiar *fold* operators in functional programming, are called *catamorphisms*. In this paper, we give conditions under which an optimisation problem can be solved by a catamorphism. Our results are a natural generalisation of earlier work by Jeuring [5, 6], who considered the same problem in a slightly less abstract setting. The main contribution of the paper is to show how several seemingly disparate results about subsequences, permutations, sequence partitions, subtrees, and so on, can all be captured in a single theorem.

The specification of an optimisation problem is usually split into three components, called the *generator*, *filter*, and *selector*, respectively. Consider, for example, the construction of a longest ascending subsequence of a sequence of numbers. This problem might be specified as

$$las = max(\#) \cdot up\triangleleft \cdot subs.$$

The function $subs$ is the generator and returns the set of all subsequences of a list. The filter $up\triangleleft$ retains only those subsequences which are ascending. The remaining term $max(\#)$ is the selector; it is a relation such that $a(max(\#))x$ holds if a is a longest sequence in x.

We can develop an algorithm from a specification like the above in two steps. First, we express the composition of the filter and generator as a catamorphism. The conditions under which this is possible are given by a generalisation of Malcolm's *promotion* theorem [8]. Second, we take the resulting catamorphism g and try to

* Research supported by the Dutch Organisation for Scientific Research, grant NFI 62–518.

express the composition $max(\#) \cdot g$ as a second catamorphism. This is also achieved by promotion but, as we shall see, the conditions are much more restrictive.

Both the generator and the result of the first step are catamorphisms returning sets of values. Using the fact that set-valued functions are isomorphic to relations, we can express many set-valued catamorphisms as *relational* catamorphisms. Briefly, these are catamorphisms to algebras in which the operators are arbitrary relations rather than functions. It turns out that this observation allows us to simplify the formal treatment substantially.

The structure of the rest of the paper is as follows. In Section 2, we discuss the basic framework of sets and functions, and define the notion of catamorphism formally. We also state and prove the promotion theorem. In Section 3 we exploit the isomorphism between relations and set-valued functions to extend this material to relations. In Section 4 we define the notion of a relational catamorphism. Examples of such catamorphisms are given in Section 5. The next step in the development is to study the properties of the relation $max(R)$, and we do this in Section 6. In particular, we will need a weak form of monotonicity, called *maxotonicity*. With this machinery we can state and prove the main theorem of the paper. In the last section we show how the theorem applies to one particular example, namely the derivation of Kruskal's algorithm for minimum cost spanning trees.

2 Preliminaries

Throughout, we assume that we are working in one of two categories: the category *Fun* of sets and total functions, and (later on) the category *Rel* of sets and relations. The restriction to set theory is for purely expository reasons, as we could have started with a suitable topos and its category of relations instead of *Fun* and *Rel* [9]. It is possible to phrase the axioms of such a topos in terms of elementary identities between relations [4], and most of our results can be derived from these identities by equational reasoning. However, rather than taking such an abstract course, we will explain ideas using set-theoretic concepts. In particular, only a passing acquaintance with category theory is assumed, and we shall review all definitions essential to the exposition.

Functions and functors. Total functions will be denoted by lower case letters. We write $f : A \to B$ to indicate that f has source type A and target type B. The identity function of type $A \to A$ is written id_A, though we usually omit the subscript. Functional composition is denoted by a dot, so $(f \cdot g)\, a = f(g\, a)$.

We shall also need the notion of a *functor*. Functors, which will be denoted by sans serif letters, are homomorphisms between categories that preserve identity arrows and composition. For example, $\mathsf{F} : Fun \to Fun$ is a functor if $\mathsf{F}A$ is an object (i.e. set or 'type') of *Fun* for each object A of *Fun* and $\mathsf{F}f : \mathsf{F}A \to \mathsf{F}B$ for each arrow (or 'function') $f : A \to B$ of *Fun*. Moreover, $\mathsf{F}id = id$ and $\mathsf{F}(f \cdot g) = \mathsf{F}f \cdot \mathsf{F}g$. Functor composition is denoted simply by juxtaposition, so we have $(\mathsf{FG})f = \mathsf{F}(\mathsf{G}f)$. Usually, we write $\mathsf{FG}f$ without any brackets.

One important example of a functor is the *power* functor P. Here, $\mathsf{P}A$ denotes the power set of A, and for $f : A \to B$ the function $\mathsf{P}f : \mathsf{P}A \to \mathsf{P}B$ is defined by

$$(\mathsf{P}f)x = \{f\ a \mid a \in x\}.$$

In words, $\mathsf{P}f$ applies f to every element of a set. The verification that P is indeed a functor is straightforward.

Products. The *product* of two sets B and C is a set $B \times C$ together with two *projection* functions, $\pi_1 : B \times C \rightarrow B$ and $\pi_2 : B \times C \rightarrow C$, and a binary operator $\langle \text{-}, \text{-} \rangle$ called *split*. Given two functions $f : A \rightarrow B$ and $g : A \rightarrow C$, we have $\langle f, g \rangle : A \rightarrow B \times C$. These operators satisfy

$$h = \langle f, g \rangle \equiv (\pi_1 \cdot h = f \ \wedge \ \pi_2 \cdot h = g)$$

for all $h : A \rightarrow B \times C$. This property characterises product up to isomorphism, so we are free to choose any representative from the isomorphism class determined by the definition. In set theory, the canonical choice for $B \times C$ is cartesian product, and the split operator is defined by

$$\langle f, g \rangle a = (f\ a, g\ a).$$

The operator $\times$ on sets can also be defined on functions: $f \times g = \langle f \cdot \pi_1, g \cdot \pi_2 \rangle$. This turns $\times$ into a functor (strictly speaking, a *bifunctor*), because $\times$ preserves composition and identities:

$$(f \times g) \cdot (h \times k) = (f \cdot h) \times (g \cdot k) \quad \text{and} \quad id_A \times id_B = id_{A \times B}\ .$$

Coproducts. Analogous to the definition of *product*, we can define the notion of *coproduct*. The coproduct of two sets B and C is a set $B + C$ together with two *injection* functions $\iota_1 : B \rightarrow (B + C)$ and $\iota_2 : C \rightarrow (B + C)$, and a binary operator $[\text{-}, \text{-}]$ called *join*. Given two functions $f : B \rightarrow A$ and $g : C \rightarrow A$, the function $[f, g]$ has type $B + C \rightarrow A$. These operators satisfy

$$h = [f, g] \equiv (h \cdot \iota_1 = f \ \wedge \ h \cdot \iota_2 = g).$$

This property characterises coproduct up to isomorphism. In set theory, the canonical choice for $B + C$ is the disjoint sum, defined by

$$B + C \ = \ \{ (1, b) \mid b \in B \} \cup \{ (2, c) \mid c \in C \}\ .$$

The join operator is defined by

$$[f, g](1, b) = f\ b \text{ and } [f, g](2, c) = g\ c.$$

Just like $\times$, we can make $+$ into a functor by defining

$$f + g = [\iota_1 \cdot f,\ \iota_2 \cdot g].$$

Polynomial Functors. The class of *polynomial* functors in *Fun* $\rightarrow$ *Fun* is defined inductively by the following clauses:

1. The identity functor and constant functors are polynomial.
2. If F and G are polynomial functors, then so are their composition FG, their sum $\mathsf{F} + \mathsf{G}$, and their product $\mathsf{F} \times \mathsf{G}$, where

$$(\mathsf{F} + \mathsf{G})(k) = \mathsf{F}(k) + \mathsf{G}(k)$$
$$(\mathsf{F} \times \mathsf{G})(k) = \mathsf{F}(k) \times \mathsf{G}(k).$$

As we shall see, polynomial functors enjoy a number of useful properties. One example of a functor that is not polynomial is the power functor P.

Algebras. Let $\mathcal{E}$ be a category and $\mathsf{F} : \mathcal{E} \to \mathcal{E}$ a functor. By definition, an F–*algebra* is an arrow $f : \mathsf{F}A \to A$ of $\mathcal{E}$. For example, consider the set L of finite sequences with elements of type E. Among other ways, we can construct all finite sequences with the help of two functions: the constant function $\nu : \mathbf{1} \to L$, which returns the empty sequence $[]$, and the binary operator $+\!\!\!+\!\!< : L \times E \to L$ (pronounced *snoc*), which takes a sequence and an element, and appends the element to the sequence:

$$[e_0, e_1, \ldots, e_{n-1}] +\!\!\!+\!\!< e_n = [e_0, e_1, \ldots, e_{n-1}, e_n].$$

(The set $\mathbf{1}$ in the type of ν stands for some distinguished one–element set and plays the role of *unit type*.)

The join $[\nu, +\!\!\!+\!\!<]$ therefore has type type $\mathbf{1} + (L \times E) \to L$. This join is an F–algebra, where $\mathsf{F} : \mathit{Fun} \to \mathit{Fun}$ is given by by

$$\begin{aligned} \mathsf{F}A &= \mathbf{1} + (A \times E) \\ \mathsf{F}h &= \mathit{id} + (h \times \mathit{id}) \ . \end{aligned}$$

Homomorphisms, initial algebras, and catamorphisms. Let $f : \mathsf{F}A \to A$ and $g : \mathsf{F}B \to B$ be F–algebras. By definition, an F–*homomorphism* from f to g is an arrow $h : A \to B$ such that

$$h \cdot f = g \cdot \mathsf{F}h \ .$$

The composition of two F–homomorphisms is again an F–homomorphism, so the F–algebras in $\mathcal{E}$ form the objects of a category $\mathcal{E}_\mathsf{F}$ in which the arrows are F–homomorphisms. For many functors F this category has an *initial object*, which we denote by $\mu\mathsf{F}$. To say that $\mu\mathsf{F}$ is initial means that for any other F–algebra f, there exists a unique homomorphism, which we denote by $(\!|f|\!)_\mathsf{F}$, from $\mu\mathsf{F}$ to f. Arrows of the form $(\!|h|\!)_\mathsf{F}$ are called *catamorphisms*.

The defining property of catamorphisms is called the *unique extension* property and is formalised by the equivalence

$$g = (\!|f|\!)_\mathsf{F} \equiv g \cdot \mu\mathsf{F} = f \cdot \mathsf{F}g \ .$$

¿From now on we shall write $(\!|f|\!)$ instead of $(\!|f|\!)_\mathsf{F}$, the functor F being understood from context.

An example of an initial F–algebra is the data type of lists discussed above. The so–called *fold–left* operator $\mathit{foldl}\,(\oplus)\, e$ common in functional programming can be defined by two recursion equations

$$\begin{aligned} \mathit{foldl}\,(\oplus)\, e\, [] &= e \\ \mathit{foldl}\,(\oplus)\, e\, (x +\!\!\!+\!\!< a) &= (\mathit{foldl}\,(\oplus)\, e\, x) \oplus a \ . \end{aligned}$$

These equations are equivalent to the statement that $\mathit{foldl}\,(\oplus)\, e$ is an F–homomorphism from $[\nu, +\!\!\!+\!\!<]$ to $[e, \oplus]$. Since $[\nu, +\!\!\!+\!\!<]$ is the initial F–algebra $\mu\mathsf{F}$, we can write $\mathit{foldl}\,(\oplus)\, e = (\!|[e, \oplus]|\!)$. For legibility we will omit the inner brackets in such expressions, writing $(\!|e, \oplus|\!)$ instead.

Promotion. The following theorem is called the *promotion* theorem. In its statement we use the abbreviation $f = g \pmod{h}$ for $f \cdot h = g \cdot h$.

Theorem 1. *For any functor* $\mathsf{F} : \mathit{Fun} \to \mathit{Fun}$ *with an initial algebra we have*

$$f \cdot (\!|g|\!) = (\!|h|\!) \equiv f \cdot g = h \cdot \mathsf{F}f \pmod{\mathsf{F}(\!|g|\!)}.$$

Proof. The proof is simple:

$$\begin{aligned}
& f \cdot (\!|g|\!) = (\!|h|\!) \\
\equiv \quad & \{\text{unique extension property}\} \\
& f \cdot (\!|g|\!) \cdot \mu\mathsf{F} = h \cdot \mathsf{F}(f \cdot (\!|g|\!)) \\
\equiv \quad & \{\mathsf{F} \text{ is a functor}\} \\
& f \cdot (\!|g|\!) \cdot \mu\mathsf{F} = h \cdot \mathsf{F}f \cdot \mathsf{F}(\!|g|\!) \\
\equiv \quad & \{\text{unique extension property}\} \\
& f \cdot g \cdot \mathsf{F}(\!|g|\!) = h \cdot \mathsf{F}f \cdot \mathsf{F}(\!|g|\!).
\end{aligned}$$

□

3 Relations

So far we have been working in the category *Fun* of sets and total functions. Now we turn to *Rel*, the category of sets and relations. Our objective is to see how and when the above theory of functions can be extended to relations.

Relations will be denoted by upper case letters $R, S, \ldots$. We write $R : A \to B$ to denote the fact that R has source A and target B. For such an R we write $b(R)a$ to denote the fact that b stands in the relation R to a. Functions are special kinds of relations, namely total and single–valued ones, so if f is a function, then $b(f)a$ means $b = f\ a$. We again use a dot for relational composition, so $c(R \cdot S)a$ holds if there is a b with $c(R)b$ and $b(S)a$. We will assume familiarity with the operations of the relational calculus displayed in Figure 1.

operator	pronunciation	meaning
R°	R converse	$b(R^{\circ})a = a(R)b$
$R \cap S$	R and S	$b(R \cap S)a = b(R)a \wedge b(S)a$
$R \cup S$	R or S	$b(R \cup S)a = b(R)a \vee b(S)a$
$R \Rightarrow S$	R implies S	$b(R \Rightarrow S)a = b(R)a \Rightarrow b(S)a$
R/S	R divided by S	$b(R/S)a = (\forall c : a(S)c \Rightarrow b(R)c)$
$p?$	test p	$b(p?)a = (b = a) \wedge (p\ a)$
$R\Box$	domain of R	$b(R\Box)a = (b = a) \wedge (\exists c : c(R)a)$
$\Box R$	range of R	$b(\Box R)a = (b = a) \wedge (\exists c : b(R)c)$

Fig. 1. Operators of the relational calculus.

Power Transpose. In set theory there are various representations of relations, of which the most well-known is to represent a relation $R : A \rightarrow B$ as a set of pairs (a, b) with $a \in A$ and $b \in B$. However, one can also think of a relation $R : A \rightarrow B$ as a set-valued function $A \rightarrow \mathsf{P}B$, where $\mathsf{P}B$ denotes the power set of B. We will take the view that a relation is a set of pairs, so that $b(R)a$ means $(a, b) \in R$. The operator Λ that sends a relation to the corresponding set-valued function is called *power transpose* and is defined by the equation

$$(\Lambda R)\, a = \{b \mid b(R)a\}.$$

In particular, $(\Lambda f)a = \{f\, a\}$. The function Λ is an isomorphism and so the inverse function Λ^{-1} is well-defined. For $R : A \rightarrow B$ and $f : A \rightarrow \mathsf{P}B$ we have

$$\in_B \cdot \Lambda R \;=\; R \quad \text{and} \quad \Lambda(\in_B \cdot f) \;=\; f,$$

where $\in_B : \mathsf{P}B \rightarrow B$ is the membership relation. Usually, we omit the subscript.

Existential Image. Readers familiar with category theory will recognise that power transpose defines a right adjoint of the inclusion functor *Fun* $\rightarrow$ *Rel*. This right adjoint is known as the *existential image*, and the adjunction is called the *power adjunction*. In fact, most of set theory can be derived from the existence of the power adjunction, an observation which lies at the heart of topos theory.

Less abstractly, the existential image of a relation $R : A \rightarrow B$ is a function $\mathsf{E}R : \mathsf{P}A \rightarrow \mathsf{P}B$, defined by

$$(\mathsf{E}R)\, x = \{b \mid (\exists a \in x : b(R)a)\}.$$

The existential image preserves identity and composition and, since we can take $\mathsf{E}A = \mathsf{P}A$, we have that E is a functor from *Rel* to *Fun*.

Note that $\mathsf{E}f = \mathsf{P}f$ for all functions f; in words, E and P coincide on functions. However, E is not the only functor of *Rel* that coincides with P on functions. In the next section we shall describe another and more useful one.

The power transpose Λ and the existential image E can each be defined in terms of the other. For $R : A \rightarrow B$ we have

$$\Lambda R \;=\; \mathsf{E}R \cdot \tau_A \quad \text{and} \quad \mathsf{E}R \;=\; \Lambda(R \cdot \in_A)\,,$$

where $\tau_A : A \rightarrow \mathsf{P}A$ is the function which sends an element a in A to the singleton set $\{a\}$. From the first identity we obtain the law of composition

$$\Lambda(R \cdot S) = \mathsf{E}R \cdot \Lambda S.$$

Singleton and Membership. The singleton function τ (again, we omit the subscript) can be defined in terms of power transpose: $\tau = \Lambda id$. The collection of singleton formers is 'polymorphic', a fact which is expressed by the identity $\mathsf{P}f \cdot \tau = \tau \cdot f$. Category theorists refer to this equation by saying that τ is a *natural transformation* from the identity functor to P.

The membership relation is dual to singleton in that $\in = \Lambda^{-1} id$. Membership is also a natural transformation; in symbols, $\in \cdot \mathsf{E}R = R \cdot \in$. This duality is no accident and stems from the fact that membership and singleton form the unit and counit of the power adjunction.

The 'big union' function $\bigcup$ is defined by $\bigcup = \mathsf{E}\in$. Big union is also a natural transformation, the proof being

$$\bigcup \cdot \mathsf{EE}R = \mathsf{E}\in \cdot \mathsf{EE}R = \mathsf{E}(\in \cdot \mathsf{E}R) = \mathsf{E}(R \cdot \in) = \mathsf{E}R \cdot \mathsf{E}\in = \mathsf{E}R \cdot \bigcup.$$

We also have

$$\mathsf{E}R = \mathsf{E}(\in \cdot \Lambda R) = \bigcup \cdot \mathsf{E}(\Lambda R).$$

(For those familiar with monads, $(\mathsf{E}, \tau, \bigcup)$ is the monad defined by the power adjunction.)

Filter. Many other set-theoretic operators can be defined in terms of power transpose and existential image. One important example is the *filter* operator. Given a Boolean-valued function $p : A \rightarrow \Omega$, we define $p\triangleleft : \mathsf{P}A \rightarrow \mathsf{P}A$ by $p\triangleleft = \mathsf{E}(p?)$. The advantage of such definitions is that they are easier to manipulate than traditional set comprehensions. For instance, here is the very short proof that filter distributes through union:

$$p\triangleleft \cdot \bigcup = \mathsf{E}(p?) \cdot \bigcup = \bigcup \cdot \mathsf{EE}(p?) = \bigcup \cdot \mathsf{E}(p\triangleleft).$$

4 Relators and Cross-operators

Let us now turn to the question of whether each functor F of *Fun* can be extended to a functor of *Rel*. An important consideration is that we would like $\mathsf{F} : Rel \rightarrow Rel$ to be *monotonic*, i.e.

$$R \subseteq S \Rightarrow \mathsf{F}R \subseteq \mathsf{F}S.$$

Inclusion rather than equality is the 'natural' way of comparing relations. Of course, for total functions inclusion means equality. Fortunately, we have

Proposition 2. *For each functor* $\mathsf{F} : Fun \rightarrow Fun$ *there exists at most one monotonic functor in* $Rel \rightarrow Rel$ *that coincides with* F *on functions.*

If F does have such an extension, then we say that F is a *relator* and we use the same symbol for both functors. We can characterise relators as those functors F that satisfy the condition

$$f^{\circ} \cdot g = h \cdot k^{\circ} \Rightarrow (\mathsf{F}f)^{\circ} \cdot \mathsf{F}g = \mathsf{F}h \cdot (\mathsf{F}k)^{\circ}$$

It can be shown that every monotonic functor on relations maps functions to functions, so every monotonic functor is the extension of a relator.

For example, all polynomial functors are relators. The list functor L is also a relator; this functor sends a set A to the set of finite sequences with elements from A. On functions, L is defined by the equation

$$(\mathsf{L}f)[a_1, a_2, \ldots, a_n] = [fa_1, fa_2, \ldots, fa_n].$$

Thus L is the familiar *map* operator from functional programming and is similar to P except that it acts on lists rather than sets. The explicit definition of $\mathsf{L} : Rel \rightarrow Rel$ is rather messy:

$$[b_1, \ldots, b_n](\mathsf{L}R)[a_1, \ldots, a_m] = (n = m) \;\wedge\; (\forall i : 1 \leq i \leq n : b_i(R)a_i).$$

The functor $\mathsf{P} : Fun \rightarrow Fun$ also satisfies the above conditions, although it is not polynomial. The extension $\mathsf{P} : Rel \rightarrow Rel$ is defined explicitly by

$$\begin{aligned} y(\mathsf{P}R)x = \; & (\forall a \in x : \exists b \in y : b(R)a) \;\wedge \\ & (\forall b \in y : \exists a \in x : b(R)a)\;. \end{aligned}$$

Hence $\mathsf{P}R$ corresponds to the *Egli–Milner* ordering induced by R. Note that $\mathsf{P} : Rel \rightarrow Rel$ is *not* the same as $\mathsf{E} : Rel \rightarrow Rel$ since P returns relations and is monotonic, while E returns functions and is not monotonic.

Cross-operators. Relators can be characterised in another useful way, namely in terms of the existence of so-called *cross-operators.* Cross-operators are a generalisation of the polymorphic cartesian product function.

By definition, a *cross-operator* on $\mathsf{F} : Fun \rightarrow Fun$ is a collection of functions $\mathsf{F}\dagger_A : \mathsf{FP}A \rightarrow \mathsf{PF}A$ that satisfies the following four axioms:

1. Crossing is polymorphic, i.e. $\mathsf{F}\dagger \cdot \mathsf{FP}f = \mathsf{PF}f \cdot \mathsf{F}\dagger$.
2. Crossing singletons gives a singleton, i.e. $\mathsf{F}\dagger \cdot \mathsf{F}\tau = \tau$.
3. Crossing distributes over union, i.e. $\mathsf{F}\dagger \cdot \mathsf{F}\bigcup = \bigcup \cdot \mathsf{PF}\dagger \cdot \mathsf{F}\dagger$.
4. Crossing is monotonic, i.e. $f \subseteq g \Rightarrow \mathsf{F}\dagger \cdot \mathsf{F}f \subseteq \mathsf{F}\dagger \cdot \mathsf{F}g$, where $f \subseteq g$ means pointwise inclusion of set-valued functions f and g.

Proposition 3. *For each functor* $\mathsf{F} : Fun \rightarrow Fun$ *there exists at most one cross-operator* $\mathsf{F}\dagger$. *This cross-operator exists if and only if* F *is a relator, in which case we have* $\mathsf{F}\dagger = \Lambda(\mathsf{F}\in)$.

For an example of a particular cross-operator, recall the list functor L. The cross-operator $\mathsf{L}\dagger$ is the polymorphic function that sends a sequence of sets to its cartesian product:

$$\mathsf{L}\dagger[x_1, x_2, \ldots, x_n] = \{[a_1, a_2, \ldots, a_n] \mid (\forall i : a_i \in x_i)\}.$$

It is possible to develop a little calculus of cross-operators for synthesising the cross-operator of a composite functor by simple calculation. Reluctantly, we omit a discussion of this calculus for reasons of space.

Relational Catamorphisms. So far we have only considered algebras and catamorphisms in the category *Fun* of sets and total functions. This is sufficient for many purposes, but for a satisfactory treatment of optimisation problems, and in particular for the selector $min\, R$, we would like to move to relations. So, how can the above theory be extended to algebras and catamorphisms in *Rel*? The obvious setting for relational algebras is the category Rel_{F} for some relator $\mathsf{F} : Fun \rightarrow Fun$ with an initial algebra. Fortunately, the initial algebra of $\mathsf{F} : Rel \rightarrow Rel$ coincides with the initial algebra of $\mathsf{F} : Fun \rightarrow Fun$.

Proposition 4. (Eilenberg and Wright [3]) *Let* $\mathsf{F} : \mathit{Fun} \to \mathit{Fun}$ *be a relator with initial algebra* $\mu\mathsf{F}$. *Then* $\mu\mathsf{F}$ *is also an initial algebra of the extension of* F *to* $\mathit{Rel} \to \mathit{Rel}$. *Furthermore,* $\Lambda(\!|R|\!) = (\!|\mathsf{E}R \cdot \mathsf{F}\dagger|\!)$.

We also have the following generalisation of the promotion theorem.

Theorem 5. (Backhouse et al. [1]) *Let* F *be a relator and possess an initial algebra. Then*

$$R \cdot (\!|S|\!) \subseteq (\!|T|\!) \Leftarrow R \cdot S \subseteq T \cdot \mathsf{F}R \quad (\mathrm{mod}\ \mathsf{F}(\!|S|\!)).$$

Moreover, the same implication holds when $\subseteq$ *is replaced by* $\supseteq$ *(and so also when* $\subseteq$ *is replaced by* $=$*).*

We will not prove the generalisation, but we will prove the following corollary:

Corollary 6.

$$p\triangleleft \cdot \Lambda(\!|R|\!) = \Lambda(\!|S|\!) \Leftarrow p? \cdot R = S \cdot \mathsf{F}(p?) \quad (\mathrm{mod}\ \mathsf{F}(\!|R|\!)).$$

Proof. We have

$$\begin{aligned}
& p\triangleleft \cdot \Lambda(\!|R|\!) \\
= \quad & \{\text{definition of filter}\} \\
& \mathsf{E}(p?) \cdot \Lambda(\!|R|\!) \\
= \quad & \{\text{power transpose of composition}\} \\
& \Lambda(p? \cdot (\!|R|\!)) \\
= \quad & \{\text{promotion}\} \\
& \Lambda(\!|S|\!)\ .
\end{aligned}$$

5 Examples

Let us now give some examples of relational catamorphisms. The first three are catamorphisms on the data type of finite sequences described above.

Subsequences. Consider a finite sequence x. By leaving out some of the elements of x, but retaining the original left-to-right order, we obtain a *subsequence* of x. Formally, y is a subsequence of x if

$$y\,(\!|\nu,\ \pi_1 \cup +\!\!\!+\!\!<|\!)\,x\ .$$

One can think of $(\pi_1 \cup +\!\!\!+\!\!<)$ as a non-deterministic operator which, when given a pair (x, a), either returns x or appends a to x. The catamorphism $(\!|\nu,\ \pi_1 \cup +\!\!\!+\!\!<|\!)$ is then a relation that holds between a sequence x and a subsequence of x. It follows that the function *subs*, which returns the set of all subsequences of a sequence, is defined by

$$\mathit{subs} = \Lambda(\!|\nu,\ \pi_1 \cup +\!\!\!+\!\!<|\!)\ .$$

We can use Proposition 4 to eliminate the power transpose from the right-hand side of this equation. After simplifying, we obtain a characterisation of *subs* expressed in terms of functions: $subs = (\!|\tau \cdot \nu,\ \otimes|\!)$, where $\tau \cdot \nu$ is the constant function returning $\{[\,]\}$, and

$$xs \otimes a = xs \cup \{x +\!\!\!+\!\!< a \mid x \in xs\}.$$

The relational characterisation of *subs* seems preferable simply because it is shorter.

Let us now instantiate the corollary of the promotion theorem to the function *subs*. We shall need the following property of predicates. A predicate p is said to be *prefix-closed* with *derivative* q if $p\,[\,]$ holds and $p\,(x +\!\!\!+\!\!< a) = p\,x \wedge q(x, a)$. More shortly, p is prefix-closed with derivative q if

$$\begin{aligned} p? \cdot \nu &= \nu \\ p? \cdot +\!\!\!+\!\!< &= +\!\!\!+\!\!< \cdot\, q? \cdot (p? \times id). \end{aligned}$$

Assuming this property of p we get

$$\begin{array}{ll} & p? \cdot [\nu,\ \pi_1 \cup +\!\!\!+\!\!<] \\ = & \{\text{coproduct and composition over union}\} \\ & [p? \cdot \nu,\ (p? \cdot \pi_1) \cup (p? \cdot +\!\!\!+\!\!<)] \\ = & \{p \text{ is prefix-closed}\} \\ & [\nu,\ (p? \cdot \pi_1) \cup (+\!\!\!+\!\!< \cdot\, q? \cdot (p? \times id))] \\ = & \{\text{since } p? \cdot \pi_1 = \pi_1 \cdot (p? \times id)\} \\ & [\nu,\ (\pi_1 \cup (+\!\!\!+\!\!< \cdot\, q?))(p? \times id)] \\ = & \{\text{coproduct}\} \\ & [\nu,\ \pi_1 \cup (+\!\!\!+\!\!< \cdot\, q?)] \cdot (id + (p? \times id)) \\ = & \{\text{definition of } \mathsf{F}\} \\ & [\nu,\ \pi_1 \cup (+\!\!\!+\!\!< \cdot\, q?)] \cdot \mathsf{F}(p?). \end{array}$$

Hence if p is prefix-closed with derivative q, then

$$p\triangleleft \cdot subs = \Lambda(\!|\nu,\ \pi_1 \cup (+\!\!\!+\!\!< \cdot\, q?)|\!).$$

Eliminating Λ, we get that $p\triangleleft \cdot subs = (\!|\tau \cdot \nu,\ \otimes|\!)$, where

$$xs \otimes a = xs \cup \{x +\!\!\!+\!\!< a \mid x \in xs \ \wedge\ q(x, a)\}.$$

We will use this result in our final example, and in the appendix we give a more abstract version of the above argument.

Partitions. A *partition* of a sequence x is a division of x into non-empty contiguous subsequences. For example, the sequence of sequences

$$[[1,2],[3],[4,5,6]]$$

is a partition of $[1,2,3,4,5,6]$. The function *parts*, which returns all partitions of its argument, is the power transpose of a relational catamorphism

$$parts = \Lambda(\![\nu,\ \oplus \cup \otimes]\!),$$

where $\oplus$ is a total function defined by $xs \oplus a\ =\ xs \mathbin{+\!\!\!+\!\!<} [a]$ and $\otimes$ is the partial function defined on nonempty lists by

$$(xs \mathbin{+\!\!\!+\!\!<} x) \oplus a = xs \mathbin{+\!\!\!+\!\!<} (x \mathbin{+\!\!\!+\!\!<} a)\ .$$

In words, we get a partition of $x \mathbin{+\!\!\!+\!\!<} a$ by taking some partition xs of x, and either appending $[a]$ to xs (represented by the term $xs \oplus a$), or appending a to the last component of xs (represented by the term $xs \otimes a$). The power-transpose can be eliminated in a similar manner as before, but we omit details.

Permutations. The function *perms*, which returns all permutations of a sequence, is also the power transpose of a relational catamorphism

$$perms = \Lambda(\![\nu,\ (\oplus \cdot +\!\!\!+^{\circ} \times id)]\!),$$

where $+\!\!\!+$ denotes concatenation, and $(y,z) \oplus a = y +\!\!\!+ [a] +\!\!\!+ z$. Thus,

$$((y,z),a)(+\!\!\!+^{\circ} \times id)(x,a) \text{ if } x = y +\!\!\!+ z.$$

Eliminating Λ from this expression, we get $perms = (\![\nu\ ;\tau,\ \otimes]\!)$, where $\otimes$ is defined by the set comprehension

$$xs \otimes a = \{y +\!\!\!+ [a] +\!\!\!+ z \mid (\exists x : x \in xs\ \wedge\ y +\!\!\!+ z = x)\}.$$

Tree Pruning. Finally, we briefly describe a relational catamorphism over a different data type. Consider the type of *binary labelled trees* in which both the leaves and the internal labels are natural numbers. This type is defined as the initial F-algebra, where F is the polynomial functor given by

$$\begin{aligned} \mathsf{F}A &= \mathrm{N} + (A \times \mathrm{N} \times A) \\ \mathsf{F}k &= id_{\mathrm{N}} + (k \times id_{\mathrm{N}} \times k)\ . \end{aligned}$$

We will call the components of the initial F-algebra *Tip* and *Node*, so

$$\mu\mathsf{F} = [Tip, Node]\ .$$

Now consider the operator *prune* which takes a tree and prunes away some of its subtrees in a non-deterministic fashion. Formally, we define *prune* as a relational catamorphism:

$$prune = (\![Tip, Node \cup (Tip \cdot \pi_2^3)]\!),$$

where π_2^3 is the projection function that returns the middle component of a triple. The power transpose of *prune* is the function that returns all possible prunings of its arguments. This function was used by Jeuring [5] to solve various optimisation problems on trees.

6 Minimisation

Now we consider the selection of an optimal element from a set of candidate solutions to an optimisation problem.

Let $R : A \rightharpoonup A$ be a *preorder*, so R is assumed to be reflexive (i.e. $id \subseteq R$) and transitive (i.e. $R \cdot R \subseteq R$). The relation $min R : \mathsf{P}A \rightharpoonup A$ relates a set to its minimum elements under R. Formally, we define

$$min\, R = \in \cap (R/\ni),$$

where $\ni = \in^{\circ}$. To understand this succinct definition, put $R = (\leq)$. We then get that $b(min(\leq))x$ if and only if b is both an element of x and a lower bound of x, i.e. for all a if $a \in x$, then $b \leq a$. Note that $max\, R = min\, R^{\circ}$.

¿From the assumption that R is reflexive we get $min\, R \cdot \tau = id$. We give the proof:

$$\begin{array}{ll} & min\, R \cdot \tau \\ = & \{\text{definition of } min\, R\} \\ & (\in \cap (R/\ni)) \cdot \tau \\ = & \{\text{composition (with a function) distributes over } \cap\} \\ & (\in \cdot \tau) \cap (R/\ni) \cdot \tau \\ = & \{\in \cdot \tau = id \text{ and } (R/\ni) \cdot \tau = R\} \\ & id \cap R \\ = & \{R \text{ is reflexive}\} \\ & id. \end{array}$$

¿From the assumption that R is transitive, we get

$$min\, R \cdot \bigcup \supseteq min\, R \cdot \mathsf{P} min\, R.$$

We have $\supseteq$ rather than $=$ in this assertion since not every set has a minimum element — in particular, the empty set does not have one. However, we can replace $\supseteq$ by $=$ provided we compose the left–hand side with $\mathsf{P}((min\, R)\Box)$, so excluding sets without minimum elements under R. For reasons of space, we omit proof of these assertions.

Another useful result is that minimisation operators can always be composed:

$$min\, S \cdot \Lambda(min\, R) = min\, T,$$

where $T = R \cap (R^{\circ} \Rightarrow S)$. The proof of this identity, which we omit, has been vigorously studied by a number of researchers in the proof methods community e.g. [10].

Monotonicity and distributivity. In order to combine our results about relational catamorphisms and minimisation, we need to know how $minR$ interacts with F–algebras. Consider, for example, addition of natural numbers $+ : (\mathrm{N} \times \mathrm{N}) \to \mathrm{N}$. Addition is an algebra of the functor F given by

$$\mathsf{F}A = A \times A$$
$$\mathsf{F}k = k \times k.$$

Addition is monotonic with respect to $\leq$, a fact we can express succinctly by the assertion $(+) \cdot \mathsf{F}(\leq) \subseteq (\leq) \cdot (+)$. This assertion translates as

$$c = a + b \ \wedge \ a \leq a' \ \wedge \ b \leq b' \Rightarrow c \leq a' + b'.$$

More generally, $f : \mathsf{F}A \to A$ is *monotonic* with respect to R if

$$f \cdot \mathsf{F}R \subseteq R \cdot f.$$

Monotonicity is equivalent to distributivity. For example, writing **min** for the function that returns the (unique) minimum element of a set of numbers, we have

$$\mathbf{min}\, x + \mathbf{min}\, y = \mathbf{min}\, \{a + b \mid a \in x \ \wedge \ b \in y\},$$

provided the left-hand side is well-defined. The above implication can be expressed more shortly as

$$(+) \cdot \mathsf{F}min(\leq) \subseteq min(\leq) \cdot \mathsf{P}(+) \cdot \mathsf{F}\dagger.$$

Generalising to an arbitrary preorder R, we have

$$f \cdot \mathsf{F}R \subseteq f \cdot R \equiv f \cdot \mathsf{F}minR \subseteq minR \cdot \mathsf{P}f \cdot \mathsf{F}\dagger.$$

The next proposition is generalisation of this result. It deals with a weaker property than monotonicity, called *minotonicity*. A function $f : \mathsf{F}A \to A$ is said to be minotonic (with respect to R) on the range of a set–valued function $g : B \to \mathsf{P}A$ if

$$f \cdot \mathsf{F}(minR \cdot \Box g \cdot \ni) \subseteq R \cdot f.$$

Taking F as the identity functor, this reads

$$c = f\, b \ \wedge \ b(minR)x \ \wedge \ x(\Box g)x \ \wedge a \in x \Rightarrow c(R)(f\ a).$$

Since

$$\begin{array}{cl} & minR \cdot \Box g \cdot \ni \\ \subseteq & \{\text{since } \Box g \subseteq id\} \\ & minR \cdot \ni \\ = & \{\text{definition of } minR\} \\ & (\in \cap (R/\ni)) \cdot \ni \\ \subseteq & \{\text{calculus}\} \\ & (\in \cdot \ni) \cap (R/\ni) \cdot \ni \\ \subseteq & \{\text{since } (R/S) \cdot S \subseteq R\} \\ & R \end{array}$$

we have that $\mathsf{F}(minR \cdot \Box g \cdot \ni) \subseteq \mathsf{F}R$, and so monotonicity implies minotonicity. The following result is called the *minotonicity lemma* and the proof can be found in [2].

Lemma 7. *The function $f : \mathsf{F}A \rightarrow A$ is minotonic with respect to R on the range of $g : B \rightarrow \mathsf{P}A$ if and only if*

$$f \cdot \mathsf{F} min R \subseteq min R \cdot \mathsf{P} f \cdot \mathsf{F} \dagger \quad (\text{mod } \mathsf{F} g).$$

7 Main Theorem

Now we come to the main theorem. Let F be a fixed relator in what follows, so $(\!|\text{-}|\!) = (\!|\text{-}|\!)_{\mathsf{F}}$. The theorem addresses the question of when it is possible to find a program for computing $min R \cdot \Lambda(\!|S|\!)$ that can be expressed as a catamorphism. We take the view that a program is a total function, so the task is to determine when

$$(\!|f|\!) \subseteq min R \cdot \Lambda(\!|S|\!)$$

for some total function f.

Theorem 8. *Suppose f is minotonic with respect to a preorder R on the range of $\Lambda(\!|S|\!)$. Then*

$$f \subseteq min R \cdot \Lambda S \Rightarrow (\!|f|\!) \subseteq min R \cdot \Lambda(\!|S|\!).$$

Proof. We calculate

$$\begin{aligned}
& (\!|f|\!) \subseteq min R \cdot \Lambda(\!|S|\!) \\
\Leftrightarrow \quad & \{\text{Proposition 4}\} \\
& (\!|f|\!) \subseteq min R \cdot (\!|\mathsf{E} S \cdot \mathsf{F} \dagger|\!) \\
\Leftarrow \quad & \{\text{promotion}\} \\
& f \cdot \mathsf{F} min R \subseteq min R \cdot \mathsf{E} S \cdot \mathsf{F} \dagger \quad (\text{mod } \mathsf{F}(\!|\mathsf{E} S \cdot \mathsf{F} \dagger|\!)) \\
\Leftarrow \quad & \{\text{minotonicity}\} \\
& min R \cdot \mathsf{P} f \cdot \mathsf{F} \dagger \subseteq min R \cdot \mathsf{E} S \cdot \mathsf{F} \dagger \quad (\text{mod } \mathsf{F}(\!|\mathsf{E} S \cdot \mathsf{F} \dagger|\!)) \\
\Leftarrow \quad & \{\text{logical entailment}\} \\
& min R \cdot \mathsf{P} f \subseteq min R \cdot \mathsf{E} S.
\end{aligned}$$

Now we have

$$\begin{aligned}
& min R \cdot \mathsf{P} f \\
\subseteq \quad & \{\text{given assumption, and monotonicity of } \mathsf{P}\} \\
& min R \cdot \mathsf{P}(min R \cdot \Lambda S) \\
= \quad & \{\text{since } \mathsf{P} \text{ is a functor}\} \\
& min R \cdot \mathsf{P} min R \cdot \mathsf{P}(\Lambda S) \\
\subseteq \quad & \{\text{since } R \text{ is transitive}\} \\
& min R \cdot \bigcup \cdot \mathsf{P}(\Lambda S) \\
= \quad & \{\text{since } \bigcup \cdot \mathsf{P}(\Lambda S) = \mathsf{E} S \text{ for any } S\} \\
& min R \cdot \mathsf{E} S\ .
\end{aligned}$$

8 Application: Kruskal's Algorithm

Let us now see how the above theorem applies to one particular example, namely the construction of a minimum cost spanning tree in a connected graph. Each edge of the graph has an associated cost, and we will write *cost e* for the cost of *e*. We will assume that the graph is given as a sequence of edges in ascending order of cost, an assumption that will allow us to develop Kruskal's algorithm.

The problem can be specified as follows:

$$mcst = min\, C \cdot \Lambda(max(\#)) \cdot acyclic\triangleleft \cdot subs.$$

Here, *subs* returns the set of subsequences of the given sequence of edges, $acyclic\triangleleft$ removes all subsequences containing a cycle, and $\Lambda(max(\#))$ returns the set of acyclic sequences of edges of greatest length, i.e. the spanning trees of the graph. The function # returns the length of a sequence. It is an abuse of notation to write $max(\#)$, since # is not an ordering; strictly speaking, we should write

$$max(\#^\circ \cdot (\leq) \cdot \#) .$$

With the same understanding, C is also a function and is defined as the sum of the costs of the edges in a given sequence. Putting all this together, *mcst* is a relation that holds between a minimum cost spanning tree and the graph.

As specified, the spanning tree problem involves the composition of two optimal selections. We can use an earlier result about combining minimisation operators to show that

$$min\, C \cdot \Lambda(max(\#)) = min\, R,$$

where the preorder R is given by

$$y(R)x = (\#y > \#x) \vee (\#y = \#x \wedge C\, y \leq C\, x) .$$

It follows that the spanning tree problem is solved by minimising with respect to R. Note that

$$y(R)x \Rightarrow (y +\!\!\!\!+\!\!< a)(R)(x +\!\!\!\!+\!\!< a)$$

for all a.

The next step is to instantiate the main theorem for problems about subsequences. In the appendix, we have stated a 'categorical' version of the corollary, and we have spelled out the proof in detail. Here we just state the result as it is applied in practice.

Corollary 9. *Assume that R is a preorder, and suppose p is prefix-closed with derivative q and satisfies*

$$q\,(x, a) \Rightarrow (x +\!\!\!\!+\!\!< a)(R)x .$$

Furthermore, assume that

$$q(y, a) \wedge q(x, a) \Rightarrow (y +\!\!\!\!+\!\!< a)(R)(x +\!\!\!\!+\!\!< a)$$
$$\neg q(y, a) \wedge q(x, a) \Rightarrow y(R)(x +\!\!\!\!+\!\!< a),$$

for all x and y satisfying

$$y(min R \cdot p\triangleleft \cdot subs)z \ \wedge \ x(\in \cdot p\triangleleft \cdot subs)z$$

for some z. Then $(\![\nu, \oplus]\!) \subseteq min R \cdot p\triangleleft \cdot subs$, where $x \oplus a = x +\!\!\!+\!\!< a$ if $q(x, a)$, and x otherwise.

For the spanning tree problem, we have that the predicate *acyclic* is prefix–closed with derivative q, where $q(x, a)$ holds if a does not create a cycle when added to x. The first condition on R holds because $\#y > \#x \Rightarrow y(R)x$. The second condition holds because $y(R)x \Rightarrow (y +\!\!\!+\!\!< a)(R)(x +\!\!\!+\!\!< a)$. This leaves us with the last condition. Let x, y, and z be as stated in the corollary, so y is some minimum cost spanning tree of z and x is some acyclic subsequence of z with $y(R)x$. Suppose that $x +\!\!\!+\!\!< a$ is acyclic, but $y +\!\!\!+\!\!< a$ contains a cycle. Since y is acyclic, $y +\!\!\!+\!\!< a$ contains a cycle of the form

$$\{(v_0, v_1), (v_1, v_2), \ldots, (v_{n-1}, v_n)\},$$

where $a = (v_0, v_n)$. Because $x +\!\!\!+\!\!< a$ is acyclic, there exists an edge $b = (v_{i-1}, v_i)$ in y which is not connected in x. Furthermore, $cost\ b \leq cost\ a$ since the graph is given as a sequence of edges in ascending order of cost. As both x and y are subsequences of z, and b is in y, it follows that there exist x_1 and x_2 with $x = x_1 +\!\!\!+ x_2$ and

$$(x_1 +\!\!\!+ [b] +\!\!\!+ x_2)(\in \cdot p\triangleleft \cdot subs)z.$$

Finally, we have

$$y(R)(x_1 +\!\!\!+ [b] +\!\!\!+ x_2) \ \text{and} \ (x_1 +\!\!\!+ [b] +\!\!\!+ x_2)(R)(x_1 +\!\!\!+ x_2 +\!\!\!+ [a])$$

so $y(R)(x +\!\!\!+\!\!< a)$. The the corollary is therefore applicable, and the result is Kruskal's algorithm.

The spanning tree problem illustrates the strength of our claim that the foregoing abstract theory gives clear guidance on the difficult proof obligations that have to be met to obtain efficient algorithms. Indeed, apart from the last condition, the rest of the development consists essentially of instantiating the theory to problems about subsequences. The last condition is less mechanical and is the crucial property in the derivation of Kruskal's algorithm. (It is, in fact, the verification of the *matroid* property [7].) But this does not invalidate the claim, because the goal of our research is to provide *organising principles* for algorithm design, not necessarily to *mechanise* the whole design process.

We should emphasise that the main theorem has many more applications besides algorithms on subsequences. The papers by Jeuring [5, 6] discuss applications to subtrees, permutations and partitions. In a forthcoming paper [2], we review some of those applications, and show how the present work can be extended to include certain dynamic programming algorithms.

References

1. R.C. Backhouse, P. de Bruin, G. Malcolm, T.S. Voermans, and J. van der Woude. Relational catamorphisms. In B. Möller, editor, *Proceedings of the IFIP TC2/WG2.1 Working Conference on Constructing Programs*, pages 287–318. Elsevier Science Publishers B.V., 1991.
2. R.S. Bird and O. de Moor. Inductive solutions to optimisation problems. Draft, 1991.
3. S. Eilenberg and J.B. Wright. Automata in general algebras. *Information and Control*, 11(4):452–470, 1967.
4. P.J. Freyd and A. Ščedrov. *Categories, Allegories*, volume 39 of *Mathematical Library*. North-Holland, 1990.
5. J. Jeuring. Deriving algorithms on binary labelled trees. In P.M.G. Apers, D. Bosman, and J. van Leeuwen, editors, *Proceedings SION Computing Science in the Netherlands*, pages 229–249, 1989.
6. J. Jeuring. Algorithms from theorems. In M. Broy and C.B. Jones, editors, *Programming Concepts and Methods*, pages 247–266. North-Holland, 1990.
7. B. Korte, L. Lovasz, and R. Schrader. *Greedoids*, volume 4 of *Algorithms and combinatorics*. Springer-Verlag, 1991.
8. G. Malcolm. Data structures and program transformation. *Science of Computer Programming*, 14:255–279, 1990.
9. O. de Moor. Categories, relations and dynamic programming. D.Phil. thesis. Technical Monograph PRG-98, Computing Laboratory, Oxford, 1992.
10. J.C.S.P. van der Woude. Free style spec wrestling ii: Preorders. *The Squiggolist*, 2(2):48–53, 1991.

Appendix: Instantiation for Subsequences

Here we show how the main theorem may be instantiated for problems about subsequences. The proofs are spelled out in great detail, to support our claim that the instantiation is largely mechanical. Throughout, F stands for the functor

$$\mathsf{F}A = \mathbf{1} + (A \times E)$$
$$\mathsf{F}f = id + (f \times id) ,$$

and α abbreviates the initial algebra $[\nu, +\!\!\!+\!\!<] : \mathsf{F}L \rightarrow L$, where L is the type of lists over E.

Recall that $subs = (\![\nu, +\!\!\!+\!\!< \cup \pi_1]\!)$. It is convenient to rewrite this in the form

$$subs = \Lambda(\![\alpha \cup \delta]\!) ,$$

where $\delta = [0, \pi_1]$, and 0 is the empty relation. We have

$$[R, S] \cup [U, V] = [R \cup U, S \cup V] ,$$

so the two expressions for $subs$ are equivalent. The advantage of naming δ is that we can refer to its algebraic properties. In particular, we have that δ is a collection of partial functions

$$\delta \cdot \delta^\circ \subseteq id ,$$

which furthermore satisfies

$$\delta \cdot \mathsf{F}Q = Q \cdot \delta .$$

The second property asserts that δ is a natural transformation. Both of these inclusions are straightforward to verify.

Our first task is to express $p\triangleleft \cdot subs$ in the form $p\triangleleft \cdot subs = \Lambda(\![S]\!)$ for some S. Recall Corollary 1 of the promotion theorem which says that

$$p\triangleleft \cdot \Lambda(\![R]\!) = \Lambda(\![S]\!) \Leftarrow p? \cdot R = S \cdot \mathsf{F}p? .$$

Also, recall that p is prefix–closed with derivative q if

$$p? \cdot \nu = \nu$$
$$p? \cdot +\!\!\!+\!\!< = +\!\!\!+\!\!< \cdot q? \cdot (p? \times id) .$$

These two identities can also be formulated as a single equation, namely

$$p? \cdot \alpha = \alpha \cdot r? \cdot \mathsf{F}p?$$

with $r = [true, q]$. Assuming the latter equation, we reason:

$$
\begin{aligned}
& p? \cdot (\alpha \cup \delta) \\
= \quad & \{\text{composition over union}\} \\
& (p? \cdot \alpha) \cup (p? \cdot \delta) \\
= \quad & \{\text{assumption: } p? \cdot \alpha = \alpha \cdot r? \cdot \mathsf{F}p? \ \} \\
& (\alpha \cdot r? \cdot \mathsf{F}p?) \cup (p? \cdot \delta) \\
= \quad & \{\text{naturality of } \delta\} \\
& (\alpha \cdot r? \cdot \mathsf{F}p?) \cup (\delta \cdot \mathsf{F}p?) \\
= \quad & \{\text{composition over union}\} \\
& ((\alpha \cdot r?) \cup \delta) \cdot \mathsf{F}p?
\end{aligned}
$$

Hence by Corollary 1,

$$p\triangleleft \cdot subs = \Lambda(\![(\alpha \cdot r?) \cup \delta]\!) \ .$$

Note that $(\alpha \cdot r?) \cup \delta = [\nu, (+\!\!\!+\!\!< \cdot\, q?) \cup \pi_1]$, giving the result cited in Section 5.

The next task is to expand the minotonicity condition

$$f \cdot \mathsf{F}(min R \cdot \Box\Lambda(\![S]\!) \cdot \ni) \subseteq R \cdot f$$

of the main theorem, in the case $S = (\alpha \cdot r?) \cup \delta$ and $f = [\nu, \oplus]$, where

$$x \oplus a = \begin{cases} x +\!\!\!+\!\!< a \text{ if } & q(x, a) \\ x & \text{otherwise} \end{cases}$$

We can rewrite f using the McCarthy conditional $(p \rightarrow R, S)$ defined by

$$(p \rightarrow R, S) = (R \cdot p?) \cup (S \cdot \neg p?) \ ,$$

where $\neg p? = (\neg p)?$. Recall that r was defined by $r = [true, q]$, so we have

$$f = (r \rightarrow \alpha, \delta) \ .$$

Writing $T = min R \cdot \Box\Lambda(\![(\alpha \cdot r?) \cup \delta]\!) \cdot \ni$, we need conditions under which

$$(r \rightarrow \alpha, \delta) \cdot \mathsf{F}T \subseteq R \cdot (r \rightarrow \alpha, \delta) \tag{1}$$

holds. We can rewrite (1) using the following two properties of conditionals

$$
\begin{aligned}
R \cdot (p \rightarrow S, T) &= (p \rightarrow R \cdot S, R \cdot T) \\
R \subseteq (p \rightarrow S, T) &\equiv (R \cdot p? \subseteq S) \wedge (R \cdot \neg p? \subseteq T) \ .
\end{aligned}
$$

Now, inclusion (1) translates to

$$
\begin{aligned}
(r \rightarrow \alpha, \delta) \cdot \mathsf{F}T \cdot r? &\subseteq R \cdot \alpha \\
(r \rightarrow \alpha, \delta) \cdot \mathsf{F}T \cdot \neg r? &\subseteq R \cdot \delta \ .
\end{aligned}
$$

Using the definition of conditions, these inclusions expand to four more:

$$\alpha \cdot r? \cdot \mathsf{F}T \cdot r? \subseteq R \cdot \alpha$$
$$\delta \cdot \neg r? \cdot \mathsf{F}T \cdot r? \subseteq R \cdot \alpha$$
$$\alpha \cdot r? \cdot \mathsf{F}T \cdot \neg r? \subseteq R \cdot \delta$$
$$\delta \cdot \neg r? \cdot \mathsf{F}T \cdot \neg r? \subseteq R \cdot \delta \; .$$

We shall assume that the first two conjuncts are satisfied. Given that $\alpha \cdot r? \subseteq R \cdot \delta$, we can prove the last two:

$$
\begin{array}{ll}
 & \alpha \cdot r? \cdot \mathsf{F}T \cdot \neg r? \\
\subseteq & \{\text{since } \neg r? \subseteq id\} \\
 & \alpha \cdot r? \cdot \mathsf{F}T \\
\subseteq & \{\text{assumption: } \alpha \cdot r? \subseteq R \cdot \delta\} \\
 & R \cdot \delta \cdot \mathsf{F}T \\
= & \{\text{naturality of } \delta\} \\
 & R \cdot T \cdot \delta \\
\subseteq & \{\text{since } T \subseteq R \text{ (Section 6)}\} \\
 & R \cdot R \cdot \delta \\
\subseteq & \{\text{assumption: } R \text{ transitive}\} \\
 & R \cdot \delta \; .
\end{array}
$$

The proof of the last inclusion is:

$$
\begin{array}{ll}
 & \delta \cdot \neg r? \cdot \mathsf{F}T \cdot \neg r? \\
\subseteq & \{\text{since } \neg r? \subseteq id\} \\
 & \delta \cdot \mathsf{F}T \\
= & \{\text{naturality of delta}\} \\
 & T \cdot \delta \\
\subseteq & \{\text{since } T \subseteq R \text{ (Section 6)}\} \\
 & R \cdot \delta \; .
\end{array}
$$

This completes the instantiation of minotonicity.

Finally we need to expand the hypothesis of the main theorem, namely

$$f \subseteq min R \cdot \Lambda S$$

where $f = (r \to \alpha, \delta)$ and $S = (\alpha \cdot r?) \cup \delta$. Using the definition of conditionals, the hypothesis reads

$$\alpha \cdot r? \subseteq min R \cdot \Lambda((\alpha \cdot r?) \cup \delta)$$
$$\delta \cdot \neg r? \subseteq min R \cdot \Lambda((\alpha \cdot r?) \cup \delta) \; .$$

To establish these results, we use the following equivalence (which was not needed in the main text), whose proof we omit

$$U \subseteq min R \cdot \Lambda V \equiv (U \subseteq V) \wedge (U \cdot V^{\circ} \subseteq R) .$$

The hypothesis is therefore established if we can show

$$\begin{aligned} \alpha \cdot r? &\subseteq (\alpha \cdot r?) \cup \delta \\ \delta \cdot \neg r? &\subseteq (\alpha \cdot r?) \cup \delta \\ \alpha \cdot r? \cdot ((\alpha \cdot r?) \cup \delta)^{\circ} &\subseteq R \\ \delta \cdot \neg r? \cdot ((\alpha \cdot r?) \cup \delta)^{\circ} &\subseteq R . \end{aligned}$$

The first inclusion is immediate, and the second follows from $\neg r? \subseteq id$. For the third we reason

$$\begin{array}{ll} & \alpha \cdot r? \cdot ((\alpha \cdot r?) \cup \delta)^{\circ} \\ \subseteq & \{\text{since } \alpha \cdot r? \text{ is a partial function}\} \\ & id \cup (\alpha \cdot r? \cdot \delta^{\circ}) \\ \subseteq & \{\text{assumption: } \alpha \cdot r? \subseteq R \cdot \delta\} \\ & id \cup (R \cdot \delta \cdot \delta^{\circ}) \\ \subseteq & \{\text{since } \delta \text{ is a partial function}\} \\ & id \cup R \\ = & \{\text{assumption: } R \text{ reflexive}\} \\ & R . \end{array}$$

This proves the first inequation. For the second, we reason:

$$\begin{array}{ll} & \delta \cdot \neg r? \cdot ((\alpha \cdot r?) \cup \delta)^{\circ} \\ = & \{\text{relation calculus}\} \\ & (\delta \cdot \neg r? \cdot r?^{\circ} \cdot \alpha^{\circ}) \cup (\delta \cdot \neg r? \cdot \delta^{\circ}) \\ = & \{\text{since } \neg r? \cdot r?^{\circ} = 0\} \\ & \delta \cdot \neg r? \cdot \delta^{\circ} \\ \subseteq & \{\text{since } \neg r? \subseteq id\} \\ & \delta \cdot \delta^{\circ} \\ \subseteq & \{\text{since } \delta \text{ is a partial function}\} \\ & id \\ \subseteq & \{\text{assumption: } R \text{ reflexive}\} \\ & R . \end{array}$$

This completes the instantiation of the hypothesis in the main theorem.

Let us now summarize the conditions. We have assumed R is reflexive and transitive, and so a preorder. We have supposed $p? \cdot \alpha = \alpha \cdot \mathsf{F}p? \cdot r?$, which is true when

p is prefix-closed with derivative q, for then one can take $r = [\mathit{true}, q]$. We have also supposed that

$$\alpha \cdot r? \subseteq R \cdot \delta$$
$$\alpha \cdot r? \cdot \mathsf{F}T \cdot r? \subseteq R \cdot \alpha$$
$$\delta \cdot r? \cdot \mathsf{F}T \cdot \neg r? \subseteq R \cdot \alpha \, .$$

With $r = [\mathit{true}, q]$, these are precisely the three conditions enunciated in Corollary 2 of the main theorem.

A Time-Interval Calculus

S. M. Brien

Oxford University Computing Laboratory,
Programming Research Group,
8–11 Keble Road,
Oxford OX1 3QD.

Abstract. The purpose of this paper is to introduce a notation for expressing the requirements of time-critical systems and a calculus for reasoning about them. The Actions, Events and States of a system are represented by sets of time intervals for which they hold. Firstly the timing model is introduced and the calculus is compared with Tarski's calculus of relations. Then states and duration are introduced and a case study is provided. Finally the connection with the duration calculus and other temporal logics is shown.

1 Introduction

The area of system design where there is the greatest complexity and least understanding is that of real-time systems. Such time-critical systems abound. Examples include aircraft control systems, process monitoring in industrial plants and in robotics. It is hoped that rigorous analysis at the more abstract level of requirements could help to reduce the final complexity of the implementation of these systems, while ensuring total reliability.

The following set of requirements for a lift system, based on studies in [1, 14, 19], are typical:

The lift has a a set of buttons and lights for each floor and a motor to move it from one floor to another. The operation of the lift satisfies the following:

- Pressing the button for a floor causes the lift to eventually visit that floor.
- If the lift does not visit a floor within 0.1 seconds of its button being pressed then its light turns on.
- The light turns on only within 0.1 seconds of a button press.
- The light turns off only when the floor is visited.
- The light for a floor is never on when the lift is at that floor.
- The lift should travel between floors within 10 seconds.
- The lift only leaves the floor when the motor is running.

The interval calculus considers the intervals of time representing the Actions, Events and States of a system. A notation is provided for expressing the relationships between them. Laws are provided for reasoning about properties of a system described in this way.

2 Intervals and Relations

A *time interval* is defined by the pair (b, e), where b, e are drawn from a set $\mathcal{T}$ of points in time. A set S of time intervals is a relation between the beginning and end points of those intervals in S. So, a set of intervals can be interpreted as all the occasions when an interval-based temporal predicate holds. Tarski's calculus of relations provides a convenient starting point.

The *events* of a system occur only in empty intervals; indicating that they represent points in time. A set of events corresponds to a subset of the identity relation. The set of all events is denoted by $\lceil\,\rceil$. In the lift system described above, the pressing of the buttons for the floor is denoted by the events in *But*; and the arrival of the lift at the floor by *Arrive*. The events marking the turning on and off of the light are contained in the sets *Light_On* and *Light_Off* respectively. Events can be used to model transitions from one state to another as the following requirement shows:

Req. 1. The light for a floor is turned off only when the lift arrives as the floor:

$$Light\_Off \;\subseteq\; Arrive$$

In order to develop a useful model for timing requirements, it is helpful to identify a subset of $\mathcal{U}$ the universal set of intervals. It is assumed that there is a partial-order $(\leq)$ over the points of time in $\mathcal{T}$. The intervals that obey this order are of most interest: they represent the forward passage of time.

Definition 2. A *well-formed* interval (b, e) is one for which $b \leq e$.

While it is useful to develop the theory within Tarski's [20] richer model, any requirements expressed should consider only such well formed intervals.

The actions of a system are those activities which take some definite time; hence they are modelled by sets of non-empty well-formed intervals. For example the set *Motor* in the lift requirements represents the action of the lift travelling from one floor to another.

In [20] Tarski states fifteen theorems which form the axiomatisation of the relational calculus. They provide a definition of a boolean algebra with the addition of composition(;) and converse ($R^{\smile}$) operators. The universal relation is denoted by $\mathcal{U}$. In this paper the identity relation is denoted by $\lceil\,\rceil$.

The sets $(\leq)$ and are respectively the top and bottom elements of the complete lattice of sets of well-formed intervals over the subset ordering with $\cap, \cup$ as the meet and join respectively. Within the calculus they are represented by the symbols $\top$ and $\bot$.

Definition 3. $\top \;=\; (\leq)$ and $\bot \;=\; \{\,\}$.

3 Relative Completeness

The well-formed time interval sets are closed under the set operations and under relational composition. However, the converse and complement operators do not preserve well-formedness.

3.1 Complement

A restricted form of the complement operator is defined as the complement with respect to $\top$:[1]

Definition 4. For any set S of well-formed intervals its *restricted complement* $\sim S$ is defined by the following properties:

$$S \cup \sim S = \top$$
$$S \cap \sim S = \bot$$

This acts as negation in the Boolean Algebra.

3.2 Truncation Operators

The converse operator raises other issues. The set of all well formed intervals $\top$ has the following property:

$$\top \cap \top^{\smile} = \lceil\,\rceil$$

Thus no meaningful analogue of converse can be defined which satisfies the closure property. The world of well-formed intervals is not as complete as relations because the converse operator cannot be used. In a manner similar to the definition of weakest pre-specification in [10] we wish to provide a complementary operation to composition. Hoare and He have shown that by providing such constructions as primitives the converse operator can be derived. We wish to develop a calculus of comparable expressive power but without a converse operator; we consider time to be progressing in only one direction. The truncation operators [3] provide a means of considering the remainder of one interval subtracted from another.

The *pre-truncation* of P with Q contains the final sub-interval of each interval in P whose corresponding initial sub interval is in Q. The following algebraic property defines pre-truncation:[2]

Definition 5. If P, Q are sets of well formed intervals, then for any set X of well-formed intervals, the pre-truncation of P with Q has the following property:

$$(Q \searrow P) \subseteq \sim X \Leftrightarrow P \subseteq \sim (Q \,;\, X)$$

Pre-truncation is disjunctive in both operands and has $\lceil\,\rceil$ as its right identity. The following theorem shows how pre-truncation distributes through composition on the right:

Theorem 6.

$$(Q \,;\, R) \searrow P = R \searrow (Q \searrow P)$$

---

[1] In the richer relational model this is defined by: $\sim S = \overline{S} \cap \top$.

[2] This operation has the following relational definition: $P \searrow Q = Q^{\smile} \,;\, P \cap \top$.

Proof:

$$((Q\ ;\ R)\setminus P)\ \subseteq\ \sim X$$

$$\Leftrightarrow \quad \{\text{by definition of } \setminus\}$$

$$P\ \subseteq\ \sim((Q\ ;\ R)\ ;\ X)$$

$$\Leftrightarrow \quad \{\text{by associativity of } ;\}$$

$$P\ \subseteq\ \sim(Q\ ;\ (R\ ;\ X))$$

$$\Leftrightarrow \quad \{\text{by definition of } \setminus\}$$

$$(Q\setminus P)\ \subseteq\ \sim(R\ ;\ X)$$

$$\Leftrightarrow \quad \{\text{by definition of } \setminus\}$$

$$R\setminus(Q\setminus P)\ \subseteq\ \sim X$$

The *post-truncation* of P with R holds for the initial sub-interval of an interval satisfying P whose final sub interval satisfies R.The following property defines post-truncation.

Definition 7. If P, R are sets of well formed intervals, then for any set X of well-formed intervals, the post-truncation of P with R has the following property:

$$(P/R) \subseteq \sim X \;\Leftrightarrow\; (X\ ;\ R) \subseteq \sim P$$

Post-truncation is disjunctive in both operands and has $\lceil\,\rceil$ as its left identity. It distributes through composition in a similar manner to pre-truncation.

Although they are not associative, the two truncation operations do associate with each other. The application of both pre and post truncation is the equivalent to truncating the beginning and end of an interval Q by R and P respectively:

Theorem 8.

$$(R\setminus P)/Q \;=\; R\setminus(P/Q)$$

4 States

The actions and events of a system are easily modeled by non-empty and empty intervals respectively. The states of a system require a more complicated model because of their particular properties. A State holds for non-empty intervals over which some predicate is continuously true. States are not considered to hold momentarily, but always for some non-empty interval. If a state holds for an interval then it holds for all its non-empty sub-intervals. If a state holds for two consecutive intervals the it holds for their composition. Finite variability of states makes their transitive closure well defined.

Definition 9. A *State* is a set of non-empty intervals which is closed under composition and contains all its non-empty sub-intervals and whose transitive closure is well-defined. The set of intervals S is a state if it has the following four properties:

1 $$\mathsf{S} \subseteq \sim\lceil\,\rceil$$

2 $$\mathsf{S} \cup \lceil\,\rceil \supseteq \mathsf{T} \searrow \mathsf{S} \swarrow \mathsf{T}$$

3 $$\mathsf{S} \,;\, \mathsf{S} \subseteq \mathsf{S}$$

4 $$\mathsf{S}^{+} \subseteq X \Leftrightarrow (\mathsf{S} \subseteq X \wedge X \,;\, \mathsf{S} \subseteq X)$$

For ease of recognition states will be written in sans-serif font. The sates of the lift system are Light and Floor representing the times when the light is on and when the lift is at the floor respectively.

4.1 The Complete Lattice of States

A state is a special set of intervals with the properties defined above. The interval predicate which represents a state is analogous to a boolean function on points in time which is also subject to the finite variability constraint. The boolean operations on pointwise functions can be reproduced for states by considering the meet and join of the ordering ($\subseteq$) over states.

Definition 10. The state true holds all the time and the state false never holds. So for any state X:[3]

$$\mathsf{false} \subseteq \mathsf{X} \subseteq \mathsf{true}\,.$$

Definition 11. The conjunction of states S and R is the greatest state which satisfies both of them. So for any state X:

$$\mathsf{X} \subseteq \mathsf{S} \sqcap \mathsf{R} \Leftrightarrow \mathsf{X} \subseteq \mathsf{S} \wedge \mathsf{X} \subseteq \mathsf{R}\,.$$

The state conjunction $\mathsf{S} \sqcap \mathsf{R}$ has the same value as $\mathsf{S} \cap \mathsf{R}$.

Definition 12. The disjunction of states S and R is the least state which both of them satisfy. So for any state X:

$$\mathsf{S} \sqcup \mathsf{R} \subseteq \mathsf{X} \Leftrightarrow \mathsf{S} \subseteq \mathsf{X} \wedge \mathsf{R} \subseteq \mathsf{X}\,.$$

The state $\mathsf{S} \sqcup \mathsf{R}$ holds for those intervals for which either S or R holds or in any intervals composed from them: $(\mathsf{S} \cup \mathsf{R})^{+}$.

Definition 13. The negation of the state S is the largest state which is disjoint from S. So for states X and Y:

$$\mathsf{S} \sqcap \mathsf{X} \subseteq \mathsf{Y} \Leftrightarrow \mathsf{X} \subseteq \mathsf{Y} \sqcup \neg\mathsf{S}\,.$$

The state $\neg\mathsf{S}$ holds in those non-empty intervals which have no sub-interval for which S holds: $\sim(\lceil\,\rceil \cup \mathsf{T} \,;\, \mathsf{S} \,;\, \mathsf{T})$.

The negation of a state can be used to say that something never happens; consider the following two lift requirements:

[3] In the interval model, the state true has the same value as $\sim\lceil\,\rceil$ and false $= \bot$.

Req. 14. The call light for a floor is only illuminated when the lift is not at the floor.

$$\mathsf{Light} \subseteq \neg\mathsf{Floor}$$

Req. 15. The light goes out only when the floor is reached.

$$\mathsf{Light}\ ;\ \neg\mathsf{Floor} \subseteq \mathsf{Light}$$

4.2 Reflexive Transitive Closure

The reflexive transitive closure of a state is true for any empty interval as well as whenever the transitive closure is true.

Definition 16. For any state S and predicate X:

$$\mathsf{S}^* \subseteq \mathsf{X} \Leftrightarrow (\lceil\rceil \subseteq \mathsf{X} \wedge \mathsf{X}\ ;\mathsf{S} \subseteq \mathsf{X})$$

Every interval satisfies the reflexive transitive closure of a state or its negation.

Theorem 17. *For state* S*:*

$$\mathsf{T} = (\mathsf{S} \cup \neg\mathsf{S})^*$$

The following induction rule permits the proof of hypotheses by extension over adjacent intervals.

Theorem 18. *For any predicate P and state* S*:*

$$\mathsf{T} \subseteq P \Leftrightarrow (\lceil\rceil \subseteq P \wedge \mathsf{S}\ ; P \subseteq P \wedge \sim\mathsf{S}\ ; P \subseteq P)$$

Proof:

$$\mathsf{T} \subseteq P$$

$\Leftrightarrow$ {by above theorem }

$$(\mathsf{S} \cup \sim\mathsf{S})^* \subseteq P$$

$\Leftrightarrow$ { by definition of P^* }

$$\lceil\rceil \subseteq P \wedge (\mathsf{S} \cup \sim\mathsf{S})\ ; P \subseteq P$$

$\Leftrightarrow$ { by pred calc}

$$\lceil\rceil \subseteq P \wedge \mathsf{S}\ ; P \subseteq P \wedge \sim\mathsf{S}\ ; P \subseteq P$$

By properties 2 and 3 of states it is possible to distribute the intersection of a state through the composition of non empty interval sets:

Theorem 19. *If A and B are sets of non-empty intervals then for any state* S *the following distribution law holds:*

$$(A\ ;\ B) \cap \mathsf{S} = (A \cap \mathsf{S})\ ;\ (B \cap \mathsf{S})$$

5 Expressing Requirements

The calculus defined above provides a way of expressing behavioural rather than temporal requirements. It is necessary to introduce a way of explicitly considering the length of time-intervals.

5.1 Duration

If it is assumed henceforth that points of time lie within a continuous metric structure with minimum element 0, then the "distance" between two point in time can be measured. The duration operator provides an interval predicate whose truth depends on the length of the interval within which it is evaluated. This definition is derived from that of the Duration Calculus [4, 5]. The continuous variable $\int$ is used to record the length of the current interval. Since 0 is the least element of the metric structure, the duration of all intervals is greater than or equal to 0:

Theorem 20.

$$(\int \leq 0) \;=\; \top$$

The empty intervals are those with 0 duration:

Theorem 21.

$$(\int = 0) \;=\; \lceil\,\rceil$$

The length of two composed intervals is the sum of their individual lengths.

Theorem 22. *For any values x, y:*

$$(\int = x + y) \;=\; (\int = x) \;;\; (\int = y)$$

This compositionality result for durations can be extended to states. Consider the following where $t > 0$:

$$\begin{array}{ll} & \mathsf{true} \\ \Leftrightarrow & \{\text{by definition}\} \\ & (\int > 0) \\ \Leftrightarrow & \{\text{by predicate calculus}\} \\ & (\int > t) \cup (t \geq \int > 0) \\ \Leftrightarrow & \{\text{by compositionality}\} \\ & (\int = t) \;;\; (\int > 0) \cup (t \geq \int > 0) \end{array}$$

Since conjunction with **true** is idempotent, then by the distribution rule for states we have the following result for any state S and $t > 0$:

$$\mathsf{S} \;=\; ((t = \int) \cap \mathsf{S}) \;;\; \mathsf{S} \cup ((\int \leq t) \cap \mathsf{S})$$

If the set *Motor* represents the actions of the lift motor, where each action moves the lift from one floor to an adjacent one, then the previous requirement can be stated as follows.

Req. 23. The lift should travel between floors within 10 seconds:

$$Motor \subseteq (\textstyle\int \leq 10)\,.$$

The requirements Buttons and Lights to take into account the pressing of a button for a floor while the lift is at that floor.

Req. 24. When the call button pressed then within 0.1 seconds the button is illuminated or the lift is at the floor.

$$But \;;\; (\textstyle\int = 0.1) \;\subseteq\; (\textstyle\int \leq 1) \;;\; (\mathsf{Light} \cup \mathsf{Floor})$$

From this requirement and the property that the light is never illuminated when the lift is at the floor, we have the following result:

$$But \;;\; ((\textstyle\int = 0.1) \cap \neg\mathsf{Floor}) \;\subseteq\; (\textstyle\int \leq 0.1) \;;\; \mathsf{Light}$$

Proof:

$$\begin{array}{ll}
 & But \;;\; ((\textstyle\int = 0.1) \cap \neg\mathsf{Floor}) \\
\subseteq & \quad \{\text{by monotonicity of } ;.\} \\
 & (But \;;\; (\textstyle\int = 0.1)) \cap \neg\mathsf{Floor} \\
\subseteq & \quad \{\text{by requirement and monotonicity of } ;.\} \\
 & ((\textstyle\int \leq 0.1) \;;\; (\mathsf{Light} \cup \mathsf{Floor})) \cap \neg\mathsf{Floor} \\
= & \quad \{\text{since intersection of states distributes through composition}\} \\
 & ((\textstyle\int \leq 0.1) \cap \neg\mathsf{Floor}) \;;\; ((\mathsf{Light} \cup \mathsf{Floor}) \cap \neg\mathsf{Floor}) \\
\subseteq & \quad \{\text{by monotonicity of composition}\} \\
 & (\textstyle\int \leq 0.1) \;;\; (\mathsf{Light} \cap \neg\mathsf{Floor}) \\
= & \quad \{\text{by requirement}\} \\
 & (\textstyle\int \leq 0.1) \;;\; \mathsf{Light}
\end{array}$$

The following result states that the the interval between the pressing of the button for a floor and the arrival of the lift at that floor can be characterised as an interval which is composed of a delay of less than 0.1 seconds and then a (possibly empty) period of time when the light is illuminated:

$$But \;;\; \sim\mathsf{Floor} \;;\; \mathsf{Floor} \;\subseteq\; (\textstyle\int \leq 0.1) \;;\; \mathsf{Light}^* \;;\; \mathsf{Floor}$$

Proof:

$$\begin{array}{ll}
& But \ ; \ \sim \mathsf{Floor} \ ; \ \mathsf{Floor} \\
= & \{\text{by decomposition of states}\} \\
& But \ ; \ (((\int \leq 1) \cap \sim \mathsf{Floor}) \cup ((\int = 1) \cap \sim \mathsf{Floor}) \ ; \ \sim \mathsf{Floor}) \ ; \ \mathsf{Floor} \\
= & \{\text{by disjunctivity of } ;\} \\
& But \ ; ((\int \leq 1) \cap \sim \mathsf{Floor}) \ ; \ \mathsf{Floor} \ \cup \ But \ ; ((\int = 1) \cap \sim \mathsf{Floor}) \ ; \ \sim \mathsf{Floor} \ ; \ \mathsf{Floor} \\
\subseteq & \{\text{by monotonicity of } ; \text{ and previous result}\} \\
& (\int \leq 1) \ ; \ \mathsf{Floor} \ \cup \ (\int \leq 1) \ ; \ \mathsf{Light} \ ; \ \sim \mathsf{Floor} \ ; \ \mathsf{Floor} \\
= & \{\text{by property of Light and } \sim \text{Floor}\} \\
& (\int \leq 1) \ ; \ \mathsf{Floor} \ \cup \ (\int \leq 1) \ ; \ \mathsf{Light} \ ; \ \mathsf{Floor} \\
= & \{\text{by disjunctivity of } ;\} \\
& (\int \leq 1) \ ; \ (\mathsf{Light} \cup \lceil\,\rceil) \ ; \ \mathsf{Floor} \\
= & \{\text{by definition of reflexive closure}\} \\
& (\int \leq 1) \ ; \ \mathsf{Light}^* \ ; \ \mathsf{Floor}
\end{array}$$

5.2 Initial and Final Events

An effective notation for expressing real-time requirements should be able to express assertions about the start/finish of Actions, the transition of a state, or the rise/fall of a waveform. The concepts are all linked, as they represent the boundaries of intervals ie: The very first point of the interval of time for which an action is executing, or for which a state is true or for which a waveform is high.

The set of initial events of P is the smallest subset of $\lceil\,\rceil$ which is a left identity for composition with P. This means that its domain is the same as that of P. This set is characterised by the following equivalence:

Definition 25. For any predicate P and set of events X:

$$^{<}P \subseteq X \Leftrightarrow P \subseteq X \, ; P$$

Likewise, the set of all the events corresponding to the end points of P can be defined :

Definition 26. For any predicate P and set of events X:

$$P^{>} \subseteq X \Leftrightarrow P \subseteq P \, ; X$$

The disjunctive properties of composition and intersection are maintained for $^{<}P$ and $P^{>}$: Sets of events are the only fixed points for these operations. The start and finish points of P are the left and right compositional identities for P respectively.

The requirements on the button and light behaviour are written as follows:

Req. 27. The call buttons illuminate only within 0.1 seconds of being pressed:

$$Light\_On \subseteq {}^{<}((\int \leq 0.1) ; But) .$$

This is an example of a minimum response timing constraint. Such a constraint can be weakened to say that something must eventually happen.

Req. 28. If the button for a floor is pressed then the lift will eventually call at the floor :

$$But \subseteq {}^{<}(\mathsf{T} ; \mathsf{Floor}) ,$$

5.3 State transition

The start transition event for a state is that event marking the the transition from false to true of the state.

Definition 29. For a state S:

$$\mathsf{S}\uparrow = (\sim \mathsf{S})^{>} ; {}^{<}\mathsf{S}$$

The end transition of a state is the event marking the point when the state ceases to hold.

Definition 30. For a state S:

$$\mathsf{S}\downarrow = \mathsf{S}^{>} ; {}^{<}(\sim \mathsf{S})$$

The events in *Light_On* are the start transition events for Light and *Light_Off* the end transition events:

$$Light\_On = \mathsf{Light}\uparrow$$
$$Light\_Off = \mathsf{Light}\downarrow$$

The events in *Arrive* mark the arrival of the lift at the floor:

$$Arrive = \mathsf{Floor}\uparrow$$

Req. 31. The lift only leaves a floor if the motor starts:

$$\mathsf{Floor}\downarrow \subseteq {}^{<}Motor$$

6 Other Approaches

6.1 Interval Temporal Logic

Interval Temporal Logic [9, 17] is a development of classical linear time temporal logic due to Moszkowski. An ITL formula describes behaviour on an interval of time. ITL operators include $\Box$ (always) , $\bigcirc$ (next) and ; (chop).

An interval is represented by a non-empty sequence of states. A state can be viewed as an instantaneous snapshot of a system. The length of an interval is one less than the number of states. An interval $\langle s_1 \rangle$ of length 0 is an empty interval.

Formulae in ITL are constructed from predicates in classical logic and the three above mentioned temporal operators.

Formulae are assigned truth values with respect to intervals. A model in ITL is a pair $\mathcal{M}(\Sigma, V)$ consisting of a set of states $\Sigma = \{s, t, u \ldots\}$ together with a valuation function V which maps formulae and states to truth values.

A formula which does not contain any of the temporal operators is true on an interval if it is true for the first state in the interval. The formula $\Box w$ (always w) is true on an interval if w is true for every subinterval which finishes in the final state. Such an operator is defined in the Time-Interval Calculus as follows:

$$X \subseteq \Box W$$

$$\Leftrightarrow$$

$$\top \diagdown X \subseteq W$$

Similarly $\bigcirc w$ (next w) is true on an interval if w is true from the next state on. This operator can defined as follows:

$$\bigcirc W = (\textstyle\int = 1) \; ; W$$

The chop operator allows an interval to be broken in two. The formula $w_1; w_2$ can be read as w_1 followed by w_2. This formula is true on an interval if there is a sequential partitioning of the interval into two subintervals which share a common state such that w_1 holds in the first subinterval and w_2 holds on the second:

The ***skip*** operator is true for any interval of length 1. The more general form of the operator is the *len n* operator which is true for any interval of length n.

This logic is very similar to the interval calculus; but it is more restricted. The time domain is defined in terms of discrete states, thus providing a next operator as a primitive. This can be derived for a discrete time domain in the interval calculus.

6.2 Duration Calculus

The distinctive feature of the duration calculus[4, 5] is reasoning about integrals of the durations of states within any given time interval.

The states of a real-time system are represented as as a function from reals to $\{0, 1\}$, where 1 denotes that the system is in the state and 0 denotes otherwise. Furthermore states are assumed to be of finite variability. Compound state expressions can be constructed from simple states by the usual boolean operators.

The duration calculus uses a system of interval temporal logic in which integrals of states over bounded intervals are introduced as variables.

For the purposes of comparison with the other calculi a model for the calculus is provided.

As in temporal logic a frame is a pairing of a set of points in time and an ordering relation $(T, \leq)$. This ordering is reflexive, linear and dense. Furthermore some metric is assumed so as to facilitate the integration. It can be assumed that the time domain is the real numbers. The distinctive feature is the duration operator for states which has the following properties:

$$\vdash \int 0 = 0$$

$$\vdash \int S \geq 0$$

$$\vdash \int S + \int T \ = \ \int (S \vee T) + \int (S \wedge T)$$

$$\vdash (\int S = t + u) \Leftrightarrow (\int S = t)^\frown(\int S = u)$$

The most distinctive and powerful characteristic feature of the duration calculus is the duration operator. This permits the description of proportional requirements such as: *in any two minute interval there must be no more than two second of gas leak* :

$$\int 1 = 120 \Rightarrow \int Leak \leq 2$$

The underlying semantics of the duration calculus and the interval calculus are very similar. The main difference between the duration and interval calculi is that the duration formulae are predicates about an arbitrary interval of time whereas the intervals formulae are defined for sets of intervals. There is a straightforward conversion by replacing $\Leftarrow$ and $\Leftrightarrow$ by $\subseteq$ and $=$ respectively. The chop operator $\frown$ of the duration calculus has the same properties as the composition ; of the interval calculus. The the full power of the duration operator could be added to the interval calculus by generalising the definition of the $\int$ variable.

7 Conclusion

The relational approach to requirements for time critical systems has the advantage of a well established theory behind it and the ability to express many of the different constructs used in requirements. The abstraction away from the details of individual intervals, (whether they are half open or closed etc) permits a cleaner notation and a simpler set of laws. With further work in the use of duration it could be possible to express more sophisticated requirements. The use of proof has been developed only to a small extent. Methods for checking the feasibility of a set of requirements would add to the power of the calculus.

8 Acknowledgements

I would like to thank Tony Hoare, He Jifeng, Anders Ravn and Zhou Chao Chen for their many helpful comments and suggestions. The original work on this material was conducted with the financial support of the SERC.

References

1. H. Barringer:. Up and Down The Temporal Way. The Computer Journal, **30:2** pp 143–148 (1987) .
2. J. van Bentham:. Time, Logic and Computation. G. Rozenberg, ed, REX Workshop on Temporal logic and Concurrency, Springer, 1988.
3. S. M. Brien:. A Relational Calculus of Intervals. MSc Thesis, PRG Oxford Univ. Comp. Lab., Sep 1990.
4. Z. Chaochen, M. R. Hansen:. A note on completeness of the duration Calculus. ProCos Note, 1991.
5. Z. Chaochen, C. A. R. Hoare, A. P. Ravn:. A Calculus of Durations. PRG Oxford Univ. Comp. Lab., Feb 1991.
6. B. A. Davey, H. A. Priestley:. An introduction to Lattices and Order. Cambridge University Press, 1990.
7. E. W. Dijkstra:. A Relational Summary. EWD **1047**, Austin TX, Nov 1990.
8. E. W. Dijkstra, C.S. Scholten:. Predicate Calculus and Program Semantics. Springer Verlag, 1990.
9. R. W. S. Hale:. Programming in Temporal Logic. PhD Thesis, Univ. Cambridge, Oct 1988.
10. C. A. R. Hoare, He Jifeng:. The Weakest Prespecification. Oxford Univ. Comp. Lab, Technical Monograph PRG-**44**, 1985.
11. G. E. Hughes, M. J. Creswell:. An Introduction to Modal Logic. Methuen, 1968.
12. G. E. Hughes, M. J. Creswell:. An Companion to Modal Logic. Methuen, 1984.
13. F. Jahanian, A. K-L. Mok:. Safety Analysis of Timing Properties in Real-Time Systems. IEEE Trans. SE, Vol. **SE-12(9)**, pp 890–904 (1986).
14. K. M. Jensen:. Specification of a Lift Control System. **ProCoS** report: ID/DTH KMJ **12/1** (1990).
15. R. Koymans:. Specifying Message Passing and Time-Critical Systems with Temporal Logic. Phd Thesis, Eindhoven, 1989
16. R. Koymans:. Specifying Real-Time Properties with Metric Temporal Logic. Real-time Systems, **2**, pp 255–299 (1990).
17. M. Moszkowski, Z. Manna:. Reasoning in Interval Temporal logic. LNCS **164**, Logics of Programs, pp 371–382, 1983.
18. N. Rescher, A. Urquhart:. Temporal Logic. Springer-Verlag, 1971.
19. S. Rossig:. Trace oriented specification of a single lift system. ProCoS Technical Report OLD SR **4/1**, 1990.
20. A. Tarski:. On the Calculus of Relations. Journal of Symbolic Logic, **6:3**, pp 73–88 (1941).

CONSERVATIVE FIXPOINT FUNCTIONS ON A GRAPH

J.P.H.W. van den Eijnde

Department of Computing Science, Hg. 7.16
Eindhoven University of Technology
P.O. Box 513, 5600 MB Eindhoven
(31)40-473922
wsinjvde@win.tue.nl

Abstract. In this paper we present a derivation of a general solution for a class of programming problems. In these problems a function over the vertices of a directed graph is to be computed, being defined as a least fixed point of some monotonic operator. If this operator satisfies a certain restriction with respect to its image for a differential change in its argument, it is called conservative, and an elegant general solution may be derived. It is stipulated that a strictly calculational derivation is only possible if the level of abstraction is sufficiently high. To that end a modest extension to the functional calculus is proposed, including partial functions, and a few simple high level programming constructs are introduced. The program scheme obtained is applied to a particular example, for which so far no derivation, other than informal ones, is known to exist. The solutions presented are not new, but the calculational, abstract and compact technique of deriving them is meant to improve and complement the current techniques [KNU77, REY81, REM84]. It is believed to simplify the derivations for a wider algorithm class [EIJ92] than the one treated here.

1 Introduction

From a fundamental point of view solving fixed-point equations for functions on a graph is interesting, because they represent generalizations of recursive equations "par excellence": each function value may depend on an arbitrary, possibly empty, set of other function values. These predecessor or successor sets simply define the graph. Obviously, the cyclicity of recursive equations is then reflected by the cyclicity of the corresponding graph. On the other hand, these fixed-point equations are of great practical use. They arise in many different fields: topography, digital networks, electronic circuits, compiler theory, computer graphics, and process scheduling. Recently the importance of solving fixed-point equations on graphs was recognized by Cai and Paige [CAI89], who devoted a comprehensive paper with many examples to the subject, although their objectives differ from ours, e.g. automatic code generation.

The technique used in this paper to solve graph programming problems is based on three corner-stones: heuristics, abstraction and calculus. Some twenty-five years ago, E.W. Dijkstra was one of the first to advocate the use of proper heuristics [DIJ76]. For a more up-to-date treatment the reader is referred to [GAS88]. These heuristic principles, and some others, are applied wherever appropriate in this paper, and an attempt is made to indicate at crucial points in the derivation why certain design decisions were taken. In [EIJ92] the reader may find explicitly formulated "rules of thumb", applying more in particular to set, graph or fixed-point problems.

The need for abstraction is widely accepted [REY81, CAI89], although in some modern textbooks [MEH84, COR90] graph algorithms are treated in terms of low level data structures. In my opinion, abstraction is the only way to reason about more complicated algorithms like graph algorithms, allowing the methods for solving small problems [DIJ88] to be applicable without requiring major changes. However, the way to abstract is subject to discussion. Since we are interested in computing functions the most logical conclusion seems to be to introduce abstract functions as data structures into the basic formalism, for which we choose the guarded command language (GCL) [DIJ88]. Because in repetitions we would like these functions to grow towards their final value we also introduce partial functions. Finally, we need abstract sets, in the spirit of [REM84]. The details of this **data abstraction** are summarized in section 2, together with a touch of **control abstraction** [LIS77] by means of some convenient programming structures, the most important one being a **for**–statement that repeats a statement list for all elements of a fixed set. A straightforward proof rule is expressed by the **precondition for theorem**, and section 2 is concluded with the **expression accumulation theorem**, needed to make efficient use of the **for**–statement.

The third corner-stone of our technique, a calculational style, is widely believed to be a prerogative of transformational programming, be it imperative [MOR90], recursive [BUR77], functional [BIR88] or even relational [ROL90]. One of the goals of this paper is to show that programming with conventional Hoare triples can also be done purely calculationally. The required set and function calculus rules are introduced in the beginning of section 2. The resulting style is, admittedly, hybrid, with program notation on the one hand, and derivations using predicates on the other. On some occasions, e.g. in the expression accumulation theorem or the function lift rules, the influence of functional programming can be distinguished, and used to advantage.

The problem class under consideration is defined in section 3. Many members have been treated in literature, from minimal distance problems, dating as far back as 1959 [DIJ59, TAR81], reachability problems [REY81, CAI89], and least ancestor problem [REM86], to the capacity or tunnel problem [CAR79, REM85]. The derivation presented in section 4 solves the general problem. Compared to [CAI89, REM85], it has the advantage of being purely calculational, requiring inventiveness only at certain crucial points.

Many of the problems in the class can be treated very elegantly using fixed-point theory and a regular algebra of matrices [BAC75, CAR79, TAR81]. However, the present class of problems is more general, because these matrices carry less information than the fixed-point operators on functions in section 3. Therefore, in section 5, the general problem is instantiated to an example that, as far as I know, cannot be described by such a regular algebra: the problem of ascending reachability [REM85]. It does, however, turn out to be an instance of the problem class in [KNU77], but there a non-calculational proof of the solution algorithm is only given afterwards.

2 The SAL formalism

In the program notation of SAL (Set Algorithmic Language), which is an extension of GCL, all variables are assigned a — constant — type at the beginning of the block with respect to which they are local, by a declaration of the form $v : \mathbb{T}$ (v a variable and $\mathbb{T}$ a type). Types are considered sets, supplemented with a collection of operators, satisfying certain laws.

In addition to the **simple types** $\mathbb{Z}$ (integers) and $\mathbb{B}$ (booleans), adapted from GCL, with the usual operators and laws, a separate, unstructured type $\mathbb{U}$ is introduced, the elements of which represent graph vertices. Apart from the **array type constructor** in GCL, we also have $\mathbb{P}$ (**powerset type constructor**), supporting the usual set operators and laws, $\rightarrow$ (**function type constructor**), and $\mapsto$ (**partial function type constructor**).

By convention sets are denoted with *CAPITALS*, and u, v, w, x, y, and z are understood to be vertices. Unless stated otherwise, they are of standard type $V \in \mathbb{P} \cdot \mathbb{U}$. Lower case letters d, e, f, g, and h denote partial or total function variables. On the other hand, for predicates and other mappings $\mathcal{SCRIPT}$ letters are used. Simple and generic types, and the powerset type constructor, are in $\mathbb{DOUBLE\ CAPITALS}$. Finally, program control structures are in **bold**, and statement (list) names *slanted*.

Although in principle the Eindhoven Quantifier Notation (EQN) [DIJ88] is adequate to formulate and prove the usual mathematical statements, I agree with [BAC89] that for complicated problems it tends to get very lengthy and cumbersome. So whenever possible, the dummies are abstracted from, and calculations are carried out on the function or set level. If not, we use the quantifiers $\forall$ (universal), $\exists$ (existential), $\downarrow$ (minimum) and $\uparrow$ (maximum), in an EQN-like notation, with an explicit dummy, a domain and a term, all separated by colons. The same convention is used for sets, so $\{v : \mathcal{Q} \cdot v : \mathcal{E} \cdot v\}$ denotes the set of values $\mathcal{E} \cdot v$ for those v that satisfy predicate $\mathcal{Q}$. In order to avoid confusion between set notation on the one hand, and the notation of hints in proofs and assertions in algorithms on the other, we adopt the convention of replacing the hint and assertion brackets { } by the special brackets $\prec \succ$.

Total function $f \in \mathbb{F} \rightarrow \mathbb{T}$ has domain $\mathbb{F}$ and range $\mathbb{T}$, and the so-called **domain operator** $\mathcal{D}$ satisfies $\mathcal{D} \cdot f = \mathbb{F}$. On the other hand, for a partial function $g \in \mathbb{G} \mapsto \mathbb{T}$, $\mathcal{D} \cdot g$ denotes the subset of domain $\mathbb{G}$ on which g is defined. So partial functions can be considered total functions with a variable domain; they are needed as program variables. All function notations and properties in the remainder of this section apply to total as well as partial functions, and $\rightarrow$ may denote either type constructor; $f \in \mathbb{F} \rightarrow \mathbb{T}$ and $g \in \mathbb{G} \rightarrow \mathbb{T}$ are supposed to be arbitrary functions. By convention, the domain of a named function is denoted by the corresponding name in *CAPITALS*, hence

$$F = \mathcal{D} \cdot f \ \wedge \ G = \mathcal{D} \cdot g \ .$$

Instances of functions can be denoted as sets of pairs, and $\emptyset$ is the **empty function**. The **constant function** with domain $\mathbb{S}$ and value c for all $s \in \mathbb{S}$ is denoted by $(\mathbb{S} : c)$, and if $\mathbb{S} = \{s\}$ is a singleton, this will be abbreviated to (s: c), the **singleton function**. A constant function with standard domain V will be written $\hat{c}$, or $\hat{\mathcal{E}}$ if $\mathcal{E}$ is an expression. The **lambda quantifier** offers a general notation for functions:

$$(\lambda i : Q{\cdot}i : \mathcal{E}{\cdot}i)$$

is the function which, for all values of dummy i that satisfy $Q{\cdot}i$, takes the value $\mathcal{E}{\cdot}i$. In this notation a **domain restricted function** is defined by

$$(A \vdash f) = (\lambda i : i{\in}A \cap F : f{\cdot}i)\ , \tag{0}$$

for arbitrary set A.

The operators on functions are defined in terms of the operators on their respective ranges, lifted to the function level. Let $\oplus \in \mathbf{T} \rightarrow \mathbf{T} \rightarrow \mathbf{T}$ be a binary infix operator. Then the lifted operator $\hat{\oplus}$ is defined by the **lift-meet rule**:

$$f \hat{\oplus} g \ = \ (\lambda i : i{\in}F \cap G : \ f{\cdot}i \oplus g{\cdot}i) \ \in \ F \cap G \rightarrow \mathbf{T}. \tag{1}$$

So far, rules (0) and (1) only allow us to restrain function domains. In programming, if a function is to be computed, we need to be able to extend a function domain, or to combine functions with disjoint domains. To that end we introduce the so-called **lift-split rule**, which, assuming $F \cap G \ = \ \emptyset$, is given by

$$f \check{\ } g \ = \ (\lambda i : i{\in}F \cup G : \quad \mathbf{if}\ i{\in}F \rightarrow f{\cdot}i \ [\!]\ i{\in}G \rightarrow g{\cdot}i \ \ \mathbf{fi}) \ \in \ F \cup G \rightarrow \mathbf{T}. \tag{2}$$

As it happens, the functions encountered in programming practice rarely have disjoint domains. In fact, a combination of (1) and (2) turns out to be very fruitful. An obvious definition for yet another lifted operator $\check{\oplus}$ would be

$$\begin{aligned} f \check{\oplus} g = \ & (f \hat{\oplus} g) \check{\ } ({\sim}G \vdash f) \ \check{\ } ({\sim}F \vdash g) \\ = \ & (\lambda i : i{\in}F \cup G : \mathbf{if}\ i \ \in \ F \cap G \rightarrow f{\cdot}i \oplus g{\cdot}i) \ \in \ F \cup G \rightarrow \mathbf{T}. \\ & \qquad\qquad\qquad [\!]\ i{\in}F \backslash G \rightarrow f{\cdot}i \\ & \qquad\qquad\qquad [\!]\ i{\in}G \backslash F \rightarrow g{\cdot}i \\ & \qquad\qquad\qquad \mathbf{fi} \end{aligned} \tag{3}$$

This rule is called the **lift-join rule**. Its power is that it generalizes both (1) and (2): using (0) and (1) we have

$$f \hat{\oplus} g \ = \ (F \cap G \vdash (f \check{\oplus} g)), \tag{4}$$

and if $F \cap G \ = \ \emptyset$ (3) reduces to (2); in that case the operator $\oplus$ on the left–hand side may be chosen arbitrarily, or omitted if it is irrelevant.

As for operator properties, it turns out that commutativity, associativity, idempotency and distributivity simply carry over to the lifted operators. For commutative and associative operators $\oplus$ we sometimes need a generalized version of (3), which reads

$$(\check{\oplus} k : Q{\cdot}k : h{\cdot}k) \ = \ (\lambda i : i{\in}(\cup k : Q{\cdot}k : H{\cdot}k) : (\oplus k : Q{\cdot}k \wedge i{\in}H{\cdot}k : h{\cdot}k{\cdot}i)), \tag{5}$$

where $h{\cdot}k \in H{\cdot}k \ \rightarrow \ \mathbf{T}$ is a k–indexed family of functions. Some function calculus rules, relating (0), (2) and (3), are summarized below. For arbitrary sets A and B we have the **chaining rule**

$$(A \cap B \vdash f) \ = \ (A \vdash (B \vdash f)). \tag{6}$$

If, in addition, $\ominus{\in}\mathbf{T} \rightarrow \mathbf{T} \rightarrow \mathbf{T}$ is an arbitrary binary infix operator on the common range $\mathbf{T}$ of f and g, the **term split rule** expresses

$$(A \vdash f \breve{\ominus} g) \;=\; (A \vdash f)\, \breve{\ominus}\, (A \vdash g). \tag{7}$$

For idempotent operators $\| \in \mathbf{T} \rightarrow \mathbf{T} \rightarrow \mathbf{T}$ the **domain disjunction rule**

$$(A \cup B \vdash f) \;=\; (A \vdash f)\, \breve{\|}\, (B \vdash f) \tag{8}$$

holds. If $A \cap B \;=\; \emptyset$, rule (8) is called domain split rule, and operator $\|$ may be omitted. Rules (6), (7) and (8) are easily derived from (0), (2) and (3). Detailed proofs are given in [EIJ92]. An application of (5), the **domain shift to term rule**, reads

$$(\breve{\oplus} k, i : Q{\cdot}k \wedge i{\in}W{\cdot}k : (\{i\} \vdash h{\cdot}k)) \;=\; (\breve{\oplus} k : Q{\cdot}k : (W{\cdot}k \vdash h{\cdot}k)), \tag{9}$$

with $\oplus$ commutative and associative, and $W{\cdot}k$ k–indexed sets. It follows from

$$\begin{array}{ll}
 & (\breve{\oplus} k, i : Q{\cdot}k \wedge i{\in}W{\cdot}k : (\{i\} \vdash h{\cdot}k)) \\
= & \langle (0); (5) \rangle \\
 & (\lambda i : i{\in}(\cup k, j : Q{\cdot}k \wedge j{\in}W{\cdot}k : \{j\} \cap H{\cdot}k) : \\
 & \qquad (\oplus k, j : Q{\cdot}k \wedge j{\in}W{\cdot}k \wedge i{\in}\{j\} \cap H{\cdot}k : h{\cdot}k{\cdot}i)) \\
= & \langle \text{case analysis } j{\in}H{\cdot}k\text{; property of } \cap \rangle \\
 & (\lambda i : i{\in}(\cup k, j : Q{\cdot}k \wedge j{\in}W{\cdot}k \cap H{\cdot}k : \{j\}) : \\
 & \qquad (\oplus k, j : Q{\cdot}k \wedge j{\in}W{\cdot}k \wedge i = j \wedge\; i{\in}H{\cdot}k : h{\cdot}k{\cdot}i)) \\
= & \langle (\cup j : j{\in}A : \{j\}) \;=\; A\text{; one-point rule for } j \rangle \\
 & (\lambda i : i{\in}(\cup k : Q{\cdot}k : W{\cdot}k \cap H{\cdot}k) : (\oplus k : Q{\cdot}k \wedge i{\in}W{\cdot}k \cap H{\cdot}k : h{\cdot}k{\cdot}i)) \\
= & \langle (0); (5) \rangle \\
 & (\breve{\oplus} k : Q{\cdot}k : (W{\cdot}k \vdash h{\cdot}k)).
\end{array}$$

Finally, we introduce a shorthand for a quantification over a function term, comparable to the reduce (/) from the Bird-Meertens formalism [BIR88], in the same way as function composition ($\circ$) is comparable to the map ($*$). For an associative and commutative operator $\oplus$ we define

$$\oplus{\cdot}f \;=\; (\oplus i : i{\in}F : f{\cdot}i). \tag{10}$$

We conclude with two domain split rules for $\oplus{\cdot}f$, with a straightforward proof:

$$\oplus{\cdot}(f \,\breve{}\, g) = \oplus{\cdot}f \;\oplus\; \oplus{\cdot}g, \quad \text{and} \tag{11}$$
$$\oplus{\cdot}(f \breve{\oplus} g) = \oplus{\cdot}f \;\oplus\; \oplus{\cdot}g. \tag{12}$$

Next we turn from **data abstraction** to **control abstraction** [LIS77]. Firstly, a convenient **miracle statement** is introduced. To that end we define statement

$$out :\vdash \mathcal{P} \cdot in \cdot out$$

with meaning: variable out is assigned a value such that $\mathcal{P} \cdot in \cdot out$ is validated. A detailed definition of this programming primitive is beyond the scope of this paper; the reader is referred to [MOR90].

A related programming primitive is element selection. Statement $x :\in X$ is defined using the weakest precondition $\mathcal{WP}$ for arbitrary predicate Q, by

$$\mathcal{WP}{\cdot}(x :\in X){\cdot}Q \;\equiv\; X \neq \emptyset \wedge (\forall v : v{\in}X : Q^x_v).$$

For reasons of convenience two new structured statements are introduced into SAL, in addition to concatenation, **if...fi** and **do...od**. The **as** statement is an alternative statement comparable to the **if** statement, but if none of the guards is satisfied a skip is executed, instead of an abort. Apart from being an obvious counterpart of the **if**, it has the advantage of making programs shorter and more transparent, at least for the majority of graph algorithms considered. For a single guard it is defined by

$$\textbf{as } \mathcal{B} \rightarrow SL \textbf{ sa} \quad \text{is equivalent to} \quad \textbf{if } \mathcal{B} \rightarrow SL \;[\!]\; \neg\mathcal{B} \rightarrow \mathit{skip} \textbf{ fi},$$

where $\mathcal{B}$ is a boolean expression and SL a statement list. Generalizations to more guards are obvious, and left to the reader. A proof rule in terms of Hoare triples without reference to the **if** statement is given by:

$$\begin{array}{l} \{\!\{\mathcal{P}\}\!\} \textbf{ as } \mathcal{B} \rightarrow SL \textbf{ sa } \{\!\{\mathcal{Q}\}\!\} \quad \text{is valid if and only if} \\ \{\!\{\mathcal{P} \wedge \mathcal{B}\}\!\}\; SL\; \{\!\{\mathcal{Q}\}\!\} \text{ and } \mathcal{P} \wedge \neg\mathcal{B} \Rightarrow \mathcal{Q} \text{ hold.} \end{array} \tag{13}$$

Another structured statement to be introduced in SAL is the **for** statement, which is a repetition, where a variable takes the value of all elements of a given set, in arbitrary order. It is defined by

$$\textbf{for } u \in W \rightarrow SL \textbf{ rof} \quad \text{is equivalent to} \quad \begin{array}{l} |[\;\; A : \mathbf{P}\cdot W; u : W \mid \;\; A := W; \\ \qquad \textbf{do } A \neq \emptyset \rightarrow \\ \qquad\qquad u :\in A; A := A \backslash \{u\}; SL \\ \qquad \textbf{od} \\ ]|. \end{array} \tag{14}$$

Evidently, the **for** statement not only turns out to be shorter than its GCL counterpart, but it also saves the introduction of the fresh set variable A. The occurrence of u in the for clause **for** $u \in W$ is supposed to implicitly open a block enclosing the repetition in which u is declared as a local variable. Of course u may be used in statement list SL, whereas A may not.

Application of the **for**–statement evidently offers no special advantages, if it replaces the corresponding repetition only after the latter is derived in full. Preferably, we would apply a proof rule directly to the **for**–statement itself. The following theorem offers one that is directly related to (14).

<u>Precondition for–theorem</u>

Let $W \in \mathbf{P}\cdot V$ be a program expression yielding a finite set, and let $u \in W$ and $A \in \mathbf{P}\cdot W$ be fresh variables. If there exists a predicate $\mathcal{P}$ depending on A, but not on u, and a statement list SL that does not refer to A, then

$$\begin{array}{ll} \text{the validity for all } A \text{ and } u \text{ of} & \{\!\{ u \in A \wedge \mathcal{P}^A_{A\cup\{u\}} \}\!\}\; SL\; \{\!\{\mathcal{P}\}\!\}, \\ \text{implies the validity of} & \{\!\{\mathcal{P}^A_W\}\!\} \textbf{ for } u \in W \rightarrow SL \textbf{ rof } \{\!\{\mathcal{P}^A_\emptyset\}\!\}. \end{array} \tag{15}$$

This theorem is proved in appendix A. It is generally applicable, but a drawback is the use of predicates, making application still rather elaborate. On the other hand, the theorem below addresses the special case of refining a complex assignment statement directly to a **for**–statement. It turns out to be adequate and practical for many applications. Not surprisingly, it reminds us of the so–called **reduction** from functional programming [BIR88]:

> Expression accumulation theorem
>
> Let $r \in \mathbf{T}$ be a program variable, $W \in \mathbf{P}\cdot V$ a program expression yielding a finite set, $\mathcal{G} \in \mathbf{P}\cdot V \to \mathbf{T} \to \mathbf{T}$ and $\mathcal{H} \in V \to \mathbf{T}$ mappings, and let $\oplus \in \mathbf{T} \to \mathbf{T} \to \mathbf{T}$ be a binary infix operator, for some type $\mathbf{T}$. If for all $A\in\mathbf{P}\cdot W$, $t\in\mathbf{T}$, and $u\in W\backslash A$
>
> $$\mathcal{G}\cdot(A\cup\{u\})\cdot t \;=\; \mathcal{G}\cdot A\cdot t \oplus \mathcal{H}\cdot u \tag{16}$$
>
> holds, then statement $r := \mathcal{G}\cdot W\cdot r$, with precondition $r = \mathcal{G}\cdot\emptyset\cdot r$, is refined by
>
> $$\textbf{for } u\in W \to r := r \oplus \mathcal{H}\cdot u \textbf{ rof}. \tag{17}$$

A proof outline is given in appendix A. For a detailed proof the reader is referred to [EIJ92]. Note that, contrary to the so–called accumulation function $\mathcal{G}$, it is essential that $\mathcal{H}$ does not depend on A or t.

3 Specification of the conservative fixpoint function problem

Consider a directed graph $G = \langle V, S\rangle$, with vertex set $V \in \mathbf{P}\cdot\mathbf{U}$, and successor function $S\in V\to\mathbf{P}\cdot V$, mapping each vertex to the set of its direct successor vertices. Alternatively, the predecessor function $P\in V\to\mathbf{P}\cdot V$ may be used instead of S, the relation between the two being given by

$$(\forall u, v :: v\in S\cdot u \equiv u\in P\cdot v).$$

Often S and P are generalized to functions of type $\mathbf{P}\cdot V \to \mathbf{P}\cdot V$, according to

$$S\cdot A \;=\; (\cup v : v\in A : S\cdot v), \tag{18}$$

and analogously for P. Denoting both the standard and generalized versions of S and P with the same symbol rarely causes confusion. Both S and P are strict with respect to $\emptyset$, monotonic, and uni-disjunctive (universally disjunctive).

Let $(\mathbf{T}, \sqsubseteq)$ be a totally ordered, complete lattice, with bottom $\bot$, supremum $\sqcup$ and infimum $\sqcap$ [DAV90]. We consider partial functions $V \mapsto \mathbf{T}$ on the vertices of the graph. Then, using lift-join rule (3), we define lifted versions $\check{\sqcup}$ and $\check{\sqcap}$ of supremum and infimum, respectively. These, in turn, define in a natural way a partial order on $V \mapsto \mathbf{T}$ (in fact a complete lattice [DAV90]), consequently denoted by $\check{\sqsubseteq}$, though itself not being defined by (3), but by

$$d \,\check{\sqsubseteq}\, e \;\equiv\; D \subseteq E \wedge (\forall v : v\in D : d\cdot v \sqsubseteq e\cdot v). \tag{19}$$

For this partial order $(A \vdash e)$ and $d\check{\sqcup}e$ turn out to be monotonic.

We are asked to determine the least fixed point $f \in V \mapsto \mathbf{T}$ of some operator $\mathcal{F}\in(V \mapsto \mathbf{T}) \to (V \mapsto \mathbf{T})$. So f satisfies

$$\mathcal{F}\cdot f \;=\; f, \tag{20}$$
$$(\forall e : \mathcal{F}\cdot e = e : f \check{\sqsubseteq} e). \tag{21}$$

Operator $\mathcal{F}$ is assumed to be monotonic, hence it satisfies

$$d \,\check{\sqsubseteq}\, e \Rightarrow \mathcal{F}\cdot d \check{\sqsubseteq} \mathcal{F}\cdot e. \tag{22}$$

Often recursive formulations can be derived from explicit expressions, using fixed-point theorems like that of Knaster-Tarski [DAV90]. In section 5 an instance of (20)–(21) is derived using the theorem below.

<u>Generalized fixed-point theorem</u>

Let $\mathcal{A} \in \mathsf{L} \rightarrow \mathsf{L}$ be a uni-disjunctive operator on the complete lattice $(\mathsf{L}, \sqsubseteq, \sqcup, \sqcap)$, and let e be an element of L. If r is given by

$$r \;=\; (\sqcup k : 0 \leq k : \mathcal{A}^{k}\cdot e), \tag{23}$$

then r is uniquely defined as the least (pre)fixed point of $(e\sqcup) \circ \mathcal{A}$, i.e.

$$e \sqcup \mathcal{A}\cdot r \;=\; r, \quad \text{and} \tag{24}$$
$$(\forall s : s \in \mathsf{L} : e \sqcup \mathcal{A}\cdot s \sqsubseteq s \;\Rightarrow\; r \sqsubseteq s). \tag{25}$$

Here $(e\sqcup) \circ \mathcal{A}$ corresponds to $\mathcal{F}$ in (20)-(21). A proof is given in appendix B.

For the special class of problems considered here, $\mathcal{F}$ also satisfies a restriction on its image, with respect to a "differential" change in its argument, a kind of "limited growth"–property, expressed by

$$\mathcal{F}\cdot(e\,\breve{}\,(x : c)) \,\check{\sqsubseteq}\, \mathcal{F}\cdot \mathrm{e}$$

An operator satisfying (26) is called **conservative**, and so are its fixpoints.

It is not hard to prove that, if one restricts the context-free grammars from [KNU77] to generate only graph problems, then fixed-point functions turn out to be conservative precisely if their generating functions are superior (see [KNU77]). Though restricted to graphs, the present formalism is more general with respect to operators and function domains.

The class of programming problems defined by (18)-(26) can not necessarily be mapped to a regular algebra of matrices [BAC75, CAR79, TAR81]. On the other hand, if the regular algebra satisfies the property that the unit of its multiplication is top ($\top$) of the lattice $(\mathbf{T}, \sqsubseteq)$, a property that corresponds to (26), it is an instance of the problem class defined above.

The required f computation is now accomplished by realizing postcondition

$$\mathcal{R}: \quad g \;=\; f.$$

where $g \in V \mapsto \mathbf{T}$ is a partial function variable.

4 Derivation of an algorithm

Because the quantity to be computed is a least fixed point we approximate it from below (cf. linear search) and propose the simplest invariant possible:

$\mathcal{P}0: \quad g \check{\sqsubseteq} f.$

¿From $\mathcal{P}0$ immediately follows the **first approximation** to the desired algorithm, using the miracle statement:

$g := \emptyset;$
do $g \check{\sqsubset} f \rightarrow$
$\quad d :\vdash g \check{\sqsubset} d \check{\sqsubseteq} f; \quad g := d$
od,

which is obviously (partially) correct. However, in this algorithm f is an unknown quantity, which should be eliminated. Again it has to be approximated from below, according to the derivation

$$f \;=\; \{\ (20)\ \}\ \mathcal{F}\cdot f \;\check{\sqsupseteq}\; \{\ \mathcal{P}0;\ (22)\ \}\ \mathcal{F}\cdot g, \tag{27}$$

so it appears useful to introduce variable $h \in V \mapsto \mathbf{T}$, specified by

$\mathcal{P}1: \quad h \;=\; \mathcal{F}\cdot g.$

Naturally we want to replace f everywhere in the algorithm by h. Since, from (27), this means strengthening the guard and selection statement, this has no consequences for the invariance of $\mathcal{P}0$, so we obtain as **second approximation**:

$g, h := \emptyset, \mathcal{F}\cdot\emptyset;$
do $g \check{\sqsubset} h \rightarrow$
$\qquad d :\vdash g \check{\sqsubset} d \check{\sqsubseteq} h; \; h := \mathcal{F}\cdot d; \; g := d$
od.

However, strengthening the guard implies weakening the postcondition, so we ought to make sure that $\mathcal{R}$ still holds upon termination. Negation of the guard could mean that g and h are not comparable, so in order to still conclude something useful we are forced to introduce invariant

$\mathcal{P}2: \quad g \check{\sqsubseteq} h.$

Then the postcondition remains valid, for we have

$$\begin{array}{ll} & \neg g \check{\sqsubset} h \\ \equiv & \{\ \mathcal{P}2;\ \mathcal{P}1\ \} \\ & g \;=\; \mathcal{F}\cdot g \\ \Rightarrow & \{\ (21)\ \} \\ & f \check{\sqsubseteq} g \\ \equiv & \{\ \mathcal{P}0\ \} \\ & \mathcal{R}. \end{array} \tag{28}$$

Evidently, $\mathcal{P}0$ and $\mathcal{P}1$ remain valid in the latest algorithm. The validity of $\mathcal{P}2^{g,h}_{d,\mathcal{F}\cdot d}$ just after the selection follows from $\mathcal{P}1$ and $\mathcal{F}$ monotonicity (22):

$$d \check{\sqsubseteq} h \;=\; \mathcal{F}{\cdot}g \check{\sqsubseteq} \mathcal{F}{\cdot}d, \tag{29}$$

so the abstract, second approximation turns out to be (partially) correct.

So far everything has gone smoothly, requiring hardly any inventiveness. But at this point we take an important design decision. The only drawback of the second approximation seems to be the lack of clarity about the efficiency, or for that matter, termination at all. We decide to settle this question rigorously, and strive for a solution where domain G of g is explicitly **extended with one vertex** $x \in H$, hence $x \notin G$, in each turn of the repetition, and choose for $d{\cdot}x$ the **greatest value** permitted by upper bound h for d, i.e. $h{\cdot}x$. Such a refinement would consequently result in the **third approximation**:

$g, h := \emptyset, \mathcal{F}{\cdot}\emptyset$;
do $H\backslash G \neq \emptyset \rightarrow$
  $x :\in H\backslash G;\quad c := h{\cdot}x;$
  $d := g\,\check{}\,(x : c);$
  $h := \mathcal{F}{\cdot}d;\; g := d$
od.

Note that domains G and H of functions g and h, respectively, need not be adjusted separately. Variable $c \in \mathbf{T}$ is only introduced as a shorthand. However, the above algorithm is only a true refinement of the previous one if

$$\mathcal{P}3:\;\; (G \vdash h) \;=\; g$$

holds, for this implies that $g\check{\sqsubset}h$ is equivalent to $H\backslash G \neq \emptyset$, using (19). Invariant $\mathcal{P}3$ means that h equals g on domain G of the latter, see (0).

Can we keep $\mathcal{P}3$ invariant? It holds initially, but before checking its invariance, we first mention a convenient property of d, directly following from conservativity property (26) and $\mathcal{P}1$:

$$\mathcal{F}{\cdot}d \;\check{\sqsubseteq}\; h\check{\sqcup}\hat{c}. \tag{30}$$

This also implies $c \;=\; \mathcal{F}{\cdot}g{\cdot}x \sqsubseteq \mathcal{F}{\cdot}d{\cdot}x \sqsubseteq h{\cdot}x \sqcup c \;=\; c$, proving

$$\mathcal{F}{\cdot}d{\cdot}x \;=\; c. \tag{31}$$

For the invariance of $\mathcal{P}3$ we then derive

$\quad \mathcal{P}3^{g,h}_{d,\mathcal{F}\cdot d}$
$\equiv \quad$ ⦃ $\mathcal{P}2$; substitution ⦄
$\quad (D \vdash \mathcal{F}{\cdot}d) \check{\sqsubseteq} d$
$\equiv \quad$ ⦃ $d \;=\; g\,\check{}\,(x : c)$; domain split (8); (19) ⦄
$\quad (G \vdash \mathcal{F}{\cdot}d) \check{\sqsubseteq} g \wedge \mathcal{F}{\cdot}d{\cdot}x \;\sqsubseteq\; c$
$\Leftarrow \quad$ ⦃ (30); monotonicity of $\vdash$; (31) ⦄
$\quad (G \vdash h\check{\sqcup}\hat{c}) \check{\sqsubseteq} g$
$\equiv \quad$ ⦃ $\vdash$ term split (7) ⦄
$\quad (G \vdash h) \check{\sqcup} (G :\; c) \check{\sqsubseteq} g$
$\equiv \quad$ ⦃ $\mathcal{P}3$; property of $\check{\sqcup}$ ⦄
$\quad (G :\; c) \check{\sqsubseteq} g$

$\equiv$ ⦃definition of $\check{\sqsubseteq}$ (19); reduce (10)⦄
$c \sqsubseteq \sqcap \cdot g$
$\Leftarrow$ ⦃ $c = h \cdot x$ and $x \in H \backslash G$ arbitrary; definition of $\vdash$ (0); reduce (10)⦄
$\mathcal{P}4$, (32)

where

$\mathcal{P}4$: $\sqcup \cdot (\sim G \vdash h) \sqsubseteq \sqcap \cdot g$.

In other words: all $h \cdot y$ for $y \in \sim G$ ($\sim G$ is G's complement with respect to V) should be at most all $g \cdot z$ values. Establishing in turn the invariance of $\mathcal{P}4$:

$\sqcup \cdot (\sim D \vdash \mathcal{F} \cdot d) \sqsubseteq \sqcap \cdot d$
$\Leftarrow$ ⦃ $d = g\,\check{}\,(x : c)$; reduce domain split (11); $\sim D \subseteq \sim G$⦄
$\sqcup \cdot (\sim G \vdash \mathcal{F} \cdot d) \sqsubseteq \sqcap \cdot g \sqcap c$
$\Leftarrow$ ⦃ (30); $x \in H \backslash G$, so from $\mathcal{P}4$: $c = h \cdot x \sqsubseteq \sqcap \cdot g$⦄
$\sqcup \cdot (\sim G \vdash h \dot{\sqcup} \hat{c}) \sqsubseteq c$
$\equiv$ ⦃$\vdash$ term split (7); reduce domain split (12)⦄
$\sqcup \cdot (\sim G \vdash h) \sqcup \ \sqcup \cdot (\sim G : c) \ \sqsubseteq \ c$
$\equiv$ ⦃ reduce constant term; property of $\sqcup$⦄
$\sqcup \cdot (\sim G \vdash h) \sqsubseteq c$. (33)

It follows that the invariance of $\mathcal{P}4$ in turn is guaranteed if x is chosen such that its h value c is maximal on domain $H \backslash G$! Now a maximum search is a rather trivial matter; however, since a witness x of the maximum is required it gives us the opportunity to demonstrate the use of the precondition for theorem.

If, in proof rule (15), we use for invariant $\mathcal{P}$

$\mathcal{P}$: $\sqcup \cdot (\sim G \backslash A \vdash h) \ \sqsubseteq \ h \cdot x \ \wedge \ x \in H \backslash G$,

then we conclude that for $A = \emptyset$ $\mathcal{P}$ implies the last member of (33), whereas for $A = H \backslash G$ the restricted domain function collapses, hence $x :\in H \backslash G$ suffices as initialization. Next, we evaluate precondition $y \notin A \ \wedge \ \mathcal{P}^{A}_{A \cup \{y\}}$ of the for–body statement list SL to be determined, first under the condition $h \cdot y \sqsubseteq h \cdot x$:

$y \notin A \ \wedge \ \sqcup \cdot (\sim G \backslash (A \cup \{y\}) \vdash h) \sqsubseteq h \cdot x \ \wedge \ h \cdot y \sqsubseteq h \cdot x \ \wedge \ x \in H \backslash G$
$\equiv$ ⦃ set calculus; property of $\sqcup$⦄
$y \notin A \ \wedge \ \sqcup \cdot ((\sim G \backslash A) \backslash \{y\} \vdash h) \sqcup h \cdot y \sqsubseteq h \cdot x \ \wedge \ x \in H \backslash G$
$\Rightarrow$ ⦃domain split of $\vdash$ (8) and reduce (11)⦄
$\sqcup \cdot (\sim G \backslash A \vdash h) \sqsubseteq h \cdot x \ \wedge \ x \in H \backslash G$,

which equals $\mathcal{P}$. Alternatively, under the condition $h \cdot x \sqsubseteq h \cdot y$ we derive

$y \notin A \ \wedge \ \sqcup \cdot (\sim G \backslash (A \cup \{y\}) \vdash h) \sqsubseteq h \cdot x \ \wedge \ h \cdot x \sqsubseteq h \cdot y \ \wedge \ x \in H \backslash G$
$\Rightarrow$ ⦃ transitivity of $\sqsubseteq$; property of $\sqcup$; type information on y⦄
$y \notin A \ \wedge \ \sqcup \cdot ((\sim G \backslash A) \backslash \{y\} \vdash h) \sqcup h \cdot y \sqsubseteq h \cdot y \ \wedge \ y \in H \backslash G$
$\Rightarrow$ ⦃ domain split of $\vdash$ (8) and reduce (11)⦄
$\sqcup \cdot (\sim G \backslash A \vdash h) \sqsubseteq h \cdot y \ \wedge \ y \in H \backslash G$
$\equiv$ ⦃substitution; definition of $\mathcal{P}$⦄
$\mathcal{P}^{x}_{y}$,

so it appears that if we perform $x := y$ in the second alternative, both yield $\mathcal{P}$ in their postcondition. Hence, as proof rule (13) can be applied, and picking up the pieces we consequently find the **fourth approximation**

$g, h := \emptyset, \mathcal{F}\cdot\emptyset;$
do $H\backslash G \neq \emptyset \rightarrow$
    $x :\in H\backslash G;$
    for $y \in H\backslash G \rightarrow$ **as** $h\cdot x \sqsubseteq h\cdot y \rightarrow x := y$ **sa rof**;
    $c := h\cdot x;\ h := \mathcal{F}\cdot(g\check{\ }(x : c));\ g\cdot x := c$
od,

where we eliminated d and simplified the g assignment. This algorithm is the end of the line if we have no further information about $\mathcal{F}$. Note that it is $\mathcal{O}(\#V^2)$, if the assignment to h can be performed in $\mathcal{O}(\#V)$ time.

In many problem instances $\mathcal{F}$ satisfies split property

$$\mathcal{F}\cdot(e\check{\ }(x : c)) \;=\; \mathcal{F}\cdot e \oplus (S\cdot x \vdash \mathcal{K}\cdot c\cdot x), \tag{34}$$

with $\mathcal{K}$ an indexed family of mappings, and $\mathcal{K}\cdot a\cdot u \in K\cdot a\cdot u \rightarrow \mathsf{T}$ for all $a \in \mathsf{T}$ and $u \in V$, and $\oplus$ is an associative binary infix operator. If $\mathcal{F}$ does not satisfy (34), one could say there is something wrong with the particular choice of the graph, because the successor function is supposed to precisely reflect the dependency relations between function values in all vertices.

If (34) does hold, the h assignment becomes

$$h := h \oplus (S\cdot x \vdash \mathcal{K}\cdot c\cdot x), \tag{35}$$

and this allows us to refine the assignment further, using the expression accumulation theorem with accumulation function $\mathcal{G}\cdot A\cdot t \;=\; t \,\check{\oplus}\, (A \vdash \mathcal{K}\cdot c\cdot x)$. Hence we check property (16) of the theorem:

$$\begin{array}{ll}
 & \mathcal{G}\cdot(A \cup \{y\})\cdot t \\
= & \langle\!\{ \text{definition of } \mathcal{G} \}\!\rangle \\
 & t \,\check{\oplus}\, (A \cup \{y\} \vdash \mathcal{K}\cdot c\cdot x) \\
= & \langle\!\{ \text{domain split of } \vdash (8),\ y \notin A;\ \text{choose } \oplus \text{ as arbitrary operator} \}\!\rangle \\
 & t \,\check{\oplus}\, (A \vdash \mathcal{K}\cdot c\cdot x) \,\check{\oplus}\, (\{y\} \vdash \mathcal{K}\cdot c\cdot x) \\
= & \langle\!\{ \text{definition of } \mathcal{G} \}\!\rangle \\
 & \mathcal{G}\cdot A\cdot t \,\check{\oplus}\, (\{y\} \vdash \mathcal{K}\cdot c\cdot x),
\end{array}$$

so $\mathcal{H}\cdot y$ is chosen to be $(\{y\} \vdash \mathcal{K}\cdot c\cdot x)$, which is indeed independent of A and t. Finally, $\mathcal{G}\cdot\emptyset\cdot t \;=\; t$ is a tautology, so the restriction on the precondition of (35) holds. Hence (35) is refined by the **for**–statement

$$\textbf{for } y \in S\cdot x \rightarrow h := h \oplus (\{y\} \vdash \mathcal{K}\cdot c\cdot x) \textbf{ rof}, \tag{36}$$

which may be expanded using (3) and (13). Also note that, since upon termination of the outer repetition $f = g = h$ holds, g has become superfluous, apart from its domain G. Substituting these findings into the fourth approximation we arrive at the final, **fifth approximation**

```
|[ G : P·V |
   h := F·∅; G := ∅;
   do H\G ≠ ∅ →
      |[ x : V; c : T | x :∈ H\G;
         for y∈H\G → as h·x ⊑ h·y → x := y sa rof;
         c := h·x;
         for y∈S·x →
            as y∈K·c·x ∩ H → h·y := h·y ⊕ 𝒦·c·x·y
               y∈K·c·x\H   → h·y := 𝒦·c·x·y
            sa
         rof; G := G ∪ {x}
      ]|
   od ⦃ h = f ⦄
]|.
```

Here, as in many applications, it is possible to implement a partial function using an array of size $\#V$, with "blanks" indicating vertices outside its domain, making membership tests for function domains ($y \in H,\ y \in K \cdot c \cdot x$) easy, $\mathcal{O}(1)$. Set $H \backslash G$ — remember that $G \subseteq H$! — may be implemented with a left adjusted array (a "stack"), making element selection also $\mathcal{O}(1)$, provided that in the maximum x search each new x value after $x := y$ is swapped to the back. Note that G itself is then obsolete. The details of the implementation phase are left to the reader, since the translation is a rather mechanical process. Also various optimizations, e.g. initializing h outside $\mathcal{D} \cdot (\mathcal{F} \cdot \emptyset)$ with the left unit of $\oplus$ and simplifying the central **as**–statement, take us beyond the scope of the paper.

Correcting for various differences in assumptions and notation the algorithm derived above corresponds to the solutions found in [BAC75, CAR79, TAR81, KNU77].

5 An example: ascending reachability

At this point we consider an application of the solution scheme derived in the previous section, called **ascending reachability**. This problem was proposed and solved in [REM85]. The problem is to determine the set of vertices reachable from a given set B via an ascending path. In this context "ascending path" means a path with the edge labels in ascending order, the edges being labelled by means of a weight function $t \in V \times V \rightarrow \mathbb{Z}$. On graph edges t is assumed to be finite.

It would not be fair to formally define the problem in a way most suited to formal manipulation, as one is tempted to do. Transformation of the "most natural" problem formulation to a suitable form is, in my opinion, an important part of the problem, and by no means the easiest. If possible, it should be based on heuristic principles that are more widely applicable.

To express the set of ascending reachables formally in a natural way, we need a path formalism. Paths are considered non-empty sequences of vertices, where all pairs of successive vertices x, y in such a sequence satisfy $y \in S \cdot x$. Thus $[x]$ is the zero-edge ("empty") path starting and ending in x, $[x, y]$ is the one-edge path from x to y, and so on. In the following p is understood to be a path in the graph. The vertices on path p are numbered, from 0 up to and including $\#p$, the path length, and $p \cdot i$, with $0 \leq i \leq \#p$, is the i-th vertex. If the last vertex of a path coincides with the first of another path, these paths may be concatenated with operator $+\!\!+$. Note that this is not the usual sequence concatenation! This brief summary will do for our purposes.

In terms of the path formalism the set of ascending reachables is given by

$$R = \{p : p \cdot 0 \in B \ \wedge \ \mathcal{A}sc \cdot p : \ p \cdot \ \#p\}, \tag{37}$$

with predicate $\mathcal{A}sc$ being defined recursively by

$$\mathcal{A}sc \cdot [u], \tag{38}$$
$$\mathcal{A}sc \cdot (p +\!\!+ [p \cdot \#p, u]) \equiv \mathcal{A}sc \cdot p \ \wedge \ t \cdot (p \cdot \#p) \cdot u \geq m \cdot p, \tag{39}$$

for $u \in S \cdot (p \cdot \#p)$, and function m in turn being defined by

$$m \cdot [u] = -\infty, \tag{40}$$
$$m \cdot (p +\!\!+ [p \cdot \#p, u]) = t \cdot (p \cdot \#p) \cdot u, \tag{41}$$

again for $u \in S \cdot (p \cdot \#p)$, i.e. the weight of the last edge of path p, and, if restricted to ascending paths, indeed the maximum label weight on the path.

If one wishes to introduce a regular algebra [BAC75] to describe (38)–(41), a homomorphism is needed on the data structure of paths with concatenation, corresponding to the regular algebra product. However, (39) tends to violate associativity. In any case, $t \cdot u \cdot v$ interpreted as a matrix is not adequate to model the problem, and at least tupled matrix elements (e.g. ascendingness, first edge label, last edge label) are required. In this way associativity can be restored, but then the regular algebra product lacks a proper unit, and it is hard to define a regular algebra sum operator, especially one over which the product should distribute. If there is a way out, it may be rather messy.

On the other hand, the present problem is easily modeled using Knuth's formalism [KNU77]. The interested reader may verify that "superior" functions

$$g_{uv} \cdot c = \textbf{if}\ t \cdot u \cdot v < c \rightarrow +\infty \ [\!]\ t \cdot u \cdot v \geq c \rightarrow t \cdot u \cdot v\ \textbf{fi}$$

will do the trick. The superiority is expressed by the fact that $g_{uv} \cdot c \geq c$.

Returning to our problem we note that definition (37) has two drawbacks. Firstly, it requires the complicated path formalism, and secondly, it does not have the shape of a recursive equation, which proved to be so fruitful in last section's derivation, and for other graph problems [EIJ92]. It turns out that a simple recursive equation of that kind does not even exist for R: the ascending reachability of a given vertex x cannot be expressed in that of its predecessors. Intuitively, we need the "degree of ascending reachability" of x's predecessors, in order to see whether paths to them can be extended with an edge to x. Therefore we apply a technique called **information extension** [EIJ92], and introduce a function $f \in V \rightarrow \mathbb{Z}$, defined by

$$f = (\lambda v : (\exists p :: \ \mathcal{A}p \cdot p \cdot v) : (\downarrow p : \ \mathcal{A}p \cdot p \cdot v : m \cdot p)), \tag{42}$$

using the shorthand

$$\mathcal{A}p \cdot p \cdot v \equiv p \cdot 0 \in B \ \wedge \ p \cdot \#p = v \ \wedge \ \mathcal{A}sc \cdot p. \tag{43}$$

The minimum quantor in the term of (42) carries information about the most favourable incoming path in v, while the severe domain restriction serves two purposes: considering only interesting vertices with finite f values increases the computing efficiency, and in addition R is conveniently equal to $\mathcal{D}{\cdot}f\ (= F)$. Function f contains more information than R in the sense that R can be expressed in terms of f, whereas the reverse is not possible.

Next, we try to transform the, indeed cumbersome, expression for f into a recursive fixed point equation, by first expressing it as a countably infinite supremum, so that subsequently a fixed point theorem can be applied, like the one in section 3. We derive

$$
\begin{array}{cl}
 & f \\
= & \{\ (42);\ \text{one-point rule}\ \} \\
 & (\lambda v : (\exists p, w :: \mathcal{A}p{\cdot}p{\cdot}w \ \wedge\ v\in\{w\}) : (\downarrow p, w : \mathcal{A}p{\cdot}p{\cdot}w \wedge v\in\{w\} : m{\cdot}p)) \\
= & \{\ \text{generalized lift-join rule (5), with } i := v \text{ and } k := (p, w)\ \} \\
 & (\check{\downarrow} p, w : \mathcal{A}p{\cdot}p{\cdot}w : (w : m{\cdot}p)) \\
= & \{\ \text{domain split over all path lengths } k;\ \text{introducing } \mathcal{Z} \text{ below}\ \} \\
 & (\check{\downarrow} k : 0 \le k : \mathcal{Z}{\cdot}k),
\end{array}
\tag{44}
$$

with

$$
\mathcal{Z}{\cdot}k \ = \ (\check{\downarrow} p, w : \mathcal{A}p{\cdot}p{\cdot}w \wedge \#p = k : (w : m{\cdot}p)).
\tag{45}
$$

If (44) is to be an explicit fixed–point expression like (23), $\mathcal{Z}\cdot k$ should take the form $\mathcal{A}^k\cdot(\mathcal{Z}\cdot 0)$, with $\mathcal{A}$ a uni–disjunctive operator. Substituting $k = 0$ into (45), we find $p = [w]$, evaluate $\mathcal{A}sc\cdot[w]$ and $m\cdot[w]$ using (38) and (40), use (5) with $k := w$, $\mathcal{Q}{\cdot}k := w{\in}B$, and $h{\cdot}k := (w : -\infty)$, and arrive at

$$
\begin{array}{rcl}
\mathcal{Z}{\cdot}0 & = & (\check{\downarrow} w :\ \mathcal{A}p{\cdot}[w]{\cdot}w : (w : m{\cdot}[w])) \\
 & = & (\check{\downarrow} w : w{\in}B : (w : -\infty)) \ = \ (B : -\infty).
\end{array}
\tag{46}
$$

Next, trying to express $\mathcal{Z}{\cdot}(k+1)$ in terms of $\mathcal{Z}{\cdot}k$ we derive

$$
\begin{array}{cl}
 & \mathcal{Z}{\cdot}(k+1) \\
= & \{\ (45)\ \} \\
 & (\check{\downarrow} p, w :\ \mathcal{A}p{\cdot}p{\cdot}w \ \wedge\ \#p = k+1 : (w : m{\cdot}p)) \\
= & \{\ k \ge 0 : \text{path split with dummy change } p := p \mathbin{+\!\!+} [v, w]\ \} \\
 & (\check{\downarrow} p, v, w :\ \mathcal{A}p{\cdot}(p \mathbin{+\!\!+} [v, w]){\cdot}w \wedge \#p = k \ \wedge\ w{\in}S{\cdot}v : (w : m{\cdot}(p \mathbin{+\!\!+} [v, w]))) \\
= & \{\ (39);(41);(43)\ \} \\
 & (\check{\downarrow} p, v, w :\ \mathcal{A}p{\cdot}p{\cdot}v \wedge \#p = k \ \wedge\ t{\cdot}v{\cdot}w \ge m{\cdot}p \ \wedge\ w{\in}S{\cdot}v : (w : t{\cdot}v{\cdot}w)) \\
= & \{\ \text{term independent of } p;\ \text{generalized domain disjunction}\ \} \\
 & (\check{\downarrow} v, w : (\exists p ::\ \mathcal{A}p{\cdot}p{\cdot}v \wedge \#p = k \ \wedge\ t{\cdot}v{\cdot}w \ge m{\cdot}p) \ \wedge\ w{\in}S{\cdot}v : (w : t{\cdot}v{\cdot}w)) \\
= & \{\ \text{property of } \downarrow\ \} \\
 & (\check{\downarrow} v, w :\ \ (\exists p ::\ \mathcal{A}p{\cdot}p{\cdot}v \wedge \#p = k) \ \wedge \\
 & \qquad\qquad (\downarrow p :\ \mathcal{A}p{\cdot}p{\cdot}v \wedge \#p = k : m{\cdot}p) \le t{\cdot}v{\cdot}w \ \wedge\ w{\in}S{\cdot}v : (w : t{\cdot}v{\cdot}w)) \\
= & \{\ \text{identifying domain and term of (45) using (5)}\ \} \\
 & (\check{\downarrow} v, w : v{\in}\mathcal{D}{\cdot}(\mathcal{Z}{\cdot}k) \ \wedge\ t{\cdot}v{\cdot}w \ge \mathcal{Z}{\cdot}k{\cdot}v \ \wedge\ w{\in}S{\cdot}v : (w : t{\cdot}v{\cdot}w)) \\
= & \{\ \text{introduce } \mathcal{L} \text{ below; calculus}\ \} \\
 & (\check{\downarrow} v, w : v{\in}\mathcal{D}{\cdot}(\mathcal{Z}{\cdot}k) \ \wedge\ w{\in}S{\cdot}v \cap \mathcal{L}{\cdot}(\mathcal{Z}{\cdot}k){\cdot}v : (w : t{\cdot}v{\cdot}w))
\end{array}
$$

$=\quad \not<$ shift w to the term (9); definition of $\mathcal{A}$ below $\not>$
$$\mathcal{A}\cdot(\mathcal{Z}\cdot k), \tag{47}$$

where

$$\mathcal{L}\cdot e\cdot v \;=\; \{z : t\cdot v\cdot z \geq e\cdot v : z\}, \quad \text{and} \tag{48}$$
$$\mathcal{A}\cdot e \;=\; (\downarrow v : v\in E : (S\cdot v \cap \mathcal{L}\cdot e\cdot v \vdash t\cdot v)). \tag{49}$$

¿From (44), (46) and (47) we conclude

$$f = (\downarrow k : 0 \leq k : \mathcal{A}^k\cdot(B : -\infty)). \tag{50}$$

In order to prove that $\mathcal{A}$ is uni-disjunctive we use the last–but–two expression in derivation (47), abbreviating $w\in S\cdot v$ to $\mathcal{P}$ and $(w : t\cdot v\cdot w)$ to T. We derive

$$
\begin{array}{ll}
 & \mathcal{A}\cdot(\downarrow i : \mathcal{Q}\cdot i : \mathcal{E}\cdot i) \\
= & \not<\ \text{definition of } \mathcal{A}\text{; domain of a generalized lift according to (5)}\ \not> \\
 & (\downarrow v,w : v\in(\cup i : \mathcal{Q}\cdot i : E\cdot i) \;\wedge\; t\cdot v\cdot w \geq (\downarrow i : \mathcal{Q}\cdot i : \mathcal{E}\cdot i)\cdot v \;\wedge\; \mathcal{P} : T) \\
= & \not<\ \text{property of } u\text{; term of generalized lift according to (5)}\ \not> \\
 & (\downarrow v,w : (\exists i : \mathcal{Q}\cdot i : v\in E\cdot i) \;\wedge\; t\cdot v\cdot w \geq (\downarrow i : \mathcal{Q}\cdot i \;\wedge\; v\in E\cdot i : \mathcal{E}\cdot i\cdot v) \;\wedge\; \mathcal{P} : T) \\
= & \not<\ \text{property of } \downarrow\ \not> \\
 & (\downarrow v,w : (\exists i : \mathcal{Q}\cdot i \;\wedge\; v\in E\cdot i : t\cdot v\cdot w \geq \mathcal{E}\cdot i\cdot v) \;\wedge\; \mathcal{P} : T) \\
= & \not<\ \text{generalized domain disjunction}\ \not> \\
 & (\downarrow i : \mathcal{Q}\cdot i : (\downarrow v,w : v\in E\cdot i \;\wedge\; t\cdot v\cdot w \geq \mathcal{E}\cdot i\cdot v \;\wedge\; \mathcal{P} : T)) \\
= & \not<\ \text{definition of } \mathcal{A}\ \not> \\
 & (\downarrow i : \mathcal{Q}\cdot i : \mathcal{A}\cdot(\mathcal{E}\cdot i)).
\end{array}
$$

Hence all conditions for application of the generalized fixed point theorem are satisfied, and f in (50) is of the required shape (23). It follows that f is the least fixed point of an operator $\mathcal{F}$ defined by

$$\mathcal{F}\cdot e \;=\; (B : -\infty) \downarrow \mathcal{A}\cdot e, \tag{51}$$

and as such matches (20)–(21), the starting point of the algorithm derivation.

The uni–disjunctivity of $\mathcal{A}$ implies monotonicity of $\mathcal{F}$, if we take $\downarrow$ for $\sqcup$, and $\geq$ for $\sqsubseteq$. It remains to check properties (26) and (35). We derive

$$
\begin{array}{ll}
 & \mathcal{F}\cdot(e\,\check{}\,(x : c)) \\
= & \not<\ (51); (49)\ \not> \\
 & (B : -\infty) \downarrow (\downarrow v : v\in E \cup \{x\} : (S\cdot v \cap \mathcal{L}\cdot(e\,\check{}\,(x : c))\cdot v \vdash t\cdot v)) \\
= & \not<\ \text{domain split; (48); (49); (51)}\ \not> \\
 & \mathcal{F}\cdot e \downarrow (\downarrow v : v = x : (S\cdot x \cap \mathcal{L}\cdot c\cdot x \vdash t\cdot x)) \\
= & \not<\ \text{one–point rule; chaining} \vdash \text{(6)}\ \not> \\
 & \mathcal{F}\cdot e \downarrow (S\cdot x \vdash (\mathcal{L}\cdot \hat{c}\cdot x \vdash t\cdot x)),
\end{array}
$$

so we recognize the required pattern of property (35) if we take $\downarrow$ for $\oplus$, and

$$\mathcal{K}\cdot c\cdot x \;=\; (\mathcal{L}\cdot \hat{c}\cdot x \vdash t\cdot x),$$

and at the same time, also realizing that $\mathcal{K}\cdot c\cdot x\cdot y\ (=\ t\cdot x\cdot y)$ is defined only on domain $\mathcal{L}\cdot \hat{c}\cdot x$ (i.e. for $t\cdot x\cdot y \geq c$), it follows that (26) holds as well. It seems we can instantiate the solution algorithm from the previous section with the above, refine the h initialization, from (46) and (51), by a **for**–statement, and after some minor simplifications we arrive at

```
|[ G : P·V | G := ∅;
   for x∈B → h·x := −∞ rof;
   do H\G ≠ ∅ →
      |[ x : V; c : Z | x :∈ H\G;
         for y∈H\G → as h·x ≥ h·y → x := y sa rof;
         c := h·x;
         for y∈S·x →
         as y∈H ∧ t·x·y ≥ c → h·y := h·y ↓ t·x·y
            y∉H ∧ t·x·y ≥ c → h·y := t·x·y
         sa
         rof; G := G ∪ {x};
      ]|
   od ⟨ h = f ∧ G = R ⟩
]|.
```

After the repetition we have $G = H$, and in accordance with the remark at the beginning of this section the latter is equal to R, so we need not calculate R separately after all. One more remark: if h is initialized to $+\infty$ outside B, the central as–statement reduces to

as $t \cdot x \cdot y \geq c \rightarrow h \cdot y := h \cdot y \downarrow t \cdot x \cdot y$ **sa**,

and upon termination h equals f, extended to the entire set V.

Comparing our solution to the one in [REM85], we observe that they are identical, except for the fact that Rem represents the domain part of function h with a left adjusted array, in which the order of the arguments matches the order of the vertices in $H \backslash G$. That does not save memory, however, the worst case size of H still being $\mathcal{O}(\#V)$.

The proof in [REM85] turns out to be entirely non–calculational, in fact verbal, though based on a complete set of more or less formal invariants, and it does not have the shape of a derivation. A flaw in the reasoning of the paper is the claim that the weight of any ascending path to the "minimal" vertex x in $H \backslash G$ must be at least $(\downarrow u : u \in H \backslash G : h \cdot u)$, because any such path has an initial part in G ending in a vertex of $H \backslash G$. This argument does not stick, because on $H \backslash G$ h is not necessarily equal to f, all these $h \cdot v$ values may still decrease.

It should be noted that the optimized solution using heaps in [REM85] is incomplete. Apart from restoring the heap-organized set $H \backslash G$ each time a vertex is added to H, also statement $h \cdot y := h \cdot y \downarrow t \cdot x \cdot y$ destroys the heap–order. Repairing this takes $\mathcal{O}(\log \#V)$. It can be carried out many times for one and the same vertex, but is in total bounded by the number of edges $\#E$. So the entire solution then becomes $\mathcal{O}((\#V + \#E) * \log \#V)$. Surprisingly, this is no worse than Rem claims, but then again his — faulty — solution is actually $\mathcal{O}(\#V * \log \#V + \#E)$. As stated before, our solution is $\mathcal{O}(\#V^2)$, which can be slightly worse than the heap solution, depending on the density of the graph.

6 Acknowledgements

I would like to thank Jaap van der Woude for his stimulating criticism during my lectures on graph algorithms: the abstraction he desired was always one level higher than I could offer. I also thank Lex Bijlsma and Roland Backhouse for their useful comments on the draft version of this paper. Finally, I am grateful to one of the conference referees for pointing out an error in the derivation of the general solution.

7 References

[BAC75] R.C. Backhouse and B.A. Carré, Regular Algebra Applied to Path-finding Problems, *J. Inst. Maths Applics* **15** (1975), 161-186

[BAC89] R.C. Backhouse, An Exploration of the Bird–Meertens Formalism, *Lecture Notes of the International Summer School on Constructive Algorithmics, Part I* (Hollum, Ameland, 1989)

[BIR88] R.S. Bird, *Constructive Functional Programming*, Lecture Notes for Marktoberdorf Summer School (1988)

[BUR77] R.M. Burstall and John Darlington, A Transformation System for Developing Recursive Programs, *J. Ass. Comp. Mach.* **24** (1977), 44-67

[CAI89] J. Cai and R. Paige, Program Derivation by Fixed Point Computation, *Sci. Comp. Progr.* **11** (1989), 197-261

[CAR79] B. Carré, *Graphs and Networks* (Clarendon Press, Oxford, U.K., 1979)

[COR90] T.H. Cormen, C.E. Leiserson and R.L. Rivest, *Introduction to Algorithms* (McGraw-Hill, New York, 1990)

[DAV90] B.A. Davey and H.A. Priestley, *Introduction to Lattices and Order* (Cambridge University Press, Cambridge, 1990)

[DIJ59] E.W. Dijkstra, A Note on Two Problems in Connexion with Graphs, *Numer. Math.* **1** (1959), 269-271

[DIJ76] E.W. Dijkstra, *A Discipline of Programming* (Prentice Hall, Englewood Cliffs, 1976)

[DIJ88] E.W. Dijkstra and W.H.J. Feijen, *A Method of Programming* (Addison-Wesley, Reading, Mass., 1988)

[EIJ92] J.P.H.W. van den Eijnde, *Program derivation in acyclic graphs and related problems* (Computing Science Notes 92/04, Technical Report, Eindhoven University of Technology, dept. Comp. Sci., 1992)

[GAS88] A.J.M. van Gasteren, *On the Shape of Mathematical Arguments* (Ph.D. Thesis, Eindhoven University of Technology, 1988)

[KNU77] D.E. Knuth, A generalization of Dijkstra's algorithm, *Inf. Proc. Lett.* **6** (1977), 1-5

[LIS77] B. Liskov et al., Abstraction Mechanisms in CLU, *Comm. ACM* **20** (1977), 564-576

[MEH84] K. Mehlhorn, *Data Structures and Algorithms 2: Graph Algorithms and NP Completeness* (Springer-Verlag, Berlin, 1984)

[MOR90] C. Morgan, *Programming from Specifications* (Prentice Hall, London, 1990)

[REM84] M. Rem, Small Programming Exercises 5, *Sci. Comp. Progr.* **4** (1984), 323-333

[REM85] M. Rem, Small Programming Exercises 7, *Sci. Comp. Progr.* **5** (1985), 219-229

[REM86] M. Rem, Small Programming Exercises 13, *Sci. Comp. Progr.* **7** (1986), 243-248

[REY81] J.C. Reynolds, *The Craft of Programming* (Prentice-Hall, Englewood Cliffs, N.J., 1981)

[ROL90] Proceedings of the Ameland Workshop on Constructive Algorithmics: *The Role of Relations in Program Development* (Utrecht Univ., 1990)

[TAR81] R.E. Tarjan, A Unified Approach to Path Problems, *J. ACM* **28** (1981), 577-593

Appendix A: Proof of the precondition for-theorem

Use $\mathcal{P}$ as invariant of the corresponding do statement in (14), and $\#A$ as variant function, then the correctness of (15) follows easily from the annotated version of the algorithm below.

```
|[ A : P·W; u : W;      ⦃ P^A_W ⦄
   A := W;              ⦃ P ⦄
   do A ≠ ∅ →           ⦃ A ≠ ∅ ∧ P ⦄
        u :∈ A;         ⦃ u ∈ A ∧ P ⦄
        A := A\{u};     ⦃ u ∉ A ∧ P^A_{A∪{u}} ⦄
        SL              ⦃ P ⦄
   od                   ⦃ P ∧ A = ∅, hence P^A_∅ ⦄
]|.
```

All transitions are trivial, only when applying the definition of $u :\in A$ the independence of $\mathcal{P}$ on x is crucial, while the decrease of the variant function follows from $\#(A\backslash\{u\}) = \#A - 1$ (for $u \in A$), provided that *SL* does not refer to A.

Proof outline for the expression accumulation theorem

It is easy to see that the desired refinement, say *FOR*, of the assignment statement $r := \mathcal{G}\cdot W\cdot r$ should satisfy

$$\{\!\!\{ r = \mathcal{G}\cdot\emptyset\cdot\rho \wedge W = \Omega \}\!\!\} \quad FOR \quad \{\!\!\{ r = \mathcal{G}\cdot\Omega\cdot\rho \}\!\!\} \tag{A0}$$

for all values of the logical variables ρ and Ω. The validity of the precondition requirement has been taken into account. Hoare triple (A0) suggests a refinement of *FOR* by a repetition with invariant

$$\mathcal{I}: \; r = \mathcal{G}\cdot(\Omega\backslash A)\cdot\rho, \tag{A1}$$

with a set variable A, initially set to W, shrinking to $\emptyset$. The correctness of the repetition below follows from the annotation in brackets:

```
|[ A : P·W; u : W |    ⦃ r = G·∅·ρ ∧ W = Ω ⦄
   A := W;             ⦃ r = G·(Ω\A)·ρ ⦄
   do A ≠ ∅ →          ⦃ A ≠ ∅ ∧ r = G·(Ω\A)·ρ ⦄
       u :∈ A;         ⦃ u∈A ∧ r = G·(Ω\A)·ρ ⦄
       A := A\{u};     ⦃ u∉A ∧ r = G·(Ω\(A ∪ {u}))·ρ ⦄
       r := r ⊕ H·u    ⦃ r = G·(Ω\A)·ρ ⦄
   od                  ⦃ A = ∅ ∧ r = G·(Ω\A)·ρ, so r = G·Ω·ρ ⦄
]|.
```

The reader is invited to carefully examine all steps. Now, applying the definition of the **for**–statement (14), we find that (17) is indeed a correct refinement of *FOR*, and hence of $r := \mathcal{G}\cdot W\cdot r$.

Appendix B: Proof of the generalized fixed point theorem

First we prove that r, as defined in (23), satisfies (24). We derive

$$\begin{array}{ll} & e \sqcup \mathcal{A}\cdot r \\ = & \langle\!\langle (23);\ \mathcal{A} \text{ is uni-disjunctive} \rangle\!\rangle \\ & e \sqcup (\sqcup k : 0 \le k : \mathcal{A}\cdot(\mathcal{A}^k\cdot e)) \\ = & \langle\!\langle \text{definition of exponentiation; dummy } k := k-1 \rangle\!\rangle \\ & \mathcal{A}^0\cdot e \sqcup (\sqcup k : 1 \le k : \mathcal{A}^k\cdot e) \\ = & \langle\!\langle \text{domain split; (23)} \rangle\!\rangle \\ & r. \end{array}$$

Secondly, it must be shown that r satisfies (25). Using a property of $\sqcup$ and definition (23) of r, we have, for arbitrary s:

$$r \sqsubseteq s \;\equiv\; (\forall k : 0 \le k : \mathcal{A}^k\cdot e \sqsubseteq s). \qquad \text{(B0)}$$

The right–hand side of (B0) can be proven with induction to k, assuming the validity of $e \sqcup \mathcal{A}\cdot s \sqsubseteq s$. The base, $\mathcal{A}^0\cdot e = e \sqsubseteq s$, follows trivially. For the induction step it follows that

$$\begin{array}{ll} & \mathcal{A}^{k+1}\cdot e \\ = & \langle\!\langle \text{definition of exponentiation} \rangle\!\rangle \\ & \mathcal{A}\cdot(\mathcal{A}^k\cdot e) \\ \sqsubseteq & \langle\!\langle \mathcal{A} \text{ is monotonic; induction hypothesis } \mathcal{A}^k\cdot e \sqsubseteq s \rangle\!\rangle \\ & \mathcal{A}\cdot s \\ \sqsubseteq & \langle\!\langle \text{assumption; property of } \sqcup \rangle\!\rangle \\ & s. \end{array}$$

Finally, the uniqueness of the solution of (24) and (25) is to be shown. Let $r0$ and $r1$ both satisfy (24) and (25). Application of (25), for $r0$, taking s to be $r1$, we find $r0 \sqsubseteq r1$, since $r0$ satisfies (24), and hence $e \sqcup \mathcal{A}\cdot r0 \sqsubseteq r0$. The reverse $r1 \sqsubseteq r0$ follows analogously. Antisymmetry then implies $r0 = r1$.

An Algebraic Construction of Predicate Transformers

Paul Gardiner[1], Clare Martin[2] and Oege de Moor[1]

[1] Oxford University Computing Laboratory, Programming Research Group, Keble Road, Oxford OX1 3QD, UK

[2] University of Buckingham, Buckingham MK18 1EG, UK

Abstract. In this paper we present an algebraic construction of the category of monotonic predicate transformers from the category of relations which is similar to the standard algebraic construction of the integers from the natural numbers. The same construction yields the category of relations from the category of total functions. This provides a mechanism through which the rich type structure of the category of total functions can be promoted to successively weaker ones in the categories of relations and predicate transformers. In addition, it has exposed two complete rules for the refinement and composition of specifications in Morgan's refinement calculus.

1 Introduction

When solving problems in mathematics it is often useful to work among a larger set of values than that in which a solution must be found. For example, the most obvious way to test whether a polynomial equation has a natural number solution is to solve it, which may require the use of integers, rationals or even complex numbers. The same is true of computer science: when calculating correct programs from their specifications it is usually difficult to work solely within the confines of the implementation language. For example, the derivation of programs in efficient functional programming languages often involves the use of relations [4], which do not correspond to programs in such languages. Monotonic predicate transformers are used in a similar way during the derivation of non-deterministic programs [26, 1, 27]. In this paper we consider some mathematical properties underlying the use of predicate transformers, using a categorical construction which is similar to the algebraic construction of the integers from the natural numbers. When applied to the category of sets and total functions once, it yields a category isomorphic to the category of sets and relations; a second application yields a category isomorphic to the category of monotonic predicate transformers. This hierarchy cannot be extended further: the category of total functions is not itself an instance of the categorical construction, and can only be extended by it twice.

In the algebraic construction of the set of integers from the set of natural numbers, the set of all pairs of natural numbers is formed and those which are equivalent with respect to an appropriate equivalence relation are identified. The rationals are generated from the integers similarly, and the associated ordering relation and algebraic operations are defined accordingly. The construction of the category of relations from the category of total functions using *spans* [21] is similar. A span is a pair of arrows with joint source. The category of relations can be constructed by forming the class of spans of all total functions, and identifying those which are equivalent with respect to an appropriate equivalence relation. The advantage of using categories instead of sets is that it is not necessary to provide an additional definition of the associated ordering and composition: both are incorporated into the definition of the span category. The new construction which unifies the categories of relations and predicate transformers is a slight variation of the standard span construction, which we call a skew span category. It differs from the standard construction by taking into consideration an ordering relation associated with the category from which it is constructed.

Both ordinary span categories and skew span categories have a factorization property which resembles the familiar categorical notion of epi/monic factorization. In both cases the factorization has an associated uniqueness property, which in the latter case is similar to the uniqueness condition normally associated with epi/monic factorization. We therefore use this analogy to motivate the factorization properties of skew span categories. The reason that such properties are interesting is that they provide necessary and sufficient conditions for the characterization of skew span categories.

The skew span construction of predicate transformers has some potential applications to programming and specification languages which have not yet been fully explored. At present, languages for program derivation such as the refinement calculus of [25, 1, 27], the semantics of which is given by monotonic predicate transformers, are untyped. The skew span construction of relations and predicate transformers provides a mechanism whereby certain aspects of the rich type structure of the category of total functions can be promoted to predicate transformers. In particular, the promotion of functors and adjunctions to lax functors and local adjunctions in span categories respectively [16] is still valid for skew span categories. Furthermore, recent research [24] has shown that data types defined by initial algebras in the category of total functions are transformed to final coalgebras under the embedding into the category of monotonic predicate transformers. Such results could prove useful for the future incorporation of types into the refinement calculus, and will be discussed in more detail in Sect. 5. A different application of this construction of predicate transformers concerns a recent result on the completeness of data refinement [10], which was shown in [9] to generalize to the categorical setting. The application on which we will focus at the end of this paper was suggested by Hoare [15], who has interpreted the findings in the context of specifications involving Hoare logic. When translated to the refinement calculus [26] this provides new single complete rules for refinement and composition.

This paper is organised as follows. In Sect. 2 we discuss the factorization properties of relations and predicate transformers which facilitate their skew span representation. The properties are described in the broader context of order enriched categories. In Sect. 3 we introduce skew span categories, and in Sect. 4 we show how such categories may be identified by their factorization properties. In Sect. 5 we briefly discuss the relevance of this work to the future development of a typed version of the refinement calculus, and in Sect. 6 we show how it provides the existing calculus with two extra proof rules.

2 Order Enriched Categories

We assume familiarity with the basic definitions of category theory, but will begin by summarizing some other standard definitions. We use the notation $(p;q)$ to denote the composite of each pair of arrows $p : A \to B$, $q : B \to C$. Functional composition is denoted by juxtaposition, and associates to the right.

A *preorder* $(P, \sqsubseteq)$ is a special case of a category: its objects are the members of a set P, and arrows are exactly the pairs (p, q) for which $p \sqsubseteq q$, where $\sqsubseteq$ is a reflexive and transitive relation. Every category determines a preorder on its class of objects, where for any pair of objects A and B, $A \sqsubseteq B$ precisely when there exists an arrow with source A and target B. An *equivalence relation* is a preorder which is symmetric. There is an equivalence relation $(\equiv)$ associated with every preorder which is defined by:

$$p \equiv q \Leftrightarrow (p \sqsubseteq q \text{ and } q \sqsubseteq p) \ .$$

Any equivalence relation on a set X gives rise to a partition of X into disjoint equivalence classes of the form $(\!|a|\!) = \{x \in X | x \equiv a\}$. The set of all equivalence classes of $(P, \sqsubseteq)$ under the equivalence relation associated with $\sqsubseteq$ forms a *partial order* $((\!|P|\!), \leq)$; a partial order (or *poset*) is a preorder which is anti-symmetric. The partial order $\leq$ on $(\!|P|\!)$ is defined by

$$(\!|p|\!) \leq (\!|q|\!) \Leftrightarrow p \sqsubseteq q \ .$$

A *preorder enriched category* $(\mathcal{C}, \sqsubseteq)$ is a category $\mathcal{C}$ with a preorder $\sqsubseteq$ defined on homsets, with respect to which the categorical composition ; is monotonic:

$$(p \sqsubseteq q \text{ and } r \sqsubseteq s) \quad \Rightarrow \quad p;r \sqsubseteq q;s$$

for all for all $p, q : A \to B$ and $r, s : B \to C$. *Poset enriched* categories are defined in the same way, except that instead of a preorder, each homset is a poset. Both these kinds of order enriched category can be thought of as a '2-dimensional' category: the arrows of $\mathcal{C}$ represent the horizontal dimension and the those of the order relation the vertical dimension. This property is formalized by the notion of a 2-*category* [11], of which preorder enriched categories are a special case. Functors between such categories must preserve both the horizontal and vertical structure, which in this case means that functors must be monotonic.

As most of the following examples illustrate, order enriched categories are well suited for modelling programming languages with types and scopes. The object structure matches the type structure of strictly typed programming languages, and composition represents sequential execution. Furthermore, the refinement relation on programs imposes a preorder enriched structure on such categories. One program p is refined by another q if for any purpose whatsoever and in any context of use q will perform at least as well, and perhaps better than p. The associated relation is clearly reflexive, and it is transitive since if the behaviour of the program p is in all respects as good as that of the program q which in turn is as good as r, then p too is as good as r.

Examples 1.

1. **$(\mathcal{M}, \leq)$ a commutative monoid.** A *monoid* is a triple $\langle \mathcal{M}, \oplus, e \rangle$ where $\mathcal{M}$ is a set, and $\oplus$ is an associative binary operator $\oplus : \mathcal{M} \times \mathcal{M} \to \mathcal{M}$ with identity e. A commutative monoid is a monoid where the operator $\oplus$ is also commutative. A monoid determines a category with one object, usually denoted $*$. The arrows of the category are the elements of $\mathcal{M}$, with $*$ as source and target; composition is the binary operator $\oplus$. Since this category has only one object it has just one homset; the preorder is defined for all $m, n \in \mathcal{M}$ by

$$m \leq n \;\Leftrightarrow\; (\exists k \in \mathcal{M} \bullet m \oplus k = n) .$$

The operator $\oplus$ is monotonic in both arguments with respect to this ordering, so $(\mathcal{M}, \leq)$ is a preorder enriched category.

2. **(Set,=), the category of total functions** whose class of objects is the class of all sets; whose homsets $\mathbf{Set}(A, B)$ are the sets of total functions from A to B; and whose composition is the usual composition of functions. Two total functions from A to B are comparable if and only if they are equal, so the partial order on each homset is the trivial equality relation.
This category can be used to model typed functional programming laguages in which all programs are terminating and deterministic. One such program refines another only if they compute the same function, so the refinement ordering is correctly represented by equality.

3. **(Rel, $\subseteq$), the category of relations** whose class of objects is the class of all sets; whose homsets $\mathbf{Rel}(A, B)$ are the sets of subsets of $A \times B$ regarded as relations from A to B; and whose composition is the usual composition of relations. Subset inclusion is a reflexive, transitive and anti-symmetric relation, so each homset $(\mathbf{Rel}(A, B), \subseteq)$ is a partial order. Relational composition is monotonic with respect to subset inclusion, so $(\mathbf{Rel}, \subseteq)$ is a poset enriched category.
This category is the natural setting in which to embed models of imperative programming languages which exhibit (angelic) non-determinism. Programs are represented by relations, and sequential composition by relational composition.

4. **(Pow, $\sqsubseteq$), the category of monotonic predicate transformers** whose class of objects is the class of all powersets; whose homsets $\mathbf{Pow}(\mathbb{P}A, \mathbb{P}B)$ are the sets of total monotonic functions from $\mathbb{P}A$ to $\mathbb{P}B$; and whose composition is the usual composition of functions. The subset inclusion relation determines a partial order on each object of this category, which is promoted to the pointwise order on arrows: if $p, q : \mathbb{P}A \to \mathbb{P}B$, then

$$p \sqsubseteq q \Leftrightarrow (p\,X \subseteq q\,X \text{ for all } X \subseteq A) .$$

The resulting partial order on each homset $(\mathbf{Pow}(\mathbb{P}A, \mathbb{P}B), \sqsubseteq)$ is preserved by functional composition since all the functions are monotonic, so $(\mathbf{Pow}, \sqsubseteq)$ is a poset enriched category.
The name of this category derives from the fact that each arrow in **Pow** is equivalent to a total function from predicates to predicates, usually called a predicate transformer [8]. For if $\mathcal{P}A$ denotes the set of predicates of classical logic with free variables drawn from the set A, then the poset $(\mathcal{P}A, \Rightarrow)$ can

be identified with $(\mathbb{P}A, \subseteq)$ by interpreting each predicate in $\mathcal{P}A$ as the set of values in A satisfying it.
Dijkstra originally introduced predicate transformers [8] in order to provide an elegant semantics for his programming language. Their strength lies in the fact that they can model non-deterministic and non-terminating behaviour in terms of total functions, rather than relations. Not all monotonic predicate transformers represent programs in Dijkstra's language: only a subset satisfying certain 'healthiness' conditions. However, the language has various extensions which can be used to describe a broader range of behaviours than is possible using relations, and each of them can be given a semantics in some larger set of predicate transformers. In particular, in Sect. 5 we will be concerned with the refinement calculus in which all monotonic predicate transformers represent programs or specifications.

Maps

Order enriched categories can accomodate a far greater range of constructs than ordinary categories, since all the standard categorical definitions can be weakened by substituting inequalities for equalities. An example of such a definition is that of a *map* [7] which is a weak analogue of the standard concept of an isomorphism. It is introduced here in order to define the factorization property shared by relations and predicate transformers.

An arrow $p : A \to B$ of a preorder enriched category $(\mathcal{C}, \sqsubseteq)$ is called a map if there exists another arrow $p^* : B \to A$ such that

$$I_A \sqsubseteq p\,;p^* \tag{1}$$

$$p^*\,;p \sqsubseteq I_B \quad . \tag{2}$$

In a trivially preorder enriched category $(\mathcal{C}, =)$ this definition coincides with the definition of isomorphism, and in the same way that isomorphisms are preserved by functors, maps are preserved by monotonic functors. Clearly all identities are maps, and the composites of maps are maps, so the maps form a subcategory of $\mathcal{C}$, which is usually denoted by $Map(\mathcal{C})$. An arrow p^* that satisfies the conditions of the above definition will be referred to here as a *comap*. Each map p determines up to equivalence its comap p^*, which is immediate from the following property of maps: let $p : C \to D$, $q : A \to B$ be maps in $\mathcal{C}$, then for all $r : A \to C$, $s : B \to D$ in $\mathcal{C}$

$$r\,;p \sqsubseteq q\,;s \quad \Leftrightarrow \quad q^*\,;r \sqsubseteq s\,;p^* \quad . \tag{3}$$

In (**Set**,=), the subcategory of maps simply consists of all the bijections; the last two categories of Examples 1 provide more interesting examples.

Examples 2.

1. $Map(\mathbf{Rel}) = \mathbf{Set}$
   The relational converse of a relation r is denoted by r°. All total functions $f : A \to B$ in **Rel** are maps, since

$$\begin{array}{ll} I_A \subseteq f; f^\circ & \text{total} \\ f^\circ; f \subseteq I_B & \text{functional} \; . \end{array}$$

Conversely, any relation which satisfies the first map axiom (1) must be total, and if it also satisfies (2) it must be functional.

2. $Map(\mathbf{Pow}) \cong \mathbf{Rel}$

A well-known property of monotonic functions between complete lattices [17] implies that an arrow $p : \mathbb{P}A \to \mathbb{P}B$ is a map in **Pow** if and only if it preserves arbitrary unions, which is to say that for all $S \subseteq \mathbb{P}A$,

$$p(\cup S) = \cup\{p\ X \mid X \in S\}\ . \tag{4}$$

It is easy to check that all maps in **Pow** have this property, and conversely any p satisfying (4) has a comap p^* defined for all $Y \in \mathbb{P}B$ by

$$p^*\ Y \mathrel{\hat=} \cup\{X \in \mathbb{P}A \mid p\ X \subseteq Y\}\ .$$

There is an isomorphism between the subcategory of **Pow** containing all functions which preserve arbitrary unions, and the category **Rel**. The isomorphism is defined by the *existential image* $\langle\_\rangle$ and *relational image* Δ functors:

$$\mathbf{Rel} \underset{\Delta}{\overset{\langle\_\rangle}{\rightleftarrows}} Map(\mathbf{Pow})$$

where for all $r : A \to B$ in **Rel**, $p : \mathbb{P}A \to \mathbb{P}B)$ in $Map($**Pow**, $X \in \mathbb{P}A$, $a \in A$ and $b \in B$,

$$\begin{array}{lcl} \langle A\rangle \mathrel{\hat=} \mathbb{P}A & \text{and} & \Delta\mathbb{P}A \mathrel{\hat=} A \\ \langle r\rangle X \mathrel{\hat=} \{b \mid \exists a \in X \bullet a(r)b\} & & a(\Delta p)b \Leftrightarrow b \in p\{a\}\ . \end{array}$$

The comap of the map $\langle r\rangle$ corresponding to each relation r is called the *universal image* of r, and will be written here as $[r]$.

Map Factorization

We will now introduce the factorization property associated with maps and comaps. Although it is motivated by analogy with the standard concept of epi/monic factorization, it is only the map factorization property that will be used in the rest of the paper.

One of the nice properties of **Set** is that every arrow $A \xrightarrow{p} B$ has a factorization

$$A \xrightarrow{p} B = A \xrightarrow{e} C \xrightarrow{m} B$$

where e is a surjection, m is an injection. This is an instance of the the more general notion of epi/monic factorization. Arrows in **Rel** do not have epi/monic factorization, but every relation $A \xrightarrow{r} B$ does have a factorization

$$A \xrightarrow{r} B = A \xrightarrow{c} C \xrightarrow{m} B$$

where c is an inverse of a total function and m is a total function; for let R denote that subset of the cartesian product $A \times B$ determined by r. The associated projections are given by:

$$\begin{array}{lcl} \pi : R \to A & \text{and} & \mu : R \to B \\ \pi(a,b) = a & & \mu(a,b) = b\ . \end{array}$$

Both projections are total functions, so $c = \pi^\circ$, $m = \mu$ is such a factorization of r. We will call this a *map factorization* since it was shown in Examples 2 that all total functions are maps in **Rel**. More generally, we say that a category C *has map factorization* if every arrow $A \xrightarrow{p} B$ has a factorization $p = c\,;m$ where c is a comap and m is a map. The following example of map factorization appears in [13].

Example 3.

Pow has map factorization: one such factorization is constructed using the union function $\cup_A : \mathbb{P}\mathbb{P}A \to \mathbb{P}A$ which sends each subset of $\mathbb{P}A$ to its union. We will write the union function as the existential image of the membership relation $\ni_A : \mathbb{P}A \to A$ which relates its left hand argument to each of its member elements: $\cup_A = \langle \ni_A \rangle$. Its comap $[\ni_A] : \mathbb{P}A \to \mathbb{P}A$ is the powerset function that sends each set in $\mathbb{P}A$ to the set of its subsets. Let $p : \mathbb{P}A \to \mathbb{P}B$ be an arrow in **Pow**. The effect of applying p to some subset X of A is equivalent to mapping X to its powerset $\mathbb{P}X$, applying p to each element of the powerset, and taking the union of the resulting set of sets. So the map factorization of p is given by

$$p = [\ni_A]\,;\langle p\,;\ni_B \rangle \ .$$

Uniqueness of Factorization

Not only do all arrows in **Set** have an epi/monic factorization; this factorization is *unique* up to isomorphism. The uniqueness condition can be defined in terms of an equivalence relation ($\equiv$) on epi/monic pairs: if (e, m) and (e', m') are two such pairs, then $(e, m) \equiv (e', m')$ if there exists an isomorphism h such that

$$e\,;h = e' \quad \text{and} \quad m = h\,;m' \ .$$

This is equivalent to saying that the following diagram commutes:

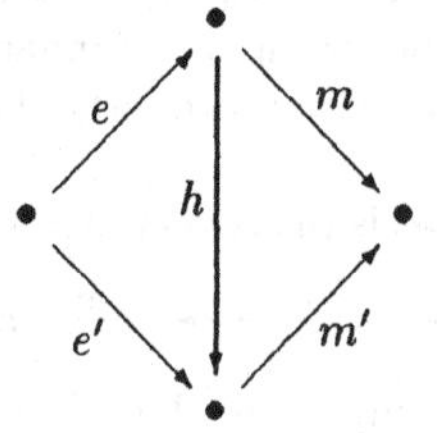

Since isomorphism is a reflexive, transitive and symmetric relation, the relation $\equiv$ is indeed an equivalence relation. If two epi/monic pairs are equivalent with respect to this relation, then it is obvious from the diagram that their composites are equal:

$$(e, m) \equiv (e', m') \quad \Rightarrow \quad e\,;m = e'\,;m' \ .$$

Epi/monic factorization is defined to be *unique up to isomorphism* provided that the converse holds: whenever an arrow p has two factorizations

$$\bullet \xrightarrow{p} \bullet \; = \; \bullet \xrightarrow{e} \bullet \xrightarrow{m} \bullet \; = \; \bullet \xrightarrow{e'} \bullet \xrightarrow{m'} \bullet$$

then $(e,m) \equiv (e',m')$. The epi/monic factorization in **Set** has this uniqueness property.

The map factorization in **Rel** has a similar uniqueness property which can be generalized to arbitrary preorder enriched categories $(\mathcal{C}, \sqsubseteq)$. Uniqueness of map factorization is defined in terms of a preorder, rather than an equivalence. The preorder relation between two comap/map pairs (c,m) and (c',m') is expressed by: $(c,m) \preceq (c',m')$ if there exists a map h such that

$$c\,;h \sqsubseteq c' \quad \text{and} \quad m \sqsubseteq h\,;m' \ .$$

This can be expressed by saying that the following diagram semi-commutes, which is indicated by the inequality signs.

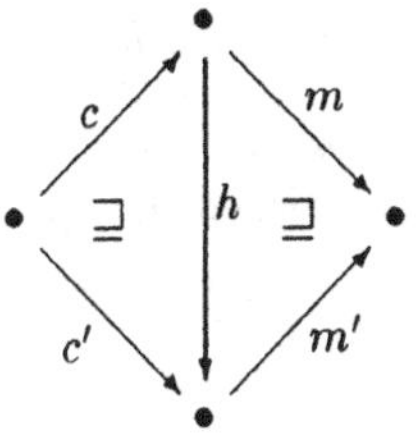

It is easy to check that this relation is reflexive and transitive. The reason that the preorder is defined in this way, and not with the inequalities reversed, is that if two map/comap pairs are related by the preorder $\preceq$ then their composites are related by the preorder $\sqsubseteq$:

$$(c,m) \preceq (c',m') \ \Rightarrow \ c\,;m \sqsubseteq c'\,;m' \ .$$

We will say that map factorization is ***unique up to equivalence*** provided that the converse holds:

$$c\,;m \sqsubseteq c'\,;m' \ \Rightarrow \ (c,m) \preceq (c',m') \ .$$

It follows that whenever an arrow p has two factorizations

$$\bullet \xrightarrow{p} \bullet \;=\; \bullet \xrightarrow{c} \bullet \xrightarrow{m} \bullet \;=\; \bullet \xrightarrow{c'} \bullet \xrightarrow{m'} \bullet$$

then the pairs (c,m) and (c',m') are equivalent with respect to the preorder $\preceq$. If the preorder $\sqsubseteq$ is a partial order we will simply refer to this property as ***unique map factorization***.

The definitions of uniqueness of factorization are not always very easy to apply to concrete examples. In the case of epi/monics there are various sufficient conditions for uniqueness that can be applied more easily [12]. An alternative statement of the uniqueness of map factorization is provided by an analogue of a well-known property of monics: if $m : A \to C$ and $m' : B \to C$ are monic arrows in some category $\mathcal{C}$, then for all $r : A \to B$,

$$m = r\,;m' \ \Rightarrow \ r \text{ is monic} \ .$$

The equivalent property does not hold for maps, but the following lemma shows how an appropriate weakening of it guarantees the uniqueness of map factorization.

Lemma 1. *If a preorder enriched category* $(\mathcal{C}, \sqsubseteq)$ *has map factorization, then the factorization is unique up to equivalence if and only if whenever* $m : A \to C$ *and* $m' : B \to C$ *are maps in* $\mathcal{C}$*, then for all* $r : A \to B$

$$m \sqsubseteq r\,;m' \;\Rightarrow\; \exists k \in Map(\mathcal{C}) \;\bullet\; (k \sqsubseteq r \text{ and } m \sqsubseteq k\,;m') \;. \tag{5}$$

Proof. First suppose that $\mathcal{C}$ has unique map factorization. Let $m : A \to C$, $m' : B \to C \in Map(\mathcal{C})$, and suppose that $m \sqsubseteq r\,;m'$ for some $r : A \to B \in \mathcal{C}$. Let $p^*\,;q$ be a map factorization of r, so

$$m \sqsubseteq p^*\,;q\,;m' \;.$$

By uniqueness of map factorization there exists a map h such that

$$h \sqsubseteq p^* \;\text{ and }\; m \sqsubseteq h\,;q\,;m' \;.$$

So $k = h\,;q$ is a map which establishes (5).

Conversely, suppose that (5) is satisfied. Let (p^*, q) and (r^*, s) be composable pairs of arrows which satisfy

$$\begin{array}{rcl} p^*\,;q \sqsubseteq r^*\,;s & \Rightarrow & \{\,(3)\,\} \\ q \sqsubseteq p\,;r^*\,;s & \Rightarrow & \{\,(5)\,\} \\ \exists k \in Map(\mathcal{C}) \;\bullet\; (k \sqsubseteq p\,;r^* \text{ and } q \sqsubseteq k\,;s) & \Rightarrow & \{\,(3)\,\} \\ \exists k \in Map(\mathcal{C}) \;\bullet\; (p^*\,;k \sqsubseteq r^* \text{ and } q \sqsubseteq k\,;s) & & \end{array}$$

which establishes the uniqueness of factorization.

□

Examples 4.

1. **Rel** has unique map factorization. For suppose that $m \subseteq r\,;m'$ in **Rel** where $m : A \to C$ and $m' : B \to C$ are maps and $r : A \to B$. We require a map k such that

   $$k \subseteq r \text{ and } m \subseteq k\,;m' \;.$$

   Since k is a map in **Rel** it must be a total function. It is constructed as follows: let $a \in A$, and take c such that $(a, c) \in m \subseteq r\,;m'$. By definition of composition there exists b such that

   $$(a, b) \in r \text{ and } (b, c) \in m' \;.$$

   Defining $k(a) = b$ establishes the result. Note that this construction of k requires the axiom of choice. Indeed, to say that **Rel** has unique map factorization is to assert the axiom of choice.
2. **Pow** has unique map factorization. For suppose that $m \sqsubseteq p\,;m'$ in **Pow**, where $m : \mathbb{P}A \to \mathbb{P}C$ and $m' : \mathbb{P}B \to \mathbb{P}C$ are maps and $p : \mathbb{P}A \to \mathbb{P}B$. We require an arrow $k : \mathbb{P}A \to \mathbb{P}B$ which satisfies the conditions of Lemma 1. Let

   $$k\,X = \cup\{p\{x\} \mid x \in X\} \text{ for all } X \in \mathbb{P}A$$

then clearly k preserves arbitrary unions, so it is a map in **Pow**. Since p is monotonic, $k \sqsubseteq p$. So it remains to check that $m \sqsubseteq k\,;m'$. Let $X \in \mathbb{P}A$, then

$$\begin{array}{ll} m\,X & \\ = & \{\ m \in Map(\mathbf{Pow})\ \} \\ \cup\{m\{x\}|x \in X\} & \\ \subseteq & \{\ m \sqsubseteq p\,;m'\ \} \\ \cup\{m'p\{x\}|x \in X\} & \\ = & \{\ m' \in Map(\mathbf{Pow})\ \} \\ m' \cup \{p\{x\}|x \in X\} & \\ = & \{\ \text{definition of } k\ \} \\ m'kX & \end{array}$$

which establishes the result.

3 Skew Span Categories

We will now introduce the skew span category which will be shown to provide the uniform construction of the categories of relations and predicate transformers in Sect. 4.

There are two categorical techniques for constructing the category of relations from the category of total functions. One is known as the *category of relations* and the other is known as the locally posettal reflection of the *category of spans* [6]. Both of these constructions can only be carried out for categories with certain properties. The second construction is much simpler than the first one, to the point of being naive. It was this simplicity that first directed our attention to the categorical theory of relations [21]. Here we consider a generalization of that construction which can be applied to any preorder enriched category $(\mathcal{C}, \sqsubseteq)$ provided that $\mathcal{C}$ has an asymmetric kind of weak pullback. If the preorder $\sqsubseteq$ in $\mathcal{C}$ is the equality relation $=$, then the new construction is identical to that in [21].

Preordered Spans

Let $(\mathcal{C}, \sqsubseteq)$ be a preorder enriched category. A *span* in $\mathcal{C}$ is an ordered pair of arrows with a common source

$$(n : E \rightarrow N, m : E \rightarrow M)\ .$$

For example, let $(\mathbb{N}, \leq_{\mathbb{N}})$ be the preorder enriched category represented by the monoid of natural numbners under addition $(\mathbb{N}, +, 0)$. Since all arrows in a monoid have the same source, a span is just a pair of natural numbers (n, m).

The preorder in $\mathcal{C}$ induces one on spans similar to the preorder associated with the uniqueness of map factorization; we define it in terms of *skew span morphisms*. Let $R = (n : E \rightarrow N, m : E \rightarrow M)$ and $R' = (n' : E' \rightarrow N, m' : E' \rightarrow M)$ be spans in $\mathcal{C}$. We define a skew span morphism to be an arrow $(h : E \rightarrow E')$ in $\mathcal{C}$ such that

$$m \sqsubseteq h; m' \quad \text{and} \quad h; n' \sqsubseteq n\ .$$

We will use the notation $(h : R \Rightarrow R')$ to denote a skew span morphism from R to R'. This definition is illustrated by the semi-commuting diagram below:

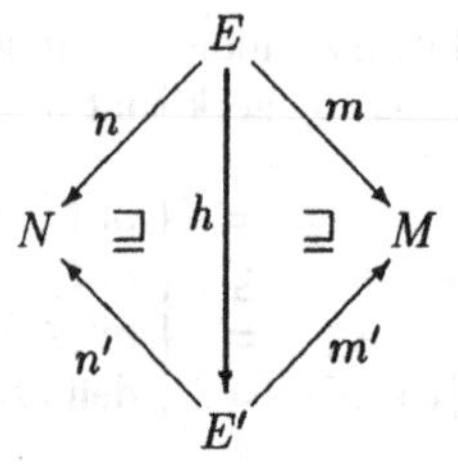

Alternatively, the same diagram with reversed inequalities could have been chosen as the definition of a skew span morphism, but this one is better suited to our examples. The standard definition of a span morphism, as used in [21, 6] is retrieved by substituting equality for inequality in this definition. Thus it is a special case of this one, in which the category $\mathcal{C}$ is considered as a preorder enriched category with respect to the equality relation $(\mathcal{C}, =)$. The preorder on spans is given by: $R \leq R'$ if and only if there is a skew span morphism from R to R'. Reflexivity is obtained by taking for h the identity arrow, which is a skew span morphism. Transitivity follows from the fact that the composition of two skew span morphisms is again a skew span morphism. We will write $(\!| n, m |\!)$ for the class of spans that are equivalent to (n, m) with respect to this preorder. The preorder on spans gives rise to a partial order on equivalence classes of spans in the usual way.

Examples 5.

1. The category of natural numbers $(\mathbb{N}, \leq_{\mathbb{N}})$ provides a simple illustration of the skew span preorder. If each span (n, m) in $\mathbb{N}$ is used to represent the integer $m-n$, then
$$(n, m) \leq (n', m') \;\Rightarrow\; (m - n) \leq_{\mathbb{N}} (m' - n') \;.$$
In contrast, the preorder on pairs of natural numbers induced by standard span morphisms implies that the corresponding integers are equal.
2. Consider any poset enriched category $(\mathcal{C}, \sqsubseteq)$ with map factorization. To say that the map factorization in $\mathcal{C}$ is unique is the same as saying that for all spans $(n : E \rightarrow N, m : E \rightarrow M)$, $(n' : E' \rightarrow N, m' : E' \rightarrow M)$ in $Map(\mathcal{C})$
$$(n, m) \leq (n', m') \;\Leftrightarrow\; n^* ; m \sqsubseteq n'^* ; m' \;.$$

Pullovers

One way that two spans $(n : E \rightarrow N, m : E \rightarrow M)$ and $(n' : D \rightarrow M, m' : D \rightarrow M')$ can be combined into a single span is by taking the pullback of the inner pair of arrows $(m : E \rightarrow M, n' : D \rightarrow M)$. The pullback is the usual composition operator on spans [3], but it cannot be used here because it is not necessarily monotonic with respect to the preorder defined by skew span morphisms. Instead we introduce the notion of a *pullover* in a preorder enriched category, which is a weak analogue of a pullback.

We define the pullover of a pair of arrows with common target $(m : E \to M, n : D \to M)$ to be a span (r, q) such that for all spans (r', q')

$$q'; n \sqsubseteq r'; m \quad \Leftrightarrow \quad (r', q') \leq (r, q) .$$

Diagramatically, a pullover is represented by

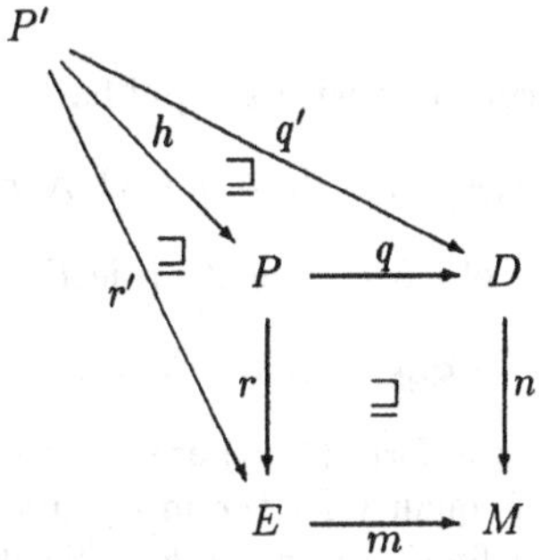

Whereas pullbacks are defined up to isomorphism, pullovers are only determined up to equivalence of spans. Their definition is related to that of *subequalizers* which were introduced by Lambek [18]. If subequalizers are substituted for equalizers in the construction of pullbacks from products and equalizers, then the result is nearly the same as a pullover. The following lemma is similar to a result about pullbacks in in [7], and provides a simple method for calculating pullovers in certain categories, including **Rel** and **Pow**.

Lemma 2. *Let* $(\mathcal{C}, \sqsubseteq)$ *be a category with unique map factorization. Then the semicommuting diagram*

$$
\begin{array}{ccc}
P & \xrightarrow{\ q\ } & D \\
{\scriptstyle r}\downarrow & \sqsupseteq & \downarrow{\scriptstyle n} \\
E & \xrightarrow{\ m\ } & M
\end{array}
$$

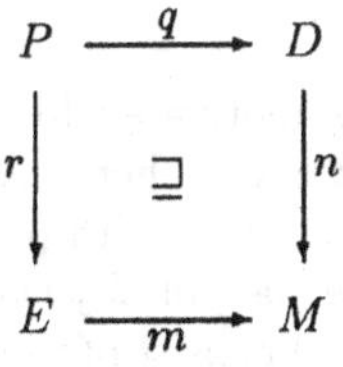

is a pullover diagram in $Map(\mathcal{C})$ *if and only if*

$$r^* ; q = m ; n^* . \tag{6}$$

Proof. Suppose that m, n, q, r satisfy (6) and that $(r' : P' \to E, q' : P' \to D)$ is another span of maps, then

$$
\begin{array}{lll}
q' ; n \sqsubseteq r' ; m & \Leftrightarrow & \{ \,(3)\, \} \\
r'^* ; q' \sqsubseteq m ; n^* & \Leftrightarrow & \{ \,(6)\, \} \\
r'^* ; q' \sqsubseteq r^* ; q & \Leftrightarrow & \{ \text{Examples } 5(2) \} \\
(r', q') \leq (r, q) .
\end{array}
$$

Hence (r,q) is the pullover of (m,n). Since any two pullovers of the same pair of arrows are equivalent with respect to the skew span preorder, by Examples 5 (2) all pullovers of (m,n) must satisfy (6).

□

The first two of the following examples can be verified using this lemma.

Examples 6.

1. In (**Set**,=) the pullover coincides with the pullback:

$$P_{\mathbf{Set}} = \{ \ (x,y) \ | \ x \in E \ \wedge \ y \in D \ \wedge \ ny = mx \ \} \ .$$

   The span (r,q) is the pair of left and right projections

$$(\pi : P_{\mathbf{Set}} \to E, \mu : P_{\mathbf{Set}} \to D) \ .$$

   Pullovers in categories of the form $(\mathcal{C},=)$ are not necessarily the same as pullbacks: the substitution of equality for the inequality $\sqsubseteq$ in the definition of the pullover yields only the definition of a *weak* pullback [21]. A weak pullback is like a pullback except that the universal arrow is not required to be unique.
2. Although pullbacks do not exist in **Rel**, it does have pullovers: one form of the pullover in **Rel** is given by

$$P_{\mathbf{Rel}} = \{ \ (x,y) \ | \ x \in \mathbb{P}E \ \wedge \ y \in \mathbb{P}D \ \wedge \ \langle n \rangle y \subseteq \langle m \rangle x \ \}.$$

   Here the span (r,q) is the pair of projections

$$((\pi \,; \ni) : P_{\mathbf{Rel}} \to E, (\mu \,; \ni) : P_{\mathbf{Rel}} \to D) \ .$$

3. Pullovers do not exist in **Pow** [20].

The skew span category *Span*($\mathcal{C}$)

The definition of the pullover was motivated by the desire to find a monotonic composition operator on spans that makes them into an preorder enriched category. Let $(\mathcal{C},\sqsubseteq)$ be a preorder enriched category with pullovers. The *skew span* category $(\mathit{Span}(\mathcal{C}),\trianglelefteq)$ is the category whose class of objects consists of all the objects in $\mathcal{C}$, and whose homsets $(\mathit{Span}(\mathcal{C}))(N,M)$ consist of the the equivalence classes of spans

$$(\!| n : E \to N, m : E \to M |\!)$$

for all such n, E, m in $\mathcal{C}$. Composition in $\mathit{Span}(\mathcal{C})$ is defined as follows: let

$$(\!| n,m |\!) : N \to M \quad \text{and} \quad (\!| n',m' |\!) : M \to K$$

be arrows in $\mathit{Span}(\mathcal{C})$. Their composition is

$$(\!| n,m |\!) \,; (\!| n',m' |\!) \quad = \quad (\!| r;n, q;m' |\!)$$

where (r,q) is a pullover of (m,n'). In a diagram:

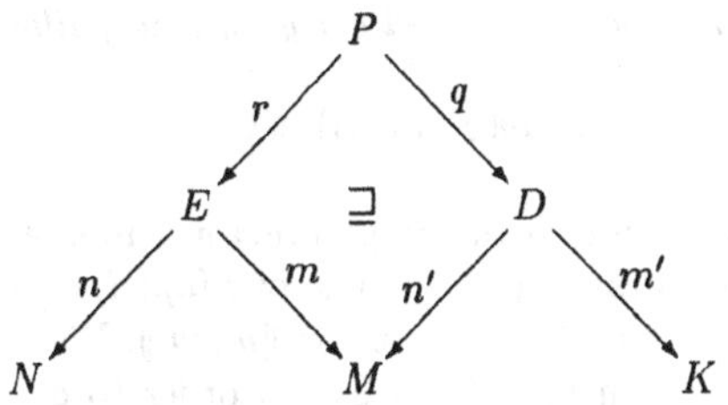

Composition is associative and monotonic and $(\!|I_N, I_N|\!)$ is the identity arrow associated with each object N. Since skew span morphisms define a partial order on objects in $Span(\mathcal{C})$, the monotonicity of composition is sufficient to ensure that it is also well-defined. The ordering $\trianglelefteq$ on homsets in $Span(\mathcal{C})$ is the partial order on equivalence classes associated with the preorder defined by skew span morphisms.

4 Characterization Theorem

We will now substantiate the claim that skew span categories, like ordinary span categories can be characterized in terms of their factorization properties alone. It is then immediate that the categories of relations and monotonic predicate transformers are isomorphic to skew span categories.

It is well-known that the span category of any suitable category $\mathcal{C}$ has map factorization, and that its subcategory of maps is isomorphic to $\mathcal{C}$. We will begin by showing that the same is true of skew span categories. The functor which embeds a category into its span category is defined as follows: for any category $(\mathcal{C}, \sqsubseteq)$ with pullovers the *graph* functor $\mathsf{G} : \mathcal{C} \to Span(\mathcal{C})$ is defined for each object M and arrow $m : E \to M$ in $\mathcal{C}$ by

$$\begin{aligned} \mathsf{G}M &= M \\ \mathsf{G}m &= (\!|I, m|\!). \end{aligned}$$

The graph functor has the property that for all arrows $m, m' : E \to M$ in $\mathcal{C}$,

$$m \sqsubseteq m' \;\Leftrightarrow\; (\mathsf{G}m \trianglelefteq \mathsf{G}m') \;.$$

This is indicative of the way that the skew span category is really a 2-categorical concept, since the graph functor preserves not only the arrows of $\mathcal{C}$ but also the corresponding order relation. It also shows that whenever $(\mathcal{C}, \sqsubseteq)$ is a poset enriched category the graph functor is injective. For each arrow m in $\mathcal{C}$, $\mathsf{G}m$ is a map in $Span(\mathcal{C})$: its comap is given by

$$(\mathsf{G}m)^* = (\!|m, I|\!) \;.$$

Moreover, every arrow $(\!|n, m|\!)$ in $Span(\mathcal{C})$ can be teased apart into the following map factorization:

$$(\!|n, m|\!) = (\mathsf{G}n)^* \,; \mathsf{G}m \;.$$

The following theorem shows that all maps in $Span(\mathcal{C})$ can be expressed in the form $\mathsf{G}m$ for some $m \in \mathcal{C}$.

Theorem 3. *If* $(\mathcal{C}, \sqsubseteq)$ *is a poset enriched category with pullovers then*

$$Map(Span(\mathcal{C})) \cong \mathcal{C} \ .$$

Proof. We have already seen that the graph functor is monotonic and injective, so it remains to show that its range is the whole of $Map(Span(\mathcal{C}))$. Let $M = (\!|n, m|\!)$ be a map in $Span(\mathcal{C})$ and let its comap $M^* = (\!|n', m'|\!)$. We must show that there exists some arrow $x \in \mathcal{C}$ such that $M = \mathsf{G}x$. In order to do so we first form the composite $M\,;M^*$: let (r, q) be a pullover of (m, n') in $\mathcal{C}$, then

$$q\,;n' \sqsubseteq r\,;m \tag{7}$$

and $M\,;M^* = (\!|r\,;n, q\,;m'|\!)$. Since (1) states that $I \trianglelefteq M\,;M^*$, there exists a skew span morphism $\eta : I \Rightarrow (r\,;n, q\,;m')$, i.e.

$$I \sqsubseteq \eta\,;q\,;m' \tag{8}$$

$$\eta\,;r\,;n \sqsubseteq I \ . \tag{9}$$

We will show that $M = \mathsf{G}(\eta\,;q\,;n')$ by proving inequality in both directions. It follows from (7) and (9) that $(\eta\,;r) : (I, \eta\,;q\,;n') \Rightarrow (n, m)$, so

$$\mathsf{G}(\eta\,;q\,;n') \trianglelefteq M \ .$$

Conversely,

$$\begin{array}{lll}
M & \trianglelefteq & \{ \text{(8) and monotonicity of G and ; } \} \\
\mathsf{G}(\eta\,;q\,;m')\,;M & = & \{ \text{G is a functor } \} \\
\mathsf{G}(\eta\,;q)\,;\mathsf{G}m'\,;M & \trianglelefteq & \{ \text{(1) and monotonicity of ; } \} \\
\mathsf{G}(\eta\,;q)\,;\mathsf{G}n'\,;(\mathsf{G}n')^*\,;\mathsf{G}m\,;M & = & \{ \text{G is a functor } \} \\
\mathsf{G}(\eta\,;q\,;n')\,;(\mathsf{G}n')^*\,;\mathsf{G}m'\,;M & = & \{ \text{map factorization } \} \\
\mathsf{G}(\eta\,;q\,;n')\,;M^*\,;M & \trianglelefteq & \{ \text{(2) and monotonicity of ; } \} \\
\mathsf{G}(\eta\,;q\,;n') & &
\end{array}$$

and therefore $M = \mathsf{G}(\eta\,;q\,;n')$ as required.

□

It follows trivially from this theorem that if $\mathcal{C}$ is a category such that $\mathcal{C} \cong Span(\mathcal{D})$ for some $\mathcal{D}$, then $\mathcal{C} \cong Span(Map(\mathcal{C}))$. This suggests that under appropriate conditions the the *Span* and *Map* functions might be mutually inverse operators on categories. The condition under which this is the case is the familiar unique map factorization property.

Theorem 4 Characterization Theorem. $(\mathcal{C}, \sqsubseteq)$ *is a poset enriched category with unique map factorization if and only if*

$$Span(Map(\mathcal{C})) \cong \mathcal{C} \ .$$

Proof. We have already seen that every arrow in $Span(Map(\mathcal{C}))$ category has at least one map factorization given for all $n : E \to N, m : E \to M \in Map(\mathcal{C})$ by

$$(\!|n, m|\!) = (\mathsf{G}n)^* \,;\mathsf{G}m \ .$$

The uniqueness of this factorization follows from Examples 5 (2) together with the fact that G is an isomorphism.

Conversely, suppose $\mathcal{C}$ is a poset enriched category with unique map factorization. Then by Lemma 2, $Map(\mathcal{C})$ has pullovers, so the category $Span(Map(\mathcal{C}))$ certainly exists. In order to show that it is isomorphic to $\mathcal{C}$ it is necessary to define a monotonic functor from one category to the other and show that it is a bijection. The obvious functor $F : Span(Map(\mathcal{C})) \to \mathcal{C}$is defined by

$$\begin{array}{ll} FN \mathrel{\hat{=}} N & \text{for all objects } N \in Map(\mathcal{C}) \\ F(\!|n, m|\!) \mathrel{\hat{=}} n^* \,;m & \text{for all } m : E \to M, n : E \to N \in Map(\mathcal{C}) \ . \end{array}$$

Since $\mathcal{C}$ has map factorization, F is clearly surjective. It is also monotonic, well-defined and bijective, since Examples 5 (2) showed that the preorder on spans is the same as that on map factorizations. It remains to show that it is a functor: suppose that $(\!|n, m|\!) : N \to M$ and $(\!|n', m'|\!) : M \to M'$ are classes of spans in $Map(\mathcal{C})$, then by Lemma 2 their composite is given by

$$(\!|n, m|\!) \,;(\!|n', m'|\!) = (\!|r \,;n, q \,;m'|\!) \qquad (10)$$

where $r^* \,;q = m \,;n'^*$. Hence

$$\begin{array}{lll} F((\!|n, m|\!) \,;(\!|n', m'|\!)) & = & \{ \ (10) \ \} \\ F(\!|r \,;n, q \,;m'|\!) & = & \{ \text{ definition of } F \ \} \\ (r \,;n)^* \,;(q \,;m') & = & \{ \text{ contravariance of }^* \ \} \\ n^* \,;r^* \,;q \,;m' & = & \{ \ r^* \,;q = n \,;m'^* \ \} \\ n^* \,;m \,;n'^* \,;m' & = & \{ \text{ definition of } F \ \} \\ F(\!|n, m|\!) \,;F(\!|n', m'|\!) \ . & & \end{array}$$

Therefore since F is monotonic and preserves identities it is a functor, which establishes the result.

□

So we have finally established that skew span categories and categories with unique map factorization are essentially the same. It is therefore obvious both that the following well-known relationship holds:

$$(\mathbf{Rel}, \subseteq) \cong Span(\mathbf{Set}, =)$$

and also, since the skew span construction preserves isomorphism, that we have the additional relationship

$$(\mathbf{Pow}, \sqsubseteq) \cong Span(\mathbf{Rel}, \subseteq) \ .$$

The category **Set** is not itself isomorphic to a skew span category, since its only maps are bijections. Moreover, it is not possible to construct the skew span category of predicate transformers, since **Pow** does not have pullovers. So this hierarchy of span categories cannot be extended further in either direction. The relationship between predicate transformers and relations can be generalized however[22]: if $\mathbf{Rel}_{\mathcal{C}}$ is the category of relations overs an arbotrary topos $\mathcal{C}$, then the skew span category of $\mathbf{Rel}_{\mathcal{C}}$ exists and is isomorphic to a category of predicate transformers.

5 Extending Functors

Associated with the construction of skew span categories from categories with pullovers is an extension of functors and adjunctions to lax functors and local adjunctions similar to that associated with ordinary span categories [16]. In this section we will first describe this extension of functors in the context of categories with unique map factorization, and then outline its application to programming languages.

Suppose that $(\mathcal{C}, \leq)$ and $(\mathcal{D}, \sqsubseteq)$ are poset enriched categories where $\mathcal{C}$ has unique map factorization. Let $F : Map(\mathcal{C}) \to Map(\mathcal{D})$ be a monotonic functor. Then F has a well-defined extension to $F' : \mathcal{C} \to \mathcal{D}$ where for all $n : E \to N, m : E \to M \in Map(\mathcal{C})$,

$$F'(n^* ;m) \mathrel{\widehat{=}} (Fn)^* ;Fm \ .$$

This extension is a monotonic graph morphism, but it is not necessarily a functor, only a *lax* functor: a lax functor $H : C \to D$ is a monotonic graph morphism such that for all composable pairs of arrows p, q and objects A in $\mathcal{C}$,

$$HI_A \sqsubseteq I_{HA}$$
$$H(p;q) \sqsubseteq Hp;Hq \ .$$

If F' is a functor, then it is the unique extension of F in the sense that if $G : \mathcal{C} \to \mathcal{D}$ is another monotonic functor that agrees with F on $Map(\mathcal{C})$, then $G = F'$. This is immediate from the fact that monotonic functors preserve maps and comaps: for all $p \in Map(\mathcal{C})$, $Gp^* = (Gp)^*$. The following theorem shows that even if F' is only a lax functor it is unique in the sense that it is the least extension of F.

Theorem 5. *Suppose that $(\mathcal{C}, \leq)$ and $(\mathcal{D}, \sqsubseteq)$ are poset enriched categories where $\mathcal{C}$ has unique map factorization. Let $F : Map(\mathcal{C}) \to Map(\mathcal{D})$ be a functor and $G : \mathcal{C} \to \mathcal{D}$ be a lax functor that agrees with F on $Map(\mathcal{C})$, that is $Fp = Gp$ for all $p \in Map(\mathcal{C})$. Then $F' \sqsubseteq G$.*

Proof. Suppose $(n^* ;m) \in \mathcal{C}$, then

$$\begin{array}{lll}
F'(n^* ;m) & = & \{ \text{ definition of } F' \ \} \\
(Fn)^* ;Fm & = & \{ \ F \text{ and } G \text{ agree on } Map(\mathcal{C}) \ \} \\
(Fn)^* ;Gm & \sqsubseteq & \{ \ (2) \text{ and monotonicity of } G \ \} \\
(Fn)^* ;G(n ;n^* ;m) & \sqsubseteq & \{ \ G \text{ is a lax functor } \} \\
(Fn)^* ;Gn ;G(n^* ;m) & = & \{ \ F \text{ and } G \text{ agree on } Map(\mathcal{C}) \ \} \\
(Fn)^* ;Fn ;G(n^* ;m) & \sqsubseteq & \{ \ (1) \ \} \\
G(n^* ;m) \ . & &
\end{array}$$

□

These observations show that every endofunctor F in **Rel** is the unique extension of an endofunctor in **Set** and has a canonical extension to **Pow** which is isomorphic to the lax endofunctor $F^\star$ defined by

$$F^\star p \mathrel{\widehat{=}} [F \ni] ; \langle F(p ; \ni) \rangle$$

where this definition uses the map factorization of predicate transformers given in Examples 3.

The technique of extending functors has applications to the theory of typed non-deterministic programming languages. Whereas deterministic languages like the lambda calculus have a definitive type structure provided by the cartesian closed category of sets and total functions, that of non-deterministic languages with either a relational or a predicate transformer semantics is less clearly defined, since neither category is cartesian closed. Therefore, when seeking to endow such languages with a natural type structure, it is useful to know that there is only one way to extend functors, and hence type constructors, from total functions to relations and predicate transformers. This fact was exploited in [23], where a relational calculus for program derivation was developed as a canonical extension of an established calculus for the derivation of functional programs [5]. Using the resulting calculus it was possible to unify a varied collection of dynamic programming algorithms.

The data types used in the functional calculus of [5] and its extensions are defined in [19] as fixed points of endofunctors in **Set**. Each data type occurs as either the initial algebra or final coalgebra of a functor, and the familiar functional programming operators such as fold or reduce are defined as homomorphisms on these data types. The reason that the calculus can be generalized so effectively to relations is that initial algebras are preserved under the extension of functors from total functions to relations. It has now been shown [24] that the extension of functors from relations to predicate transformers simply dualizes the universal properties of initial algebras by transforming them to final coalgebras. This means that the theoretical foundations for the future development of a corresponding calculus of predicate transformers are fully established. Such a calculus would combine all the benefits of the refinement calculus with the additional capability of reasoning about types.

6 An Application

In [15], it is shown how the description of specifications in terms of Hoare logic [14] can clarify the intuition behind the skew span representation of predicate transformers. Specification statements are represented as spans of relations, and the inference rules of Hoare logic are used to show that the refinement ordering on specifications is equivalent to the preorder on spans. Similarly, the inference rule for composition is used to show that the composition operator for spans would also be a suitable operator for specifications. In this section we will discuss the implications of these observations on the laws of the refinement calculus of [26]. In particular, they provide single complete rules for the refinement and composition of certain commands.

Until now we have only discussed the mathematical model of the refinement calculus: the monotonic predicate transformers. The language itself consists of a notation and a set of refinement laws for deriving programs from their specifications. It takes its inspiration from the earlier methods of Hoare and Dijkstra, but is innovative in incorporating both programs and specifications within a single framework: each command of the refinement calculus represents either a guarded command from Dijkstra's programming language [8] or a non-executable specification. The transition from each specification to executable code is broken down into a sequence of development steps, each of which is justified by one of the refinement laws. Therefore it is important to know whether the set of laws is complete, which is to say that every valid refinement can be verified using only the given laws. We will now show how the span representation of predicate transformers provides two such completeness results.

One of the most general forms of specification statement in the refinement calculus is the following:

$$|[\mathbf{con}\ X : E \bullet [pre, post]]| \ . \tag{11}$$

We will call such statements *transition* statements. As we will see, it is possible to express any statement of the refinement calculus as a transition specification. Any program satisfying specification (11) will, when activated in a state satisfying *pre*, establish a final state satisfying *post*. The two predicates *pre* and *post* have free variables drawn from a set of program variables V and a set of logical variables E; the purpose of the logical variables is to relate initial and final values of the program variables. They are assumed to denote the same value when they occur in the precondition *pre* as they do in the postcondition *post*, and therefore must not be altered by programs in order to meet the specification. A simple example of a transition specification is that met by the program which increments the value of the program variable x by 1:

$$|[\mathbf{con}\ X : E \bullet [x = X, x = X + 1]]| \ .$$

The results of Sect. 4 show that the meaning of any command in the refinement calculus can be expressed not only as a monotonic predicate transformer, but also as an equivalence class of spans of relations. The predicates *pre* and *post* in (11) both describe a relationship between the two sets of variables E and V so they can be considered alternatively as relations from E to V, as it is observed in [15]. By doing so, and using the laws of the refinement calculus to calculate the meaning of (11), we find that it has the following simple expression as a class of spans of relations:

$$(\!|post, pre|\!) \ .$$

So every monotonic predicate transformer, and therefore any statement of the refinement calculus, corresponds to an equivalence class of transition specifications, associated with which is a notion of refinement and composition. The refinement ordering on specifications is defined to be the pointwise ordering on the corresponding predicate transformers, which is equivalent to the skew span preorder on spans of relations. Therefore, the latter can be translated into a single complete rule for the refinement of transition specifications. Let *pre* and *post* be a pair of logical formulae with free variables drawn from the set of the program variables V and the set of logical variables E. Similarly for the pair *pre'* and *post'* with respect to the sets V and E'. When it is necessary to make the set of free variables associated with a predicate explicit we will use subscripts to do so: for example *pre* would be written as pre_{EV}. The definition of the preorder on spans of relations translates to the following preorder relation between specifications:

Span Preorder :

$$|[\mathbf{con}\ X : E \bullet [pre, post]]| \sqsubseteq |[\mathbf{con}\ X' : E' \bullet [pre', post']]|$$

$$\widehat{=}$$

$$\exists h_{EE'} \bullet (pre \Rightarrow (\exists X' : E' \bullet h \wedge pre')) \wedge ((\exists X' : E' \bullet h \wedge post') \Rightarrow post) \ .$$

This rule can be broken down into the following three more familiar rules, which are together equivalent to it. For simplicity suppose now that unless indicated otherwise, all predicates share the same two sets of free variables E and V. Then the Span Preorder rule is equivalent to the following:

1. *Weaken Precondition*: If $pre \Rightarrow pre'$ then

$$|[\mathbf{con}\ X : E \bullet [pre, post]]| \sqsubseteq |[\mathbf{con}\ X : E \bullet [pre', post]]| \ .$$

2. *Strengthen Postcondition*: If $post' \Rightarrow post$ then

$$|[\mathbf{con}\ X : E \bullet [pre, post]]| \sqsubseteq |[\mathbf{con}\ X : E \bullet [pre, post']]| \ .$$

3. *Change logical variable*:

$$|[\mathbf{con}\ X' : E' \bullet [(\exists X : E \bullet h_{EE'} \wedge pre), (\exists X : E \bullet h_{EE'} \wedge post)]]|$$

$$\sqsubseteq$$

$$|[\mathbf{con}\ X : E \bullet [pre, post]]| \ .$$

Both the first two of these laws occur among the laws of the refinement calculus. Law (3) is not listed explicitly, but is derivable within the calculus, so this shows that the laws of the refinement calculus are complete for proving refinements of transition specifications.

Similar reasoning can be used to translate the skew span composition operator into a composition law for transition specifications. For simplicity we will just consider the case where the specifications being composed share the same set of logical variables and program variables E and V respectively. The composition becomes:

Span Composition :

$$|[\mathbf{con}\ X : E \bullet [pre, mid]]| \,;|[\mathbf{con}\ X : E \bullet [mid', post]]|$$

$$\widehat{=}$$

$$|[\mathbf{con}\ Y : D \bullet [(\exists X : E \bullet h_{DE} \wedge pre), (\exists X : E \bullet h'_{DE} \wedge post]]|$$

where (h, h') is any pair of predicates such that for all predicates ψ such that Y does not occur free in ψ,

$$(\exists Y : D \bullet (h \wedge (\forall X : E \bullet h' \Rightarrow \psi)) = (\forall x : V \bullet mid \Rightarrow (\exists X : E \bullet mid' \wedge \psi)) \ .$$

One suitable choice for h and h' is the following: let $\mathcal{P}E$ denote the set of predicates with free variables drawn from E, and let $D = \mathcal{P}E \times \mathcal{P}E$. Then for all $Y : D$ there exist predicates ϕ_E and ψ_E such that $Y = (\phi, \psi)$, so we can define h and h' as follows

$$\begin{aligned} mids_D &\mathrel{\widehat=} (\exists X : E \bullet mid \wedge \phi) \Rightarrow (\exists X : E \bullet mid' \wedge \psi) \\ h &\mathrel{\widehat=} mids \wedge \psi \\ h' &\mathrel{\widehat=} mids \wedge \phi \ . \end{aligned}$$

This definition of composition is rather unwieldy, and unlike the law for refinement, it cannot be proved from the laws of the refinement calculus, although it is possible to show that the expression on the right of the definition is refined by that on the left. Therefore the laws are complete for proving refinements of specifications into smaller pieces. The laws are usually only used in this way, since the process of derivation usually involves breaking a specification down into manageable parts, each of which is independently refined to code. However, it is conceivable that it might be useful sometimes to have a rule which does the opposite, and combines two specifications into a larger one. For example, it might be possible to substitute a library procedure which meets the composite of two specifications, but neither individually. Or, more probably, it might be useful once derivation is completed, to calculate the loosest specification satisfied by the composite of two programs whose separate specifications are known.

Acknowledgements

The authors would like to thank Tony Hoare and Carroll Morgan for their comments and encouragement. We gratefully acknowledge the financial support of the SERC and British Petroleum International, which made our cooperation possible.

References

1. Back, R.J.R.: A Calculus of Refinements for Program Derivation. Acta Informatica **25** (1988) 593-624
2. Barr, M.: Relational Algebras. In Reports of the Midwest Category Seminar *IV*, Springer-Verlag Lecture Notes in Mathematics **137** (1069) 435-477
3. Benabo, J.: Introduction to Bicategories. Springer-Verlag Lecture Notes in Mathematics **47** (1967)
4. Bird, R.S.,de Moor, O.: Solving Optimisation Problems with Catamorphisms. (1992) This volume.
5. Bird, R.S.: Lectures on Constructive Functional Programming. In M. Broy, editor. Constructive Methods in Computing Science, Springer-Verlag NATO ASI series F **55** (1989) 151-216
6. Carboni, A.,Kasangian, S.: Bicategories of Spans and Relations. Journal of Pure and Applied Algebra **33** (1984) 259-267
7. Carboni, A., Kelly, M., and Wood, R.: A 2-Categorical Approach to Geometric Morphisms I. Sydney Pure Mathematics Research Reports **89-19** (1989). Department of Pure Mathematics, University of Sydney, NSW 2006, Australia.
8. Dijkstra, E.W.: A Discipline of Programming. Prentice-Hall, Englewood Cliffs (1976)

9. Gardiner, P.H.B.: Data Refinement of Maps. Draft (1990)
10. Gardiner, P.H.B., Morgan, C.C.: A Single Complete Rule for Data Refinement. Technical Report PRG-TR-7-89, Programming Research Group, 11 Keble Road, Oxford OX1 3QD, 1989.
11. Gray, J.W.: Formal Category Theory: Adjointness for 2-categories. Springer-Verlag Lecture Notes in Mathematics **391** (1974)
12. Herrlich H., Strecker G.E.: Category Theory. Allyn and Bacon Inc., Boston (1973)
13. Hesselink, W.H.: Modalities of Nondeterminancy. In Beauty is Our Business: A Birthday Salute to Edsgar W. Dijkstra. Springer-Verlag, (1990) 182-193
14. Hoare, C.A.R.: An Axiomatic Basis for Computer Programming. Communications of the ACM **12(10)** (1969) 576-580,583
15. Hoare, C.A.R. Handwritten note (1991)
16. Jay, C.B.: Local Adjunctions. Journal of Pure and Applied Algebra **53** (1988) 227-238
17. Johnstone, P.T.: Stone Spaces Cambridge University Press (1982)
18. Lambek, J.: Subequalizers. Bulletin Amer. Math. Soc. **13(3)** (1970) 337-349
19. Malcolm, G.: Data Structures and Program Transformation. Science of Computer Programming **14** (1990) 255-279
20. Martin, C.E.: Preordered Categories and Predicate Transformers. D. Phil Thesis. Programming Research Group, 11 Keble Road, Oxford OX1 3QD. (1991)
21. de Moor, O.: Categories, Relations and Dynamic Programming. PRG Technical Report PRG-TR-18-90, Programming Research Group, 11 Keble Road, Oxford OX1 3QD. (1990)
22. de Moor, O., Gardiner, P.H.B., and Martin, C.E.: Factorizing Predicate Transformers in a Topos. Draft (1991)
23. de Moor, O.: Categories, Relations and Dynamic Programming. D. Phil Thesis. PRG Technical Monograph PRG-98, Programming Research Group, 11 Keble Road, Oxford OX1 3QD. (1992)
24. de Moor, O.: Inductive Data Types for Predicate Transformers. Information Processing Letters (to appear)
25. Morgan, C.C.: Programming From Specifications. Prentice-Hall (1990)
26. Morgan, C.C., Robinson, K. and Gardiner, P.H.B.: On the Refinement Calculus. PRG Technical Monograph PRG-70. Programming Research Group, 11 Keble Road, Oxford OX1 3QD. (1988)
27. Morris, J.M.: Laws of Data Refinement. Acta Informatica **26** (1989) 287-308

Upwards and downwards accumulations on trees

Jeremy Gibbons

Dept of CS, University of Auckland, Private Bag 92019, Auckland, New Zealand.
Email: jeremy@cs.aukuni.ac.nz

Abstract. An *accumulation* is a higher-order operation over structured objects of some type; it leaves the shape of an object unchanged, but replaces each element of that object with some accumulated information about the other elements. Upwards and downwards accumulations on trees are two instances of this scheme; they replace each element of a tree with some function—in fact, some homomorphism—of that element's descendants and of its ancestors, respectively. These two operations can be thought of as passing information up and down the tree.
We introduce these two accumulations, and show how together they solve the so-called prefix sums problem.

1 Introduction

The value of being able to calculate computer programs formally from their specifications is now widely recognized. This ability to calculate can only be achieved with the aid of mathematically precise and concise tools. In this paper we look at the calculation of solutions to problems about trees. Trees are important in computing because they capture the idea of hierarchical structure. They permit fast parallel collection and dissemination of information among their elements; indeed, it could be argued that *all* algorithms that take logarithmic time, whether sequentially or in parallel, do so because of an underlying tree structure.

We introduce two tools for reasoning with problems about trees, namely, *upwards* and *downwards accumulations* on trees. These accumulations embody the notions of passing information up a tree, from the leaves towards the root, and down, from the root towards the leaves. We use these accumulations in the derivation of a fast algorithm for the prefix sums problem.

The workbench on which we use these tools is the *Bird-Meertens formalism* [12, 3, 4, 2]. We give a crash course in the relevant notation below.

2 Notation

The identity function is written id ; the constant function always returning a is written $!a$. Function application is written with an infix $\cdot$, so $!a \cdot b = a$. Application is tightest binding, and *right* associative, so $f \cdot g \cdot a$ parses as $f \cdot (g \cdot a)$; we find this more useful than left associative application. Function composition is backwards, written with an infix $\circ$, and is weakest binding:

$$(f \circ g) \cdot a = f \cdot g \cdot a$$

The Bird-Meertens formalism makes free use of infix binary operators. Such operators are turned into unary functions by *sectioning*:

$$(a\oplus)\cdot b = a \oplus b = (\oplus b)\cdot a$$

The *converse* of a binary operator $\oplus$ is written $\tilde{\oplus}$ and satisfies

$$a \mathbin{\tilde{\oplus}} b = b \oplus a$$

The type judgement 'a has type A' is written $a \in A$ (this is not intended to mean that types are sets). The function type former is written $\rightarrow$. The cartesian product of two types is denoted by $\parallel$, and their sum by $\mid$. Indeed, $\parallel$ and $\mid$ are bifunctors, acting on functions as well as types: if $f \in A \rightarrow C$ and $g \in B \rightarrow D$ then

$$f \parallel g \in A \parallel B \rightarrow C \parallel D$$
$$f \mid g \in A \mid B \rightarrow C \mid D$$

We write A^2 and f^2 as abbreviations for $A \parallel A$ and $f \parallel f$. The product and sum morphisms 'fork' and 'join' are denoted by $\curlywedge$ and $\curlyvee$, the shapes suggesting processes splitting and recombining as they 'move down the page'; if $f \in A \rightarrow B$ and $g \in A \rightarrow C$, and $h \in B \rightarrow A$ and $k \in C \rightarrow A$, then

$$f \curlywedge g \in A \rightarrow B \parallel C$$
$$h \curlyvee k \in B \mid C \rightarrow A$$

The projections for products are

$$\ll\ \in A \parallel B \rightarrow A$$
$$\gg\ \in A \parallel B \rightarrow B$$

We do not need injections for sums in this paper.

Data types are constructed as the least fixed points of polynomial functors, that is, as the initial algebras in the appropriate categories of algebras. We will just use the ideas informally here, not having the space to present them formally; the reader is referred to Malcolm [11] or Hagino [8] for the details. For functor F, an F-algebra is a pair (X, τ) such that $\tau \in F\cdot X \rightarrow X$. The trees we will be considering, *homogeneous binary trees*, are defined by the equation

$$tree\cdot A = \triangle\cdot A \mid tree\cdot A \perp_A tree\cdot A$$

Informally, this says that the algebra $(tree\cdot A, \triangle \curlyvee \perp)$ of trees with elements of type A is the least fixed point of the functor T that maps X to $A \mid (X \parallel A \parallel X)$; the constructors $\triangle$ (pronounced 'leaf') and $\perp$ (a corruption of the Chinese ideogram pronounced 'moo' and meaning 'wood' or 'tree') have types

$$\triangle \in A \rightarrow tree\cdot A$$
$$\perp \in tree\cdot A \parallel A \parallel tree\cdot A \rightarrow tree\cdot A$$

Note that $\perp$ is a *ternary* operator; its middle argument is written as a subscript. For example, the tree

$$\vartriangle \cdot b \perp_a (\vartriangle \cdot d \perp_c \vartriangle \cdot e)$$

corresponds to the tree

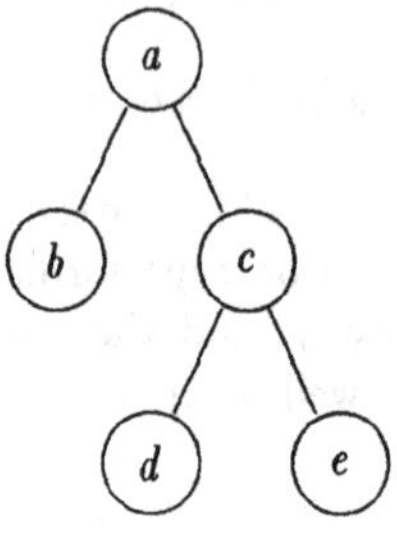

We will call this tree *five* , and use it as a running example.

The operation on types that sends A to $tree \cdot A$ is a functor, and so its action on functions—written with a postfix $*$—respects identity and composition. Thus, if $f \in A \to B$, then $f* \in tree \cdot A \to tree \cdot B$, and $id* = id$ and $(f \circ g)* = f* \circ g*$.

We say that a function h is (f, g) *F-promotable* if

$$h \circ f = g \circ F \cdot h$$

The important fact about the initial algebra in a category of algebras, the *unique extension property*, states that, if (X, τ) is the initial F-algebra, then for a given f there is a *unique* function that is (τ, f) F-promotable. This function is called a *catamorphism* and is written $\llparenthesis F{:}\, f \rrparenthesis$; if the functor F is clear from context, we write simply $\llparenthesis f \rrparenthesis$. In the case of trees, the catamorphism $\llparenthesis f \triangledown \odot \rrparenthesis$ satisfies

$$\llparenthesis f \triangledown \odot \rrparenthesis \cdot \vartriangle \cdot a = f \cdot a$$
$$\llparenthesis f \triangledown \odot \rrparenthesis \cdot (x \perp_a y) = \llparenthesis f \triangledown \odot \rrparenthesis \cdot x \odot_a \llparenthesis f \triangledown \odot \rrparenthesis \cdot y$$

and is the unique function that is $(\vartriangle \triangledown \perp, f \triangledown \odot)$ T-promotable. For example, the functions returning the numbers of leaves and branches of a tree are catamorphisms, given by

$$leaves = \llparenthesis !1 \triangledown \oplus \rrparenthesis \qquad \text{where} \quad u \oplus_a v = u + v$$
$$branches = \llparenthesis !0 \triangledown \otimes \rrparenthesis \qquad \text{where} \quad u \otimes_a v = u + 1 + v$$

Catamorphisms are important because they are 'eminently manipulable'.

One corollary of the unique extension property is called the *promotion theorem*; the proof is in Malcolm's thesis, among other places.

Theorem 1. *If h is (f, g) F-promotable, then* $h \circ \llparenthesis F{:}\, f \rrparenthesis = \llparenthesis F{:}\, g \rrparenthesis$.

For example, the popular student exercise in structural induction of showing that

$$\langle 1+ \rangle \circ branches = leaves$$

follows from the promotion theorem, because $\langle 1+ \rangle$ is $(!0 \triangledown \otimes, !1 \triangledown \oplus)$ T-promotable.

3 Upwards accumulations

We now turn to the main topic of this paper, namely accumulations. Functional programmers will be familiar with the function *inits* on (non-empty) lists, which returns a list of all the initial segments of its argument; for example,

$$inits \cdot [a, b, c] = [[a], [a, b], [a, b, c]]$$

Upwards and downwards accumulations arise from generalizing this concept to trees.

One way of thinking about *inits* is as replacing every element of a list with its *predecessors*, that is, with the initial segment of the list ending with that element. By analogy, the function *subtrees* on trees replaces every element of a tree with its *descendants*, that is, with the subtree rooted at that element. For example, applying *subtrees* to the tree *five* yields the tree of trees

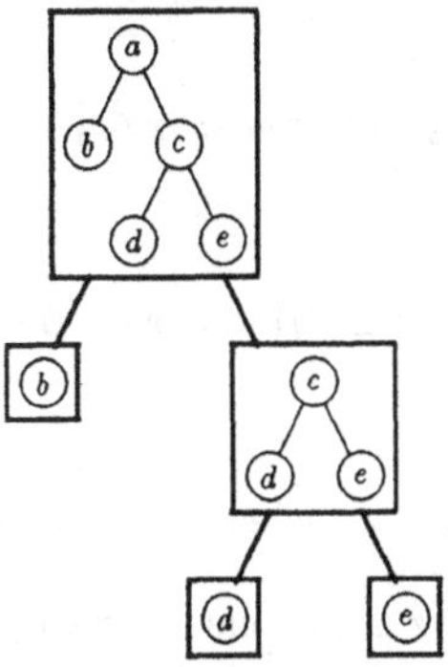

The equations characterizing *subtrees* are

$$subtrees \cdot \triangle \cdot a = \triangle \cdot \triangle \cdot a$$
$$subtrees \cdot (x \perp_a y) = subtrees \cdot x \perp_{x \perp_a y} subtrees \cdot y$$

We can see by case analysis that

$$root \circ subtrees = id$$

and so

$$subtrees \cdot (x \perp_a y) = subtrees \cdot x \oplus_a subtrees \cdot y$$

where

$$u \oplus_a v = u \perp_z v \qquad \text{where} \quad z = root \cdot u \perp_a root \cdot v$$

That is, *subtrees* is a catamorphism, $(\!|(\triangle \circ \triangle) \curlyvee \oplus|\!)$.

Functions that 'pass information upwards', from the leaves of a tree towards the root, are characterized in terms of *subtrees* :

Definition 2. Functions of the form $g* \circ subtrees$ are called *upwards passes*.

Suppose that h is an upwards pass, and that $h = g* \circ subtrees$, so that

$$h \cdot \triangle \cdot a = \triangle \cdot g \cdot \triangle \cdot a$$
$$h \cdot (x \perp_a y) = h \cdot x \perp_b h \cdot y \qquad \text{where} \quad b = g \cdot (x \perp_a y)$$

This does not yield a quick way of computing h, even if we have one for g; for example, if g takes time proportional to the depth of the tree, then h will take parallel time proportional to the square of the depth.

However, suppose further that g is a tree catamorphism, so that

$$g \cdot (x \perp_a y) = g \cdot x \odot_a g \cdot y$$

for some $\odot$. Now, we already have

$$root \circ h = g$$

and so

$$g \cdot (x \perp_a y) = root \cdot h \cdot x \odot_a root \cdot h \cdot y$$

Thus,

$$h \cdot (x \perp_a y) = h \cdot x \oplus_a h \cdot y$$

where .

$$u \oplus_a v = u \perp_b v \qquad \text{where} \quad b = root \cdot u \odot_a root \cdot v$$

The important point is that $h \cdot (x \perp_a y)$ can be computed from $h \cdot x$ and $h \cdot y$ using only *one* more application of $\odot$, and so h can be computed in parallel time proportional to the depth of the tree times the time taken by $\odot$. Such functions are what we mean by *upwards accumulations*:

Definition 3. Functions of the form $(\!| f \curlyvee \odot |\!)* \circ subtrees$ are called *upwards accumulations*, and are written $(f, \odot)\Uparrow$.

The phrase above about the time taken to compute h is rather unwieldy; we shall say instead that a function on trees can be computed *quickly up to* $\odot$, or just *quickly* when the $\odot$ is understood, if it can be computed in parallel time proportional to the depth of the tree times the time taken by $\odot$.

One simple example of an upwards accumulation is the function *sizes*, which replaces every element of a tree with the number of descendants it has:

$$sizes = size* \circ subtrees$$

Since *size* is a catamorphism,

$$size = (\!| !1 \curlyvee \odot |\!) \qquad \text{where} \quad u \odot_a v = u + 1 + v$$

we have a quick algorithm

$$sizes = (!1, \odot)\Uparrow$$

for *sizes*

The equation

$$(\!| f \curlyvee \odot |\!) * \circ \mathit{subtrees} = (f, \odot)\Uparrow$$

can be seen as an efficiency-improving transformation, when used from left to right. It can also, of course, be used from right to left, when it forms a 'manipulability-improving' transformation; catamorphisms, maps and *subtrees* enjoy many useful properties, and the left hand side may be more amenable to calculation than the right. This choice between manipulability and efficiency is a characteristic of accumulations.

4 Downwards accumulations

We have just discussed upwards accumulation, which captures the notion of passing information up through a tree from the leaves towards the root. We turn now to *downwards* accumulation, which corresponds to passing information in the opposite direction, from the root towards the leaves.

As upwards accumulations arose by considering the function *subtrees*, which replaces every element of a tree with its descendants, so downwards accumulations arise by considering the function *paths*, which replaces every element of a tree with its *ancestors*. The ancestors of an element in a tree themselves form a special kind of tree, called a *thread*: a tall thin tree with that element as its one and only leaf. For example, the ancestors of the element d in the tree *five* form the three-element thread

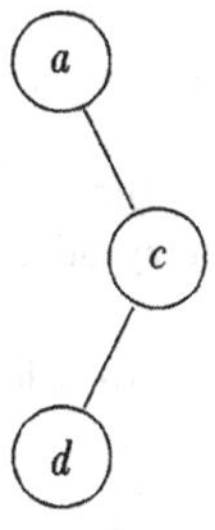

and applying *paths* to *five* yields the tree of threads

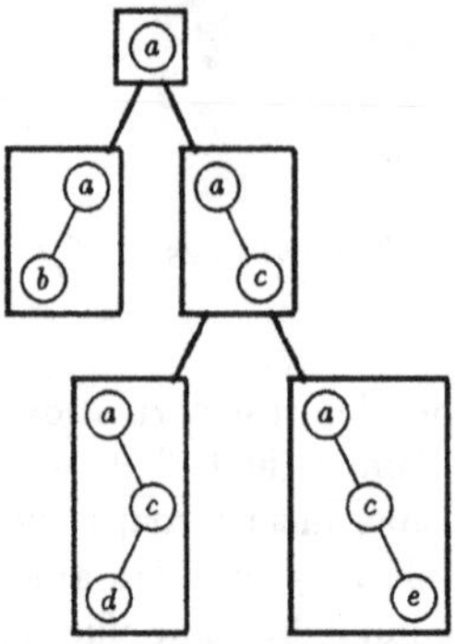

Threads are the least fixed point of the functor H which sends the 'base type' X to $A \mid (X \parallel A) \mid (X \parallel A)$; they are given by the type equation

$$thread \cdot A = \diamond \cdot A \mid thread \cdot A \swarrow A \mid thread \cdot A \searrow A$$

and the three-element thread given above is written $(\diamond \cdot a \searrow c) \swarrow d$. The constructors $\swarrow$ and $\searrow$ could be pronounced 'left snoc' and 'right snoc', and threads thought of as 'snoc' lists with two different colours of constructor.

Now, *paths* is a catamorphism, and is given by the equations

$$\begin{aligned} paths \cdot \triangle \cdot a &= \triangle \cdot \diamond \cdot a \\ paths \cdot (x \perp_a y) &= (a \oslash) * \cdot paths \cdot x \ \perp_{\diamond \cdot a} \ (a \obslash) * \cdot paths \cdot y \end{aligned} \tag{1}$$

where

$$\begin{aligned} a \oslash p &= (\!|H: (\diamond \cdot a \swarrow) \curlyvee \swarrow \curlyvee \searrow|\!) \cdot p \\ a \obslash p &= (\!|H: (\diamond \cdot a \searrow) \curlyvee \swarrow \curlyvee \searrow|\!) \cdot p \end{aligned}$$

Informally, $\oslash$ and $\obslash$ 'cons' elements to threads; for example,

$$a \obslash (\diamond \cdot c \swarrow d) = (\diamond \cdot a \searrow c) \swarrow d$$

Because threads have three constructors, thread catamorphisms involve a 'three-way join', which should be considered a ternary operator rather than two applications of a binary one.

Functions that 'pass information downwards' are characterized by the following definition.

Definition 4. Functions of the form $g* \circ paths$ are called *downwards passes.*

As before, downwards passes need not be quick; however, if the 'multiplier' g is a thread catamorphism then $g* \circ paths$ is more tractable.

Definition 5. Functions of the form $(\!|H: f \curlyvee \oplus \curlyvee \otimes|\!)* \circ paths$ are called *downwards accumulations* and are written $(f, \oplus, \otimes)\Downarrow$.

The downwards accumulation $(f,\oplus,\otimes)\Downarrow$ can be computed *quickly up to* $\oplus$ *and* $\otimes$, that is, quickly up to the more expensive of $\oplus$ and $\otimes$, since

$$
\begin{array}{cl}
 & (f,\oplus,\otimes)\Downarrow\cdot(x \perp_a y) \\
= & \qquad \{ \Downarrow \} \\
 & (\!|f \curlyvee \oplus \curlyvee \otimes|\!)*\cdot paths\cdot(x \perp_a y) \\
= & \qquad \{ \mathit{paths} \} \\
 & (\!|f \curlyvee \oplus \curlyvee \otimes|\!)*\cdot(\langle a\oslash\rangle*\cdot paths\cdot x \perp_{\diamond\cdot a} \langle a\obslash\rangle*\cdot paths\cdot y) \\
= & \qquad \{ \text{promotion (see below)} \} \\
 & (\!|\langle f\cdot a\oplus\rangle \curlyvee \oplus \curlyvee \otimes|\!)*\cdot paths\cdot x \perp_{f\cdot a} (\!|\langle f\cdot a\otimes\rangle \curlyvee \oplus \curlyvee \otimes|\!)*\cdot paths\cdot y \\
= & \qquad \{ \Downarrow \} \\
 & (\langle f\cdot a\oplus\rangle,\oplus,\otimes)\Downarrow\cdot x \perp_{f\cdot a} (\langle f\cdot a\otimes\rangle,\oplus,\otimes)\Downarrow\cdot y
\end{array}
$$

and the recursion on *paths* can be made in parallel after only one more application each of $\oplus$ and $\otimes$. (The promotion property mentioned in the calculation is that

$$(\!|f \curlyvee \oplus \curlyvee \otimes|\!) \circ \langle a\oslash\rangle = (\!|\langle f\cdot a\oplus\rangle \curlyvee \oplus \curlyvee \otimes|\!)$$

and similarly for $\langle a\obslash\rangle$, both of which are consequences of the definitions of $\oslash$ and $\obslash$ and properties of catamorphisms.) For example, we see that *paths* can be computed quickly up to $\swarrow$ and $\searrow$, that is, in parallel time proportional to the depth of the tree, because *paths* is itself a downwards accumulation:

$$
\begin{array}{cl}
 & paths \\
= & \qquad \{ \text{identity catamorphism} \} \\
 & (\!|\diamond \curlyvee \swarrow \curlyvee \searrow|\!)* \circ paths \\
= & \qquad \{ \Downarrow \} \\
 & (\diamond,\swarrow,\searrow)\Downarrow
\end{array}
$$

Notice, however, that downwards accumulations are not in general catamorphic: the accumulation $(f,\oplus,\otimes)\Downarrow$ of a branch $x \perp_a y$ depends on different accumulations $(\langle f\cdot a\oslash\rangle,\oplus,\otimes)\Downarrow$ and $(\langle f\cdot a\obslash\rangle,\oplus,\otimes)\Downarrow$ of the children.

The characterization (1) we gave for *paths* is an instance of the frequently-occurring idiom

$$
\begin{aligned}
h\cdot\triangle\cdot a &= \triangle\cdot f\cdot a \\
h\cdot(x \perp_a y) &= \langle a\ominus\rangle*\cdot h\cdot x \perp_{f\cdot a} \langle a\oslash\rangle*\cdot h\cdot y
\end{aligned}
\tag{2}
$$

This looks like a downwards pass, in the sense that every element of the tree is replaced with some function of its ancestors, but it is not immediately obvious how it matches the pattern $g* \circ paths$. The following theorem makes the correspondence clear.

Theorem 6. *If* h *satisfies (2) then* h *is a downwards pass.*

Consider the type *daerht*, pronounced 'dirt'; this is the least fixed point of the functor D sending X to $A \mid (A \parallel X) \mid (A \parallel X)$, and is given by the equation

$$daerht \cdot A = \diamond \cdot A \mid A \nearrow daerht \cdot A \mid A \nwarrow daerht \cdot A$$

(We make no apology for using the same symbol $\diamond$ for singleton threads and for singleton daerhts.) Informally, daerhts are to threads as cons lists are to snoc lists. More formally, the correspondence between the two is given by the isomorphism

$$td = (\!|H \colon \diamond \curlyvee \oslash \curlyvee \circledS |\!) \in thread \cdot A \rightarrow daerht \cdot A$$

where

$$p \oslash a = (\!|D \colon (\nearrow \diamond \cdot a) \curlyvee \nearrow \curlyvee \nwarrow|\!) \cdot p$$
$$p \circledS a = (\!|D \colon (\nwarrow \diamond \cdot a) \curlyvee \nearrow \curlyvee \nwarrow|\!) \cdot p$$

The operators $\oslash$ and $\circledS$ effectively 'snoc' elements to daerhts.

It turns out that the function h of Theorem 6 satisfies

$$h = (\!|D \colon f \curlyvee \ominus \curlyvee \oslash|\!) * \circ td * \circ paths$$

To show this we will call upon the following lemma, concerning the operations $\oslash$ and $\circledS$ from the definition of *paths*.

Lemma 7.

$$td \circ (a \oslash) = (a \nearrow) \circ td$$
$$td \circ (a \circledS) = (a \nwarrow) \circ td$$

Proof. The lemma is a consequence of the promotion theorem, since $\nearrow$ and $\nwarrow$ each *associate with* $\oslash$ and $\circledS$, that is, $(a \nearrow)$ and $(a \nwarrow)$ are both $(\oslash, \oslash)$ and $(\circledS, \circledS)$ F-promotable, where $F \cdot X = X \parallel A$.

We are now equipped to prove Theorem 6. Let $\circledast$ satisfy

$$u \circledast_a v = (a \ominus) * \cdot u \perp_{f \cdot a} (a \oslash) * \cdot v$$

so that $h = (\!|(\triangle \circ f) \curlyvee \circledast|\!)$. We will show that

$$(\!|D \colon f \curlyvee \ominus \curlyvee \oslash|\!) * \circ td * \circ paths = (\!|(\triangle \circ f) \curlyvee \circledast|\!)$$

too.

Proof of Theorem 6. On leaves we have

$$(\!|f \curlyvee \ominus \curlyvee \oslash|\!) * \cdot td * \cdot paths \cdot \triangle \cdot a = \triangle \cdot f \cdot a$$

while on branches we get

$$
\begin{array}{ll}
 & \llparenthesis f \curlyvee \ominus \curlyvee \oslash \rrparenthesis * \cdot td * \cdot paths \cdot (x \perp_a y) \\
= & \quad \{ \ paths \ \} \\
 & \llparenthesis f \curlyvee \ominus \curlyvee \oslash \rrparenthesis * \cdot td * \cdot (\langle a \varoslash \rangle * \cdot paths \cdot x \ \perp_{\diamond \cdot a} \ \langle a \varobslash \rangle * \cdot paths \cdot y) \\
= & \quad \{ \text{ Lemma 7 } \} \\
 & \llparenthesis f \curlyvee \ominus \curlyvee \oslash \rrparenthesis * \cdot (\langle a \nearrow \rangle * \cdot td * \cdot paths \cdot x \ \perp_{\diamond \cdot a} \ \langle a \searrow \rangle * \cdot td * \cdot paths \cdot y) \\
= & \quad \{ \text{ catamorphisms } \} \\
 & \langle a \ominus \rangle * \cdot \llparenthesis f \curlyvee \ominus \curlyvee \oslash \rrparenthesis * \cdot td * \cdot paths \cdot x \ \perp_{f \cdot a} \ \langle a \oslash \rangle * \cdot \llparenthesis f \curlyvee \ominus \curlyvee \oslash \rrparenthesis * \cdot td * \cdot paths \cdot y \\
= & \quad \{ \ \circledast \ \} \\
 & (\llparenthesis f \curlyvee \ominus \curlyvee \oslash \rrparenthesis * \cdot td * \cdot paths \cdot x) \ \circledast_a \ (\llparenthesis f \curlyvee \ominus \curlyvee \oslash \rrparenthesis * \cdot td * \cdot paths \cdot y)
\end{array}
$$

Combining the two we get

$$\llparenthesis f \curlyvee \ominus \curlyvee \oslash \rrparenthesis * \circ td * \circ paths \circ (\vartriangle \curlyvee \perp) = ((\vartriangle \circ f) \curlyvee \circledast) \circ T \cdot (\llparenthesis f \curlyvee \ominus \curlyvee \oslash \rrparenthesis * \circ td * \circ paths)$$

and therefore

$$\llparenthesis f \curlyvee \ominus \curlyvee \oslash \rrparenthesis * \circ td * \circ paths = \llparenthesis (\vartriangle \circ f) \curlyvee \circledast \rrparenthesis = h$$

We have just seen that functions satisfying (2) are downwards passes. Under what conditions are they accumulations? That is, under what conditions is $\llparenthesis f \curlyvee \ominus \curlyvee \oslash \rrparenthesis \circ td$ a thread catamorphism? This question is answered next.

Definition 8. We say $(f, \ominus, \oslash)$ *inverts to* $(f, \oplus, \otimes)$ if

$$b \ominus f \cdot a = f \cdot b \oplus a$$
$$b \oslash f \cdot a = f \cdot b \otimes a$$

and $\ominus$ and $\oslash$ each associate with both $\oplus$ and $\otimes$. We say $(f, \ominus, \oslash)$ *is top down* if there exist $\oplus$ and $\otimes$ such that $(f, \ominus, \oslash)$ inverts to $(f, \oplus, \otimes)$.

Theorem 9. *If* $(f, \ominus, \oslash)$ *inverts to* $(f, \oplus, \otimes)$ *then*

$$\llparenthesis D : f \curlyvee \ominus \curlyvee \oslash \rrparenthesis \circ td = \llparenthesis H : f \curlyvee \oplus \curlyvee \otimes \rrparenthesis$$

Proof. According to the promotion theorem, we need only show that $\llparenthesis f \curlyvee \ominus \curlyvee \oslash \rrparenthesis$ is $(\varoslash, \oplus)$ and $(\varobslash, \otimes)$ promotable, that is, that the two equations

$$\llparenthesis f \curlyvee \ominus \curlyvee \oslash \rrparenthesis \cdot (p \varoslash a) = \llparenthesis f \curlyvee \ominus \curlyvee \oslash \rrparenthesis \cdot p \oplus a$$
$$\llparenthesis f \curlyvee \ominus \curlyvee \oslash \rrparenthesis \cdot (p \varobslash a) = \llparenthesis f \curlyvee \ominus \curlyvee \oslash \rrparenthesis \cdot p \otimes a$$

hold. We prove the first of these; the second is symmetric.

$$
\begin{array}{ll}
 & \llparenthesis f \curlyvee \ominus \curlyvee \oslash \rrparenthesis \cdot (p \varoslash a) = \llparenthesis f \curlyvee \ominus \curlyvee \oslash \rrparenthesis \cdot p \oplus a \\
= & \quad \{ \ \varoslash \ \} \\
 & \llparenthesis f \curlyvee \ominus \curlyvee \oslash \rrparenthesis \cdot \llparenthesis \langle \nearrow \diamond \cdot a \rangle \curlyvee \nearrow \curlyvee \searrow \rrparenthesis \cdot p = \llparenthesis f \curlyvee \ominus \curlyvee \oslash \rrparenthesis \cdot p \oplus a \\
= & \quad \{ \text{ promotion, catamorphisms } \} \\
 & \llparenthesis \langle \ominus f \cdot a \rangle \curlyvee \ominus, \oslash \rrparenthesis \cdot p = \llparenthesis f \curlyvee \ominus \curlyvee \oslash \rrparenthesis \cdot p \oplus a
\end{array}
$$

$$
\begin{array}{ll}
\Leftarrow & \{ \text{ condition on } \ominus \text{ and } \oplus \} \\
& b \ominus f\cdot a = f\cdot b \oplus a \ \wedge\ (\!|((\oplus a) \circ f) \curlyvee \ominus \curlyvee \oslash|\!) = (\oplus a) \circ (\!| f \curlyvee \ominus \curlyvee \oslash |\!) \\
\Leftarrow & \{ \text{ promotion } \} \\
& b \ominus f\cdot a = f\cdot b \oplus a \ \wedge\ (\oplus a) \text{ is } (\ominus, \ominus) \text{ and } (\oslash, \oslash) \text{ promotable} \\
= & \{ \text{ definition } \} \\
& b \ominus f\cdot a = f\cdot b \oplus a \ \wedge\ (\oplus a) \text{ associates with } \ominus \text{ and with } \oslash
\end{array}
$$

Corollary 10. *If* h *satisfies (2), and* $(f, \ominus, \oslash)$ *inverts to* $(f, \oplus, \otimes)$, *then*

$$h = (f, \oplus, \otimes)\Downarrow$$

For example, consider the function *depths*, which replaces every element of a tree with its depth in that tree; it satisfies

$$
\begin{aligned}
depths \cdot \triangle \cdot a &= \triangle \cdot 1 \\
depths \cdot (x \perp_a y) &= \langle 1+\rangle * \cdot depths \cdot x \ \perp_1 \ \langle 1+\rangle * \cdot depths \cdot y
\end{aligned}
$$

That is, *depths* satisfies (2), with f being $!1$ and $\ominus$ and $\oslash$ both being the $\odot$ such that $a \odot u = 1 + u$; hence, by Theorem 6, *depths* is a downwards pass and a catamorphism. Computed naively, *depths* will take parallel time quadratic in the depth of the tree, but $(!1, \odot, \odot)$ inverts to $(!1, \tilde{\odot}, \tilde{\odot})$ and so

$$depths = (!1, \tilde{\odot}, \tilde{\odot})\Downarrow$$

and *depths* can be computed quickly too.

5 Parallel prefix

Accumulations on trees provide a valuable tool for abstraction, as we hope to show by the following example.

The *prefix sums* problem is a problem on non-empty lists, which are given by the equation

$$list \cdot A = \square \cdot A \mid list \cdot A +\!\!+ list \cdot A$$

modulo the law that $+\!\!+$ is associative. The list catamorphism $(\!| f \curlyvee \odot |\!)$ satisfies the equations

$$
\begin{aligned}
(\!| f \curlyvee \odot |\!) \cdot \square \cdot a &= f \cdot a \\
(\!| f \curlyvee \odot |\!) \cdot (x +\!\!+ y) &= (\!| f \curlyvee \odot |\!) \cdot x \ \odot \ (\!| f \curlyvee \odot |\!) \cdot y
\end{aligned}
$$

where the operator $\odot$ must also be associative. One example of a list catamorphism is the function *last*, which returns the last element of a list:

$$last = (\!| id \curlyvee \gg |\!)$$

Another is the function *inits* mentioned earlier:

$$inits = (\!| (\square \circ \square) \curlyvee \oplus |\!) \qquad \text{where} \quad u \oplus v = u +\!\!+ \langle last \cdot u +\!\!+ \rangle * \cdot v$$

The prefix sums problem is to evaluate the 'running totals' $(\!|f \curlyvee \odot|\!)* \circ \mathit{inits}$ of a list. The operator $\odot$ must be associative; we also assume that it has a unit, e . For example, applied to the list $[a_1, \ldots, a_n]$, the problem is to compute

$$[f \cdot a_1,\ f \cdot a_1 \odot f \cdot a_2,\ \ldots,\ f \cdot a_1 \odot \cdots \odot f \cdot a_n]$$

This problem encapsulates a very common pattern of computation on lists; it has applications in, among other places, the evaluation of polynomials, compiler design, and numerous graph problems including minimum spanning tree and connected components [1].

It might appear from the above example that the problem inherently takes linear time to solve, even in parallel; the structure of the result seems to preclude any faster solution. However, Ladner and Fischer [10], reworking earlier results by Kogge and Stone [9] and Estrin [5], show that the evaluation can be performed in logarithmic time on a linear number of processors acting in parallel. It turns out that their 'parallel prefix' algorithm, which we derive here, is naturally expressed in terms of accumulations on trees.

The problem is to evaluate

$$ps = (\!|f \curlyvee \odot|\!)* \circ \mathit{inits}$$

The first step in the derivation is to change the problem from one on lists to one on trees; the motivation for this is that trees often lead to logarithmic algorithms, whereas lists rarely do. So, we are looking for a quick function tps , which evaluates prefix sums on a tree, satisfying

$$\mathit{fringe} \circ tps = ps \circ \mathit{fringe} \tag{3}$$

where *fringe* is the function returning the leaves of a tree as a list:

$$\mathit{fringe} = (\!|\square \curlyvee \oplus|\!) \qquad \text{where} \quad u \oplus_a v = u +\!\!+ v$$

We can calculate immediately the result of applying tps to a leaf, since

$$\mathit{fringe} \cdot tps \cdot \triangle \cdot a = \square \cdot f \cdot a$$

and hence

$$tps \circ \triangle = \triangle \circ f$$

because *fringe* is injective on leaves. Letting $s = (\!|f \curlyvee \odot|\!) \circ \mathit{fringe}$, we have on branches

$$\begin{array}{ll} & \mathit{fringe} \cdot tps \cdot (x \perp_a y) \\ = & \qquad \{ \text{ specification of } tps \ \} \\ & ps \cdot \mathit{fringe} \cdot (x \perp_a y) \\ = & \qquad \{ \ \mathit{fringe} \ \} \\ & ps \cdot (\mathit{fringe} \cdot x +\!\!+ \mathit{fringe} \cdot y) \\ = & \qquad \{ \ ps,\ \mathit{inits} \ \} \\ & (\!|f \curlyvee \odot|\!)* \cdot (\mathit{inits} \cdot \mathit{fringe} \cdot x \ +\!\!+ \ (\mathit{fringe} \cdot x +\!\!+)* \cdot \mathit{inits} \cdot \mathit{fringe} \cdot y) \end{array}$$

$$
\begin{aligned}
&= \quad \{ *, \text{catamorphisms} \} \\
&\quad (\!|f \curlyvee \odot|\!)*\cdot inits\cdot fringe\cdot x \;+\!\!+\; (s\cdot x\odot)*\cdot(\!|f \curlyvee \odot|\!)*\cdot inits\cdot fringe\cdot y \\
&= \quad \{ ps;\ \text{specification of } tps \} \\
&\quad fringe\cdot tps\cdot x \;+\!\!+\; (s\cdot x\odot)*\cdot fringe\cdot tps\cdot y
\end{aligned}
$$

This does not completely determine tps on branches, since $fringe$ is not injective on branches, but it is 'sweetly reasonable' to suppose that

$$tps\cdot(x \perp_a y) = tps\cdot x \perp_b (s\cdot x\odot)*\cdot tps\cdot y$$

for some b; certainly, this supposition satisfies the indirect specification (3) of tps. The calculation can tell us nothing about b, the root of $tps\cdot(x \perp_a y)$, because $fringe$ throws branch labels away.

This gives us now a direct—that is, executable—specification of tps:

$$
\begin{aligned}
tps\cdot\triangle\cdot a &= \triangle\cdot f\cdot a \\
tps\cdot(x \perp_a y) &= tps\cdot x \perp_b (s\cdot x\odot)*\cdot tps\cdot y
\end{aligned}
\tag{4}
$$

for some b. Executing this specification requires parallel time quadratic in the depth of the tree; we show next how to improve this to linear parallel time, by exploiting the freedom in the choice of value for b.

Suppose that $b = s\cdot x$, that is, that

$$tps\cdot(x \perp_a y) = tps\cdot x \perp_{s\cdot x} (s\cdot x\odot)*\cdot tps\cdot y$$

Intuitively, this allows the computation of $tps\cdot(x \perp_a y)$ from $tps\cdot x$ and $tps\cdot y$ to be split into two parts, the first bringing $s\cdot x$ to the root of the tree and the second mapping $(s\cdot x\odot)$ over the right child. More formally, suppose that

$$
\begin{aligned}
up\cdot(x \perp_a y) &= up\cdot x \perp_{s\cdot x} up\cdot y \\
down\cdot(u \perp_b v) &= down\cdot u \perp_b (b\odot)*\cdot down\cdot v
\end{aligned}
$$

whence

$$down\cdot up\cdot(x \perp_a y) = down\cdot up\cdot x \perp_{s\cdot x} (s\cdot x\odot)*\cdot down\cdot up\cdot y$$

so $down \circ up$ follows the same pattern as tps. An inductive proof shows that

$$(tps = down \circ up) \Leftarrow (down \circ up \circ \triangle = tps \circ \triangle)$$

(We cannot use the unique extension property because we do not yet know whether tps is a catamorphism.) The premise of this implication is satisfied if

$$
\begin{aligned}
up\cdot\triangle\cdot a &= \triangle\cdot f\cdot a \\
down\cdot\triangle\cdot b &= \triangle\cdot b
\end{aligned}
$$

We have not yet improved the efficiency; up and $down$ both take parallel time quadratic in the depth of the tree. However, as the names suggest, up and $down$ are upwards and downwards passes, and we know how to make such functions quick: we turn them into accumulations.

Let $sl = root \circ up$, so

$$sl \cdot \triangle \cdot a = f \cdot a$$
$$sl \cdot (x \perp_a y) = s \cdot x$$

Now,

$$up = sl* \circ subtrees$$

so up is indeed an upwards pass. It is not an accumulation, because sl is not a catamorphism: $sl \cdot (x \perp_a y)$ depends on $s \cdot x$ and not just on $sl \cdot x$. However, as this suggests, $s \curlywedge sl$ *is* a catamorphism,

$$s \curlywedge sl = (\!|(f \curlywedge f) \curlyvee ((\odot \curlywedge \ll) \circ \ll^2)|\!)$$

(In fact, sl is a *zygomorphism* [11]: a function which, although not catamorphic by itself, becomes catamorphic when tupled with another function—in this case, s—that is itself a catamorphism.) This means that

$$\begin{array}{ll} & up \\ = & \quad \{ \text{ above } \} \\ & sl* \circ subtrees \\ = & \quad \{ \text{ pairs } \} \\ & \gg * \circ (s \curlywedge sl)* \circ subtrees \\ = & \quad \{ \ s \curlywedge sl \text{ is a catamorphism } \} \\ & \gg * \circ (\!|(f \curlywedge f) \curlyvee ((\odot \curlywedge \ll) \circ \ll^2)|\!)* \circ subtrees \\ = & \quad \{ \Uparrow \} \\ & \gg * \circ (f \curlywedge f, (\odot \curlywedge \ll) \circ \ll^2)\Uparrow \end{array}$$

which can be evaluated quickly up to $\odot$.

So much for up; what about $down$? We have

$$down \cdot \triangle \cdot a = \triangle \cdot a$$
$$down \cdot (u \perp_b v) = down \cdot u \ \perp_b \ (b\odot)* \cdot down \cdot v$$

and so by Theorem 6 $down$ is a downwards pass. Again, it is not an accumulation, because $(id, \gg, \odot)$ is not top down. For, suppose $(id, \gg, \odot)$ were to invert to $(id, \oplus, \otimes)$; then by Theorem 9, the two functions

$$f = (\!|H\colon id \curlyvee \gg \curlyvee \odot|\!)$$
$$g = (\!|D\colon id \curlyvee \oplus \curlyvee \otimes|\!) \circ td$$

would be equal. Consider now the three threads

$$p = a \searrow \diamond \cdot b$$
$$q = a \searrow (b \swarrow \diamond \cdot c)$$
$$r = (a \odot b) \swarrow \diamond \cdot c$$

with a and c such that $a \odot c$ differs from c. We have

$$\begin{array}{ll} f \cdot p = a \odot b & g \cdot p = a \otimes b \\ f \cdot q = a \odot c & g \cdot q = (a \otimes b) \oplus c \\ f \cdot r = c & g \cdot r = (a \odot b) \oplus c \end{array}$$

If f and g are to be equal, we see that $\otimes$ and $\odot$ must also be equal, in which case g returns the same values for q and r, whereas f returns different values.

So, *down* is not an accumulation. Consider, though, the fork of thread catamorphisms

$$\llparenthesis !e \curlyvee \gg \curlyvee \odot \rrparenthesis \curlywedge \llparenthesis id \curlyvee \gg \curlyvee \odot \rrparenthesis$$

This is itself a catamorphism [6]:

$$\llparenthesis !e \curlyvee \gg \curlyvee \odot \rrparenthesis \curlywedge \llparenthesis id \curlyvee \gg \curlyvee \odot \rrparenthesis = \llparenthesis (!e \curlywedge id) \curlyvee \gg \curlyvee \oslash \rrparenthesis$$

where

$$a \oslash (b, c) = (a \odot b, a \odot c)$$

Moreover, it is top down—it inverts to $\llparenthesis (!e \curlywedge id) \curlyvee \oplus \curlyvee \otimes \rrparenthesis$ where

$$\begin{aligned} (b, c) \oplus d &= (b, b \odot d) \\ (b, c) \otimes d &= (c, c \odot d) \end{aligned}$$

Thus,

$$\begin{array}{cl} & down \\ = & \quad \{ \text{ Theorem 6 } \} \\ & \llparenthesis id \curlyvee \gg \curlyvee \odot \rrparenthesis * \circ td * \circ paths \\ = & \quad \{ \text{ pairs } \} \\ & \gg * \circ \llparenthesis (!e \curlywedge id) \curlyvee \gg \curlyvee \oslash \rrparenthesis * \circ td * \circ paths \\ = & \quad \{ \text{ Theorem 9 } \} \\ & \gg * \circ (!e \curlywedge id, \oplus, \otimes) \Downarrow \end{array}$$

This gives us the promised efficient algorithm for *tps*, an upwards accumulation followed by a downwards accumulation, with maps after each accumulation to 'tidy up':

$$tps = \gg * \circ (!e \curlywedge id, \oplus, \otimes) \Downarrow \circ \gg * \circ (f \curlywedge f, (\odot \curlywedge \ll) \circ \ll^2) \Uparrow$$

6 Conclusion

We have presented two kinds of accumulation on binary trees: upwards accumulation, which captures the notion of passing information up from the leaves of a tree towards the root, and downwards accumulation, which corresponds to passing information in the other direction, from the root towards the leaves. We have given conditions under which the accumulations are both catamorphic and quick, that is, requiring parallel time proportional to the depth of the tree times the time to perform the individual operations. We have shown how these accumulations neatly provide a solution to the prefix sums problem.

O'Donnell [15] has presented a derivation similar to ours, but only went as far as producing the characterization (4) of *tps* , then developing a quick implementation of it as a single monolithic function without separating out the two phases. The result is a 'sweep' operation consisting of a tree of processes, each of which 'sends information in both directions on each data path'. Accumulations make the data flow much clearer, cleanly separating the two phases, and provide a systematic way of deriving an efficient solution to the problem.

The material presented here is covered in much greater depth in the author's thesis [7], but the interested reader is warned that the notation differs in places: different names have been used for the same concept, and even the same name for different concepts.

One interesting question that remains to be answered is, under what extra conditions on the individual operators can the accumulations can be evaluated in logarithmic parallel time, even if the tree has greater than logarithmic depth? That is, when can we use an algorithm like parallel tree contraction [14]?

Another topic to explore is the generalization of accumulations to arbitrary initial data types. Meertens [13] gives a general construction yielding the subtrees of a tree, and upwards accumulations are a special case of his *paramorphisms*, but it is by no means obvious how to generalize downwards accumulations to other data types.

The author is grateful to Richard Bird, Geraint Jones, Lambert Meertens, David Skillicorn, and the anonymous referees, all of whom have made suggestions to improve the presentation of this paper.

References

1. Selim G. Akl. *Design and Analysis of Parallel Algorithms.* Prentice-Hall, 1989.
2. Roland Backhouse. An exploration of the Bird-Meertens formalism. In *International Summer School on Constructive Algorithmics, Hollum, Ameland.* STOP project, 1989. Also available as Technical Report CS 8810, Department of Computer Science, Groningen University, 1988.
3. Richard S. Bird. An introduction to the theory of lists. In M. Broy, editor, *Logic of Programming and Calculi of Discrete Design*, pages 3–42. Springer-Verlag, 1987. Also available as Technical Monograph PRG-56, from the Programming Research Group, Oxford University.
4. Richard S. Bird. Lectures on constructive functional programming. In Manfred Broy, editor, *Constructive Methods in Computer Science.* Springer-Verlag, 1988. Also available as Technical Monograph PRG-69, from the Programming Research Group, Oxford University.

5. G. Estrin. Organization of computer systems—the fixed plus variable structure computer. In *Proceedings Western Joint Computer Conference*, pages 33–40, May 1960.
6. Maarten M. Fokkinga. Tupling and mutumorphisms. *The Squiggolist*, 1(4):81–82, June 1990.
7. Jeremy Gibbons. *Algebras for Tree Algorithms.* D. Phil. thesis, Programming Research Group, Oxford University, 1991. Also available as Technical Monograph PRG-94 from the Programming Research Group, Oxford.
8. Tatsuya Hagino. *A Categorical Programming Language.* PhD thesis, Laboratory for the Foundations of Computer Science, Edinburgh, September 1987.
9. Peter M. Kogge and Harold S. Stone. A parallel algorithm for the efficient solution of a general class of recurrence equations. *IEEE Transactions on Computers*, C-22(8):786–793, August 1973.
10. Richard E. Ladner and Michael J. Fischer. Parallel prefix computation. *Journal of the ACM*, 27(4):831–838, October 1980.
11. Grant Malcolm. *Algebraic Data Types and Program Transformation.* PhD thesis, Rijksuniversiteit Groningen, September 1990.
12. Lambert Meertens. Algorithmics: Towards programming as a mathematical activity. In J. W. de Bakker, M. Hazewinkel, and J. K. Lenstra, editors, *Proc. CWI Symposium on Mathematics and Computer Science*, pages 289–334. North-Holland, 1986.
13. Lambert Meertens. Paramorphisms. Technical Report CS-R9005, CWI, Amsterdam, 1990.
14. Gary L. Miller and John H. Reif. Parallel tree contraction and its application. In *26th FOCS*, pages 478–489, 1985.
15. John T. O'Donnell. Derivation of fine-grain algorithms. Presentation at IFIP Working Group 2.8 meeting, Rome, 1990.

Distributing a class of sequential programs

H. Peter Hofstee [1]
Department of Computer Science, 256-80
California Institute of Technology
Pasadena, CA 91125

Abstract. A class of sequential programs is distributed through a series of program transformations. To construct a concurrent solution, a sequential solution is given first. A decision is made about the distribution of the variables and the sequential solution is transformed so that guards at the outermost level can be evaluated using variables that will be allocated to one process only. Next we introduce processes and communication. The resulting distributed algorithm does not terminate, but it will become quiescent, and in this state the original postcondition will hold. The distributed algorithm is highly nondeterministic and not network specific. A synchronization primitive, the nonblocking channel, is introduced, and used to generalize the first distributed solution to a larger class of networks.
We give two examples of problems that can be solved with this approach. First we show how a more general version of the load-balancing algorithm of [7] can be derived as an instance of this class. Next we instantiate our solution to arrive at an algorithm for distributed sorting. Finally we refine this solution to arrive at a terminating distributed sorting algorithm.

1 Introduction

The algorithms we present in this paper have been constructed with a specific architecture in mind; a message passing multicomputer with many small nodes. Unlike single instruction multiple data (SIMD) machines, synchronization is not implied, but has to be programmed. It is unlikely that all processors will perform the same task in synchrony; if they would, a SIMD machine would be more suitable for the task. One would like to develop

[1] H.Peter Hofstee is supported by an IBM graduate fellowship.

algorithms that can be used in a wide variety of contexts. The algorithm should therefore not depend on detailed analysis of a particular network, or on information about other tasks performed by the processors. These considerations make the cost of synchronization hard to predict, therefore synchronization is kept to a minimum in the algorithms developed here. It also seems unlikely that deterministic programs, whose execution does not depend on properties of the network, or the workload of the processors, are most efficient. The algorithms presented in this paper are therefore highly nondeterministic.

We will make only weak assumptions about the communication network that connects the nodes. We assume that we can use information about the network, and perhaps about the complete programs to make our final solution more deterministic and perhaps more efficient. In our experience, this turns out to be a good choice. The algorithm for distributed sorting presented here allows us to make use of more connections in a network than previous versions that were more deterministic, but also restricted to a specific communication topology. This property has caused it to outperform the previous solutions, even on relatively small networks. For the load-balancing example we did not have a deterministic solution to compare it to, but because nondeterminism is even more inherent to this problem, we do not doubt that the choice was appropriate for this problem as well.

The remainder of this paper focuses on the construction, and even though the above gives direction to the derivation, we try to make the formalism work for us as much as possible.

2 Transformations of a Sequential Program

We use an extended subset of Edsger W. Dijkstra's guarded command language [1]. We allow any commutative infix operator as a quantifier in a quantified expression. The meaning of such an expression is the meaning of the expression after expanding the expression for each value in its range separated by the operator.

[htbp]

$\{\forall(i :: v_i = V_i)\}$

do $[\!]\ (i,j : iRj : c(v_i, v_j) \rightarrow S(v_i, v_j)\ )$ **od**

$\{\ Q\ \}$

Fig. 1.

Our task is to find a distributed implementation for the program in

Figure 1. In this program iRj describes a finite directed acyclic graph; iRj is *true* just when the graph has a directed edge from node i to node j. Condition $c(v_i, v_j)$ is understood not to modify any variables. Condition c and statement S operate on their arguments only. The type of the variables v_i is left unspecified. We assume the program is correct, there exist an invariant P and a bound function bf that satisfy

- $\forall(i :: v_i = V_i) \Rightarrow P$
- $\forall(i,j : iRj :$
  $P \wedge c(v_i, v_j) \wedge bf = BF \Rightarrow \text{wp}(S(v_i, v_j), P \wedge bf < BF))$
- $P \wedge \forall(i,j : iRj : \neg c(v_i, v_j)) \Rightarrow Q$
- $bf > constant$

We intend to allocate variable v_i to process i. For any additional variables that we introduce, we will honor the convention that the first subscript denotes the process to which the variable will be allocated. However, we delay the introduction of processes and communications as long as possible, since properties of sequential programs are easier to prove. As a first step we introduce variables v_{ij} to replace the variables v_j. This allows us to evaluate the guard locally. We could have tried to maintain the invariant $\forall(i,j : iRj : v_{ij} = v_j)$. Maintaining this invariant would require statements that involve variables of more than two processes, which might make the distributed solution inefficient. A solution that involves variables from two processes in each alternative only is given in Figure 2.

[htbp]

$\{\forall(i :: v_i = V_i)\}$

do $[\!]\ (i,j : iRj : c(v_i, v_{ij}) \rightarrow$ **if** $c(v_i, v_j) \rightarrow S(v_i, v_j)$

$[\!]\ \neg c(v_i, v_j) \rightarrow v_{ij} := v_j$

fi

)

$[\!]\ \ [\!]\ (i,j : iRj : v_j \neq v_{ij} \rightarrow v_{ij} := v_j)$

od

Fig. 2.

It is easily verified this program will establish the same postcondition as the original program. The bound function is now a two tuple, $(bf, N(i,j : iRj : v_{ij} \neq v_j))$. This bound function decreases (lexicographic ordering) on every iteration and is bounded from below, therefore the second program also terminates. Curiously enough the variables v_{ij} need not appear in the invariant: the invariant is the same as the invariant of the original program. Therefore their initial value need not be specified.

We now remark that the alternative that we have added on the outermost level of the program has guard $v_j \neq v_{ij}$ which cannot be evaluated locally. We solve this problem by introducing variables ov_{ji} for which we maintain the invariant $\forall(i,j : iRj : ov_{ji} = v_{ij})$. This invariant can be maintained through statements involving variables of two processes only. The resulting program can be found in Figure 3.

[htbp]

$\{\forall(i :: v_i = V_i)\}$

$;\ (i,j : iRj : ov_{ji} := v_{ij})\ ;$

do $[\!]\ (i,j : iRj : c(v_i, v_{ij}) \rightarrow$ **if** $c(v_i, v_j) \rightarrow S(v_i, v_j)$

$[\!]\ \neg c(v_i, v_j) \rightarrow v_{ij}, ov_{ji} := v_j, v_j$

fi)

$[\!]\ \ [\!]\ (i,j : iRj : v_j \neq ov_{ji} \rightarrow v_{ij}, ov_{ji} := v_j, v_j)$

od

Fig. 3.

In order to establish the new conjunct in the invariant we have been forced to make a choice for the initial value of the ov_{ji}. The bound function for the program in Figure 2 can also be used to show that the algorithm in Figure 3 will terminate. The extended invariant and the negation of all guards implies the original postcondition. Therefore this program is a correct refinement as well.

Since the statements in the alternatives in Figure 3 contain references to variables that are not local to one process, communication will have to be introduced at this point. When we inspect the program in Figure 3, we see that for a pair of processes (i,j) that satisfy iRj a communication can be initiated both by process i and by process j. Obviously this might create deadlock; i may be committed to a communication with j to jointly execute the first alternative, whereas j is dedicated to a communication with i to jointly execute the second alternative. One solution to this problem is to merge both alternatives and use the same channels, which implies that the communication will succeed if both processes are dedicated to communicating with each other, irrespective of which process initiated the communication. The algorithm where both alternatives have been merged is given in Figure 4.

[htbp]

$\{\forall(i :: v_i = V_i)\}$

$; (i, j : iRj : ov_{ji} := v_{ij}) ;$

$\mathbf{do} \; [\!] \; (i, j : iRj : c(v_i, v_{ij}) \rightarrow \; S_0(v_i, v_j); \; v_{ij}, ov_{ji} := v_j, v_j \;)$

$[\!] \;\; [\!] \; (i, j : iRj : v_j \neq ov_{ji} \rightarrow \; S_0(v_i, v_j); \; v_{ij}, ov_{ji} := v_j, v_j \;)$

od

where $S_0(v_i, v_j) \equiv \mathbf{if} \; c(v_i, v_j) \rightarrow S(v_i, v_j) \; [\!] \; \neg c(v_i, v_j) \rightarrow \mathbf{skip} \; \mathbf{fi}$

Fig. 4.

This algorithm also maintains the invariant $P \wedge \forall(i, j : iRj : ov_{ji} = v_{ij})$ and also decreases $(bf, N(i, j : iRj : v_{ij} \neq v_j))$ on every iteration. Because we have not changed the guards, the algorithm in Figure 4 is a correct refinement as well.

3 Processes and Communication

Now that all guards of the loop can be evaluated locally, we are ready to introduce processes and communication. For two statements S and T, $S \| T$ denotes their parallel execution. The semantics of the message passing primitives is as described by A.J. Martin in [4]. The main difference with C.A.R. Hoare's proposal in [2] is in the naming of channels rather than processes. In [3], the same author proposes to name channels instead of processes in communication commands, but differs from our notation by using one name per channel instead of our two: output command $R!E$ in one process is paired with input command $L?v$ in another process by declaring the pair (R, L) to be a channel between the two processes. Each channel is between two processes only. When declaring (R, L) to be a channel, we write the name on which the output actions are performed first and the name on which the input actions are performed last. We allow the channels to be probed (cf. [5]) on one side. In this paper the side of the channel that is probed is the side on which the input actions are performed. The semantics of the **if** statement is reinterpreted (cf. [2]). If all guards in an **if** statement evaluate to *false*, its execution results in suspension of the execution rather than **abort**. When the guards contain no shared variables, an **if** statement that is suspended remains suspended forever and therefore this definition is compatible with the original semantics.

Our next task is to split the statement S_0 into two statements, each involving variables of one process only. The construction is given in figure 5.

[htbp]

$$
\begin{array}{ll}
 & S_0(v_i, v_j);\;\; v_{ij}, ov_{ji} := v_j, v_j \\
= & \{\text{new variables } x_i \text{ and } x_j, \\
 & \quad \text{require } x_i = v_j \;\wedge\; x_j = v_i \;\wedge\; i \neq j \Rightarrow \\
 & \qquad S_0'(v_i, x_i) \parallel S_0''(x_j, v_j) = S_0(v_i, v_j);\;\; x_i, x_j := v_j, v_i\} \\
 & (x_i := v_j \qquad\qquad\qquad\qquad \parallel x_j := v_i)\;; \\
 & ((\; S_0'(v_i, x_i);\;\; v_{ij} := x_i\;) \qquad \parallel (S_0''(x_j, v_j);\;\; ov_{ji} := v_j)\;) \\
= & \{\text{implement first two assignments and synchronization as com-} \\
 & \quad \text{munications using channels } (down_{ij}, down_{ji}) \text{ and } (up_{ji}, up_{ij})\;\} \\
 & \quad (\;(up_{ij}?x_i \parallel down_{ij}!v_i)\;;\;\; S_0'(v_i, x_i);\;\; v_{ij} := x_i\;) \\
 & \parallel (\;(up_{ji}!v_j \parallel down_{ji}?x_j)\;;\;\; S_0''(x_j, v_j);\;\; ov_{ji} := v_j\;)
\end{array}
$$

Fig. 5.

A natural choice for S_0' and S_0'' that satisfies the requirement would be to have them differ from S_0 only in their arguments.

The only statement in Figure 4 that remains to be distributed is the initialization of the variables ov_{ji}. The invariant merely requires $ov_{ji} = v_{ij}$, hence we can choose **any** initial value for these variables as long as it is the same for corresponding v_{ij} and ov_{ji}.

We have chosen not to worry about termination at this stage. One problem for which we have used this solution, a distributed load-balancing algorithm, was intended not to terminate. We arrive at the solution in Figure 6. We have renamed the dummy in the first and fourth alternative in order to textually separate alternatives that result in a communication with a predecessor in R from alternatives that result in a communication with a successor.

The first two alternatives jointly implement the first alternative of Figure 4. Since the original guard is evaluated in one of the two processes only, a new guard has to be found for the other process that becomes *true* just when the first one is. Since the body of the alternative starts with communication actions, probing one of the channels gives us just the right condition. Following [5], $\overline{c}$ (read c-probe) evaluates to *true* just when a communication on c is pending.

We have replaced a terminating algorithm by a nonterminating one, and therefore cannot expect that this algorithm implements our original one. We do expect the following:

(1) When all processes are quiescent, and this condition is stable, the original postcondition is established.

(2) After finite time all processes are quiescent and stay quiescent.

Quiescence is defined as the state in which a process is suspended. We now show that the first requirement is **not** met by the program in Figure 6. The relation iRj does not allow directed cycles, but it does allow cycles. Since the communications may go in either direction any cycle may cause deadlock when all processes in such a cycle are suspended on a communication with a successor in the cycle. If the graph represented by iRj is cycle free, the above is a correct implementation of the original algorithm. Because we do not want to restrict R, we opt for a different solution. We do not give a proof of correctness of the algorithm in Figure 6 for restricted iRj, but we will prove correctness of the more general version.

4 Nonblocking Channels

We require that all communications that do not follow the direction of the graph terminate. This requires the introduction of a special kind of channel (cf. [7]) with the property that any input action on the channel is followed by an output action of the channel. We do **not** require, however, that the number of outputs matches the number of inputs, or make assumptions about the values the channel produces.

With, for a channel $c = (c_i, c_o)$,

- $\mathbf{r}c \equiv$ number of $c_o?$ actions that have not been followed by a $c_i!$ action

and, following [4],

- $\mathbf{c}$A $\equiv$ number of completed A-actions
- $\mathbf{q}$A $\equiv$ 'an A-action is pending'

a nonblocking channel is characterized by

- $\neg \mathbf{q}c_i!$
- $\mathbf{c}c_o? \leq \mathbf{c}c_i!$
- $0 \leq \mathbf{r}c \leq \mathit{slack}$
- $\mathbf{q}c_o? \Rightarrow \mathbf{r}c = 0$

In the above the nonblocking channel is represented by a pair (c_i, c_o) and two operations, $c_i!$ and $c_o?$, which satisfy the properties given. Another example of a construct that satisfies these axioms is a special kind of semaphore, represented by a single variable, with the usual V-operation that

[htbp]

$$
\begin{array}{l}
\{\forall(i :: v_i = V_i)\} \\
\|(i :: \\
\quad ;(j : iRj : v_{ij} := V) ; \\
\quad ;(h : hRi : ov_{ih} := V) ; \\
\quad \textbf{do } \mathit{true} \rightarrow \\
\qquad \textbf{if } [\!]\ (j : iRj : c(v_i, v_{ij}) \rightarrow (up_{ij}?x_i \| \ down_{ij}!v_i) ; \\
\qquad\qquad\qquad\qquad\qquad\qquad S_0'(v_i, x_i);\ \ v_{ij} := x_i\ ) \\
\qquad [\!]\ [\!]\ (h : hRi : \overline{down_{ih}} \ \rightarrow (up_{ih}!v_i \| \ down_{ih}?x_i) ; \\
\qquad\qquad\qquad\qquad\qquad\qquad S_0''(x_i, v_i);\ \ ov_{ih} := v_i\ ) \\
\qquad [\!]\ [\!]\ (h : hRi : v_i \neq ov_{ih} \rightarrow (up_{ih}!v_i \| \ down_{ih}?x_i) ;) \\
\qquad\qquad\qquad\qquad\qquad\qquad S_0''(x_i, v_i);\ \ ov_{ih} := v_i\ ) \\
\qquad [\!]\ [\!]\ (j : iRj : \overline{up_{ij}} \qquad \rightarrow (up_{ij}?x_i \| \ down_{ij}!v_i) \\
\qquad\qquad\qquad\qquad\qquad\qquad S_0'(v_i, x_i);\ \ v_{ij} := x_i\ ) \\
\qquad \textbf{fi} \\
\quad \textbf{od} \\
)
\end{array}
$$

Fig. 6.

increases the semaphore, but a P-operation that suspends if the semaphore is 0, but sets it to 0 and completes if it is greater than 0. In this case $slack = 1$.

Figure 7 gives an implementation of a nonblocking channel with $slack = 2$ that does not share variables between processes. When the values communicated are irrelevant, they are omitted.

Since both of the alternatives in the alternative statement of the first process are probed, and since both alternatives involve no further communications, both are guaranteed to terminate, hence no output action on the channel is ever suspended. The local boolean ensures that the number of communications on *out* never exceeds the number of communications on *in* by more than the slack of the local channel (*inti*, *into*), and ensures the first alternative in the first process is chosen eventually. We assume the slack is zero; since both processes of the nonblocking channel can reside on the same processor, this does not present complications when implementing the channel. We can get rid of this assumption by exchanging *inti*! and *into*? in the program, but we have not done so because some implementa-

[htbp]

process *nbchan*(*in*, *out*) $\equiv$

$b := false;$

do *true* $\rightarrow$

if $\overline{in} \rightarrow in?;\ b := true$

$[]\ \overline{into}\ \wedge\ b \rightarrow into?;\ b := false$

fi

od

$\|$

do *true* $\rightarrow$ *inti*!; *out*! **od**

Fig. 7.

tions of concurrent programming languages do not allow the input side of a channel to be probed. At most two output actions can follow an input action, hence the $slack = 2$ for this nonblocking channel. If the last action on the channel was an input action, either $b = true$ or the second process is suspended on an output action. In either case an output action will not suspend.

5 A Solution with Nonblocking Channels

The new channel allows us to modify the solution in Figure 6. For a pair of processes (i, j) for which iRj holds, the code for the alternatives that implement a communication initiated by process i remains unchanged. Process j however, uses a nonblocking channel to process i to request a joint action. The action by such a process j on the nonblocking channel does not lead to progress. That problem is resolved using the boolean variables $sent_{ij}$. We use the additional channels (to_{ij}, to'_{ij}) and $(from_{ij}, from_{ji})$ to communicate with the nonblocking channel processes. Finally, we have changed the order of the statements in such a way that if alternatives that textually precede others are more likely to be executed, long chains of processes waiting for each other are less likely to occur. Correctness of this algorithm, and the following ones, does not depend on that order: the semantics of the **if** statements merely states that one of the alternatives for which the guard evaluates to *true* is executed. The resulting algorithm is given in Figure 8.

We prove the program in Figure 8 satisfies the following properties:

(1) When the program is quiescent, the original postcondition is established.

[htbp]

$$
\begin{array}{l}
\{\forall(i :: v_i = V_i)\} \\
\|(i :: \\
\quad ;(j : iRj : v_{ij} := V) ; \\
\quad ;(h : hRi : ov_{ih} := V \ ; \ sent_{ih} := false) ; \\
\quad \textbf{do } true \rightarrow \\
\qquad \textbf{if } [\!]\ (h : hRi : \overline{down_{ih}} \ \rightarrow \ sent_{ih} := false \ ; \\
\qquad\qquad (up_{ih}!v_i \| \ down_{ih}?x_i) \ ; \\
\qquad\qquad S_0''(x_i, v_i) \ ; \ ov_{ih} := v_i) \\
\qquad [\!] \ [\!]\ (j : iRj : c(v_i, v_{ij}) \ \rightarrow \ (up_{ij}?x_i \| \ down_{ij}!v_i) \ ; \\
\qquad\qquad S_0'(v_i, x_i) \ ; \ v_{ij} := x_i) \\
\qquad [\!] \ [\!]\ (j : iRj : \overline{from_{ij}} \rightarrow \ from_{ij}?; \ (up_{ij}?x_i \| \ down_{ij}!v_i) \ ; \\
\qquad\qquad S_0'(v_i, x_i) \ ; \ v_{ij} := x_i) \\
\qquad [\!] \ [\!]\ (h : hRi : v_i \neq ov_{ih} \ \wedge \ \neg sent_{ih} \rightarrow \\
\qquad\qquad to_{ih}!; \ sent_{ih} := true \ ) \\
\qquad \textbf{fi} \\
\quad \textbf{od} \\
\ ) \\
\| \\
\|(i, j : jRi : nbchan(to'_{ij}, from_{ij}))
\end{array}
$$

Fig. 8.

(2) The program becomes quiescent in finite time.

Proof of (1):

By definition, quiescence occurs when all processes are suspended. Processes can be suspended either because all guards in the **if** statement are *false* or because a process is suspended on a communication. We show that if any process is suspended on a communication, not all processes are suspended. A process cannot be suspended on any of the communications in the first alternative; $\overline{down_{ij}}$ becomes *true* only when another process is committed to a set of communications matching those in the first alternative. A process cannot be suspended on the communication in the fourth alternative either because it is a communication on a non-blocking channel. Thus a process i can be suspended only on one of the communications in the second or third alternatives, i.e. a communication with a process

j such that iRj holds. It follows that process j must be suspended on a communication, since at least one of the guards in its alternative statement ($\overline{down_{ij}}$) is true. ¿From the finiteness of R and from the fact that the transitive closure of R is irreflexive, it follows that a process must exist that is suspended on a communication with a process that is not suspended on a communication, and, since one of its guards is *true*, not suspended at all, which is a contradiction.

It follows that quiescence occurs only when all guards in the **if** statements of all processes are *false*. We first prove $\neg sent_{ij}$. If $sent_{ji}$ is *true* and all processes are quiescent a message has been sent on to_{ji} that has not been followed by the receipt of a message on $from_{ij}$. Had it been followed by such a message, the resulting communication between process i and j must have been completed since all processes are suspended, thus $sent_{ji}$ had been set to *false* again. The specification of the nonblocking channel guarantees us that a communication on *to* will result in $\overline{from}$ becoming *true*, therefore the situation in which $sent_{ij}$ is *true* precludes the processes from staying quiescent. $\forall(i,j : iRj : \neg c(v_i, v_{ij}) \ \wedge \ v_j = ov_{ji})$ and the invariant $\forall(i,j : iRj : v_{ij} = ov_{ji})$ implies $\forall(i,j : iRj : \neg c(v_i, v_j))$ which suffices to show that the old postcondition is established.

Proof of (2):

The four tuple $(bf, N(i,j : iRj : v_{ij} \neq v_j), N(i,j : jRi : sent_{ij} = (v_i = ov_{ij}), N(i,j : jRi : cto_{ij}! - cfrom_{ji}?))$ decreases for the forth alternative and for each other matching pair of alternatives in the program and is bounded from below. Checking this is not difficult, and it is left to the reader.

One last remark: using the nonblocking channel as a primitive, one can translate the algorithm in Figure 3 directly. We chose to refine the algorithm in Figure 6, because it gives a solution with fewer channels.

6 A Load-Balancing Algorithm

In this section we show how a generalized version of the load-balancing algorithm of Hofstee, Lukkien, and van de Snepscheut [7] can be derived with this approach. A sequential program that establishes the required balance is given in Figure 9.

In this program p_i represents the number of tasks in node i, and T a threshold. This program does not have the desired form, because iRj inserted in the range would be true for all pairs (i,j) and thus not represent a directed acyclic graph. We therefore choose a subset for our relation. If the subset is chosen in such a way that iRj represents a **rooted** directed acyclic graph that spans the original graph, we can rewrite the algorithm as in Figure 10.

Upon termination, a predecessor of a node with less than T tasks will have less than T tasks. By induction all predecessors of a node will have less than T tasks if that node has less than T tasks. Since the root is a

[htbp]

$\{\forall(i) :: p_i = P_i\}$

do $[\!]\ (i,j :: p_i < T\ \wedge\ p_j > T \rightarrow\ p_i, p_j := p_i + 1, p_j - 1)$ **od**

$\{\forall(i :: p_i \geq T) \vee \forall(i :: p_i \leq T)\}$

Fig. 9.

[htbp]

$\{\forall(i) :: p_i = P_i\}$

do $[\!]\ (i,j : iRj : p_i < T\ \wedge\ p_j \geq T \rightarrow\ p_i, p_j := p_i + 1, p_j - 1)$

$[\!]\quad [\!]\ (i,j : iRj : p_i > T\ \wedge\ p_j \leq T \rightarrow\ p_i, p_j := p_i - 1, p_j + 1)$

od

$\{\forall(i :: p_i \geq T) \vee \forall(i :: p_i \leq T)\}$

Fig. 10.

predecessor of all other nodes in the graph, it will have less than T tasks upon termination if any node has less than T tasks upon termination. Similarly, the root will have more than T tasks if any node has more than T tasks. Since the two conditions are mutually exclusive, the original postcondition is established when all guards are false. The bound function for this program is $\sum_i w_i\ |p_i - T|$ where w_i is the longest distance to the root. The function is bound by 0 and decreases on every iteration, hence the algorithm terminates.

We now have two alternatives, whereas the desired program only has one. Therefore we rewrite the algorithm once more to get the desired form. The inner **do**-statement may be replaced by an **if**-statement, but that is likely to be less efficient.

Now that the algorithm has the right form, we can apply the transformation to obtain the distributed solution in Figure 12.

The fifth alternative that we have added represents the consumption or generation of tasks by another process. Of course we can now no longer prove that the load will be balanced eventually, but it follows from our proofs that the load will become balanced across the network if the fifth alternative is no longer executed, and that actions that do not involve the fifth alternative lead to progress towards a balanced state. A complete solution should communicate tasks whenever the p_i are changed, the extension is trivial The solution derived here is more general than the solution pre-

[htbp]

$$
\begin{array}{l}
\{\forall(i) :: p_i = P_i\} \\
\textbf{do} \ [\!] \ (i,j : iRj : (p_i < T \ \wedge \ p_j \geq T) \vee (p_i > T \ \wedge \ p_j \leq T) \rightarrow \\
\qquad \textbf{do} \ p_i < T \ \wedge \ p_j \geq T \rightarrow \ p_i, p_j := p_i + 1, p_j - 1 \\
\qquad [\!] \ \ p_i > T \ \wedge \ p_j \leq T \rightarrow \ p_i, p_j := p_i - 1, p_j + 1 \\
\qquad \textbf{od} \\
\textbf{od} \\
\{\forall(i :: p_i \geq T) \vee \forall(i :: p_i \leq T)\}
\end{array}
$$

Fig. 11.

sented in [7] since it allows the communication graph to contain cycles (but no directed cycles). The network can now be chosen in such a way that the root is not a bottleneck, even for very inhomogeneous problems. The solution can be generalized to more thresholds in exactly the same manner as the solution presented in [7].

7 Distributed Sorting

A program that sorts bags according to the relation R is given in Figure 13.

We see that it has the shape we required in Figure 1, and therefore the algorithm in Figure 14 is a correct refinement.

In this program we have made a slightly different choice for S_0' and S_0'' than in the load balancing example to avoid updating both bags in both nodes. It is easily verified that this choice satisfies the requirement on S_0' and S_0'' in Figure 5 in section 3. Inspection of the algorithm in Figure 14 reveals that even though whole bags are communicated, only their minimum or only their maximum is relevant. Thus it suffices to send a minimum or a maximum only. We modify the program accordingly and arrive at the solution in Figure 15.

Obviously the program can be improved by introducing variables to maintain the minimum and maximum of a bag. Also, we can leave out one of the communications in S_0' and S_0'', because the value communicated is never used. The program can even be modified to avoid both communications in the statements S_0' and S_0''. These communications can be left out if the statements S_0' and S_0'' and the following assignments to b_{ij} and ob_{ij} are reversed in order. The resulting program is then a refinement of the program in Figure 4 where the order of the statements S_0 and the multiple assignment directly following it has been reversed. The program is then still

[htbp]

$\|(i ::$

$; (j : iRj : p_{ij} := 0)$;

$; (h : hRi : op_{ih} := 0 ;\ sent_{ih} := false)$;

do *true* $\rightarrow$

$$
\begin{array}{lll}
\textbf{if}\ [\!]\ (h : hRi : \overline{down_{ih}} & \rightarrow & sent_{ih} := false\ ; \\
 & & (up_{ih}!p_i \| \ down_{ih}?x_i)\ ; \\
 & & S_0''(x_i, p_i)\ ;\ op_{ih} := p_i) \\
[\!]\ [\!]\ (j : iRj : c(p_i, p_{ij}) & \rightarrow & (up_{ij}?x_i \| \ down_{ij}!p_i)\ ; \\
 & & S_0'(p_i, x_i)\ ;\ p_{ij} := x_i) \\
[\!]\ [\!]\ (j : iRj : \overline{from_{ij}} \rightarrow & & from_{ij}?;\ (up_{ij}?x_i \| \ down_{ij}!p_i)\ ; \\
 & & S_0'(p_i, x_i)\ ;\ p_{ij} := x_i) \\
[\!]\ [\!]\ (h : hRi : p_i \neq op_{ih} & \wedge & \neg sent_{ih} \rightarrow \\
 & & to_{ih}!;\ sent_{ih} := true\ ) \\
[\!]\ \overline{P}_i \rightarrow P_i?p_i & & \\
\textbf{fi} & &
\end{array}
$$

od)

$\|\ \|(i, j : jRi : nbchan(to'_{ij}, from_{ij}))$

$$
\begin{array}{l}
c(p_i, p_{ij}) \ = \ (p_i < T \ \wedge \ p_{ij} \geq T) \vee (p_i > T \ \wedge \ p_{ij} \leq T) \\
S_0'(p_i, p_j) \ = \ S_0''(p_i, p_j) \ = \ S_0(p_i, p_j) \ = \\
\textbf{if}\ c(p_i, p_j) \rightarrow \textbf{do}\ p_i < T \ \wedge \ p_j \geq T \rightarrow \ p_i, p_j := p_i + 1, p_j - 1 \\
\qquad\qquad\qquad\quad [\!]\ \ p_i > T \ \wedge \ p_j \leq T \rightarrow \ p_i, p_j := p_i - 1, p_j + 1 \\
\qquad\qquad\qquad\quad \textbf{od} \\
[\!]\ \neg c(p_i, p_j) \rightarrow \ \textbf{skip} \\
\textbf{fi} \\
= \\
\textbf{do}\ p_i < T \ \wedge \ p_j \geq T \rightarrow \ p_i, p_j := p_i + 1, p_j - 1 \\
[\!]\ \ p_i > T \ \wedge \ p_j \leq T \rightarrow \ p_i, p_j := p_i - 1, p_j + 1 \\
\textbf{od}
\end{array}
$$

Fig. 12.

[htbp]

$\{\forall(i) :: b_i = B_i\}$

do $[\!] (i,j : iRj : max(b_i) > min(b_j) \rightarrow$

$b_i, b_j := b_i - max(b_i) + min(b_j), b_j - min(b_j) + max(b_i))$

od

Fig. 13.

a refinement of the program in Figure 1. It has the same bound function and maintains the invariant of the corresponding refinements. However, the program will go through more iterations before terminating or becoming quiescent because the second part of the bound function increases by 1 more, as compared to the algorithm given, whenever the first part decreases.

As a final transformation we add variables that allow us to detect termination. This transformation is specific for the sorting problem. We assume that the bags are numbered from 0 to N and that the pairs $(i, i+1)$ for $0 \leq i < N$ form a subset of R, that is, we assume we want a total sort. Obviously, a termination condition for a process must be a stable condition, which presses us to look for monotonic variables. The reasoning is similar to the reasoning in [6]. The minimum and maximum of any individual bag may change nonmonotonically. However, we can prove that for any union of bags $lb_i = +(k : 0 \leq k \leq i \leq N : b_k)$ the maximum decreases monotonically, and for any union of bags $rb_i = +(k : 0 \leq i \leq k \leq N : b_k)$ the minimum increases monotonically. We can easily verify this by inspecting Figure 13. The minima and maxima change by exchanging elements between bags only. If both bags are in one of the sets described above, or both are not, its minimum, or maximum, does not change. If one bag is in the set, and the other is not, then if $i < j$ only (i, j) can be in R since adding (j, i) to R would create a directed cycle. This implies that in that case the minimum of bag j only increases and the maximum of bag i only decreases.

We introduce variables that approximate the maxima and minima of the sets introduced above. The following invariant is maintained: $\forall(i,j : iRj : LM_{ji} = PM_{ij} \geq max(lb_i)) \ \wedge \ \forall(i,j : jRi : rm_{ji} = pm_{ij} \leq min(rb_i))$. We arrive at the solution in Figure 16.

Verifying the equalities in the invariant is easy, they hold initially, and every assignment to PM or pm is followed by a pair of communications that restores the invariant. The other two relations in the invariant hold because none of the PM_{ij} or pm_{ij} can be assigned a value that does not satisfy the relation, unless one of the other variables PM_{ij} or pm_{ij} did not

[htbp]

$\{\forall(i :: b_i = B_i)\}$

$\|(i ::$

$\quad ; (j : iRj : b_{ij} := \emptyset) \; ;$

$\quad ; (h : hRi : ob_{ih} := \emptyset; sent_{ih} := false) \; ;$

$\quad$ **do** $true \rightarrow$

$\quad\quad$ **if** $[\!] \, (h : hRi : \overline{down_{ih}} \rightarrow sent_{ih} := false \; ;$

$\quad\quad\quad\quad (up_{ih}!b_i \| \; down_{ih}?x_i) \; ;$

$\quad\quad\quad\quad S_0''(x_i, b_i) \; ; \; ob_{ih} := b_i)$

$\quad\quad [\!] \; [\!] \, (j : iRj : c(b_i, b_{ij}) \rightarrow (up_{ij}?x_i \| \; down_{ij}!b_i) \; ;$

$\quad\quad\quad\quad S_0'(b_i, x_i) \; ; \; b_{ij} := x_i)$

$\quad\quad [\!] \; [\!] \, (j : iRj : \overline{from_{ij}} \rightarrow from_{ij}?; \; (up_{ij}?x_i \| \; down_{ij}!b_i) \; ;$

$\quad\quad\quad\quad S_0'(b_i, x_i) \; ; \; b_{ij} := x_i)$

$\quad\quad [\!] \; [\!] \, (h : hRi : b_i \neq ob_{ih} \; \wedge \; \neg sent_{ih} \rightarrow$

$\quad\quad\quad\quad to_{ih}!; \; sent_{ih} := true \;)$

$\quad\quad$ **fi**

$\quad$ **od**

)

$\|$

$\|(i, j : jRi : nbchan(to'_{ij}, from_{ij}))$

$$c(b_i, b_{ij}) = max(b_i) > min(b_{ij})$$

$$S_0'(b_i, x_i) = \textbf{if}\, c(b_i, x_i) \rightarrow b_i := b_i - max(b_i) + min(x_i)$$
$$[\!] \neg c(b_i, x_i) \rightarrow \textbf{skip}$$
$$\textbf{fi}; \; (up_{ij}?x_i \| \; down_{ij}!b_i)$$

$$S_0''(x_i, b_i) = \textbf{if}\, c(x_i, b_i) \rightarrow b_i := b_i - min(b_i) + max(x_i)$$
$$[\!] \neg c(x_i, b_i) \rightarrow \textbf{skip}$$
$$\textbf{fi}; \; (up_{ih}!b_i \| \; down_{ih}?x_i)$$

Fig. 14.

[htbp]

$\{\forall(i :: b_i = B_i)\}$

$\|(i ::$

$; (j : iRj : b_{ij} := \infty)\ ;$

$; (h : hRi : ob_{ih} := \infty;\ sent_{ih} := false)\ ;$

do $true \rightarrow$

if $[\!]\ (h : hRi : \overline{down_{ih}} \rightarrow sent_{ih} := false\ ;$

$(up_{ih}!min(b_i) \| down_{ih}?x_i)\ ;$

$S_0''(x_i, b_i)\ ;\ ob_{ih} := min(b_i))$

$[\!]\ [\!]\ (j : iRj : c(b_i, b_{ij}) \rightarrow (up_{ij}?x_i \| down_{ij}!max(b_i))\ ;$

$S_0'(b_i, x_i)\ ;\ b_{ij} := x_i)$

$[\!]\ [\!]\ (j : iRj : \overline{from_{ij}} \rightarrow from_{ij}?;\ (up_{ij}?x_i \| down_{ij}!max(b_i))\ ;$

$S_0'(b_i, x_i)\ ;\ b_{ij} := x_i)$

$[\!]\ [\!]\ (h : hRi : min(b_i) \neq ob_{ih} \wedge \neg sent_{ih} \rightarrow$

$to_{ih}!;\ sent_{ih} := true\ )$

fi

od

)

$\|$

$\|(i, j : jRi : nbchan(to'_{ij}, from_{ij}))$

$$
\begin{aligned}
c(b_i, b_{ij}) &= max(b_i) > b_{ij} \\
S_0'(b_i, x_i) &= \textbf{if } max(b_i) > x_i \rightarrow b_i := b_i - max(b_i) + x_i \\
&\quad [\!] \neg max(b_i) \leq x_i \rightarrow \textbf{skip} \\
&\quad \textbf{fi};\ (up_{ij}?x_i \| down_{ij}!max(b_i)) \\
S_0''(x_i, b_i) &= \textbf{if } x_i > min(b_i) \rightarrow b_i := b_i - min(b_i) + x_i \\
&\quad [\!]\ x_i \leq min(b_i) \rightarrow \textbf{skip} \\
&\quad \textbf{fi};\ (up_{ih}!min(b_i) \| down_{ih}?x_i)
\end{aligned}
$$

Fig. 15.

[htbp]

$\{\forall(i :: b_i = B_i)\}$

$\|(i ::$

$; (j : iRj : b_{ij}, rm_{ij}, PM_{ij} := +\infty, -\infty, +\infty)$;

$; (h : hRi : ob_{ih}, sent_{ih}, LM_{ih}, pm_{ih} := +\infty, false, +\infty, -\infty)$;

do $\exists(j : jRi : LM_{ij} > pm_{ij}) \vee \exists(j : iRj : PM_{ij} > rm_{ij}) \rightarrow$

if $[\!] (h : hRi : \overline{down_{ih}} \rightarrow sent_{ih} := false$;

$pm_{ih} := min(b_i + +(k : iRk : rm_{ik}))$;

$(up_{ih}!(min(b_i), pm_{ih}) \| down_{ih}?(x_i, LM_{ih}))$;

$S_0''(x_i, b_i)$; $ob_{ih} := min(b_i))$

$[\!]\ [\!] (j : iRj : max(b_i) > b_{ij} \wedge PM_{ij} > rm_{ij} \rightarrow$

$PM_{ij} := max(b_i + +(k : kRi : LM_{ik}))$;

$(up_{ij}?(x_i, rm_{ij}) \| down_{ij}!(max(b_i), PM_{ij}))$;

$S_0'(b_i, x_i)$; $b_{ij} := x_i)$

$[\!]\ [\!] (j : iRj : \overline{from_{ij}} \wedge PM_{ij} > rm_{ij} \rightarrow from_{ij}?$;

$PM_{ij} := max(b_i + +(k : kRi : LM_{ik}))$;

$(up_{ij}?(x_i, rm_{ij}) \| down_{ij}!(max(b_i), PM_{ij}))$;

$S_0'(b_i, x_i)$; $b_{ij} := x_i)$

$[\!]\ [\!] (h : hRi : min(b_i) \neq ob_{ih} \wedge \neg sent_{ih} \rightarrow$

$to_{ih}!$; $sent_{ih} := true$)

$[\!]\ [\!] (j : iRj : max(b_i + +(k : kRi : LM_{ik})) \neq PM_{ij} \wedge PM_{ij} > rm_{ij} \rightarrow$

$PM_{ij} := max(b_i + +(k : kRi : LM_{ik}))$;

$(up_{ij}?(x_i, rm_{ij}) \| down_{ij}!(max(b_i), PM_{ij}))$;

$S_0'(b_i, x_i)$; $b_{ij} := x_i)$

$[\!]\ [\!] (h : hRi : min(b_i + +(k : iRk : rm_{ik})) \neq pm_{ih} \wedge \neg sent_{ih} \rightarrow$

$to_{ih}!$; $sent_{ih} := true$)

fi

od

)

$\|\ \|(i, j : jRi : nbchan(to'_{ij}, from_{ij}))$

S_0' and S_0'' as before

Fig. 16.

satisfy the relation. This follows from the form of the assignments to the pm and PM. Since all pm and PM are guaranteed to satisfy the relations in the invariant initially, the invariant is maintained.

Given this invariant, and the old invariant, we prove that the final program satisfies the original specification:

(1) Termination of all processes implies the old postcondition.

(2) Not all processes are suspended. (Unless they have all terminated.)

(3) Each matching pair of alternatives decreases a bound function.

Proof of (1):

$$\begin{aligned}
& \forall(i,j : jRi : LM_{ij} \leq pm_{ij}) \ \wedge \ \forall(i,j : iRj : PM_{ij} \leq rm_{ij}) \\
\Rightarrow \ & \{\text{Invariant} : PM_{ij} = LM_{ji}\} \\
& \forall(i,j : iRj : LM_{ji} \leq rm_{ij}) \\
\equiv \ & \{\text{Invariant} : LM_{ji} \geq max(lb_i) \ \wedge \ rm_{ij} \leq min(rb_j)\} \\
& \forall(i,j : iRj : max(lb_i) \leq LM_{ji} \leq rm_{ij} \leq min(rb_j)) \\
\Rightarrow \ & \{max(b_i) \leq max(lb_i), min(rb_j) \leq min(b_j)\} \\
& \forall(i,j : iRj : max(b_i) \leq min(b_j))
\end{aligned}$$

Proof of (2); Not all processes are suspended or terminated, unless they have all terminated:

First we claim that a process i that has not terminated and is blocked on a communication has a successor j such that iRj holds and j has the same property. Since R is acyclic and finite, this implies that no process is blocked on a communication. The reasoning is virtually identical to the reasoning in section 5, with one important exception: because processes may now terminate, we also have to show that no process is committed to a communication with a process that has terminated. The conjuncts $PM_{ij} > rm_{ij}$ with the invariant $PM_{ij} = LM_{ji} \ \wedge \ rm_{ij} = pm_{ji}$ in the second, third, and fifth alternative guarantee just that. The rest of the proof is left to the reader.

Next we show that if not all bags are sorted, not all processes are suspended or have terminated:

$$\exists(i,j : iRj : max(b_i) > min(b_j))$$
$$\Rightarrow \{\text{Invariants}\}$$
$$\exists(i,j : iRj : PM_{ij} \geq max(lb_i) \geq max(b_i) > min(b_j) \geq min(rb_j) \geq pm_{ji})$$
$$\Rightarrow \{\text{Invariants}\}$$
$$\exists(i,j : iRj : PM_{ij} > rm_{ij} \ \wedge \ LM_{ji} > pm_{ji}$$
$$\wedge \ (max(b_i) > b_{ij} \vee ob_{ji} \neq min(b_j)))$$
$$\Rightarrow \{\text{guards}\}$$

process i and process j not terminated and

(process i not suspended or process j not suspended or $sent_{ji} = true$)

We had concluded before that the situation in which $sent_{ji}$ is true and process i is not terminated eventually leads to process i not being suspended.

The following step is to show that if all bags are sorted and all processes are suspended or have terminated, the property $max(b_i) = max(b_i + +(k : kRi : LM_{ik}))$ holds for all processes that have not terminated. We prove it holds for all processes by induction on the graph without directed cycles that R represents. The induction step requires us to prove that a node for which all predecessors satisfy the property, satisfies the property also:

$$\forall(j : jRi : LM_{ij} \leq pm_{ij} \vee$$
$$(LM_{ij} > pm_{ij} \ \wedge \ max(b_j) = max(b_j + +(k : kRj : LM_{jk}))))$$
$$\Rightarrow \{\text{Invariants}\}$$
$$\forall(j : jRi : LM_{ij} \leq pm_{ij} \vee$$
$$(PM_{ji} > rm_{ji} \ \wedge \ max(b_j) = max(b_j + +(k : kRj : LM_{jk}))))$$
$$\Rightarrow \{ \text{ process } j \text{ is suspended, guard of fifth alternative } \}$$

$$\forall(j : jRi : LM_{ij} \leq pm_{ij} \vee$$
$$(PM_{ji} > rm_{ji} \ \wedge \ max(b_j) = max(b_j + +(k : kRj : LM_{jk})) = PM_{ji}))$$
$$\Rightarrow \{\text{Invariants, sortedness}\}$$
$$\forall(j : jRi : LM_{ij} \leq pm_{ij} \leq min(b_i) \vee LM_{ij} = max(b_j) \leq min(b_i))$$
$$\Rightarrow$$
$$max(b_i) = max(b_i + +(j : jRi : LM_{ij}))$$

Finally we show that if all bags are sorted, and if all processes that have not terminated are suspended and have the above property, then any process i that has not terminated has a successor j such that iRj holds, and j has not terminated. The fact that R is finite and acyclic then implies that no process is suspended, thus all processes have terminated, which is what we set out to prove.

$$\exists(j : iRj : PM_{ij} > rm_{ij}) \vee \exists(j : jRi : LM_{ij} > pm_{ij})$$

$\Rightarrow$ { inv., all proc. suspended ($sent = false$, guard of 6th alternative }

$$\exists(j : iRj : PM_{ij} > rm_{ij}) \vee$$
$$\exists(j : jRi : PM_{ji} > rm_{ji} \ \wedge \ pm_{ij} = min(b_i + +(k : iRk : rm_{ik}))$$

$\Rightarrow$ { invariant, all proc. suspended, guard of fifth alternative }

$$\exists(j : iRj : PM_{ij} > rm_{ij}) \vee$$
$$\exists(j : jRi : PM_{ji} > rm_{ji} \ \wedge \ pm_{ij} = min(b_i + +(k : iRk : rm_{ik})) \ \wedge$$
$$PM_{ji} = max(b_j + +(k : kRi : LM_{ik})))$$

$\Rightarrow$ { invariant, $max(b_i) = max(b_i + +(k : kRi : LM_{ik}))$, sortedness

$$\exists(j : iRj : PM_{ij} > rm_{ij}) \vee$$
$$\exists(j : jRi : LM_{ij} = max(b_j) > pm_{ij} = min(b_i + +(k : iRk : rm_{ik}))$$
$$\wedge \ max(b_j) \leq min(b_i))$$

$\Rightarrow$

$$\exists(j : iRj : PM_{ij} > rm_{ij}) \vee \exists(k : iRk : min(b_i) > rm_{ik})$$

$\Rightarrow$ $\{PM_{ik} \geq max(lb_i) \geq max(b_i) \geq min(b_i)\}$

$$\exists(j : iRj : PM_{ij} > rm_{ij})$$

Proof of (3); Each matching pair of alternatives decreases a bound function:

The four tuple

$$(old - bf,$$
$$N(i,j : iRj : v_{ij} \neq v_j) + +(i,j : iRj : PM_{ij}) - +(i,j : jRi : pm_{ji}),$$
$$N(i,j : jRi : \neg sent_{ij} \ \wedge \ \neg(min(b_i) = ob_{ij})) +$$
$$N(i,j : jRi : \neg sent_{ij} \ \wedge \ \neg(min(b_i + +(k : iRk : rm_{ik})) = pm_{ij})),$$
$$N(i,j : jRi : cto_{ij}! - cfrom_{ji}?))$$

decreases for the fourth and fifth alternative in the program and for each of the three matching pairs of alternatives. For the sake of this variant function we have to read $+\infty$ and $-\infty$ in the program as 'some finite number bigger than anything else' and 'some finite number smaller than everything else' respectively.

8 Discussion

This paper was written in an attempt to unify the derivation of a distributed algorithm for load balancing [7], with the derivation of a new series of distributed sorting algorithms [8] that the paper and a previous paper on distributed sorting [6] inspired. Essential to the derivation is realizing that since we do not put a limit on the number of processes that can interact with any process and change its variables, information in any process about any other process is bound to be volatile. This problem can be overcome by imposing a high degree of synchronization, but, unless one is willing to make further assumptions about the specified problem, such a solution may be inefficient. Assuming that information is volatile, has the further advantage that introducing other processes that modify variables in one or more of the processes, as in the load balancing algorithm [7], does not present any difficulties. For the proof that quiescence is ultimately achieved in that case, one must make the assumption that these other processes are quiescent at some point in time. Other than that the proofs remain the same.

Since the load balancing algorithm does not terminate, we chose to treat termination as a late refinement step. The nonterminating version of the sorting algorithm is not without merit either, it can most likely be used, for instance, in a distributed database application to keep a changing collection of bags sorted 'on the fly'. Since the specification in Figure 1 is quite general, it seems likely that several other problems can be cast in this form as well.

9 Related Work

The sequential program that specifies the problem we try to distribute, is similar to a class of programs known as 'Action Systems' [10] or 'Unity Programs' [9].

The motivation for and goals of the first two refinements steps, those that maintain a sequential program, differs from [10] only in that we require the guards to involve variables local to one process only. This is related to the fact that we do not allow output guards in our distributed programs. In general, output guards cannot be implemented efficiently. More stringent conditions on the guards and structure of the sequential programs can ensure that a translation into communicating sequential processes can be generated automatically [11]. This method has the advantage of staying in the domain of sequential programs throughout the whole derivation. It is unclear, however, to what extent the restrictions on the final sequential program affect the efficiency of the final distributed algorithm or the ability of the programmer to find an efficient solution.

Quiescence of our distributed algorithms is related to the fixpoint of a 'Unity Program'. Methods for fixpoint detection described in [9] will therefore be helpful when trying to detect quiescence and thus create terminating algorithms.

Acknowledgements
I thank my advisor, Jan van de Snepscheut, for encouraging me to do something more methodical than another example of a derivation; I hope I have succeeded. Jan, Klaas Esselink, and Rustan Leino did most of the formalization of the nonblocking channel. Johan Lukkien invented the implementation of the nonblocking channels on the spot when, in a group meeting at Caltech, I expressed my doubts that it could be done without sharing variables between processes. I also thank Jan, Johan and Alain Martin, for work on papers that inspired the present one. I thank Ralph-Johan Back for lectures and discussions on program refinement. Audiences at T. U. Eindhoven, Koninklijke/Shell-Laboratorium Amsterdam, and Groningen University had several questions and comments that have led to improvements in the paper. Finally, I thank Jan, Rustan, Ulla Binau and Wim Feijen for reading the manuscript and many helpful comments.

References

1. E. W. Dijkstra, *A discipline of programming* (Prentice-Hall, Englewood Cliffs, NJ, 1976).
2. C.A.R. Hoare, Communicating sequential processes, *Comm. ACM* (1978) 666-677.
3. C.A.R. Hoare, *Communicating sequential processes* (Prentice-Hall International Series in Computer Science, 1985).
4. A.J. Martin, An axiomatic definition of synchronization primitives, *Acta Informatica* **16** (1981) 219-235.
5. A.J. Martin, The probe: an addition to communication primitives, *Information Processing Letters*, **20** (1985) 125-130.
6. H. P. Hofstee, A. J. Martin and J. L. A. van de Snepscheut, Distributed sorting, *Science of Computer Programming* **15** (1990) 119-133.
7. H. P. Hofstee, J. J. Lukkien and J. L. A. van de Snepscheut, A distributed implementation of a taskpool, In *Research Directions in High-Level Parallel Programming Languages*, J.B. Banâtre and D. Le Métayer (Eds.), LNCS 574 (1992) 338-348.
8. H. P. Hofstee, Distributed sorting revisited, *unpublished*.
9. K. Mani Chandy and J. Misra, *Parallel program design, a foundation* (Addison-Wesley, Reading, MA, 1988).

10. R.-J.R. Back and R. Kurki-Suonio, Decentralization of process nets with centralized control, *Distributed Computing*, (1989) 3:73-87.
11. R.-J.R. Back and K. Sere, Deriving an occam implementation of action systems. In *Proc. of the 3rd Refinement Workshop BCS FACS*, J.M. Morris and R. C. Shaw (Eds.), (Springer Verlag Workshops in Computing Series, 1991).

(Relational) Programming Laws in the Boom Hierarchy of Types

Paul F. Hoogendijk*

Department of Mathematics and Computing Science,
Eindhoven University of Technology,
P.O. Box 513, 5600 MB Eindhoven, The Netherlands

Abstract. In this paper we demonstrate that the basic rules and calculational techniques used in two extensively documented program derivation methods can be expressed, and, indeed, can be generalised within a relational theory of datatypes. The two methods to which we refer are the so-called "Bird-Meertens formalism" (see [22]) and the "Dijkstra-Feijen calculus" (see [15]).
The current paper forms an abridged, though representative, version of a complete account of the algebraic properties of the Boom hierarchy of types [19, 18]. Missing is an account of extensionality and the so-called cross-product.

1 Introduction

The "Bird-Meertens formalism" (to be more precise, our own conception of it) is a calculus of total functions based on a small number of primitives and a hierarchy of types including trees and lists. The theory was set out in an inspiring paper by Meertens [22] and has been further refined and applied in a number of papers by Bird and Meertens [8, 9, 11, 12, 13]. Its beauty derives from the small scale of the theory itself compared with the large scale of applications.

Essentially there are just three primitive operators in the theory - "reduce", "map" and "filter". (Actually, the names used by Meertens for the first two of these operators were "inserted-in" and "applied-to-all" in line with Backus [7]; Iverson [20] used the name "reduce". Moreover, just the first two are primitive since filter is defined in terms of reduce and map.) These operators are defined at each level of a hierarchy of types called the "Boom hierarchy" [2] after H.J. Boom to whom Meertens attributes the concept.

The basis of this hierarchy is given by what Meertens calls "D-structures". A D-structure, for given type D, is formed in one of two ways: there is an embedding

* The investigations were (partly) supported by the Foundation for Computer Science in the Netherlands (SION) with financial support from the Netherlands Organization for Scientific Research (NWO)

[2] For the record: Doaitse Swierstra appears to have been responsible for coining the name "Bird-Meertens Formalism" when he cracked a joke comparing "BMF" to "BNF" — Backus-Naur Form — at a workshop in Nijmegen in April, 1988. The name "Boom hierarchy" was suggested to Roland Backhouse by Richard Bird at the same workshop.

function that maps an element of D into a D-structure, and there is a binary join operation that combines two D-structures into one. Thus, a D-structure is a full binary tree with elements of D at the leaves. (By "full" we mean that every interior node has exactly two children.) The embedding function and the join operation are called the *constructors* of the type. Other types in the hierarchy are obtained by adding extra algebraic structure. Trees — binary but non-full — are obtained by assuming that the base type D contains a designated **nil** element which is a left and right unit of the join operation. Lists, bags and sets are obtained by successively introducing the requirements that join is associative, symmetric and idempotent.

Meertens describes the D-structures as "about the poorest (i.e., in algebraic laws) possible algebra" and trees as "about the poorest-but-one possible algebra". Nevertheless, in [3] we exploit the power of abstraction afforded by the notion of a so-called relator (a relator is a generalization of a functor) to add several more levels to the Boom hierarchy each of which is "poorer" than those considered by Meertens. Each level is characterised by a class of relators that specialises the class at the level below it. In decreasing order of abstraction these are the "sum" relators, "grounded" and "polymorphically grounded" relators, "monadic" relators and "pointed" relators. The reason for introducing these extra levels is organisational: the goal is to pin down as clearly as possible the minimum algebraic structure necessary to be able to, first, define the three operators of the Bird-Meertens formalism and, second, establish each of the basic properties of the operators. In the present paper, we start with (an instantiation of) pointed relators and give the definition of map, reduce and filter. For further discussion we refer to the paper [3].

The unconventional nature (and perhaps also the conciseness) of the notations used in the Bird-Meertens formalism makes the formalism difficult to comprehend for many groups. The program calculations carried out within the formalism are, however, strongly related to calculations within other systems. In particular there is a strong link between a certain combination of the three basic operators of the formalism and the quantifier expressions used for many years in the Eindhoven school of program development, this link being expressed via a correspondence between the basic laws of the two systems. For the benefit of those familiar with the Eindhoven calculus we use the opportunity to point out elements of this correspondence. What emerges is that there are typically more laws in the Bird-Meertens formalism than the quantifier calculus but the Bird-Meertens formalism exhibits a much better developed separation of concerns.

The theorems presented in the current paper are more general than those in the publications of Bird and Meertens since their work is restricted to total functions. (Meertens [22] does discuss the issue of indeterminacy but this part of his paper — we regret to have to say — is in our view the least satisfactory.) A danger of generalisation is that it brings with it substantial overhead making a theory abstruse and unworkable. At this stage in our work, however, the generalisation from (total) functions to relations has been very positive bringing to mind a parallel with the extension of the domain of real numbers to complex numbers. The fact of the matter is that we are rarely aware of working with relations rather than functions. The following pages are intended to provide some justification for that claim.

In [1] a rigorous discussion of the "Bird-Meertens formalism" can be found. This report was the starting point for the present paper. We start with a brief introduction to a relational calculus of datatypes as described in [4, 6]. Thereafter, we define the so-called binary structures; also we define the map, reduce and filter operators in our system. Before we start with the original Boom hierarchy where laws play an important rôle, we first define what it means for a relation to be associative, symmetric or idempotent; also we give a definition for sectioning and units. Next we define the original Boom-hierarchy in our system. Many properties from E. Voermans' paper [24] will be used. That paper describes how laws can be incorporated into the relational calculus. Finally, we relate the formalism to the quantifier calculus. We prove rules like range translation, trading, range splitting, etc., within our formalism.

1.1 Plat Calculus

In this section we summarise the axiom system in which we conduct our calculations. For pedagogic reasons we prefer to decompose the algebra into three layers with their interfaces and two special axioms. The algebra is, nevertheless, well known and can also be found in, for example, [23].

Let $\mathcal{A}$ be a set, the elements of which are to be called *specs* (from *spec*ification). We use identifiers R, S, etc., to denote specs. On $\mathcal{A}$ we impose the structure of a complete, completely distributive, complemented lattice ($\mathcal{A}$, $\sqcap$, $\sqcup$, $\neg$, $\top\!\!\top$, $\bot\!\!\bot$) where "$\sqcap$" and "$\sqcup$" are associative and idempotent binary infix operators with unit elements "$\top\!\!\top$" and "$\bot\!\!\bot$", respectively, and "$\neg$" is the unary prefix operator denoting complement (or negation). We assume familiarity with the standard definition of a lattice given above. We call such a structure a *plat*, the "p" standing for power set and "lat" standing for lattice.

The second layer is the monoid structure for composition: ($\mathcal{A}$, $\circ$, I) where $\circ$ is an associative binary infix operator with unit element I. The interface between these two layers is: $\circ$ is coordinatewise universally "cup-junctive". I.e. for $V, W \subseteq \mathcal{A}$, $(\sqcup V)\circ(\sqcup W) = \sqcup(\,R,S \;:\; R\in V \wedge S\in W \;:\; R\circ S\,)$

The third layer is the reverse structure: ($\mathcal{A}$, $\cup$) where "$\cup$" is a unary postfix operator such that it is its own inverse. The interface with the first layer is that "$\cup$" is an isomorphism of plats. I.e. for all $R, S \in \mathcal{A}$, $R\cup \sqsupseteq S \equiv R \sqsupseteq S\cup$. The interface with the second layer is that "$\cup$" is a contravariant monoid isomorphism: $(R\circ S)\cup \;=\; S\cup \circ R\cup$.

A model for this axiom system is the set of binary relations over some universe, with $\sqcup$, $\sqcap$ and $\sqsupseteq$ interpreted as set union, set intersection and set containment, respectively.

1.2 Operator precedence

Some remarks on operator precedence are necessary to enable the reader to parse our formulae. First, operators in the metalanguage ($\equiv$, $\Leftarrow$ and $\Rightarrow$ together with $\vee$ and $\wedge$) have lower precedence than operators in the object language. Next, the operators in the object language "$=$", "$\sqsupseteq$" and "$\sqsubseteq$" all have equal precedence; so do "$\sqcup$" and "$\sqcap$"; and, the former is lower than the latter. Composition "$\circ$" has a yet higher precedence than all operators mentioned thus far. Finally, all unary operators in the object language, whether prefix or postfix, have the same precedence which is the highest of all. Parentheses will be used to disambiguate expressions where necessary.

1.3 The Middle Exchange Rule and the Cone Rule

To the above axioms we now add an axiom relating all three layers:
Middle Exchange Rule: $X \sqsupseteq R \circ Y \circ S \ \equiv\ \neg Y \sqsupseteq R^{\cup} \circ \neg X \circ S^{\cup}$
Our last axiom, which is sometimes referred to as "Tarski's Rule", we call the
Cone Rule: $\top\!\!\top \circ R \circ \top\!\!\top = \top\!\!\top \ \equiv\ R \neq \perp\!\!\perp$

1.4 Imps, totality, co-imps and surjectivity

The notions of functionality, totality, injectivity and surjectivity can be defined using the operations introduced above. To avoid confusion between object language and metalanguage we use the term *imp* instead of *function*:

Definition 1

(a) R is an *imp* iff $R \circ R^{\cup} \sqsubseteq I$
(b) R is *total* iff $R^{\cup} \circ R \sqsupseteq I$
(c) R is a *co-imp* iff $R^{\cup}$ is an imp
(d) R is *surjective* iff $R^{\cup}$ is total
□

The term *imp* is derived from the word *imp*lementation. By interpretation, the definition of imps says that R is zero- or single-valued; the definition of totality means that R always returns some result.

1.5 Domains

We say that spec A is a *monotype* iff $I \sqsupseteq A$. Monotypes may be interpreted as an implementation of sets: element x is contained in the set corresponding to monotype A iff $x\langle A\rangle x$.

We need to refer to the "domain" and "co-domain" (or "range") of a spec. In order to avoid unhelpful operational interpretations we use the terms *left-domain* and *right-domain* instead. These are denoted by "<" and ">", respectively, and defined by:

Definition 2 *domains*

(a) $R< \ = \ R \circ R^{\cup} \sqcap I$
(b) $R> \ = \ R^{\cup} \circ R \sqcap I$
□

Note that domains are monotypes. Moreover, $R<$ is the smallest monotype satisfying the following equation:

(3) $X :: \ R = X \circ R$

One of the main considerations of the development of the theory described in [6] was that the notion of domain is a part of the system itself: it is not something defined outside the system. This means that we can use type-information in our calculations in a very natural way. So, type considerations are a part of a particular law itself rather than being expressed in some overall context within which the law is applicable.

1.6 Polynomial Relators

In categorical approaches to type theory a parallel is drawn between the notion of type constructor and the categorical notion of "functor", thereby emphasising that a type constructor is not just a function from types to types but also comes equipped with a function that maps arrows to arrows. In [4, 6] an extension to the notion of functor is given, the so-called "relator":

Definition 4 A *relator* is a function, F, from specs to specs such that

(a) $I \sqsupseteq F.I$ (b) $R \sqsupseteq S \Rightarrow F.R \sqsupseteq F.S$
(c) $F.(R \circ S) = F.R \circ F.S$ (d) $F.(R\cup) = (F.R)\cup$
□

One of the main properties of a relator is that it commutes with the domain operators, $F.(R>) = (F.R)>$. Another is that relators respect imps.

1.7 Disjoint Sum and Cartesian Product

We begin by postulating the existence of four specs: the two projections $\ll$ (pronounced "project left") and $\gg$ (pronounced "project right") for cartesian product and the two injections $\hookrightarrow$ (pronounced "inject left",) and $\hookleftarrow$ (pronounced "inject right") for disjoint sum. Further, experience leads us to introduce four binary operators on specs: for cartesian product $\vartriangle$ (pronounced "split", commonly the notation $\langle\_,\_\rangle$ is used) and $\times$ (pronounced "times"), and for disjoint sum $\triangledown$ (pronounced "junc", commonly the notation $[\_,\_]$ is used) and $+$ (pronounced "plus"), defined in terms of the projection and injection specs as follows:

(5) $P \vartriangle Q = (\ll\cup \circ P) \sqcap (\gg\cup \circ Q)$
(6) $P \triangledown Q = (P \circ \hookrightarrow\cup) \sqcup (Q \circ \hookleftarrow\cup)$
(7) $P \times Q = (P \circ \ll) \vartriangle (Q \circ \gg)$
(8) $P+Q = (\hookrightarrow \circ P) \triangledown (\hookleftarrow \circ Q)$

The relational model that we envisage assumes that the universe is a term algebra formed by closing some base set under three operators: the binary operator mapping the pair of terms x, y to the term (x, y), and two unary operators $\hookrightarrow$ and $\hookleftarrow$ mapping the term x to the terms $\hookrightarrow.x$ and $\hookleftarrow.x$, respectively. The interpretation of $\ll$ and $\gg$ is that they project a pair onto its left and right components. The operators defined above have a higher precedence than composition "$\circ$". In [4, 6] the axioms are given for the projections and injections such that cartesian product and disjoint sum have the desired properties. One of these properties is that cartesian product and disjoint sum are (binary) relators. Furthermore, $\ll$, $\gg$ are imps and $\hookrightarrow$, $\hookleftarrow$ are imps and co-imps. Also, we have the following properties:

(9) $P \vartriangle Q \circ R \sqsubseteq (P \circ R) \vartriangle (Q \circ R)$
(10) $P \vartriangle Q \circ f = (P \circ f) \vartriangle (Q \circ f)$, for imp f
(11) $R \circ P \triangledown Q = (R \circ P) \triangledown (R \circ Q)$

Property (10) is called split-imp fusion and property (11) is called spec-junc fusion.

1.8 Unit

The last axiom we add is the existence of a unit type. The unit type (denoted $\mathbb{1}$), when viewed as a set of pairs, consists of at most one pair, the two components of which are identical:

Unit: $\perp\!\!\!\perp \neq \mathbb{1} \;\wedge\; I \sqsupseteq \mathbb{1} \circ \top\!\!\!\top \circ \mathbb{1}$

One of the main properties of $\mathbb{1}$ is that it is a monotype and that it is an atom.

2 Binary Structures

Throughout this paper we consider the following so-called "pointed relator":

(12) $F.X \;=\; I + (\mathbb{1} + X \times X)$

For this relator F we define μF to be the least solution of $X :: X \;=\; F.X$. One of the main properties of μF is that μF is a monotype (see [5, 6]). One can view μF as the recursively defined type corresponding to the relator F. For μF we have the following four constructors:

(13) $\tau \;=\; \mu F \circ \hookrightarrow \;=\; \hookrightarrow$
(14) $\eta \;=\; \mu F \circ \hookleftarrow \;=\; \hookleftarrow \circ \; \mathbb{1} + \mu F \times \mu F$
(15) $\Box \;=\; \mu F \circ \hookleftarrow \circ \hookrightarrow \;=\; \hookleftarrow \circ \hookrightarrow \circ \mathbb{1}$
(16) $+\!\!+ \;=\; \mu F \circ \hookleftarrow \circ \hookleftarrow \;=\; \hookleftarrow \circ \hookleftarrow \circ \; \mu F \times \mu F$

Informally stated, τ is the singleton constructor: from an element of the universe I it constructs a singleton containing that element. $\Box$ is the unit constructor and $+\!\!+$ is the join operator: from two elements of μF it constructs their join. Note that

(17) $\eta = \Box \triangledown +\!\!+$

So, η is just the combination of two other constructors. Elements of μF we call bins (short for binary structures); bins correspond to the D-structures which Meertens describes as "about the poorest possible algebra".

Corresponding with μF we can define the so-called catamorphism operator denoted by $(\!|\_|\!)$. Catamorphism $(\!|R|\!)$ is defined to be the least solution of

$$X :: X \;=\; R \circ F.X$$

Catamorphisms can be viewed as a recursively defined specs which follow the same recursion pattern as the elements of μF. For a further account of catamorphisms see [5, 6].

We may assume without loss of generality that every catamorphism corresponding with the relator F can be written as $(\!|R \triangledown (S \triangledown \otimes)|\!)$ with $\otimes\!> \;\sqsubseteq\; I \times I$. We call specs with right domains contained in $I \times I$ *binary specs*. From now on $\otimes$ denotes a binary spec.

For the constructors defined above we have the following computation rules:

Theorem 18 (Computation Rule)

(a) $(\![R \triangledown (S \triangledown \otimes)]\!) \circ \tau \;=\; R$
(b) $(\![R \triangledown (S \triangledown \otimes)]\!) \circ \square \;=\; S \circ \mathbb{1}$
(c) $(\![R \triangledown (S \triangledown \otimes)]\!) \circ +\!\!+ \;=\; \otimes \circ (\![R \triangledown (S \triangledown \otimes)]\!) \times (\![R \triangledown (S \triangledown \otimes)]\!)$
□

For later use we define the following shorthand:

Definition 19 (Empty Tree)

$$\varepsilon \;=\; \square \circ \top\!\!\top$$

□

ε can be viewed as the constant function always returning $\square<$, i.e. the unit element. Functions like ε are so-called points. The definition of a point is:

Definition 20 (Point) Spec x is a *point* iff

$$x \text{ is an imp} \quad \wedge \quad x = x \circ \top\!\!\top$$

□

As mentioned above, points are constant functions: applied to an arbitrary element they always return the same element.

We also need the following shorthands:

(21) $R \otimes S \;=\; \otimes \circ R \vartriangle S$
(22) $R \underline{\otimes} S \;=\; (R \circ \ll) \otimes (S \circ \gg) \;=\; \otimes \circ R \times S$

For $R \otimes S$ we have by split-imp fusion (10):

(23) $R \otimes S \circ f \;=\; (R \circ f) \otimes (S \circ f)$, for imp f

Note that the computation rule for $+\!\!+$ can now be rewritten as:

(24) $(\![R \triangledown (S \triangledown \otimes)]\!) \circ +\!\!+ \;=\; (\![R \triangledown (S \triangledown \otimes)]\!) \underline{\otimes} (\![R \triangledown (S \triangledown \otimes)]\!)$

2.1 Map and Reduce

Because relator F is of the shape $I\oplus$, i.e. $F.X \;=\; I \oplus X$ where $\oplus$ is the binary relator defined by $R \oplus S \;=\; R + (\mathbb{1} + S \times S)$, we can define a map operator according to section 5 of [5] or section 7 of [6]. Instantiating the definition given there with our relator, we get:

Theorem 25 (Map) The function $*$ from specs to specs defined by

$$*R \;=\; (\![R + (\mathbb{1} + I \times I)]\!)$$

is a relator.
□

The function $*$ defines a family of monotypes, namely the monotypes $*B$ where B ranges over monotypes. In particular, $*I = \mu F$. For each spec R, the spec $*R$ has left domain $*(R{<})$ and right domain $*(R{>})$. In addition, for monotypes A and B and imps $f \in A \longleftarrow B$, $*f \in *A \longleftarrow *B$. An instance of such a relator is the List relator. In functional programming texts $*f$ is commonly called "map f" (and sometimes written that way too) and denotes a function from lists to lists that "maps" the given function f over the elements of the argument list (i.e. constructs a list of the same length as the argument list, where the elements are obtained by applying f to each of the elements of the argument list). This then is the origin of the name "map" for $*$.

We will mostly use a different but equivalent definition for map that exploits the particular structure of the relator $\oplus$. That definition is obtained by using the following theorem:

Theorem 26 (Map Fusion)

$$(\!|P \triangledown Q|\!) \circ *R = (\!|(P \circ R) \triangledown Q|\!)$$

□

Theorem 27 (Map – Alternative Definition)

$$*R = (\!|(\tau \circ R) \triangledown \eta|\!)$$

□

The reason why we sometimes prefer this definition is that catamorphisms of the shape $(\!|R \triangledown \eta|\!)$ enjoy many properties.

Instantiating the computation rule (18) with the above definition of $*$ we obtain the following computation rules:

$$\begin{aligned} *R \circ \tau &= \tau \circ R \\ *R \circ \square &= \square \\ *R \circ +\!\!\!+ &= +\!\!\!+ \circ *R \times *R \end{aligned}$$

One can view $*R$ as a spec which, when applied to an element of μF, applies R to the ground elements but does not destroy the original structure.

The second primitive in the Bird-Meertens formalism is called "reduce" and is denoted by the symbol "/". In the context of our work, reduce is a function from specs to specs. We shall adopt the same symbol but use it as a prefix operator in order to be consistent with our convention of always writing function and argument in that order. Thus we write $/S$ and read "reduce with S" or just "reduce S". (In choosing to write reduce as a prefix operator we are turning the clock back to Backus' Turing award lecture [7] rather than following the example of Bird and Meertens. In the context of Bird and Meertens' original work reduce was a binary infix operator with argument a pair consisting of a binary operator, say $\oplus$, and a list, say x, thus giving $\oplus/x$. In the course of time it was recognised that calculations and laws could be made more compact by working with the *function* $(x \mapsto \oplus/x)$ rather than the *object* $\oplus/x$. To achieve the compactness the notation $\oplus/$ (or sometimes $(\oplus/)$) was adopted for the function, the process of abstracting one of the arguments of a binary operator being commonly referred to as "sectioning". By this development, presumably, they came to the convention of using "/" as a postfix operator. Since our concern is to profit from what has been learnt rather than repeat the learning process we shall not adopt their notation in its entirety.) The idea behind reduce is that it should have a complementary behaviour to map. Recall that map, applied to an element of μF, leaves the structure unchanged but applies its argument to the ground elements. Reduce should do the opposite: leave the ground elements unchanged but destroy the structure. Since a catamorphism does both (modifies the ground elements and the structure) we formulate the requirement on reduce as being that every catamorphism is factorisable into a reduce composed with a map. I.e. for all specs R and S,

$$/S \circ *R \;=\; (\!|R \triangledown S|\!)$$

Let us try to calculate a suitable definition for $/S$.

$$\begin{array}{ll} & /S \circ *R \\ = & \quad \{ \text{ We try to express } /S \text{ as a catamorphism} \\ & \quad\;\; \bullet\;\; /S = (\!|P \triangledown Q|\!) \;\} \\ & (\!|P \triangledown Q|\!) \circ *R \\ = & \quad \{ \text{ map fusion: theorem 26 } \} \\ & (\!|(P \circ R) \triangledown Q|\!) \\ = & \quad \{ \text{ choose } P = I \text{ and } Q = S \;\} \\ & (\!|R \triangledown S|\!) \end{array}$$

Thus if we define the reduce operator by:

(28) $/S \;=\; (\!|I \triangledown S|\!)$

then we have established the following factorisation property:

Lemma 29

$$/S \circ *R \;=\; (\!|R \triangledown S|\!)$$

□

A special reduce is $/\eta$ (for list-structures this is the "flattening" catamorphism; it maps a list of lists to a list). For this catamorphism there exist two well-known leapfrog properties:

Theorem 30 ($/\eta$ **Leapfrog**)

(a) $/S \circ /\eta \;=\; /S \circ */S$
(b) $*R \circ /\eta \;=\; /\eta \circ **R$

□

2.2 Conditionals

Conditionals (**if-then-else** statements) are, of course, a well-established feature of programming languages. Several publications have already appeared documenting the algebraic properties of conditionals, the most comprehensive account that we know of being given by Hoare *et al* [17].

In order to be able to define conditionals we need to have the complement of monotypes. We achieve this in a slightly roundabout way. That is to say, we consider the right domains of so-called right conditions. For right conditions we have the following definition:

Definition 31 (Right Condition) We call spec p a right condition if

$$p = \top\!\top \circ p$$

□

(Of course we have a dual definition for left conditions.) We adopt the convention that lower case letters p, q and r denote right conditions.

In the relational model right conditions may be used to represent boolean tests: the predicate b may be represented by the right condition p where for all elements x and y

$$x\langle p\rangle y \;\equiv\; b.y$$

Now the complement of the monotype $p{>}$ is just $(\neg p){>}$, i.e. we have the properties $p{>} \sqcup (\neg p){>} = I$ and $p{>} \sqcap (\neg p){>} = \bot\!\bot$. Right conditions are closed under union, intersection, negation, and right composition. In order to give the properties of conditionals a more familiar appearance we shall write $p \wedge q$ instead of $p \sqcap q$, and $p \vee q$ instead of $p \sqcup q$. Other familiar boolean operators can then be defined on right conditions, such as $p \Rightarrow q \;=\; \neg p \vee q$. We shall also sometimes write *true* instead of $\top\!\top$ and *false* instead of $\bot\!\bot$. We define:

Definition 32 (Conditional) For all right conditions p we define the binary operator $\lhd p \rhd$ by:

$$R \lhd p \rhd S \;=\; R \circ p{>} \;\sqcup\; S \circ (\neg p){>}$$

□

Notes: In [16] Hoare defines conditionals, although for propositions, as follows,

$$P \lhd Q \rhd R \;=\; (P \sqcap Q) \sqcup (R \sqcap \neg Q)$$

Note that our definition corresponds to Hoare's definition because

$$R \circ p{>} \sqcup S \circ (\neg p){>} = (R \sqcap p) \sqcup (S \sqcap \neg p)$$

In [3] an alternative definition of conditionals is proposed that eliminates the need to restrict p to a right condition. Moreover, a more comprehensive account of their properties is included.
End of Notes.

The conditional $R \lhd p \rhd S$ can be viewed as a spec which applies R to those elements for which condition p holds and applies S to the other ones.

For conditionals we have the commonly known properties which one can find for instance in [17]. The subset of those properties that we use in the present paper is:

Theorem 33 For all specs R, S, T, conditions p and q, and imp f:

(a) $R \lhd true \rhd S \;=\; R$
(b) $R \lhd false \rhd S \;=\; S$
(c) $R \lhd p \rhd R \;=\; R$
(d) $R \lhd \neg p \rhd S \;=\; S \lhd p \rhd R$
(e) $R \lhd p \rhd (S \lhd p \rhd T) \;=\; R \lhd p \rhd T \;=\; (R \lhd p \rhd S) \lhd p \rhd T$
(f) $R \lhd (p \wedge q) \rhd S \;=\; (R \lhd p \rhd S) \lhd q \rhd S$
(g) $R \lhd (p \vee q) \rhd S \;=\; R \lhd p \rhd (R \lhd q \rhd S)$
(h) $(R \lhd p \rhd S) \lhd q \rhd T \;=\; (R \lhd q \rhd T) \lhd p \rhd (S \lhd q \rhd T)$
(i) $(R \lhd p \rhd S) \vartriangle T \;=\; (R \vartriangle T) \lhd p \rhd (S \vartriangle T)$
(j) $T \circ R \lhd p \rhd S \;=\; (T \circ R) \lhd p \rhd (T \circ S)$
(k) $R \lhd p \rhd S \circ f \;=\; (R \circ f) \lhd (p \circ f) \rhd (S \circ f)$
□

Note that for property (k) the condition that f is an imp is necessary. Property (k) is called the range translation rule for conditionals.

2.3 Definition of Filters

The definition of filter is borrowed directly from the work of Meertens [22] and Bird [10]:

Definition 34 (Filter) For right-condition p,

$$\lhd p \;=\; /\eta \circ *(\tau \lhd p \rhd \varepsilon)$$

□

Note that from the fact that τ and ε are imps and the fact that conditionals, junc and catamorphisms respect imps it follows that $\lhd p$ is an imp.

In this section we explore some algebraic properties of the filter operation. The properties that we seek are motivated by the relationship between the Bird-Meertens formalism and the so-called quantifier calculus, which relationship will be clarified in section 5.

By design $\lhd true$ is the identity function on specs of the correct type:

Theorem 35

$$\lhd true \;=\; *I$$

□

Now we consider whether two filters can be fused into one. Since $\lhd p$ is a catamorphism of the form $/\eta \circ *\overline{p}$ where $\overline{p} \;=\; \tau \lhd p \rhd \varepsilon$ it pays to begin by exploring whether a map can be fused with a filter. Indeed it can.

Lemma 36

(a) $*R \circ \lhd p \;=\; /\eta \circ *((\tau \circ R) \lhd p \rhd \varepsilon)$
(b) $/\eta \circ *R \circ \lhd p \;=\; /\eta \circ *(R \lhd p \rhd \varepsilon)$
□

A direct consequence of lemma 36b is:

Theorem 37 ($\triangleleft$ distribution)

$$\triangleleft p \circ \triangleleft q \;=\; \triangleleft(p \wedge q)$$

□

Yet another fusion property for filters is

Theorem 38 (Filter Range Translation) For all imps f,

$$\triangleleft p \circ *f \;=\; *f \circ \triangleleft(p \circ f) \circ *f\!>$$

□

3 Sectioning and Laws

In this section we give definitions for left- and right- sectioning, symmetry, associativity, idempotency and left- and right units. For each of these notions we derive a number of properties.

3.1 Sectioning

One important notion we know for binary functions is sectioning, i.e. we can construct a unary function from a binary function, by fixing one of the arguments. For instance (1+) is defined to be the function such that

$$\forall(x :: (1+).x = 1+x)$$

In the relational setting, the fixed argument does not necessarily have to be a point; we can generalize it to a left condition because left conditions can be viewed as the relational generalization of constants. We define left-sectioning by

Definition 39 (Left-Sectioning) For p a left condition,

$$p\otimes \;=\; p \otimes I$$

□

For the spec $p\otimes$ we then have the desired property:

Theorem 40 (Left-Sectioning)

$$p\otimes \circ R \;=\; p \otimes R$$

□

Note that by taking $R := I$ it follows that $p\otimes$ is the only spec for which property (40) holds for arbitrary spec R.

Similarly, we define right-sectioning by:

Definition 41 (Right-Sectioning) For left condition q,

$$\otimes q \;=\; I \otimes q$$

□

For which we have the property:

Theorem 42 (Right-Sectioning)

$$\otimes q \circ R \;=\; R \otimes q$$

□

3.2 Symmetry

We define symmetry by:

Definition 43 (Symmetry) $\otimes$ is symmetric iff

$$\forall (R, S :: R \otimes S \;=\; S \otimes R)$$

□

There is an other characterization of symmetry:

Theorem 44 (Symmetry) $\otimes$ is symmetric iff

$$\otimes \;=\; \otimes \circ \gamma$$

where γ is the natural transformation such that

$$R \times S \circ \gamma \;=\; \gamma \circ S \times R \quad \wedge \quad \gamma{>} \;=\; I \times I$$

□

3.3 Associativity

We define associativity by:

Definition 45 (Associativity) $\otimes$ is associative iff

$$\forall (R, S, T :: R \otimes (S \otimes T) \;=\; (R \otimes S) \otimes T))$$

□

There is another characterization of associativity:

Theorem 46 (Associativity) $\otimes$ is associative iff

$$\otimes \circ I \times \otimes \;=\; \otimes \circ \otimes \times I \circ \beta$$

where β is the natural transformation such that

$$(R \times S) \times T \circ \beta \;=\; \beta \circ R \times (S \times T) \quad \wedge \quad \beta{>} \;=\; I \times (I \times I)$$

□

For the definition of the natural transformations β and γ see [4, 6].

3.4 Idempotency

For idempotency we must be careful. Were we to define idempotency by

$$\forall(R :: R \otimes R \; = \; R)$$

we would have in particular $I \otimes I \; = \; I$. Since $I \otimes I \; = \; \otimes \circ I \vartriangle I$ this implies the very strong surjectivity condition $\otimes< \; = \; I$. Fortunately, a simple definition is at hand:

Definition 47 (Idempotency) $\otimes$ is idempotent iff,

$$I \otimes I \;\; \sqsubseteq \;\; I$$

□

For idempotency, we have the following property

Theorem 48 (Idempotency) $\otimes$ is idempotent iff

$$\forall(f : \mathit{imp}.f : f \otimes f = (\otimes \circ I \vartriangle I)> \circ f)$$

□

3.5 Units

We define left units by

Definition 49 (Left Unit) $l_\otimes$ is a left unit of $\otimes$ iff

$$\begin{array}{ll} & \mathit{left\_condition}.l_\otimes \;\; \wedge \;\; (\otimes \circ \ll^{\cup})> \;\; \sqsupseteq \;\; l_\otimes< \\ \wedge & \\ & l_\otimes \otimes \;\; = \;\; (\otimes \circ \gg^{\cup})> \end{array}$$

□

We demand the first conjunct because otherwise $l_\otimes \otimes$ is not defined. The second conjunct expresses that $l_\otimes \otimes$ should not contain junk outside the domain of the left argument of $\otimes$. The third conjunct expresses that $l_\otimes$ is the identity on the domain of the right argument of $\otimes$.

Of course, we can give similar definitions for $r_\otimes$ being a right unit of $\otimes$:

Definition 50 (Right Unit) $r_\otimes$ is a right unit of $\otimes$ iff,

$$\begin{array}{ll} & \mathit{left\_condition}.r_\otimes \;\; \wedge \;\; (\otimes \circ \gg^{\cup})> \;\; \sqsupseteq \;\; r_\otimes< \\ \wedge & \\ & \otimes r_\otimes \;\; = \;\; (\otimes \circ \ll^{\cup})> \end{array}$$

□

For arbitrary binary spec $\otimes$ nothing is known about the uniqueness of the left- or right unit. But if we know that $\otimes$ has a left and a right unit we have:

$$
\begin{array}{ll}
& r_\otimes \\
= & \qquad \{ \text{ definition left-, right-unit: } I \sqsupseteq l_\otimes \otimes \sqsupseteq r_\otimes < \ \} \\
& l_\otimes \otimes \circ r_\otimes \\
= & \qquad \{ \text{ sectioning } \} \\
& \otimes r_\otimes \circ l_\otimes \\
= & \qquad \{ \text{ definition left-, right-unit: } I \sqsupseteq \otimes r_\otimes \sqsupseteq l_\otimes < \ \} \\
& l_\otimes
\end{array}
$$

So they are the same and hence unique. This leads us to the following definition:

Definition 51 (Unit)
$1_\otimes$ is the unique unit of $\otimes$ iff $1_\otimes$ is a left and a right unit of $\otimes$.
□

Note: Very often we are interested in the reduce $/(1_\otimes \triangledown \otimes)$; that is why we allow ourself to write just $/\otimes$ instead of the cumbersome $/(1_\otimes \triangledown \otimes)$.

3.6 Conditionals

Because we now have the notation $\_ \otimes \_$ we can state two more properties about conditionals:

Theorem 52 ($\_ \otimes \_$-Conditional Abides Law)

(a)
$$
\begin{array}{l}
(R \triangleleft p \triangleright S) \otimes (P \triangleleft p \triangleright Q) \\
= \\
(R \otimes P) \triangleleft p \triangleright (S \otimes Q)
\end{array}
$$

(b)
$$
\begin{array}{l}
(R \triangleleft p \triangleright S) \otimes (P \triangleleft q \triangleright Q) \\
= \\
((R \otimes P) \triangleleft q \triangleright (R \otimes Q)) \triangleleft p \triangleright ((S \otimes P) \triangleleft q \triangleright (S \otimes Q))
\end{array}
$$

□

4 The Boom Hierarchy

The Boom-Hierarchy consists of four levels: trees, lists, bags and sets. A tree can be represented by a bin. Note that many bins can represent the same tree. For instance bin x and $\varepsilon +\!\!\!+ x$ represent the same tree. Also, we can represent a list by a tree if we forget about the brackets, i.e. $x +\!\!\!+ (y +\!\!\!+ z)$ and $(x +\!\!\!+ y) +\!\!\!+ z$ represent the same list. A bag can be represented by a list if we forget about the ordering of the elements: $x +\!\!\!+ y$ and $y +\!\!\!+ x$ represent the same bag. Similarly, a set can be represented by a bag if we forget about the number of occurrences of an element in a bag: $x +\!\!\!+ x$ and x represent the same set. In the paper by E. Voermans [24] a construction is given for an equivalence relation on μF such that the equivalence classes of that equivalence relation are just those bins (trees, lists, bags) that represent the same tree (list, bag, set). It is beyond the scope of the current paper to give that construction in detail; only the results are relevant to us. We denote the four equivalence relations by *Tree*, *List*, *Bag* and *Set*. Let $\mathcal{L}$ be one of these relations; we define the following constructors

$$
\begin{aligned}
&(53)\ \tau_{\mathcal{L}} &= \mathcal{L} \circ \tau \\
&(54)\ \Box_{\mathcal{L}} &= \mathcal{L} \circ \Box \\
&(55)\ +\!\!+_{\mathcal{L}} &= \mathcal{L} \circ +\!\!+
\end{aligned}
$$

So, for instance spec τ_{List} is the singleton-list constructor, i.e. when we apply τ_{List} to an element we get as result the whole equivalence class of *List* of which each bin represents the same singleton-list.

For these three constructors we have the following desired properties:

Theorem 56 (Boom-hierarchy)

$$
\begin{aligned}
&(a)\quad +\!\!+_{Tree} \circ \varepsilon_{Tree} \vartriangle Tree = Tree \\
&(b)\quad +\!\!+_{Tree} \circ Tree \vartriangle \varepsilon_{Tree} = Tree \\
&(c)\quad +\!\!+_{List} \circ +\!\!+_{List} \times I = +\!\!+_{List} \circ I \times +\!\!+_{List} \circ \beta \\
&(d)\quad +\!\!+_{Bag} = +\!\!+_{Bag} \circ \gamma \\
&(e)\quad +\!\!+_{Set} \circ I \vartriangle I = Set
\end{aligned}
$$

where β and γ are the natural transformations as used for the definition of associativity (46) and symmetry (44).
□

For $\mathcal{L}$ equal to *Tree*, *List*, *Bag* or *Set* we have the so-called absorption rule:

$$(57)\ \mathcal{L} \circ +\!\!+ = \mathcal{L} \circ +\!\!+ \circ \mathcal{L} \times \mathcal{L} = \mathcal{L} \circ +\!\!+ \circ I \times \mathcal{L} = \mathcal{L} \circ +\!\!+ \circ \mathcal{L} \times I$$

By unfolding the definition of $+\!\!+_{\mathcal{L}}$ and $\varepsilon_{\mathcal{L}}$ and using the absorption rule, we can shift the type information entirely to the left, which gives us:

Theorem 58 (Property constructors)

$$
\begin{aligned}
&(a)\quad Tree \circ \varepsilon +\!\!+ = Tree \\
&(b)\quad Tree \circ +\!\!+ \varepsilon = Tree \\
&(c)\quad List \circ +\!\!+ \circ +\!\!+ \times I = List \circ +\!\!+ \circ I \times +\!\!+ \circ \beta \\
&(d)\quad Bag \circ +\!\!+ \circ \gamma = Bag \circ +\!\!+ \\
&(e)\quad Set \circ +\!\!+ \circ I \vartriangle I = Set
\end{aligned}
$$

where β and γ are the natural transformations as used for the definition of associativity (46) and symmetry (44).
□

For example, property (d) expresses the fact that bins constructed by the functions $+\!\!+$ and $+\!\!+ \circ \gamma$ represent the same bag. In general it is the case that if

$$\mathcal{L} \circ R = \mathcal{L} \circ S$$

then R and S are the same up to the choice of the representatives of the equivalence classes resulting from the application of R and S to an element.

The properties of the constructors take on a more familiar form if we compose both left and right sides with an appropriate number of argument specs. For example, property (a) becomes, for all specs R,

(59) $\mathit{Tree} \circ \varepsilon \mathbin{+\!\!+} R \;=\; \mathit{Tree} \circ R$

and similarly property (b) becomes

(60) $\mathit{Tree} \circ R \mathbin{+\!\!+} \varepsilon \;=\; \mathit{Tree} \circ R$

If we compose both left and right sides of (e) with imp f, property (e) becomes, for all imps f,

(61) $\mathit{Set} \circ f \mathbin{+\!\!+} f \;=\; \mathit{Set} \circ f$

If more bins represent the same element of a particular type, the equivalence classes become bigger. Formally stated, we have the following lattice ordering which follows from the fact that the four types are the least solutions of a specification and that each type higher in the hierarchy satisfies the conditions of the types lower in the hierarchy:

(62) $\mu F \sqsubseteq \mathit{Tree} \sqsubseteq \mathit{List} \sqsubseteq \mathit{Bag} \sqsubseteq \mathit{Set}$

Also, we have the following properties:

(63) $\mathit{Tree} = \mathit{Tree} \circ \mu F$
(64) $\mathit{List} = \mathit{List} \circ \mathit{Tree} = \mathit{List} \circ \mu F$
(65) $\mathit{Bag} = \mathit{Bag} \circ \mathit{List} = \mathit{Bag} \circ \mathit{Tree} = \mathit{Bag} \circ \mu F$
(66) $\mathit{Set} = \mathit{Set} \circ \mathit{Bag} = \mathit{Set} \circ \mathit{List} = \mathit{Set} \circ \mathit{Tree} = \mathit{Set} \circ \mu F$

We say that X *respects* $\mathcal{L}$ if

$$X \circ \mathcal{L} \;=\; X$$

this condition expresses that spec X does not differentiate between representatives of the same equivalence class of $\mathcal{L}$. For instance, if $R \circ \mathit{Set} = R$ holds then spec R yields the same result when applied to bins x and $x \mathbin{+\!\!+} x$.

For each level of the Boom-hierarchy we have the following conditions for a catamorphism to be of that type.

Theorem 67 (Type of Catamorphism)
In order to state the theorem let us introduce the following abbrevations. First let $\mathcal{C}$ be a shorthand for $(\!|R \triangledown (S \triangledown \otimes)|\!)$. Then let us consider the following five properties, which we refer to later by their labels:

(a) $\otimes \circ (S \circ \top\!\top) \vartriangle I \circ \mathcal{C} \;=\; \mathcal{C}$
(b) $\otimes \circ I \vartriangle (S \circ \top\!\top) \circ \mathcal{C} \;=\; \mathcal{C}$
(c) $(\mathcal{C} \underline{\otimes} \mathcal{C}) \underline{\otimes} \mathcal{C} \;=\; \mathcal{C} \underline{\otimes} (\mathcal{C} \underline{\otimes} \mathcal{C}) \circ \beta$
(d) $\mathcal{C} \underline{\otimes} \mathcal{C} \;=\; \mathcal{C} \underline{\otimes} \mathcal{C} \circ \gamma$
(e) $\otimes \circ \mathcal{C} \vartriangle \mathcal{C} \;=\; \mathcal{C}$

(Note that (c), respectively (d), holds if $\otimes$ is associative, respectively symmetric).

Now we can state the theorem, which is:

$\mathcal{C}$ respects *Tree* $\equiv$ (a) $\wedge$ (b).
$\mathcal{C}$ respects *List* $\equiv$ $\mathcal{C}$ respects *Tree* $\wedge$ (c).
$\mathcal{C}$ respects *Bag* $\equiv$ $\mathcal{C}$ repects *List* $\wedge$ (d).
$\mathcal{C}$ respects *Set* $\equiv$ $\mathcal{C}$ respects *Bag* $\wedge$ (e).
□

4.1 Map and Filter

In this section we derive properties for map and filter concerning laws. From these properties it follows that the type of a catamorphism remains the same if it is composed with a filter or a map (in the case of *Set* only if the argument of the map is an imp). An important theorem is:

Theorem 68 (Type-η Fusion) For $\mathcal{L}$ equal to *Tree*, *List*, *Bag* or *Set*,

$$\mathcal{L} \circ (\!|R \triangledown \eta|\!) \;=\; (\!|(\mathcal{L} \circ R) \triangledown (\mathcal{L} \circ \eta)|\!)$$

□

Using this property we can prove the following type judgements:

Theorem 69 For $\mathcal{L}$ equal to *Tree*, *List* or *Bag*, and imp f,

(a) $\mathcal{L} \circ (\!|R \triangledown \eta|\!)$ respects $\mathcal{L}$
(b) $Set \circ (\!|f \triangledown \eta|\!)$ respects Set
(c) $\mathcal{L} \circ *R$ respects $\mathcal{L}$
(d) $Set \circ *f$ respects Set
(e) $\mathcal{L} \circ \triangleleft p$ respects $\mathcal{L}$
(f) $Set \circ \triangleleft p$ respects Set
□

The importance of theorem (69) is that if a spec R does not differentiate between representatives of, say, the same bag, i.e. spec R respects *Bag*, then for instance $R \circ \triangleleft p$ also respects *Bag*. Thus using these theorems we can move around the type information in our formulae.

5 Comparison with Quantifier Notation

In this section we will make a link to the Quantifier Notation. We define

(70) $\otimes(i : p.i : R.i) \;=\; /\!\otimes \circ *R \circ \triangleleft p$

where $R.i$ is defined by

(71) $R.i \;=\; R \circ i\bullet$

with $i\bullet$ being a point "pointing at i", i.e. $i\bullet< \;=\; \{(i,i)\}$.

It is vital to note that the expression $\otimes(i : p.i : R.i)$ denotes a function. This appears to be counter to normal usage; that it is not so is explained by the fact that the domain of the dummy i is always left implicit in the quantifier notation. We can make it explicit by writing $\otimes(i \in A : p.i : R.i)$; if we now suppose that $\mathcal{A}$ is an enumeration of the elements of the type A and $\otimes$ is both associative and symmetric then we make the definition

(72) $\otimes(i \in A : p.i : R.i) \;=\; /\otimes \,\circ\, *R \,\circ\, \lhd p \,\circ\, \mathcal{A}$

Note that for the quantification over $\otimes$ to be meaningful it is always assumed that $\otimes$ is both associative and symmetric. We will not demand this unless it is necessary.

Two rules known for the quantifier notation hold at the level of bins. The first rule is trading; the laws govern the interchange of expressions between the range and the function part of the quantification. Using theorem (36a) and (30) it is easy to prove that:

(73) $/\otimes \,\circ\, *R \,\circ\, \lhd p \;=\; /\otimes \,\circ\, *(R\lhd p\rhd 1_{\otimes})$

The well-known trading rules for universal and existential quantification are immediate corollaries of (73).

(74) $/\mathbf{and} \,\circ\, *\overline{q} \,\circ\, \lhd p \;=\; /\mathbf{and} \,\circ\, *(\overline{p\Rightarrow q})$

(75) $/\mathbf{or} \,\circ\, *\overline{q} \,\circ\, \lhd p \;=\; /\mathbf{or} \,\circ\, *(\overline{p\wedge q})$

where p and q are right-conditions, and $\overline{q}$ is a total imp to $\mathbb{B}$ defined by

$$\overline{q} \;=\; \mathbf{true}\lhd q\rhd\mathbf{false}$$

with **true** and **false** the points "pointing" at true and false ($\mathbb{B}$ is a monotype containing two elements: false and true) and **and** and **or** the total imps from $\mathbb{B}\times\mathbb{B}$ to $\mathbb{B}$ corresponding to disjunction and conjunction.

To derive (74) and (75) it suffices to observe that

(76) $\overline{q}\lhd p\rhd\mathbf{true} \;=\; \overline{p\Rightarrow q}$

and

(77) $\overline{q}\lhd p\rhd\mathbf{false} \;=\; \overline{p\wedge q}$

which follows directly from the definition of the bar and the properties of conditionals. As an example we prove (76):

$$
\begin{array}{ll}
 & \overline{q}\lhd p\rhd\mathbf{true} \\
= & \quad\{ \text{ property conditionals (33d) } \} \\
 & \mathbf{true}\lhd\neg p\rhd\overline{q} \\
= & \quad\{ \text{ definition bar } \} \\
 & \mathbf{true}\lhd\neg p\rhd(\mathbf{true}\lhd q\rhd\mathbf{false})
\end{array}
$$

$$
\begin{array}{ll}
= & \{ \text{ property conditionals (33g) } \} \\
 & \mathbf{true} \lhd \neg p \vee q \rhd \mathbf{false} \\
= & \{ \text{ definition } \Rightarrow \text{, definition bar } \} \\
 & \overline{p \Rightarrow q}
\end{array}
$$

Less familiar consequences of (73) are the trading rules for equivalence and inequivalence.

(78) $/\!\equiv \circ *\overline{q} \circ \lhd p \;=\; /\!\equiv \circ *(\overline{p \Rightarrow q})$
(79) $/\!\not\equiv \circ *\overline{q} \circ \lhd p \;=\; /\!\not\equiv \circ *(\overline{p \wedge q})$

These follow because, like conjunction, the unit of equivalence is **true** and, like disjunction, the unit of inequivalence is **false**.

The literal translation of (74) and (75) into the quantifier notation yields

$$\forall(i : \, p.i : \, q.i) \;=\; \forall(i :: \, p.i \Rightarrow q.i)$$

and

$$\exists(i : p.i : q.i) \;=\; \exists(i :: p.i \wedge q.i)$$

The trading rules in the quantifier calculus are, however, slightly more general; specifically:

$$\forall(i : \, p.i \wedge r.i : \, q.i) \;=\; \forall(i : \, r.i : \, p.i \Rightarrow q.i)$$

and

$$\exists(i : p.i \wedge r.i : q.i) \;=\; \exists(i : r.i : p.i \wedge q.i)$$

But these properties stated in our formalism follow directly from theorem (37):

(80) $\lhd(p \wedge r) \;=\; \lhd p \circ \lhd r$

The second rule we have already derived is range translation. From the filter range translation rule (38) and compositionality of map it follows that, for imp f,

(81) $/\!\otimes \circ *R \circ \lhd p \circ *f \;=\; /\!\otimes \circ *(R \circ f) \circ \lhd(p \circ f) \circ *f_>$

In the quantifier notation this would be expressed as

(82) $\otimes(i \in t.A : p.i : f.i) \;=\; \otimes(j \in A : p.(t.j) : f.(t.j))$

In the following sections we derive the other well-known rules like the unit rule, range splitting and range disjunction. We try to derive these properties as early as possible in the Boom-hierarchy.

5.1 Unit Law

In this section we concentrate on trees. Using the results of the previous section and the property of the constructors (58a) we can prove the unit rule. We have the following property:

Theorem 83 (Unit)

$$\mathit{Tree} \circ \lhd \mathit{false} \;=\; \mathit{Tree} \circ \varepsilon \circ \mu F$$

□

Note that the added type information is crucial: without it, the equality does not hold. For instance, $\lhd \mathit{false}$ transforms $a \mathbin{+\!\!+} (b \mathbin{+\!\!+} c)$ into $\varepsilon \mathbin{+\!\!+} (\varepsilon \mathbin{+\!\!+} \varepsilon)$ and the latter is only equal to ε if we consider trees. The added "$\circ\ \mu F$" is necessary because the right domain of $\lhd \mathit{false}$ is μF.

Using theorem (83) we can prove:

Corollary 84 (Unit Rule) If $(\!| R \triangledown (S \triangledown \otimes) |\!)$ respects *Tree* then

$$(\!| R \triangledown (S \triangledown \otimes) |\!) \circ \lhd \mathit{false} \;=\; S \circ \top\!\!\top \circ \mu F$$

Proof

$$\begin{array}{ll}
 & (\!| R \triangledown (S \triangledown \otimes) |\!) \circ \lhd \mathit{false} \\
= & \quad \{ (\!| R \triangledown (S \triangledown \otimes) |\!) \text{ respects } \mathit{Tree} \} \\
 & (\!| R \triangledown (S \triangledown \otimes) |\!) \circ \mathit{Tree} \circ \lhd \mathit{false} \\
= & \quad \{ \text{theorem (83)} \} \\
 & (\!| R \triangledown (S \triangledown \otimes) |\!) \circ \mathit{Tree} \circ \varepsilon \circ \mu F \\
= & \quad \{ (\!| R \triangledown (S \triangledown \otimes) |\!) \text{ respects } \mathit{Tree} \} \\
 & (\!| R \triangledown (S \triangledown \otimes) |\!) \circ \varepsilon \circ \mu F \\
= & \quad \{ \text{definition } \varepsilon \} \\
 & (\!| R \triangledown (S \triangledown \otimes) |\!) \circ \Box \circ \top\!\!\top \circ \mu F \\
= & \quad \{ \text{computation rule } \Box \} \\
 & S \circ \top\!\!\top \circ \mu F
\end{array}$$

□

The above calculation is a nice example of how we treat type information in our system. By the assumption the type information pops up at the right place and having used the type information we can get rid of it by using the assumption again.

Examples of the unit rule are

$$\begin{array}{lcl}
/+ \circ *R \circ \lhd \mathit{false} & = & 0\bullet \circ \mu F \\
/\times \circ *R \circ \lhd \mathit{false} & = & 1\bullet \circ \mu F \\
/\mathbf{and} \circ *R \circ \lhd \mathit{false} & = & \mathbf{true} \circ \mu F \\
/\mathbf{or} \circ *R \circ \lhd \mathit{false} & = & \mathbf{false} \circ \mu F.
\end{array}$$

because $0\bullet$, $1\bullet$, **true** and **false** are the units of $+$, $\times$, **and** and **or**. These are written in the quantifier form as follows

$$\begin{array}{lcl} \Sigma(i : false : R.i) & = & 0 \\ \Pi(i : false : R.i) & = & 1 \\ \forall(i : false : R.i) & = & true \\ \exists(i : false : R.i) & = & false \end{array}$$

Note that the proof given here is valid for trees (and thus also if the catamorphism respects lists, bags or sets). This is an improvement over the one in [2] since there the range splitting rule was used but that rule is only valid for bags and hence also for sets.

5.2 Associativity and Symmetry

Throughout this section we assume the existence of an associative and symmetric binary spec $\otimes$ and the existence of a spec U, such that

$$U = U \otimes U$$

An obvious candidate for U is $1_\otimes$ if it exists.

First we prove a property of $\otimes$:

Theorem 85 ($\underline{\otimes}$-$\otimes$ Abides Law)

$$(R \underline{\otimes} S) \otimes (P \underline{\otimes} Q) = (R \otimes P) \underline{\otimes} (S \otimes Q)$$

Proof

$$\begin{array}{ll} & (R \underline{\otimes} S) \otimes (P \underline{\otimes} Q) \\ = & \quad \{ \text{ property (22) } \} \\ & ((R \circ \ll) \otimes (S \circ \gg)) \otimes ((P \circ \ll) \otimes (Q \circ \gg)) \\ = & \quad \{ \otimes \text{ associative and symmetric } \} \\ & ((R \circ \ll) \otimes (P \circ \ll)) \otimes ((S \circ \gg) \otimes (Q \circ \gg)) \\ = & \quad \{ \ll, \gg \text{ imps, split-imp fusion (23) } \} \\ & (R \otimes P \circ \ll) \otimes (S \otimes Q \circ \gg) \\ = & \quad \{ \text{ property (22) } \} \\ & (R \otimes P) \underline{\otimes} (S \otimes Q) \end{array}$$

□

For this spec $\otimes$ and U we can prove:

Theorem 86 (Asso-Sym Rule)

$$(\!|(R \otimes S) \triangledown (U \triangledown \otimes)|\!) = (\!|R \triangledown (U \triangledown \otimes)|\!) \otimes (\!|S \triangledown (U \triangledown \otimes)|\!)$$

□

We can now prove, because a filter is an imp:

Theorem 87

$$\begin{array}{ll} & (\!|(R \otimes S) \triangledown (U \triangledown \otimes)|\!) \circ \triangleleft p \\ = & \\ & ((\!|R \triangledown (U \triangledown \otimes)|\!) \circ \triangleleft p) \otimes ((\!|S \triangledown (U \triangledown \otimes)|\!) \circ \triangleleft p) \end{array}$$

Proof

$$
\begin{array}{ll}
& (\!|(R \otimes S) \triangledown (U \triangledown \otimes)|\!) \circ \triangleleft p \\
= & \quad \{ \text{ theorem (86) } \} \\
& (\!|R \triangledown (U \triangledown \otimes)|\!) \otimes (\!|S \triangledown (U \triangledown \otimes)|\!) \circ \triangleleft p \\
= & \quad \{ \triangleleft p \text{ imp, split-imp fusion (23) } \} \\
& ((\!|R \triangledown (U \triangledown \otimes)|\!) \circ \triangleleft p) \otimes ((\!|S \triangledown (U \triangledown \otimes)|\!) \circ \triangleleft p)
\end{array}
$$

□

Theorem (87) is the direct analogue of the associativity and symmetry rule in [2], expressed in the quantifier notation as follows:

$$\otimes(i : p.i : f.i \otimes g.i) = \otimes(i : p.i : f.i) \otimes \otimes(i : p.i : g.i)$$

By definition $Bag \circ +\!\!+$ is associative and $Bag \circ \varepsilon$ is the unit of $Bag \circ +\!\!+$, thus theorem (86) holds with $U := Bag \circ \Box$ and $\otimes := Bag \circ +\!\!+$. This results in:

Theorem 88 (Asso-Sym Rule)

$$Bag \circ (\!|(R +\!\!+ S) \triangledown \eta|\!) = Bag \circ (\!|R \triangledown \eta|\!) +\!\!+ (\!|S \triangledown \eta|\!)$$

□

For the proof of Range Splitting and Range Disjunction in the next section we will use theorem (88) to move the join operator inside the catamorphism, then we use property (52a) or (52b) to move it inside the arguments of the conditionals and finally, with some shuffling around, we use the properties of the constructors to remove it entirely.

Theorem 89 (Range Splitting)

$$Bag \circ \triangleleft p +\!\!+ \triangleleft \neg p = Bag$$

Proof

First we prove:

$$
\begin{array}{ll}
& Bag \circ (\tau \triangleleft p \triangleright \varepsilon) +\!\!+ (\tau \triangleleft \neg p \triangleright \varepsilon) \\
= & \quad \{ \text{ property conditionals (33d) } \} \\
& Bag \circ (\tau \triangleleft p \triangleright \varepsilon) +\!\!+ (\varepsilon \triangleleft p \triangleright \tau) \\
= & \quad \{ +\!\!+ \text{ abides with } \langle p \rangle \text{ (52a) } \} \\
& Bag \circ (\tau +\!\!+ \varepsilon) \triangleleft p \triangleright (\varepsilon +\!\!+ \tau) \\
= & \quad \{ \text{ property conditonals (33j) } \} \\
& (Bag \circ \tau +\!\!+ \varepsilon) \triangleleft p \triangleright (Bag \circ \varepsilon +\!\!+ \tau) \\
= & \quad \{ \text{ property constructors (58ab) } \} \\
& (Bag \circ \tau) \triangleleft p \triangleright (Bag \circ \tau) \\
= & \quad \{ \text{ property conditionals (33a) } \} \\
& Bag \circ \tau
\end{array}
$$

And from this the theorem follows:

$$
\begin{array}{ll}
 & Bag \circ \lhd p \mathbin{+\!\!+} \lhd \neg p \\
= & \quad \{ \text{ definition filter } \} \\
 & Bag \circ (\!|(\tau \lhd p \rhd \varepsilon) \triangledown \eta|\!) \mathbin{+\!\!+} (\!|(\tau \lhd \neg p \rhd \varepsilon) \triangledown \eta|\!) \\
= & \quad \{ \text{ asso-sym rule (88) } \} \\
 & Bag \circ (\!|((\tau \lhd p \rhd \varepsilon) \mathbin{+\!\!+} (\tau \lhd \neg p \rhd \varepsilon)) \triangledown \eta|\!) \\
= & \quad \{ \text{ type-}\eta \text{ fusion (68) } \} \\
 & (\!|(Bag \circ (\tau \lhd p \rhd \varepsilon) \mathbin{+\!\!+} (\tau \lhd \neg p \rhd \varepsilon)) \triangledown (Bag \circ \eta)|\!) \\
= & \quad \{ \text{ calculation above } \} \\
 & (\!|(Bag \circ \tau) \triangledown (Bag \circ \eta)|\!) \\
= & \quad \{ \text{ type-}\eta \text{ fusion (68) } \} \\
 & Bag \circ (\!|\tau \triangledown \eta|\!) \\
= & \quad \{ (\!|\tau \triangledown \eta|\!) = \mu F, Bag \circ \mu F = Bag \} \\
 & Bag
\end{array}
$$

□

Using the range splitting rule, we can prove:

Theorem 90 If $(\!|R \triangledown (S \triangledown \otimes)|\!)$ respects *Bag* then,

$$
\begin{array}{ll}
 & (\!|R \triangledown (S \triangledown \otimes)|\!) \circ \lhd q \\
= & \\
 & ((\!|R \triangledown (S \triangledown \otimes)|\!) \circ \lhd (p \wedge q)) \otimes ((\!|R \triangledown (S \triangledown \otimes)|\!) \circ \lhd (\neg p \wedge q))
\end{array}
$$

Proof

$$
\begin{array}{ll}
 & (\!|R \triangledown (S \triangledown \otimes)|\!) \circ \lhd q \\
= & \quad \{ (\!|R \triangledown (S \triangledown \otimes)|\!) \text{ respects } Bag, \text{ range splitting } \} \\
 & (\!|R \triangledown (S \triangledown \otimes)|\!) \circ \lhd p \mathbin{+\!\!+} \lhd \neg p \circ \lhd q \\
= & \quad \{ \text{ definition } \_\mathbin{+\!\!+}\_, \text{ computation rule } \mathbin{+\!\!+} \} \\
 & ((\!|R \triangledown (S \triangledown \otimes)|\!) \circ \lhd p) \otimes ((\!|R \triangledown S \triangledown \otimes|\!) \circ \lhd \neg p) \circ \lhd q \\
= & \quad \{ \lhd q \text{ imp, split-imp fusion } \} \\
 & ((\!|R \triangledown (S \triangledown \otimes)|\!) \circ \lhd p \circ \lhd q) \otimes ((\!|R \triangledown S \triangledown \otimes|\!) \circ \lhd \neg p \circ \lhd q) \\
= & \quad \{ \text{ filter-distribution } \} \\
 & ((\!|R \triangledown (S \triangledown \otimes)|\!) \circ \lhd (p \wedge q)) \otimes ((\!|R \triangledown S \triangledown \otimes|\!) \circ \lhd (\neg p \wedge q))
\end{array}
$$

□

Expressed in the quantifier notation, theorem (90) takes the form:

$$
\begin{array}{ll}
 & \otimes(i : q.i : f.i) \\
= & \\
 & \otimes(i : p.i \wedge q.i : f.i) \;\otimes\; \otimes(i : \neg p.i \wedge q.i : f.i)
\end{array}
$$

5.3 Adding Idempotency

In this section we add idempotency, i.e. we consider sets. We know that $Set = Set \circ Bag$ so we may use the asso-sym rule (88). Furthermore, we know that $Set \circ \mathbin{+\!\!+}$ is idempotent, which gives us:

Theorem 91 (Range Disjunction)

$$Set \circ \triangleleft p + \!\! + \triangleleft q \;=\; Set \circ \triangleleft(p \vee q)$$

Proof

First we prove:

$$\begin{array}{ll} & Set \circ (\tau \lhd p \rhd \varepsilon) + \!\! + (\tau \lhd q \rhd \varepsilon) \\ = & \quad \{ \text{ property conditionals (52b) } \} \\ & Set \circ ((\tau + \!\! + \tau) \lhd q \rhd (\tau + \!\! + \varepsilon)) \lhd p \rhd ((\varepsilon + \!\! + \tau) \lhd q \rhd (\varepsilon + \!\! + \varepsilon)) \\ = & \quad \{ \text{ property conditionals (33j), constructors (58abe) } \} \\ & Set \circ (\tau \lhd q \rhd \tau) \lhd p \rhd (\tau \lhd q \rhd \varepsilon) \\ = & \quad \{ \text{ property conditionals (33a) } \} \\ & Set \circ \tau \lhd p \rhd (\tau \lhd q \rhd \varepsilon) \\ = & \quad \{ \text{ property conditionals (33g) } \} \\ & Set \circ \tau \lhd p \vee q \rhd \varepsilon \end{array}$$

Then we have:

$$\begin{array}{ll} & Set \circ \triangleleft p + \!\! + \triangleleft q \\ = & \quad \{ \text{ definition filter } \} \\ & Set \circ (\!| \tau \lhd p \rhd \varepsilon \triangledown \eta |\!) + \!\! + (\!| \tau \lhd q \rhd \varepsilon \triangledown \eta |\!) \\ = & \quad \{ \text{ theorem (88) } \} \\ & Set \circ (\!| ((\tau \lhd p \rhd \varepsilon) + \!\! + (\tau \lhd q \rhd \varepsilon)) \triangledown \eta |\!) \\ = & \quad \{ \text{ theorem (68), calculation above } \} \\ & Set \circ (\!| \tau \lhd p \vee q \rhd \varepsilon \triangledown \eta |\!) \\ = & \quad \{ \text{ definition filter } \} \\ & Set \circ \triangleleft(p \vee q) \end{array}$$

□

And again we can combine this with a catamorphism:

Theorem 92 If $(\!| R \triangledown (S \triangledown \otimes) |\!)$ respects *Set* then,

$$\begin{array}{ll} & (\!| R \triangledown (S \triangledown \otimes) |\!) \circ \triangleleft(p \vee q) \\ = & \\ & ((\!| R \triangledown (S \triangledown \otimes) |\!) \circ \triangleleft p) \otimes ((\!| R \triangledown (S \triangledown \otimes) |\!) \circ \triangleleft q) \end{array}$$

Proof

$$\begin{array}{ll} & (\!| R \triangledown (S \triangledown \otimes) |\!) \circ \triangleleft(p \vee q) \\ = & \quad \{ \; (\!| R \triangledown (S \triangledown \otimes) |\!) \text{ respects } Set, \text{ range disjunction } \} \\ & (\!| R \triangledown (S \triangledown \otimes) |\!) \circ \triangleleft p + \!\! + \triangleleft q \\ = & \quad \{ \text{ definition } \_ + \!\! + \_, \text{ computation rule } + \!\! + \} \\ & ((\!| R \triangledown (S \triangledown \otimes) |\!) \circ \triangleleft p) \otimes ((\!| R \triangledown (S \triangledown \otimes) |\!) \circ \triangleleft q) \end{array}$$

□

Theorem (92) stated in the quantifier notation yields:

$$\otimes(i : p.i \vee q.i : f.i) = \otimes(i : p.i : f.i) \otimes \otimes(i : q.i : f.i)$$

6 Conclusion

By now the so-called "Boom hierarchy of types" is very familiar (if perhaps not under that name) and the substantial majority of properties established in this paper have been published elsewhere (in texts on APL, in Backus's Turing award lecture and in Meertens' paper). The contribution of this paper has principally been to organise the rules, making clear at which level of the hierarchy each rule becomes valid. The paper has also generalised the rules from a strictly typed, functional framework to a polymorphic, relational framework. Two surprising outcomes are the absence of appeals to extensionality and the low incidence of properties that rely on functionality.

A major distinguishing feature of the spec calculus is that it tries to capture *true* polymorphism rather than *parameterised* polymorphism, which is expressed by naturality properties in category theory. A consequence of this design decision is that type considerations occasionally enter into equational laws rather than being expressed in some overall context within which the law is applicable. This difference is illustrated by the split operator. In category theory the arrow $R \vartriangle S$ is only defined if R and S have the same right domain; if this is the case one has the computation rule

(93) $\ll \circ\ R \vartriangle S \ = \ R$

In the spec calculus split is a *total* operator: $R \vartriangle S$ is a spec for all specs R and S irrespective of whether R and S have the same right domain, and the rule (93) takes the form

(94) $\ll \circ\ R \vartriangle S \ = \ R \circ S{>}$

Our experience is that this has been a fortuitious design decision: one is rarely hindered by the types that occasionally crop up in calculations, and calculations proceed more smoothly because one is not continually nervous about unconscious omission of some type restriction, the laws themselves containing the reminder that such is necessary.

This aspect of the theory is particularly emphasised in this paper: a great many of our calculations are prefaced by "*List* $\circ$" or "*Bag* $\circ$" etc. It may indeed seem that the decision to develop a calculus in which types are part and parcel of the equational laws was misguided: here, after all, is the evidence how cumbersome it can be. But one must remember the whole content of this paper is the discussion of which laws are valid for Trees, which for Lists etc. and it is because of this that the type constraints are ever present.

7 Acknowledgements

We would like to thank Roland Backhouse, Ed Voermans and Netty van Gasteren for their contributions to this work.

Preparation of this paper was expedited by the use of the proof editor developed by Paul Chisholm [14].

References

1. R.C. Backhouse. An exploration of the Bird-Meertens formalism. Technical Report CS 8810, Department of Computing Science, Groningen University, 1988.
2. R.C. Backhouse. Program construction and verification. Prentice-Hall, 1986.
3. Roland Backhouse and Paul Hoogendijk. Elements of a Relational Theory of Datatypes. To appear in: *Proceedings of IFIP summerschool*, Brazil 1992.
4. R.C. Backhouse, P. de Bruin, P. Hoogendijk, G. Malcolm, T.S. Voermans, and J. van der Woude. Polynomial relators. To appear in: *Proceedings of the 2nd Conference on Algebraic Methodology and Software Technology*, May 22-25, 1991.
5. R.C. Backhouse, P. de Bruin, G. Malcolm, T.S. Voermans, and J van der Woude. Relational Catamorphisms. In Möller B., editor, *Proceedings of the IFIP TC2/WG2.1 Working Conference on Constructing Programs*, pages 287-318. Elsevier Science Publishers B.V., 1991.
6. R.C. Backhouse, T.S. Voermans, J. v.d. Woude. A Relational Theory Of Datatypes. *Proceedings of the EURICS on Calculational Theories of Program Structure, Ameland, The Netherlands*, September 1991.
7. J. Backus. Can programming be liberated from the von Neumann style? A functional style and its algebra of programs. *Communications of the ACM*, 21(8):613-641, 1978.
8. R.S. Bird. Transformational Programming and the Paragraph Problem. *Science of Computing Programming*, 6:159-189, 1986.
9. R.S. Bird. An introduction to the Theory of Lists. In M. Broy, editor, *Logic of Programming and Calculi of Discrete Design* . Springer-Verlag, 1987. NATO ASI Series, vol. F36.
10. R.S. Bird. Lectures on constructive functional programming. Oxford University, 1988.
11. R.S. Bird and J. Gibbons and G. Jones. Formal Derivation of a Pattern Matching Algorithm. Technical report, Programming Research Group, Oxford University, 11, Keble Road, Oxford, OX1 3QD, U.K., 1988.
12. R.S. Bird. A Calculus of Functions for Program Derivation. Technical report, Programming Research Group, Oxford University, 11, Keble Road, Oxford, OX1 3QD, U.K., 1988.
13. R.S. Bird and L. Meertens. Two exercises found in a book on algorithmics. In L.G.L.T. Meertens, editor, *Program Specification and Transformations*, pages 451-457. Elsevier Science Publishers B.V., North Holland, 1987.
14. P. Chisholm. Calculation by Computer. In *Third International Workshop Software Engineering and its Applications*, pages 713-728, Toulouse, France, December 3-7 1990. EC2.
15. E.W. Dijkstra, W.H.J. Feijen. Een methode van programmeren. Academic Service, 1984.
16. C.A.R. Hoare. A couple of novelties in the propositional calculus. *Zeitschr. fuer Math. Logik und Grundlagen der Math.*, 31(2):173-178, 1985.
17. C.A.R. Hoare *et al.* Laws of Programming. *Communications of the ACM*, 30(8):672-686, 1987.
18. P.F. Hoogendijk. A Hierarchy of Freebies. *Proceedings of the EURICS on Calculational Theories of Program Structure, Ameland, The Netherlands*, September 1991.
19. P.F. Hoogendijk. The Boom hierarchy. *Proceedings of the EURICS on Calculational Theories of Program Structure, Ameland, The Netherlands*, September 1991.

20. K. Iverson. A Programming Language. John Wiley & Sons New York, 1962.
21. G. Malcolm. Algebraic data types and program transformation. PhD thesis, Groningen University, 1990.
22. L. Meertens. Algorithmics – towards programming as a mathematical activity. In *Proceedings of the CWI Symposium on Mathematics and Computer Science*, pages 289-334. North-Holland, 1986.
23. G. Schmidt and T. Ströhlein. Relationen und Grafen. Springer-Verlag, 1988.
24. E. Voermans, Pers as Types, Inductive Types and Types with Laws. To appear in: *PHOENIX seminar and workshop on declarative programming, Sasbachwalden 1991*. Workshops in computing, Springer-Verlag.

A Logarithmic Implementation of Flexible Arrays

Rob R. Hoogerwoord

Eindhoven University of Technology, department of Mathematics and Computing Science, postbus 513, 5600 MB Eindhoven, The Netherlands

Abstract. In this paper we derive an implementation of so-called ***flexible arrays***; a flexible array is an array whose size can be changed by adding or removing elements at either end. By representing flexible arrays by so-called ***Braun trees***, we are able to implement all array operations with logarithmic —in the size of the array— time complexity.
Braun trees can be conveniently defined in a recursive way. Therefore, we use functional programming to derive (recursive) definitions for the functions representing the array operations. Subsequently, we use these definitions to derive (iterative) sequential implementations.

0 Introduction

A flexible array is an array the size of which can be changed by adding or removing elements at either end. In 1983 W. Braun and M. Rem designed an implementation of flexible arrays by means of balanced binary trees, which are used in such a way that all array operations can be performed in logarithmic time. Examples of programs in which flexible arrays are used can be found in [1].

The original presentation of this design by Braun and Rem is, however, rather complicated [0]. In this paper we use functional programming to derive this implementation in a more straightforward way. We do so in three steps. First, the binary trees used to represent arrays are defined as a recursive data type. Second, we derive recursive definitions for the functions implementing the array operations; this is relatively easy. Finally, we use these definitions to derive iterative sequential implementations of the array operations, where the Braun trees are represented by means of nodes linked together by pointers.

1 Functional Programs

1.0 Specifications

For the sake of simplicity of presentation, we assume that all (flexible) arrays have 0 as their lower bound. Then, an array of size d, $0 \le d$, is a function of type $[0,d) \rightarrow A$, where A denotes the element type of the array. As usual for functions, we call $[0,d)$ the *domain* of the array and we call values of type A *elements*. The size of an array is an attribute of that array's value, not of its type —as in Pascal— . Arrays being flexible means that different arrays may have different sizes; yet, they are all of the same type.

We denote the size of array x by $\#x$. The function $\#$ is one of the operations to be implemented. Another important operation on arrays is *element selection*, which is the same as function application: for array x and $i, 0 \leq i < \#x$, element $x{\cdot}i$ is the value of x in point i of its domain. Notice that an array is completely determined by its size and its elements. In what follows we use this without explicit reference.

By means of *element replacement*, an array can be modified in a single point of its domain. For array x, natural $i, 0 \leq i < \#x$, and element b, we use $x{:}i, b$ to denote the array y that satisfies:

$$\#y = \#x \quad \wedge \quad y{\cdot}i = b \quad \wedge \quad (\forall j \; : \; 0 \leq j < \#x \wedge j \neq i \; : \; y{\cdot}j = x{\cdot}j) \quad .$$

Notice that element selection and replacement are only meaningful for nonempty arrays.

The functions *le* and *he* represent the operations to *extend* an array with an additional element at the *lower* or *higher* end of its domain; that is, for element b and array x, arrays $le{\cdot}b{\cdot}x$ and $he{\cdot}b{\cdot}x$ are the arrays y and z that satisfy:

$$\#y = \#x{+}1 \quad \wedge \quad y{\cdot}0 = b \quad \wedge \quad (\forall j : 0 \leq j < \#x : y{\cdot}(j{+}1) = x{\cdot}j) \quad , \quad \text{and}$$
$$\#z = \#x{+}1 \quad \wedge \quad z{\cdot}(\#x) = b \quad \wedge \quad (\forall j : 0 \leq j < \#x : z{\cdot}j = x{\cdot}j) \quad .$$

Finally, the (partial) inverses of *le* and *he* are the functions *lr* and *hr*; they can be used to *remove* the extreme elements of an array. For array x satisfying $\#x \geq 1$, arrays $lr{\cdot}x$ and $hr{\cdot}x$ are the arrays y and z satisfying:

$$\#y = \#x{-}1 \quad \wedge \quad (\forall j : 0 \leq j < \#y : y{\cdot}j = x{\cdot}(j{+}1)) \quad , \quad \text{and}$$
$$\#z = \#x{-}1 \quad \wedge \quad (\forall j : 0 \leq j < \#z : z{\cdot}j = x{\cdot}j) \quad .$$

¿From these specifications it follows, for instance, that $lr{\cdot}(le{\cdot}b{\cdot}x) = x$ and that $hr{\cdot}(he{\cdot}b{\cdot}x) = x$.

1.1 Braun Trees

For the sake of the required (logarithmic) efficiency, we use a divide-and-conquer approach. The unique array of size 0 can be represented by the unique element of a unit type. An array x of size $d{+}1, 0 \leq d$, can be represented as follows. One element, namely $x{\cdot}0$, is kept separate; the remaining elements $x{\cdot}(j{+}1), 0 \leq j < d$, are partitioned according to the *parity* of j. That is, we distinguish the elements $x{\cdot}(2{*}i{+}1)$, $0 \leq 2{*}i < d$, and the elements $x{\cdot}(2{*}i{+}2), 0 \leq 2{*}i{+}1 < d$. The ranges $0 \leq 2{*}i < d$ and $0 \leq 2{*}i{+}1 < d$ can be rewritten into $0 \leq i < d\,\mathrm{div}\,2 + d\,\mathrm{mod}\,2$ and $0 \leq i < d\,\mathrm{div}\,2$ respectively. Hence, the two collections of elements thus obtained can be considered as arrays again, of sizes $d\,\mathrm{div}\,2 + d\,\mathrm{mod}\,2$ and $d\,\mathrm{div}\,2$.

So, an array of size $d{+}1$ can be represented by a triple consisting of an element and two arrays of sizes $d\,\mathrm{div}\,2 + d\,\mathrm{mod}\,2$ and $d\,\mathrm{div}\,2$. By applying the same trick to these two subarrays, we obtain a recursive data-type the elements of which we call *Braun trees*. We represent them by tuples, using the empty tuple $\langle\rangle$ to represent the empty array and using the triple $\langle a, s, t\rangle$ to represent the array consisting of element a and subarrays s and t.

We devote the remainder of this section to a formal definition of Braun trees and of how they represent flexible arrays. Trees are defined recursively by:

$\langle\rangle$ is a tree , and
$\langle a, s, t\rangle$ is a tree , for element a and tree s, t .

The fact that the two subtrees of $\langle a, s, t\rangle$ represent arrays of almost equal sizes gives rise to trees that are *balanced*, which can be formalised as follows. We introduce a predicate *bal* on the set of trees and we define the Braun[0] trees as those trees that satisfy *bal*. This predicate is defined in terms of the sizes of the subtrees; therefore, we denote the size of tree s by $\#s$:

$$\begin{aligned} bal\cdot\langle\rangle &\equiv \mathsf{true} \\ bal\cdot\langle a,s,t\rangle &\equiv \#t \le \#s \ \wedge\ \#s \le \#t{+}1 \ \wedge\ bal\cdot s \ \wedge\ bal\cdot t \\ \#\langle\rangle &= 0 \\ \#\langle a,s,t\rangle &= 1 + \#s + \#t \end{aligned}$$

Throughout the rest of this paper we only consider Braun trees and we use the term "tree" for "Braun tree".

We now define how trees represent arrays, by defining how the size and the elements of the array depend on the tree representing it. First, we have:

the size of the array represented by tree s equals $\#s$.

Second, we denote element i of the array represented by tree s by $s!i$; for element a and tree s, t, the elements of the array represented by $\langle a, s, t\rangle$ are defined as follows:

$$\begin{aligned} \langle a,s,t\rangle\,!0 &= a \\ \langle a,s,t\rangle\,!(2*i{+}1) &= s!i \ , \quad \text{for } i : 0 \le i < \#s \\ \langle a,s,t\rangle\,!(2*i{+}2) &= t!i \ , \quad \text{for } i : 0 \le i < \#t \end{aligned}$$

Notice that the domain of the array thus represented is $[0, d{+}1)$, where $d = \#s{+}\#t$; the tripartitioning $\{0\} \cup \{i : 0 \le i < \#s : 2*i{+}1\} \cup \{i : 0 \le i < \#t : 2*i{+}2\}$ exactly characterizes this domain. That trees are balanced plays a crucial role here, as is reflected by the following property.

Property 0:

$$bal\cdot\langle a,s,t\rangle \ \wedge\ \#\langle a,s,t\rangle = d{+}1 \ \Rightarrow\ \#s = d\,\mathsf{div}2 + d\,\mathsf{mod}2 \ \wedge\ \#t = d\,\mathsf{div}2$$

□

The size of a tree equals the size of the array represented by it; computing the size of the array in this way requires a linear amount of time, though. It is possible and sufficient to record the size of the whole array separately. As a consequence of Property 0, it is not necessary to record the size of *each* subtree together with that subtree.

As a consequence of its balance, the *height* of a tree is proportional to the logarithm of its size. This is the reason why all operations on trees require an amount of time that is at most logarithmic in the size of the tree.

[0] We use the term "Braun tree" because "balanced" is too general a notion here: Braun trees are trees that are so neatly balanced that they admit the operation of element selection, as defined in the next paragraph.

1.2 Element Selection and Replacement

In the previous section we have defined element selection. This definition is recursive and it can be considered as a (functional) program right away. We repeat it here:

$$\begin{array}{lll} \langle a,s,t\rangle\,!\,0 & = & a \\ \langle a,s,t\rangle\,!\,(2{*}i{+}1) & = & s!i \quad , \quad \text{for } i: 0 \le i < \#s \\ \langle a,s,t\rangle\,!\,(2{*}i{+}2) & = & t!i \quad , \quad \text{for } i: 0 \le i < \#t \end{array}$$

Element replacement is so similar to element selection that there is hardly anything to derive; the difference lies only in the value produced:

$$\begin{array}{lll} \langle a,s,t\rangle : 0, b & = & \langle b,s,t\rangle \\ \langle a,s,t\rangle : (2{*}i{+}1), b & = & \langle\, a\,,\,(s{:}i,b)\,,\,t\,\rangle \quad , \quad \text{for } i: 0 \le i < \#s \\ \langle a,s,t\rangle : (2{*}i{+}2), b & = & \langle\, a\,,\,s\,,\,(t{:}i,b)\,\rangle \quad , \quad \text{for } i: 0 \le i < \#t \end{array}$$

1.3 Intermezzo on Bag Insertions

Arrays as well as trees can be considered as (representations of) *bags* of elements. In terms of bags, the two array extension operations, *le* and *he*, both boil down to insertion of an element into a bag. To separate our concerns, we first investigate bag insertion in isolation.

We denote the bag represented by tree s by $[\![s]\!]$; a recursive definition for function $[\![\cdot]\!]$ is —where $\{\}$ denotes the *empty bag* and $+$ denotes *bag summation*—:

$$\begin{array}{lll} [\![\langle\rangle]\!] & = & \{\} \\ [\![\langle a,s,t\rangle]\!] & = & \{a\} + [\![s]\!] + [\![t]\!] \end{array}$$

We now derive definitions for a function *ins*, where for element b and tree s the tree $ins{\cdot}b{\cdot}s$ is a solution of the equation[1] (with unknown u):

$$u \;:\; [\![u]\!] = \{b\} + [\![s]\!] \;\wedge\; bal{\cdot}u \quad .$$

We use the first conjunct of this equation to guide our derivation, whereas the second conjunct remains as an a posteriori proof obligation. By induction over the size of the trees we derive, starting with the case that s is the empty tree:

$$\begin{array}{ll} & [\![u]\!] = \{b\} + [\![\langle\rangle]\!] \\ \equiv & \{\ \{\}\text{ is the identity of } +\text{; definition of } [\![\cdot]\!]\ \} \\ & [\![u]\!] = \{b\} + [\![\langle\rangle]\!] + [\![\langle\rangle]\!] \\ \equiv & \{\ \text{definition of } [\![\cdot]\!]\ \} \\ & [\![u]\!] = [\![\langle b,\langle\rangle,\langle\rangle\rangle]\!] \\ \Leftarrow & \{\ \text{Leibniz}\ \} \\ & u = \langle b,\langle\rangle,\langle\rangle\rangle \quad . \end{array}$$

[1] The word "equation" is used here for arbitrary predicates, not just equalities.

The first step of this derivation may look like a rabbit, but it is not: the only way to solve an equation of the form $u: [\![u]\!] = E$ is to transform E into an expression of the form $[\![F]\!]$ and then to apply "Leibniz", as we did in the last step. In view of the definition of $[\![\cdot]\!]$ and the occurrence of $\{b\}$, the only thing we can do is to work towards a formula of the shape $[\![\langle b, ?, ?\rangle]\!]$.

Notice that in the above derivation in all steps except the last one, only the right-hand side of the equality is manipulated. In order to avoid the continued rewriting of the constant left-hand side, we shall carry out the calculation with the right-hand expression only. That is, in order to solve an equation of the form $u: f\cdot u = E$, we transform E into an equivalent expression $f\cdot F$ in isolation, after which we conclude that F is a solution of the equation.

For the composite tree $\langle a, s, t\rangle$ we derive, in the same "goal-driven" way:

$$\{b\} + [\![\langle a, s, t\rangle]\!]$$
$$= \quad \{ \text{ definition of } [\![\cdot]\!] \ \}$$
$$\{b\} + \{a\} + [\![s]\!] + [\![t]\!]$$
$$= \quad \{ \text{ specification of } ins, \text{ by induction hypothesis (see below) } \}$$
$$\{b\} + [\![ins\cdot a\cdot s]\!] + [\![t]\!]$$
$$= \quad \{ \text{ definition of } [\![\cdot]\!] \ \}$$
$$[\![\langle b\, ,\, ins\cdot a\cdot s\, ,\, t\rangle]\!] \ .$$

The second step of this derivation represents a choice out of many possibilities; because bag summation is symmetric and associative, we have 8 possibilities here: a and b may be interchanged, s and t may be interchanged, and the recursive application of ins may be taken as the "right" or as the "left" subtree in the resulting tree. Thus, we obtain 8 different definitions for a function ins satisfying the first conjunct of the above specification.

Regarding the remaining proof obligation we observe that, by the definition of bal, $bal\cdot\langle b, \langle\rangle, \langle\rangle\rangle$ holds. For the 8 alternatives a simple calculation reveals that, generally, $bal\cdot\langle a, s, t\rangle \Rightarrow bal\cdot u$ only holds when s and t satisfy an additional precondition, as follows —notice that we need write down 4 cases only, since the relative positions of a and b are irrelevant—:

$$\begin{array}{llll}
bal\cdot\langle a,s,t\rangle & \Rightarrow bal\cdot\langle b\, ,\, ins\cdot a\cdot s\, ,\, t\rangle & , \text{ if } & \#s = \#t \\
bal\cdot\langle a,s,t\rangle & \Rightarrow bal\cdot\langle b\, ,\, ins\cdot a\cdot t\, ,\, s\rangle & , \text{ if } & \mathsf{true} \\
bal\cdot\langle a,s,t\rangle & \Rightarrow bal\cdot\langle b\, ,\, s\, ,\, ins\cdot a\cdot t\rangle & , \text{ if } & \#s = \#t+1 \\
bal\cdot\langle a,s,t\rangle & \Rightarrow bal\cdot\langle b\, ,\, t\, ,\, ins\cdot a\cdot s\rangle & , \text{ if } & \mathsf{false}
\end{array}$$

Apparently, the last alternative is never useful; thus, only 6 out of the 8 alternatives can be used when the trees are to remain balanced. As for the sizes of the trees, ins has the following property.

Property 1: for all b and s we have $\#(ins\cdot b\cdot s) \ = \ \#s+1$

□

The definitions we shall derive for the array extension operations turn out to correspond to some of the recursive schemes discussed here. That is, these operations are refinements of the above functions *ins*. As a consequence, we have done away with the proof obligations regarding the size and the balance of the trees.

1.4 Low Extension and Removal

We recall the specification of functon *le*, but now reformulated in terms of trees. For element b and tree s, the value $le{\cdot}b{\cdot}s$ is the tree u satisfying:

$$\#u = \#s{+}1 \;\wedge\; u!0 = b \;\wedge\; (\forall j : 0 \leq j < \#s : u!(j{+}1) = s!j) \quad .$$

According to the analysis in the previous section, we have only one option for $le{\cdot}b{\cdot}\langle\rangle$; fortunately, it satisfies all requirements. So, we define:

$$le{\cdot}b{\cdot}\langle\rangle \;=\; \langle b, \langle\rangle, \langle\rangle\rangle \quad .$$

For the composite tree $\langle a,s,t\rangle$ of size $d{+}1$, the value $le{\cdot}b{\cdot}\langle a,s,t\rangle$ is the tree u satisfying:

$$\#u = d{+}2 \;\wedge\; u!0 = b \;\wedge\; (\forall j : 0 \leq j < d{+}1 : u!(j{+}1) = \langle a,s,t\rangle!j) \quad .$$

The term $\langle a,s,t\rangle!j$ and the definition of element selection suggest a 3-way case analysis in the range $0 \leq j < d{+}1$; so, using the definition of !, we rewrite this as:

$$\begin{array}{l} \#u = d{+}2 \;\wedge\; u!0 = b \;\wedge\; u!1 = a \;\wedge \\ (\forall i : 0 \leq i < \#s : u!(2{*}i{+}2) = s!i) \;\wedge\; (\forall i : 0 \leq i < \#t : u!(2{*}i{+}3) = t!i) \quad . \end{array}$$

This provides an explicit definition of the elements of u in terms of a, b, s, t. In view, again, of the definition of !, and observing that u will be a composite tree, we are forced to distinguish between $u!0$, $u!(2{*}i{+}1)$, and $u!(2{*}i{+}2)$. By comparing this with the above requirement we conclude that we must choose $u = \langle b,v,s\rangle$, where v is the tree satisfying:

$$\#v = \#t{+}1 \;\wedge\; v!0 = a \;\wedge\; (\forall i : 0 \leq i < \#t : v!(i{+}1) = t!i) \quad .$$

This specification of v is precisely the specification of the tree $le{\cdot}a{\cdot}t$; because t is smaller than $\langle a,s,t\rangle$ we may use this recursive application of *le*. Thus we obtain the following definition for *le*:

$$\begin{array}{lcl} le{\cdot}b{\cdot}\langle\rangle & = & \langle b, \langle\rangle, \langle\rangle\rangle \\ le{\cdot}b{\cdot}\langle a,s,t\rangle & = & \langle\, b\,,\, le{\cdot}a{\cdot}t\,,\, s\,\rangle \end{array}$$

This definition corresponds to one of the alternatives for bag insertion discussed in the previous section; as already stated there, this definition also satisfies the requirements regarding the size and the balance of the resulting tree.

Finally, we recall the specification of the function *lr*, reformulated in terms of trees. For composite tree s the value $lr{\cdot}s$ is the tree u that satisfies:

$$\#u = \#s - 1 \;\wedge\; (\forall j : 0 \leq j < \#u : u!j = s!(j{+}1)) \quad .$$

Using that $lr\cdot(le\cdot b\cdot s) = s$, we obtain from the above definition for le the following definition for lr, by means of "program inversion". The verification that this definition indeed satisfies the specification requires a calculation very much like the above derivation for le.

$$\begin{array}{lll} lr\cdot\langle a,\langle\rangle,\langle\rangle\rangle & = & \langle\rangle \\ lr\cdot\langle a,s,t\rangle & = & \langle\, s!0\,,\, t\,,\, lr\cdot s\,\rangle \quad , \text{ for } s : s\neq\langle\rangle \end{array}$$

1.5 High Extension and Removal

We recall the specification of function he, now reformulated in terms of trees. For element b and tree s, the value $he\cdot b\cdot s$ is the tree u satisfying:

$$\#u = \#s{+}1 \quad\wedge\quad u!(\#s) = b \quad\wedge\quad (\forall j : 0\leq j<\#s : u!j = s!j) \quad .$$

As in the previous section, the only possible definition for $he\cdot b\cdot\langle\rangle$ is:

$$he\cdot b\cdot\langle\rangle \;=\; \langle b,\langle\rangle,\langle\rangle\rangle \quad .$$

For the composite tree $\langle a,s,t\rangle$ we observe —calculation omitted— that the first and the third conjuncts of the above specification can be met both by:

$$\begin{array}{lll} he\cdot b\cdot\langle a,s,t\rangle & = & \langle\, a\,,\, he\cdot b\cdot s\,,\, t\,\rangle \quad , \text{ and by:} \\ he\cdot b\cdot\langle a,s,t\rangle & = & \langle\, a\,,\, s\,,\, he\cdot b\cdot t\,\rangle \quad . \end{array}$$

To investigate which one we need, we try to prove the second conjunct of the specification, where we assume $\#\langle a,s,t\rangle = d{+}1$; recall that then $\#s = d\,\mathsf{div}\,2{+}d\,\mathsf{mod}\,2$ and $\#t = d\,\mathsf{div}\,2$. For the first alternative to satisfy the second conjunct, we need:

$$\begin{array}{ll} & \langle\, a\,,\, he\cdot b\cdot s\,,\, t\,\rangle\,!\,(d{+}1) \\ = & \quad \{\text{ assume } d \text{ to be even, set } d{=}2{*}e\,;\text{ definition of } !\ \} \\ & he\cdot b\cdot s\,!\,e \\ = & \quad \{\ d{=}2{*}e\,,\text{ so } \#s{=}e\,;\text{ specification of } he,\text{ by ind. hyp.}\ \} \\ & b\quad . \end{array}$$

So, the expression $\langle\, a\,,\, he\cdot b\cdot s\,,\, t\,\rangle$ satisfies the specification if d is even, provided that we also have $bal\cdot\langle\, a\,,\, he\cdot b\cdot s\,,\, t\,\rangle$; the condition for this is $\#s{=}\#t$, which is equivalent to d being even.

Similarly, we can derive that the other expression, $\langle\, a\,,\, s\,,\, he\cdot b\cdot t\,\rangle$, satisfies the specification if d is odd; in that case we have $\#s{=}\#t{+}1$, which is exactly the condition for $bal\cdot\langle\, a\,,\, s\,,\, he\cdot b\cdot t\,\rangle$.

Thus, we obtain the following definition for he:

$$\begin{array}{lll} he\cdot b\cdot\langle\rangle & = & \langle b,\langle\rangle,\langle\rangle\rangle \\ he\cdot b\cdot\langle a,s,t\rangle & = & \mathsf{if}\ d\,\mathsf{mod}2 = 0 \;\rightarrow\; \langle\, a\,,\, he\cdot b\cdot s\,,\, t\,\rangle \\ & & []\ d\,\mathsf{mod}2 \neq 0 \;\rightarrow\; \langle\, a\,,\, s\,,\, he\cdot b\cdot t\,\rangle \\ & & \mathsf{fi}\ \ \mathsf{where}\ d = \#\langle a,s,t\rangle - 1\ \ \mathsf{end} \end{array}$$

To allow for logarithmic computation times, the value $\#\langle a,s,t\rangle$ occurring in this definition must not be computed but must be supplied as an additional parameter instead. By Property 0 this is possible.

Finally, we recall the specification of the function hr, reformulated in terms of trees. For composite tree s the value $hr{\cdot}s$ is the tree u that satisfies:

$$\#u = \#s-1 \ \wedge\ (\forall j : 0 \le j < \#u : u!j = s!j) \ .$$

Using that $hr{\cdot}(he{\cdot}b{\cdot}s) = s$, we can derive the following definition for hr from the above one for he, again by program inversion:

$$\begin{array}{lcl} hr{\cdot}\langle a,\langle\rangle,\langle\rangle\rangle & = & \langle\rangle \\ hr{\cdot}\langle a,s,t\rangle & = & \textsf{if } d\textsf{mod}2 = 0 \ \rightarrow\ \langle a\,,\, hr{\cdot}s\,,\, t\rangle \\ & & [\!]\ d\textsf{mod}2 \neq 0 \ \rightarrow\ \langle a\,,\, s\,,\, hr{\cdot}t\rangle \\ & & \textsf{fi where } d = \#\langle a,s,t\rangle - 2 \textsf{ end} \quad , \text{ for } s : s \neq \langle\rangle \end{array}$$

1.6 Summary of the Functional Programs

$$\begin{array}{lcll} \langle a,s,t\rangle\,!\,0 & = & a & \\ \langle a,s,t\rangle\,!\,(2{*}i{+}1) & = & s!i \ , & \text{for } i : 0 \le i < \#s \\ \langle a,s,t\rangle\,!\,(2{*}i{+}2) & = & t!i \ , & \text{for } i : 0 \le i < \#t \end{array}$$

$$\begin{array}{lcll} \langle a,s,t\rangle : 0, b & = & \langle b,s,t\rangle & \\ \langle a,s,t\rangle : (2{*}i{+}1), b & = & \langle a\,,\,(s{:}i,b)\,,\,t\rangle \ , & \text{for } i : 0 \le i < \#s \\ \langle a,s,t\rangle : (2{*}i{+}2), b & = & \langle a\,,\,s\,,\,(t{:}i,b)\rangle \ , & \text{for } i : 0 \le i < \#t \end{array}$$

$$\begin{array}{lcl} le{\cdot}b{\cdot}\langle\rangle & = & \langle b,\langle\rangle,\langle\rangle\rangle \\ le{\cdot}b{\cdot}\langle a,s,t\rangle & = & \langle b\,,\, le{\cdot}a{\cdot}t\,,\, s\rangle \end{array}$$

$$\begin{array}{lcl} lr{\cdot}\langle a,\langle\rangle,\langle\rangle\rangle & = & \langle\rangle \\ lr{\cdot}\langle a,s,t\rangle & = & \langle s!0\,,\, t\,,\, lr{\cdot}s\rangle \quad , \text{ for } s : s \neq \langle\rangle \end{array}$$

$$\begin{array}{lcl} he{\cdot}b{\cdot}\langle\rangle & = & \langle b,\langle\rangle,\langle\rangle\rangle \\ he{\cdot}b{\cdot}\langle a,s,t\rangle & = & \textsf{if } d\textsf{mod}2 = 0 \ \rightarrow\ \langle a\,,\, he{\cdot}b{\cdot}s\,,\, t\rangle \\ & & [\!]\ d\textsf{mod}2 \neq 0 \ \rightarrow\ \langle a\,,\, s\,,\, he{\cdot}b{\cdot}t\rangle \\ & & \textsf{fi where } d = \#\langle a,s,t\rangle - 1 \textsf{ end} \end{array}$$

$$\begin{array}{lcl} hr{\cdot}\langle a,\langle\rangle,\langle\rangle\rangle & = & \langle\rangle \\ hr{\cdot}\langle a,s,t\rangle & = & \textsf{if } d\textsf{mod}2 = 0 \ \rightarrow\ \langle a\,,\, hr{\cdot}s\,,\, t\rangle \\ & & [\!]\ d\textsf{mod}2 \neq 0 \ \rightarrow\ \langle a\,,\, s\,,\, hr{\cdot}t\rangle \\ & & \textsf{fi where } d = \#\langle a,s,t\rangle - 2 \textsf{ end} \quad , \text{ for } s : s \neq \langle\rangle \end{array}$$

2 Sequential Implementations

In this section we derive sequential implementations from the recursive function definitions in the previous section. We represent trees by data structures built from smaller units called ***nodes***, which are linked together by means of ***pointers***. For this purpose we use a program notation that is a mixture of guarded commands and Pascal.

Before doing so, however, we explain the techniques used for this transformation by means of a simple example. Readers who are sufficiently familiar with techniques for pointer manipulation and recursion elimination may wish to skip the next subsection at first reading.

2.0 A Few Simple Transformation Techniques

This subsection consists of two parts. First, we present two (well-known) instances of how tail-recursive definitions can be transformed into equivalent non-recursive programs. Second, we illustrate how a simple function on lists, namely list catenation, can be implemented as a sequential program.

*   *   *

We consider the following tail-recursive definition of a function F:

$$
\begin{array}{lllll}
F\cdot x & = & \mathsf{if} & \neg b\cdot x & \rightarrow f\cdot x \\
 & & [] & b\cdot x & \rightarrow F\cdot(g\cdot x) \\
 & & \mathsf{fi} & &
\end{array}
$$

¿From this definition we obtain the following iterative, sequential program for the computation of, say, $F\cdot X$. The correctness of this program follows from the invariance of $F\cdot X = F\cdot x$:

$$
\begin{array}{l}
\quad x := X \\
;\ \mathsf{do}\ b\cdot x \rightarrow x := g\cdot x\ \mathsf{od} \\
;\ r := f\cdot x \\
\{\ r = F\cdot X\ \}
\end{array}
$$

Next, we consider the following tail-recursive procedure P, in which x is assumed to be a value parameter:

$$
\begin{array}{l}
\mathsf{procedure}\ P(x) \\
= |[\ \mathsf{if}\ \neg b\cdot x \rightarrow S_0 \\
\quad\ \ []\ \ b\cdot x \rightarrow S_1\ ;\ P(g\cdot x) \\
\quad\ \ \mathsf{fi} \\
\quad]|
\end{array}
$$

This procedure can be transformed into the following equivalent non-recursive one:

$$
\begin{array}{l}
\mathsf{procedure}\ P(x) \\
= |[\ \mathsf{do}\ b\cdot x \rightarrow S_1\ ;\ x := g\cdot x\ \mathsf{od} \\
\quad\ ;\ S_0 \\
\quad]|
\end{array}
$$

If we wish to get rid of the procedure altogether, each call $P(X)$ may be replaced by the following program fragment; we call this ***unfolding*** $P(X)$:

$$
\begin{array}{l}
|[\ \mathsf{var}\ x\,; \\
\quad x := X \\
\ \ ;\ \mathsf{do}\ b{\cdot}x\ \rightarrow\ S_1\ ;\ x := g{\cdot}x\ \ \mathsf{od} \\
\ \ ;\ S_0 \\
]|
\end{array}
$$

* * *

We consider the function f that maps two lists onto their catenation, defined recursively as follows:

$$
\begin{array}{rl}
f{\cdot}x{\cdot}y\ =\ \mathsf{if} & x = []\ \rightarrow\ y \\
[] & x \neq []\ \rightarrow\ hd{\cdot}x\ \mathsf{cons}\ f{\cdot}(tl{\cdot}x){\cdot}y \\
\mathsf{fi} &
\end{array}
$$

The first step towards a sequential implementation is to recode this definition as a procedure definition; the reason for using a result parameter for the function result will become clear later:

$$
\begin{array}{l}
\mathsf{procedure}\ P_0(x, y;\ \mathsf{result}\ z) \\
=\ |[\ \{\ \mathsf{post:}\ \ z\ =\ f{\cdot}x{\cdot}y\ \} \\
\qquad \mathsf{var}\ h\,; \\
\qquad \mathsf{if}\ \ x = []\ \ \rightarrow z := y \\
\qquad []\ \ x \neq []\ \ \rightarrow P_0(tl{\cdot}x\ ,\ y\ ,\ h) \\
\qquad\qquad\qquad\quad ;\ z\ :=\ hd{\cdot}x\ \mathsf{cons}\ h \\
\qquad \mathsf{fi} \\
\quad]|
\end{array}
$$

In the second step we introduce the representation of lists by means of ***nodes*** and ***pointers***. A list is represented by a pointer; a pointer is either nil, or a reference to a node —Pascal: a record— consisting of an element and a pointer. The value nil has as its only property that it differs from all pointer values referring to nodes. By means of a call of the standard procedure $new(p)$, pointer variable p is assigned a value that differs from nil and that differs from all pointer values "currently in use": we call such a pointer value, and the node referred to by it, ***fresh***.

The type definitions needed to define this data structure formally are:

$$
\begin{array}{rl}
\mathsf{type}\ listp & =\ \mathsf{pointer\ to}\ node\,; \\
node & =\ \langle\, hd : element\ ;\ tl : listp\,\rangle
\end{array}
$$

For p a pointer of type $listp$ and $p \neq \mathsf{nil}$, we use $p\uparrow$ to denote the node to which p refers. We denote the two components of this node by $p\uparrow{\cdot}hd$ and $p\uparrow{\cdot}tl$; that is, we use the Pascal convention of ***field selectors*** to identify the components of a pair. As in Pascal, we admit and shall use assignments to individual components of nodes.

A pointer p represents a list $[\![p]\!]$, say, as follows:

$$
\begin{array}{lcl}
[\![\mathsf{nil}]\!] & = & [] \\
[\![p]\!] & = & p\uparrow{\cdot}hd\ \mathsf{cons}\ [\![p\uparrow{\cdot}tl]\!]\quad ,\ \ \mathsf{for}\ p : p \neq \mathsf{nil}
\end{array}
$$

(The use of nil to represent the (one and only) empty list is somewhat opportunistic, but it simplifies things a little.)

By incorporation of this list representation we obtain from the above procedure P_0 our next version; notice that $[\![p]\!] = [\,] \;\equiv\; p = \mathsf{nil}$:

$$
\begin{array}{l}
\mathsf{procedure}\ P_1(p, q;\ \mathsf{result}\ r) \\
= |[\ \{\ \mathsf{post}:\ [\![r]\!] = f\cdot[\![p]\!]\cdot[\![q]\!]\ \} \\
\quad \mathsf{var}\ h; \\
\quad \mathsf{if}\ p = \mathsf{nil} \rightarrow r := q \\
\quad [\,]\ p \neq \mathsf{nil} \rightarrow P_1(p\uparrow\cdot tl,\ q,\ h) \\
\qquad\qquad\qquad ;\ new(r) \\
\qquad\qquad\qquad ;\ r\uparrow := \langle p\uparrow\cdot hd,\ h\rangle \\
\quad \mathsf{fi} \\
]|
\end{array}
$$

Procedure P_1 entails an important design decision. In this procedure the only assignment to values of type *node* is the assigment to $r\uparrow$, which is a fresh node. As a result, existing nodes are not modified. Generally, *if the value of a node*, once it has been created and initialised, *is never changed, then this node may be freely shared* among different data structures. If such a node contains pointers to other nodes, then the "never change" condition must also hold for all nodes that are reachable from this node. By building data structures in this way, the use of sharing saves both storage space and computation time. For example, a *list assignment* can now be implemented by copying a pointer only, which is an $\mathcal{O}(1)$ operation, instead of by copying the whole list. As a result, the two lists are represented in storage by the same data structure as long as they remain equal. In procedure P_1, for instance, we have used the pointer assignment $r := q$ to implement the list assignment $z := y$ from procedure P_0. The price to be paid for this flexibility is that efficient storage management involves some form of garbage collection.

Aside: As a matter of fact, sharing is possible when, for every pointer p representing a list, the value $[\![p]\!]$ is never changed. This requirement is weaker than the requirement that $p\uparrow$ is never changed. That is, nodes may be *overwritten* as long they still represent the same abstract values —in our case: lists—. For example, node overwriting is frequently used in graph-reduction machines. In this paper, we can live with the stricter regime.

□

In the third step we employ the possibility to use assignments to individual components of nodes, to rearrange the order of the assignments in such a way that the procedure becomes tail-recursive. Thus, we obtain:

$$
\begin{array}{l}
\mathsf{procedure}\ P_2(p, q;\ \mathsf{result}\ r) \\
= |[\ \{\ \mathsf{post}:\ [\![r]\!] = f\cdot[\![p]\!]\cdot[\![q]\!]\ \} \\
\quad \mathsf{if}\ p = \mathsf{nil} \rightarrow r := q \\
\quad [\,]\ p \neq \mathsf{nil} \rightarrow new(r) \\
\qquad\qquad\qquad ;\ r\uparrow\cdot hd := p\uparrow\cdot hd \\
\qquad\qquad\qquad ;\ P_2(p\uparrow\cdot tl,\ q,\ r\uparrow\cdot tl) \\
\quad \mathsf{fi} \\
]|
\end{array}
$$

Procedure P_2 is tail-recursive but it still contains a result parameter; the transformation into iterative form discussed in the beginning of this subsection is only applicable to procedures with value parameters only. Result parameter r may be turned into a value parameter, provided that we remove all assignments to r from the procedure. (Notice that assignments to $r\uparrow$ are not assignments to r.) The assignment $r := q$ can be removed by the introduction of an additional procedure P_3 that is identical to P_2 but with its precondition strengthened with $p \neq \text{nil}$. This requires, of course, that the case $p = \text{nil}$ be dealt with separately. Similarly, the assignment $new(r)$ can be eliminated by strengthening the precondition of P_3 even further, namely with "r is fresh" . Thus, we obtain:

```
procedure P2(p, q; result r)
= |[ { post: [[r]] = f·[[p]]·[[q]] }
     if p = nil  → r := q
     [] p ≠ nil  → new(r)
                   { p ≠ nil ∧ "r is fresh" }
                   ; P3(p, q, r)
     fi
  ]|

procedure P3(p, q, r)
= |[ {  pre: p ≠ nil ∧ "r is fresh" }
     { post: [[r]] = f·[[p]]·[[q]] }
     r↑·hd := p↑·hd
   ; if p↑·tl = nil  → r↑·tl := q
     [] p↑·tl ≠ nil  → new(r↑·tl)
                      ; P3(p↑·tl, q , r↑·tl)
     fi
  ]|
```

Finally, if we now distribute the assignment $r\uparrow\cdot hd := p\uparrow\cdot hd$ over the alternatives of the succeeding selection statement, procedure P_3 can be transformed into iterative form; by unfolding its one and only call, P_3 can be eliminated altogether. This yields our final, iterative implementation of list catenation:

```
procedure P4(p, q; result r)
= |[ { post: [[r]] = f·[[p]]·[[q]] }
     var h;
     if p = nil  → r := q
     [] p ≠ nil  → new(r) ; h := r
                   { invariant: p ≠ nil ∧ "h is fresh" }
                   ; do p↑·tl ≠ nil  → h↑·hd := p↑·hd
                                     ; new(h↑·tl)
                                     ; p, h := p↑·tl, h↑·tl
                     od
                   ; h↑ := ⟨p↑·hd, q⟩
     fi
  ]|
```

2.1 Selection and Replacement

We represent trees by means of data structures composed from nodes and pointers. A tree is represented by a pointer; a pointer is either nil or a pointer to a node consisting of an element and 2 pointers. The value nil has as its only property that it differs from all pointer values referring to nodes. By means of a call of the standard procedure $new(p)$, pointer variable p is assigned a value that differs from nil and that differs from all pointer values "currently in use": we call such a pointer value, and the node referred to by it, *fresh*.

Assignments to nodes will be restricted to fresh nodes; as a result, the values of existing nodes will never be changed and we may employ ***node sharing***. (See the previous subsection, for a slightly more elaborate discussion of node sharing.)

The type definitions needed to define the data structure formally are:

$$\begin{array}{lll} \textsf{type} & treep & = \textsf{pointer to } node\,; \\ & node & = \langle\, a : element\ ;\ s,t : listp \rangle \end{array}$$

For p a pointer of type *treep* and $p \neq \textsf{nil}$, we use $p\uparrow$ to denote the node to which p refers. We denote the three components of this node by $p\uparrow\cdot a$, $p\uparrow\cdot s$, and $p\uparrow\cdot t$. We admit and shall use assignments to individual components of nodes.

A pointer p represents a tree $[\![p]\!]$, say, as follows:

$$\begin{array}{lll} [\![\textsf{nil}]\!] & = & \langle\rangle \\ [\![p]\!] & = & \langle p\uparrow\cdot a\,,\, [\![p\uparrow\cdot s]\!],\, [\![p\uparrow\cdot t]\!]\rangle \quad , \quad \text{for } p : p \neq \textsf{nil} \end{array}$$

The definition of element selection is tail-recursive. Such a definition can be transformed into an iterative program in a straightforward way. Next, the tree operations in terms of tuples are recoded in terms of node and pointer operations. Thus, we obtain the following program; to make verification of this transformation somewhat easier, we repeat the definition of element selection here with some of the syntactic sugar —the parameter patterns— removed:

```
u!i = if  i=0  → u·a
      [] i>0  → if  jmod2=0  → u·s!(jdiv2)
                 [] jmod2≠0  → u·t!(jdiv2)
                 fi where j=i−1 end
      fi

procedure EL(p,i; result e)
= |[ {  pre:  0≤i<#[[p]] }
     { post:  e = [[p]]!i (initial values of p,i) }
     do i≠0  → i:=i−1
               ; if  imod2=0  →  p,i := p↑·s , idiv2
                 [] imod2=1  →  p,i := p↑·t , idiv2
                 fi
     od
     ; e:=p↑·a
  ]|
```

For the implementation of element replacement, we introduce a procedure *REP* with the following specification:

```
procedure REP(p, i, b; result r)
= |[ { pre: 0≤i<#[[p]] }
     { post: [[r]] = [[p]] : i, b }
  ]|
```

As a first approximation, we construct the following (recursive) code from the (recursive) definiton of element replacement:

```
procedure REP(p, i, b; result r)
= |[ var h;
     if i=0 → new(r) ; r↑ := ⟨b, p↑·s, p↑·t⟩
     [] i≠0 → i := i−1
              ; if imod2=0 → REP(p↑·s, idiv2, b, h)
                             ; new(r) ; r↑ := ⟨p↑·a, h, p↑·t⟩
                [] imod2=1 → REP(p↑·t, idiv2, b, h)
                             ; new(r) ; r↑ := ⟨p↑·a, p↑·s, h⟩
                fi
     fi
  ]|
```

By means of the techniques from the previous section, this procedure can be transformed into the following iterative one:

```
procedure REP1(p, i, b; result r)
= |[ var h;
     new(r) ; h := r
   ; do i≠0 → i := i−1
              ; if imod2=0 → h↑·a := p↑·a ; h↑·t := p↑·t
                             ; new(h↑·s) ; p, i, h := p↑·s, idiv2, h↑·s
                [] imod2=1 → h↑·a := p↑·a ; h↑·s := p↑·s
                             ; new(h↑·t) ; p, i, h := p↑·t, idiv2, h↑·t
                fi
     od
   ; h↑ := ⟨b, p↑·s, p↑·t⟩
  ]|
```

Procedure REP_1 constructs a new tree that shares the majority of its nodes with the old tree. As a result, this operation requires at most logarithmic computation time.

2.2 Extension and Removal

The techniques used in the previous subsections can be used to derive iterative implementations of the extension and removal operations as well. Although these cases are slightly more complicated, this transformation offers no further surprises;

therefore, we only present the resulting programs. As is the case with element replacement, the following procedures construct, in logarithmic time, new trees that share the majority of their nodes with the old ones.

```
procedure LE(p, b; result r)
= |[ var h ;
     new(r) ; h := r
   ; do p≠nil → h↑·a := b ; h↑·t := p↑·s
              ; new(h↑·s) ; p, b, h := p↑·t , p↑·a , h↑·s
     od
   ; h↑ := ⟨b, nil, nil⟩
  ]|

procedure HE(p, d, b; result r)
= |[ { pre: d = #[[p]] }
     var h ;
     new(r) ; h := r
   ; do d≠0 → d := d−1
            ; if dmod2=0 → h↑·a := p↑·a ; h↑·t := p↑·t
                         ; new(h↑·s) ; p, d, h := p↑·s , ddiv2 , h↑·s
              [] dmod2=1 → h↑·a := p↑·a ; h↑·s := p↑·s
                         ; new(h↑·t) ; p, d, h := p↑·t , ddiv2 , h↑·t
              fi
     od
   ; h↑ := ⟨b, nil, nil⟩
  ]|

procedure HR(p, d; result r)
= |[ { pre: d = #[[p]] }
     var h ;
     if d=1 → r := nil
     [] d>1 → new(r) ; d, h := d−2 , r
            ; do d≥2 ∧ dmod2=0 → h↑·a := p↑·a ; h↑·t := p↑·t
                               ; new(h↑·s)
                               ; p, d, h := p↑·s , ddiv2 − 1 , h↑·s
              [] d≥2 ∧ dmod2=1 → h↑·a := p↑·a ; h↑·s := p↑·s
                               ; new(h↑·t)
                               ; p, d, h := p↑·t , ddiv2 − 1 , h↑·t
              od
            ; if d=0 → h↑ := ⟨p↑·a , nil , nil⟩
              [] d=1 → h↑ := ⟨p↑·a , p↑·s , nil⟩
              fi
     fi
  ]|
```

```
procedure LR(p; result r)
= |[ var h;
     if p↑·s = nil  →  r := nil
     [] p↑·s ≠ nil  →  new(r) ; h := r
                     ; do p↑·s↑·s ≠ nil  →  h↑·a := p↑·s↑·a ; h↑·s := p↑·t
                                          ; new(h↑·t) ; p, h := p↑·s , h↑·t
                       od
                     ; h↑ := ⟨p↑·s↑·a , p↑·t , nil⟩
     fi
  ]|
```

3 Concluding Remarks

The introduction of Braun trees to represent flexible arrays constitutes a major design decision. Once this decision has been taken, functional programs for the array operations can be derived smoothly. These functional programs are compact and provide a good starting point for the construction of a sequential, nonrecursive implementation of the array operations. Although the resulting programs are nontrivial they can be obtained by systematic transformation of the functional programs; this requires great care but little ingenuity.

The exercise in this paper also shows that recursive data types, such as Braun trees, can be implemented by means of nodes and pointers in a simple and systematic way. Notice that we have not employed an (elaborate) formal theory on the semantics of pointer operations; nevertheless, we are convinced of the correctness of the programs thus obtained. The programs derived in this paper employ sharing of nodes by adhering to the "never-change-an-existing-node" discipline; it is equally well possible to derive programs that, instead of building a new tree, modify the existing tree by modifying nodes.

Braun trees admit several variations and embellishments. To mention a few: instead of binary trees, k-ary trees, for some fixed k, can be used. Operations like mod2 and div2 then become mod$k$ and div$k$. Larger values of k give rise to trees of smaller height —when the size remains the same—. The price to be paid is that storage utilisation decreases: each node now contains k pointers per array element instead of 2. To compensate for this, it is possible to store several consecutive array elements, instead of a single one, per node of the tree. This reduces the height of the tree even further and improves storage utilisation (, except for very small arrays). Note, however, that in a setting where nodes are shared, large nodes are awkward, because of the time needed to make copies of nodes; this places an upper bound on what can be considered as a reasonable node size.

Acknowledgement

To Anne Kaldewaij and the members of the Kleine Club, for their constructive comments on an earlier version of this paper.

References

[0] Braun, W., Rem, M.: A logarithmic implementation of flexible arrays. Memorandum MR83/4, Eindhoven University of Technology (1983).

[1] Dijkstra, Edsger W.: A discipline of programming. Prentice-Hall, Englewood Cliffs (1976).

Designing Arithmetic Circuits by Refinement in Ruby

Geraint Jones[1] and Mary Sheeran[2]

[1] Programming Research Group, Oxford University Computing Laboratory, 11 Keble Road, Oxford OX1 3QD, England; electronic mail: Geraint.Jones@comlab.oxford.ac.uk.
[2] Informationsbehandling, Chalmers Tekniska Högskola, S 412 96 Göteborg, Sweden; electronic mail: ms@cs.chalmers.se.

Abstract. This paper presents in some detail the systematic derivation of a static bit-level parallel algorithm to implement multiplication of integers, that is to say one which might be implemented as an electronic circuit. The circuit is well known, but the derivation shows that its design can be seen as the consequence of decisions made (and explained) in terms of the abstract algorithm. The systematic derivation serves both as an explanation of the circuit, and as a demonstration that it is correct 'by construction'. We believe that the technique is applicable to a wide range of similar algorithms.

1 Introduction

We advocate a style of 'design by calculation' for the very fine-grained parallel algorithms that are implemented as regular arrays of electronic circuits. The design of such circuits is particularly difficult because the implementation medium imposes severe constraints on what is possible and what is reasonably efficient. In consequence the details of the final implementation have a pervasive influence on the whole design process, and it is regrettable that many such designs are presented with little or no abstract justification, and solely in the low-level terms of the final detailed implementation.

These same constraints, it seems to us, make it easier to apply the systematic design methods which are being used in the development of software algorithms. Our work is much influenced by that of Bird and Meertens [2, 12], although in contrast to their use of functions we choose to use relations as our abstract programs. This same choice has been made by the authors of [1] and there are close parallels [8] between the framework in which they work and ours.

We represent by relations both abstract specifications of the desired behaviour of circuits and the components from which circuits are built. Ruby [10] is a language of operations on relations in which the combining forms correspond naturally to the ways that components are connected in our target implementations. The algebra of these combining forms is a framework in which to refine the specification of an algorithm into an implementable form.

The notion of an abstraction relation is important, as is the composition of an abstraction with the inverse of another abstraction. This is a representation changer: a device to consume one representation of some abstraction, and return another representation. In this paper we show that a multiplication circuit can be specified naturally as such a representation changer, and that the specification can be refined into a regular array of components each of which is also a representation changer. These small representation changers are the sorts of devices which would be found in a hardware designer's library of standard components, and are the parts from which the circuit would be built. The multiplier is not new; we first presented it much more informally in 1985 [13]. Here we present its derivation as a new account of how such a circuit might be designed.

As well as explaining how a design came about, a derivation constitutes a proof that the implementation meets its specification. Usually such proofs are constructed after the circuit has been designed [15, 4], and independently of the method used to produce the design [17]. With that approach the design process offers no help in the proof, nor does the verification offer any assistance in the design. Even when circuits are synthesised mechanically from parametric designs, after-the-fact verification of the synthesis functions [5] gives no guidance on their construction.

2 Ruby

We model the behaviour of circuits by binary relations, denoted in Ruby by expressions which have a form that suggests the shapes of the corresponding circuits as well as their behaviour. In this paper we confine our attention to combinational circuits, operating on simple data values. Lifting to time sequences of values, to describe the behaviour of circuits with internal state, is also a part of the interpretation of the Ruby expression. In this way we our combinational designs can be transformed into very high throughput systolic designs [13], but this lifting is not covered in this paper.

2.1 Basic Structuring Functions

The most important structuring functions are relational composition and inverse (or converse):

$$a\,(R\,;S)\,c \Leftrightarrow \exists b.\, a\,R\,b\ \&\ b\,S\,c$$
$$a\,R^{-1}\,b \Leftrightarrow b\,R\,a$$

Repeated composition is denoted by exponentiation: $R^1 = R$ and $R^{n+1} = R^n\ ;R$. These operators have many simply expressed properties:

$$R\,;(S\,;T) = (R\,;S)\,;T$$
$$(R^{-1})^{-1} = R$$
$$(R\,;S)^{-1} = S^{-1}\,;R^{-1}$$
$$(R^n)^{-1} = (R^{-1})^n$$

and it is these kinds of properties that we use in deriving implementations from the specification of a required behaviour.

The pictures in this paper are drawn to a convention in which the domain of a relation is on the left and the range (or codomain) on the right, so that relational

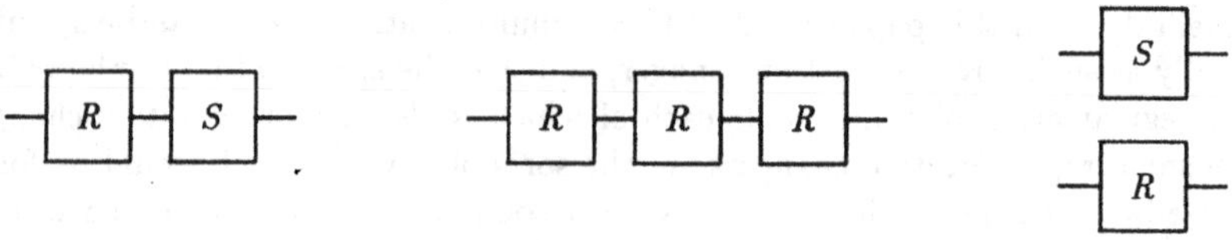

Fig. 1. $R\,;S$ and R^3 and $[R,S]$

composition is as shown in Fig. 1. Taking the inverse of a relation corresponds to flipping the picture over so that the domain and range are swapped.

The parallel composition of two circuits $[R,S]$ relates a pair of values in the domain to a pair of values in the range if and only if the components are related by the corresponding component circuits, $\langle a,b\rangle\ [R,S]\ \langle x,y\rangle \Leftrightarrow a\,R\,x\ \&\ b\,S\,y$. The identity relation id is defined by $x\,id\,y \Leftrightarrow x=y$. The functions fst and snd abbreviate frequently used instances of parallel composition: $\mathsf{fst}\,R=[R,id]$ and $\mathsf{snd}\,S=[id,S]$. Among the properties of parallel composition that we will use are that

$$\begin{aligned}
[R,S]^{-1} &= [R^{-1},S^{-1}]\\
[R,S]\,;[T,U] &= [R\,;T,S\,;U]\\
(\mathsf{fst}\,R)^{-1} &= \mathsf{fst}(R^{-1})\\
\mathsf{fst}\,R;\mathsf{fst}\,S &= \mathsf{fst}(R\,;S)\\
[R,S] &= \mathsf{fst}\,R\,;\mathsf{snd}\,S\\
&= \mathsf{snd}\,S\,;\mathsf{fst}\,R
\end{aligned}$$

2.2 Some Primitive Relations

In addition to the full identity relation id, we introduce identities on finite sets of values, including subranges of the integers. For example, $\|[0,1,2]\|$ is the identity on the set $\{0,1,2\}$. Constant relations are identities on singleton sets. The relation $\|[a\,..\,b]\|$ is the identity on integers in the range a to b inclusive, $x\,\|[a\,..\,b]\|\,y \Leftrightarrow x=y\ \&\ a\le x\le b$.

The following primitive relations are useful building blocks:

$$\begin{aligned}
\langle a,b\rangle\ \pi_1\ c &\Leftrightarrow a=c\\
\langle a,b\rangle\ \pi_2\ c &\Leftrightarrow b=c\\
\langle a,b\rangle\ swp\ \langle c,d\rangle &\Leftrightarrow a=d\ \&\ b=c\\
a\,fork\ \langle b,c\rangle &\Leftrightarrow a=b=c
\end{aligned}$$

Note that the inverse of a projection is a perfectly good relation, which will occur often in our calculations: a component of its *range* is unconstrained. For the arithmetic circuits in this paper we will use:

$$\begin{aligned}
\langle a,b\rangle + c &\Leftrightarrow a+b=c\\
\langle a,b\rangle \times c &\Leftrightarrow a\times b=c
\end{aligned}$$

Note that the symbols $+$ and $\times$ are being used as the names of relations. For binary arithmetic, two other useful components are the relation $\times 2$ that relates an integer x to the even integer $2x$, and $s+ = \mathsf{snd}\ \times 2\, ; +$.

2.3 Measuring the Domain and Range of a Relation

Despite the misleading notation, beware that $R^{-1}\, ; R$ is not always an identity relation. If f is a function from domain to range, then $f^{-1}\, ; f$ is the identity on the range of f, that is $f = f\, ; f^{-1}\, ; f$. For more general relations, however, we can only assume that R is a subrelation of $R\, ; R^{-1}\, ; R$. That is, $R^{-1}\, ; R$ may be bigger than the identity on the range of R. For example, the relation $+^{-1}\, ; +$ is the identity on the integers (written *int*), but $+\, ; +^{-1}$ relates a given pair of integers to every other pair of integers that has the same sum.

It is convenient to introduce notations for the forms $R^{-1}\, ; R$ and $R\, ; R^{-1}$:

$$R{>} = R^{-1}\, ; R$$
$${<}R = R\, ; R^{-1}$$

which we read as 'R right' and 'left R' respectively. It can be shown that $R{>} = (R{>})^{-1} = {<}(R^{-1}) = ({<}(R^{-1}))^{-1}$. Some examples: $\mathit{fork}{>}$ is the identity on pairs whose elements are equal; $\mathit{swp}{>} = [\mathit{id}, \mathit{id}]$ is the identity on all pairs.

In calculations, we often make and use assertions of the form $A = A\, ; B$, so we introduce another abbreviation: $A \vdash B \Leftrightarrow A = A\, ; B$, which we read as '$A$ guarantees B (on the right)'. Think of $A \vdash B$ as an assertion about the range of A. Some useful properties of $\vdash$ are

$$B \vdash C \Rightarrow A\, ; B \vdash C$$
$$B \vdash C\, ; D \Rightarrow B\, ; C \vdash D\, ; C$$
$$B \vdash C\, ; D \;\&\; D\, ; E \vdash F \Rightarrow B\, ; E \vdash F$$

Similarly, $\dashv$ (guarantees on the left) captures assertions about the domain of a relation: $A \dashv B \Leftrightarrow A\, ; B = B$. Because $A \dashv B \Leftrightarrow B^{-1} \vdash A^{-1}$, the properties of $\dashv$ are dual to those of the right-handed operator.

2.4 Types

Data types in Ruby are partial equivalence relations (*pers*), that is, relations P that satisfy $P \vdash P^{-1}$. If A is a per and $A \dashv R$, then we say that R has (a possible) domain type A. Similarly, if B is a per and $R \vdash B$, say that R has range type B. A relation may have many such domain and range types. We write $R : A \sim B$ as an abbreviation for $A \dashv R \& R \vdash B$. It is convenient to have data types that are relations because then types can be manipulated and reasoned about in just the same way as any other Ruby relation. The use of pers as types been developed independently by Voermans [8].

The largest type is the full relation *any* for which $x\ \mathit{any}\ y$ for any x and y. For example, the assertion $\mathsf{snd}\ \mathit{any} \dashv \pi_1$ expresses the fact that the second component of the domain value of the selector π_1 is not constrained in any way: it is not connected to anything by the selector.

We are now in a position to consider what we mean by R^0. For $R : T \sim T$, R^0 must also have type $T \sim T$, and it must be both a left type and a right type of R. Accordingly, we choose R^0 to be the type T' for which $R : T' \sim T'$ and $R : T \sim T \Rightarrow T' : T \sim T$.

2.5 Lists

We collect data values into lists, and have structuring functions which construct relations to operate on them. The repeated parallel composition $\mathsf{map}_n\, R$ relates two n-lists if their corresponding elements are related by R, that is

$$x\,(\mathsf{map}_n\, R)\,y \Leftrightarrow \#x = \#y = n \;\&\; \forall i.\, x_i\, R\, y_i$$

Triangle, another pointwise operator, relates the ith elements of two n-lists by R^i.

$$x\,(\mathsf{tri}_n\, R)\,y \Leftrightarrow \#x = \#y = n \;\&\; \forall i.\, x_i\, R^i\, y_i$$

For example, $\mathsf{map}_n \times 2$ doubles each element of a list of n integers, while $\mathsf{tri}_n \times 2$ relates the ith element of the list, x_i, to $x_i 2^i$, the relation $(\times 2)^0$ being the identity on the integers. Figure 2 shows instances of map and triangle.

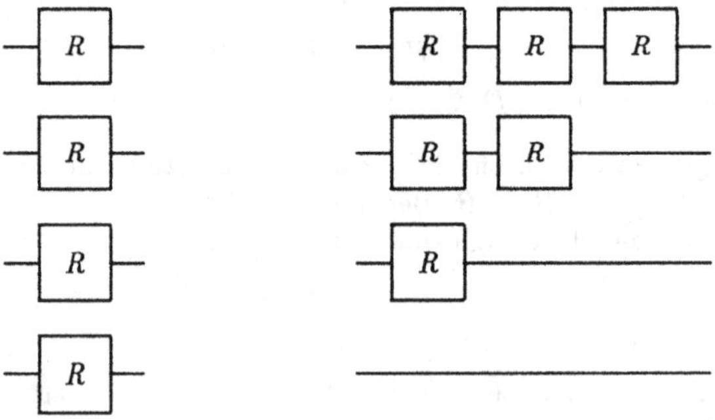

Fig. 2. $\mathsf{map}_4\, R$ and $\mathsf{tri}_4\, R$

The relation *zip* interleaves a pair of lists to give a list of pairs: $\langle x, y\rangle\; zip_n\; z \Leftrightarrow \#x = \#y = \#z = n \;\&\; \forall i.\, \langle x_i, y_i\rangle = z_i$. It satisfies

$$[\mathsf{tri}_n\, A, \mathsf{tri}_n\, B]\,;\, zip_n = zip_n\,;\, \mathsf{tri}_n[A, B]$$
$$[\mathsf{map}_n\, A, \mathsf{map}_n\, B]\,;\, zip_n = zip_n\,;\, \mathsf{map}_n[A, B]$$

2.6 Circuits Connected in Two Dimensions

To broaden the range of circuit forms that can be described, we introduce a new convention for pictures of pair-to-pair relations: they can be viewed as four-sided tiles, with the domain on the left and top edges, and the range on the bottom and right edges. This ordering of the labels is chosen to be consistent with the labelling of circuits drawn under the earlier linear convention.

The structuring function $\leftrightarrow$ (read 'beside') places two four-sided tiles side by side as shown in Fig. 3. The resulting relation is again pair-to-pair.

$$\langle a, \langle b, c\rangle\rangle \; R \leftrightarrow S \; \langle\langle d, e\rangle, f\rangle \;\Leftrightarrow\; \exists g.\, \langle a, b\rangle \; R \; \langle d, g\rangle \;\;\&\;\; \langle g, c\rangle \; S \; \langle e, f\rangle$$

Composing further relations around the edges of a beside can be done in various ways:

$$[A, [B, C]] \,;\, (R \leftrightarrow S) \,;\, [[D, E], F] = ([A, B] \,;\, R \,;\, \mathsf{fst}\, D) \leftrightarrow (\mathsf{snd}\, C \,;\, S \,;\, [E, F])$$

and a relation on the arc between R and S can be bracketed with R or with S.

$$(R \,;\, \mathsf{snd}\, A) \leftrightarrow S = R \leftrightarrow (\mathsf{fst}\, A \,;\, S)$$

Perhaps the simplest pair-to-pair relation is $[id, id]$. Placing two of these side by side gives $rsh = [id, id] \leftrightarrow [id, id]$, which is also shown in Fig. 3. From the properties of beside, it follows that

$$[A, [B, C]] \,;\, rsh = rsh \,;\, [[A, B], C]$$

which we will call 'right-shifting' of $[A, [B, C]]$.

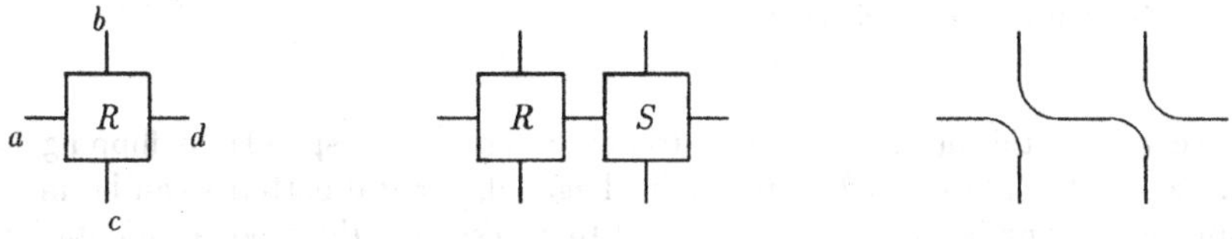

Fig. 3. $\langle a, b\rangle \; R \; \langle c, d\rangle$ and layouts suggested by $R \leftrightarrow S$ and rsh

Just as map generalises parallel composition, we can generalise the beside operator by an operator that connects a horizontal array of pair-to-pair relations:

$$\begin{aligned} &\langle a, x\rangle \; (\mathsf{row}_n \; R) \; \langle y, b\rangle \\ &\quad = \#x = \#y = n \;\;\& \\ &\qquad\qquad \exists z.\, \#z = n + 1 \;\;\&\;\; a = z_0 \;\;\&\;\; (\forall i.\, \langle z_i, x_i\rangle \; R \; \langle y_i, z_{i+1}\rangle) \;\;\&\;\; z_n = b \end{aligned}$$

If $R : [A, B] \sim [C, A]$ then $\mathsf{row}_{n+1}\, R : [A, \mathsf{map}_{n+1}\, B] \sim [\mathsf{map}_{n+1}\, C, A]$, although the corresponding result does not hold for row_0. It seems that there is no 'obviously right' meaning for the row of zero width, but we have chosen here that $\langle a, b\rangle \; (\mathsf{row}_0 \; R) \; \langle c, d\rangle$ if and only if $a = d$ and $b = c = \langle\,\rangle$, that is $\mathsf{row}_0 \; R = \mathsf{snd}\; \mathsf{map}_0 \; id \,;\, swp$.

Two properties of row used often in calculations are

$$\mathsf{snd\ map}_n\, R\,;\mathsf{row}_n\, S\,;\mathsf{fst\ map}_n\, T = \mathsf{row}_n(\mathsf{snd}\, R\,;S\,;\mathsf{fst}\, T)$$

which we call 'pushing maps into row' and

$$\mathsf{fst}\, R\,; S = T\,;\mathsf{snd}\, R \Rightarrow \mathsf{fst}\, R\,;\mathsf{row}_{n+1}\, S \;=\; \mathsf{row}_{n+1}\, T\,;\mathsf{snd}\, R$$

which is 'row induction'. Again there seems to be no right way of dealing with the empty row, which we will deal with separately if necessary.

Complex wiring relations can be built from rows of simple cells: for example, the relation $\mathsf{row}_n(\mathit{fork}{>} \leftrightarrow \mathit{swp})$ can be customised to give some useful standard wiring relations. The relation that distributes a value across a list to give a list of pairs, is

$$\begin{aligned} \mathit{dstl}_n &= \mathsf{row}_n(\mathsf{snd}\,\pi_2^{-1}\,;(\mathit{fork}{>} \leftrightarrow \mathit{swp}))\,;\pi_1 \\ &= \mathsf{row}_n(\mathit{fork}\,;\mathsf{snd}\,\pi_1)\,;\pi_1 \end{aligned}$$

From the properties of row and of *fork* it can be shown that $[A, \mathsf{map}_n\, B]\,;\mathit{dstl}_n = \mathit{dstl}_n\,;\mathsf{map}_n[A,B]$. Figure 4 shows an instance of *dstl*.

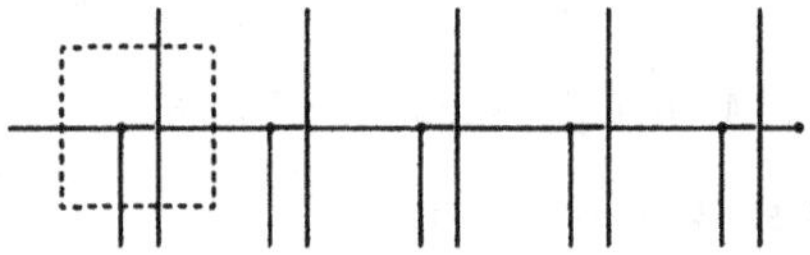

Fig. 4. $\mathit{dstl}_5 \;=\; \mathsf{row}_5(\mathit{fork}\,;\mathsf{snd}\,\pi_1)\,;\pi_1$

Since taking the inverse of a pair-to-pair relation corresponds to flipping its picture about the bottom-left to top-right diagonal, vertical patterns can be made by flipping, making horizontal patterns and then restoring the original orientation. We define $\updownarrow$ (read 'below') and col as duals of $\leftrightarrow$ and row.

$$\begin{aligned} R \updownarrow S &= (R^{-1} \leftrightarrow S^{-1})^{-1} \\ \mathsf{col}_n\, R &= (\mathsf{row}_n\, R^{-1})^{-1} \end{aligned}$$

The properties of below and column can be derived from those of beside and row. The dual of *rsh* is $\mathit{lsh} = \mathit{rsh}^{-1} = [\mathit{id}, \mathit{id}] \updownarrow [\mathit{id}, \mathit{id}]$. The inverse of $\mathsf{row}_n(\mathit{fork}{>} \leftrightarrow \mathit{swp})$ is $\mathsf{col}_n(\mathit{fork}{>} \updownarrow \mathit{swp})$ and we can make a 'distribute right' wiring relation from this.

$$\begin{aligned} \mathit{dstr}_n &= \mathsf{col}_n(\mathsf{fst}\,\pi_1^{-1}\,;(\mathit{fork}{>} \updownarrow \mathit{swp}))\,;\pi_2 \\ &= \mathsf{col}_n(\mathit{fork}\,;\mathsf{fst}\,\pi_2)\,;\pi_2 \end{aligned}$$

Not all of the relations that we want to connect in these kinds of patterns are pair-to-pair. We use degenerate cases of the structuring functions already introduced to cope with those relations that do not quite have the right numbers of connections.

For example, to make a continued version of a pair-to-value relation, we use a version of row that has some of its arcs removed.

$$\mathsf{rdl}_n\, R = \mathsf{row}_n(R\,;\pi_2^{-1})\,;\pi_2$$

This operator is familiar from functional programming, where it is often called 'reduce left' or 'fold'. For example, to sum n integers:

$$sum_n = \pi_2^{-1}\,;\mathsf{fst}\,[\![0]\!]\,;\mathsf{rdl}_n\,+$$

The $[\![0]\!]$ constrains the first component of the domain of the reduction to be zero, and the inverse projection hides this from the domain of sum.

3 Abstraction and Representation

An abstraction relates concrete values to the abstract values that they represent. We say that abs is an abstraction if $abs \vdash abs{>}$. If the assertion $abs \vdash abs{>}$ holds, then $abs{>}$ is guaranteed to be a type. We call this range type of abs the abstract type. Similarly ${<}abs$ is guaranteed to be a type, and this domain type of abs is called the concrete type. This is really only a matter of convenience as this notion of abstraction is symmetric: if abs is an abstraction, so is abs^{-1}. (An earlier paper [9] of ours on refinement in Ruby incorrectly denies this.) The relations which we call abstractions are those that are called *difunctionals* [8]: all functions are abstractions, as are all inverses of functions [14].

Since we want to design arithmetic circuits, we will be considering abstractions onto the integers and subranges of the integers. The relations $+$, $\times$, $s+$ and sum_n are all abstractions onto the integers. The relation $\times 2$ is an abstraction not onto the integers but onto the *even* integers, that is the type $(\times 2){>}$.

In the abstraction sum_n, each element of the list of integers in the domain has the same weight. We can give the elements of a list weights that are increasing powers of two by using a triangle of $\times 2$ components. The resulting relation, which we call a *ladder*, is also an abstraction onto the integers.

$$ladder_n = \mathsf{tri}_n\,\times 2\,;sum_n$$

We will be constructing circuits which operate on 'voltage levels', which are the values of the type $sig = [\![high, low]\!]$. The abstraction bi, for 'bit to integer', relates abstract voltages to numbers in the set $\{0, 1\}$.

$$\begin{aligned} bi &= \{(high, 1), (low, 0)\} \\ bi{>} &= [\![0, 1]\!] \\ {<}bi &= sig \end{aligned}$$

Similarly, we can represent the numbers 0 and -1 by signals. We call the resulting 'negative bit' a *nit*, and the abstraction is called ni for 'negative bit to integer'.

$$\begin{aligned} ni &= \{(high, -1), (low, 0)\} \\ ni{>} &= [\![-1, 0]\!] \\ {<}ni &= sig \end{aligned}$$

We can also represent numbers by pairs of signals in various ways. A pair of bits of the same weight can represent numbers in $\{0,1,2\}$. This abstraction is often used so we give it the name $bb+$.

$$bb+ = [bi, bi] \,;\, +$$
$$(bb+)> = \|[0,1,2]\|$$

This is a redundant representation; the number 1 is represented both by the pair $\langle high, low\rangle$ and by the pair $\langle low, high\rangle$. This means that $<(bb+) \neq [sig, sig]$, but rather it is the partial equivalence relation described by

$$\begin{aligned}<(bb+) = \{ & (\langle low, low\rangle, \langle low, low\rangle), \ (\langle high, high\rangle, \langle high, high\rangle),\\ & (\langle high, low\rangle, \langle high, low\rangle), (\langle high, low\rangle, \langle low, high\rangle),\\ & (\langle low, high\rangle, \langle low, high\rangle), (\langle low, high\rangle, \langle high, low\rangle) \ \}\end{aligned}$$

This is the concrete type of pairs of equal weight bits.

A pair of bits, one of weight 1 and one of weight 2, can represent numbers in $\{0,1,2,3\}$. The corresponding abstraction relation is $[bi, bi] \,;\, s+$. We call this a 'bit step' and abbreviate it to $bbs+$.

$$bbs+ = [bi, bi] \,;\, s+$$
$$(bbs+)> = \|[0,1,2,3]\|$$
$$<(bbs+) = [sig, sig]$$

A nit and a bit, each of weight one, can represent numbers in $\{-1,0,1\}$. This is a standard abstraction in arithmetic circuit design, and is usually called a *trit*.

$$ti = [ni, bi] \,;\, +$$
$$ti> = \|[-1,0,1]\|$$

Again, $<ti$ is a partial equivalence relation that is bigger than $[sig, sig]$. This is because the integer 0 is represented both by the pair $\langle high, high\rangle$ and by the pair $\langle low, low\rangle$. (We can write that assertion as $ti \,;\, \|[0]\| = fork^{-1} \,;\, sig \,;\, any \,;\, \|[0]\|$.)

If we restrict the elements of the lists in the domain of a ladder to be bits, we get $\mathsf{map}_n\ bi \,;\, ladder_n$, which is the well known n-bit binary least-significant bit at the left representation of numbers in the range 0 to $2^n - 1$. The form $\mathsf{map}_n\ abs \,;\, ladder_n$, where abs is an abstraction, appears so often that we abbreviate it to $\mathsf{L}_n\ abs$. We think of the n-bit binary abstraction as a 'ladder of bits'.

$$(\mathsf{L}_n\ bi)> = \|[0\,..\,2^n - 1]\|$$
$$\text{where } \mathsf{L}_n\ abs = \mathsf{map}_n\ abs \,;\, ladder_n$$

This is the type of numbers that can be represented as n-bit binary numerals. Because the abstraction is one-to-one, $<(\mathsf{L}_n\ bi) = \mathsf{map}_n\ sig$.

The carry-save abstraction is a ladder of pairs of equal weight bits, which we can write as $\mathsf{L}_n\ bb+$.

$$(\mathsf{L}_n\ bb+)> = \|[0\,..\,2(2^n - 1)]\|$$

Carry-save is a redundant abstraction both because $bb+$ is redundant, and because the value that can be represented at each step of the ladder is not constrained to be smaller than the radix (which is two).

A number of useful transformations of these abstractions can be derived by induction on the lengths of the ladders, starting from the properties of the underlying arithmetic. The sum of two ladders can be written as a ladder of sums, provided the two concrete lists are first interleaved (or zipped together):

$$[\mathsf{L}_n\, a, \mathsf{L}_n\, b] \,;\, + = zip_n \,;\, \mathsf{L}_n\,([a,b]\,;\,+) \tag{1}$$

Similarly, the product of an abstraction and a ladder is a ladder of products:

$$[a, \mathsf{L}_n\, b] \,;\, \times = dstl_n \,;\, \mathsf{L}_n\,([a,b]\,;\,\times) \tag{2}$$

$$[\mathsf{L}_n\, a, b] \,;\, \times = dstr_n \,;\, \mathsf{L}_n\,([a,b]\,;\,\times) \tag{3}$$

To push abstractions around in circuit descriptions, we use instances of the theorems about the structuring functions, for example by row induction

$$\mathsf{fst}\, a \,;\, R \vdash \mathsf{snd}\, a{>} \quad \Rightarrow \quad \mathsf{fst}\, a \,;\, \mathsf{row}_n\, R = \mathsf{row}_n(\mathsf{fst}\, a \,;\, R \,;\, \mathsf{snd}\, a^{-1}) \,;\, \mathsf{snd}\, a$$

3.1 Representation Changers

Many design problems can naturally be cast as the implementation of representation changers. This seems particularly to be the case for arithmetic circuits. The specification of a representation changer is the composition of an abstraction relation – which relates a given concrete representation to its abstract meaning – and the inverse of another abstraction relation – which relates that abstract value to the desired concrete value. If a and b are abstractions, then $a \,;\, b^{-1}$ relates the type $<a$ to the type $<b$. Provided $a> = b>$ the translation is complete and faithful in both directions; if they differ it is faithful only between those representations which correspond to abstract values which are in both types.

We choose to think of our circuits as representation changers because simple representation changers can often be implemented directly in hardware. For example, $[bi, bi] \,;\, \times \,;\, bi^{-1}$ is just an *and* gate: that is, it is a relation between a pair of signals and a signal which relates $\langle high, high \rangle$ to *high* and any other pair to *low*. Similarly, $bb+ \,;\, bbs+^{-1}$ is just a half-adder, thought of as a function from two input bits to a sum and carry pair.

The representation changer $[bi, ni] \,;\, s+ \,;\, ti^{-1}$ can also be implemented by a half-adder in which the carry flows from range to domain. It is a half-adder rotated through ninety degrees, so that the domain is a pair consisting of one input and the carry out, and the range is the pair of the other input and the sum output. Notice that we abstracted away from the distinction between inputs and outputs by choosing to represent circuits by relations.

A full-adder relates a bit and a pair of equal weight bits to a sum and carry pair, so it is $[bi, bb+] \,;\, + \,;\, bbs+^{-1}$. Just as we did for a half-adder, we can implement the representation changer $[bb+, ni] \,;\, s+ \,;\, ti^{-1}$ by a full-adder rotated so that the carry out is in the domain and the carry in is in the range. The easiest way to check such assertions about small relations is to enumerate the values which they relate.

3.2 Refinement

Representation changers that cannot immediately be implemented as hardware primitives must be re-expressed as networks of smaller representation changers. These smaller changers can then either be implemented directly in hardware or rewritten as networks of yet smaller changers of representation. This refinement proceeds hierarchically, and stops when we reach cells that we know how to implement directly in hardware.

We will need to know how to reorganise large changers of representation into networks of smaller ones. Because we want to refine to regular array circuits, we concentrate on ways of calculating with rows and columns. We have previously shown [9], using the Ruby equivalent of Horner's rule [3] for polynomial evaluation, that

$$\mathsf{snd}\, ladder_n \,;\, + \,;\, +^{-1} \,;\, [ladder_n, \times 2^n]^{-1} = \mathsf{row}_n(+ \,;\, s{+}^{-1}) \tag{4}$$

This allows us to calculate

$$\begin{aligned}
&\mathsf{snd}(\mathsf{L}_n\, a) \,;\, + \,;\, +^{-1} \,;\, [\mathsf{L}_n\, b, \times 2^n]^{-1} \\
= &\quad \{\text{definition of } \mathsf{L}_n\} \\
&\mathsf{snd}(\mathsf{map}_n\, a \,;\, ladder_n) \,;\, + \,;\, +^{-1} \,;\, [(\mathsf{map}_n\, b \,;\, ladder_n), \times 2^n]^{-1} \\
= &\quad \{\text{properties of inverse}\} \\
&\mathsf{snd}\,\mathsf{map}_n\, a \,;\, \mathsf{snd}\, ladder_n \,;\, + \,;\, +^{-1} \,;\, [ladder_n, \times 2^n]^{-1} \,;\, \mathsf{fst}\,\mathsf{map}_n\, b^{-1} \\
= &\quad \{\text{equation } 4\} \\
&\mathsf{snd}\,\mathsf{map}_n\, a \,;\, \mathsf{row}_n(+ \,;\, s{+}^{-1}) \,;\, \mathsf{fst}\,\mathsf{map}_n\, b^{-1} \\
= &\quad \{\text{pushing maps into row}\} \\
&\mathsf{row}_n(\mathsf{snd}\, a \,;\, + \,;\, s{+}^{-1} \,;\, \mathsf{fst}\, b^{-1})
\end{aligned} \tag{5}$$

Similarly,

$$\begin{aligned}
&\mathsf{snd}(\mathsf{L}_n\, a) \,;\, s{+} \,;\, +^{-1} \,;\, [\mathsf{L}_n\, b, \times 2^n]^{-1} \\
= &\quad \{\text{definition of } s{+} \text{ and } (\mathsf{L}_n\, a) \,;\, \times 2 = \mathsf{L}_n(a \,;\, \times 2)\} \\
&\mathsf{snd}(\mathsf{L}_n(a \,;\, \times 2)) \,;\, + \,;\, +^{-1} \,;\, [\mathsf{L}_n\, b, \times 2^n]^{-1} \\
= &\quad \{\text{equation } 5\} \\
&\mathsf{row}_n(\mathsf{snd}(a \,;\, \times 2) \,;\, + \,;\, s{+}^{-1} \,;\, \mathsf{fst}\, b^{-1}) \\
= &\quad \{\text{definition of } s{+}\} \\
&\mathsf{row}_n(\mathsf{snd}\, a \,;\, s{+} \,;\, s{+}^{-1} \,;\, \mathsf{fst}\, b^{-1})
\end{aligned} \tag{6}$$

Taking inverses on both sides of (5) and (6) and renaming variables gives

$$[\mathsf{L}_n\, a, \times 2^n] \,;\, + \,;\, +^{-1} \,;\, \mathsf{snd}(\mathsf{L}_n\, b)^{-1} = \mathsf{col}_n(\mathsf{fst}\, a \,;\, s{+} \,;\, +^{-1} \,;\, \mathsf{snd}\, b^{-1}) \tag{7}$$

and

$$[\mathsf{L}_n\, a, \times 2^n] \,;\, + \,;\, s{+}^{-1} \,;\, \mathsf{snd}(\mathsf{L}_n\, b)^{-1} = \mathsf{col}_n(\mathsf{fst}\, a \,;\, s{+} \,;\, s{+}^{-1} \,;\, \mathsf{snd}\, b^{-1}) \tag{8}$$

These four equations guide our refinements, so that if we are aiming for a column of components, for instance, we try to massage the expression to match the left hand side either of (7) or of (8).

A useful property of the composition of $+$ and its inverse, which appears often in these expressions, is that

$$+ \,;\, +^{-1} = \mathsf{snd}\,+^{-1} \,;\, rsh \,;\, \mathsf{fst}\,+$$

and from this, one can use arithmetic to show that

$$s{+} \,;\, s{+}^{-1} = \mathsf{snd}\,+^{-1} \,;\, rsh \,;\, \mathsf{fst}\,s{+}$$

4 Small Examples

As small examples of the method, we will develop some well-known adder circuits by refinement of representation changers.

To specify the addition of a carry-in bit to a binary number to give a binary number and a carry-out bit, the abstraction $[bi, \mathsf{L}_n\, bi] \,;\, +$ is composed with the inverse of $[\mathsf{L}_n\, bi, (bi \,;\, \times 2^n)] \,;\, +$. This matches the pattern of (5).

$$\begin{aligned}
&[bi, \mathsf{L}_n\, bi] \,;\, + \,;\, +^{-1} \,;\, [\mathsf{L}_n\, bi, (bi \,;\, \times 2^n)]^{-1} \\
&\quad = \quad \{\text{properties of inverse}\} \\
&\quad \mathsf{fst}\, bi \,;\, \mathsf{snd}\, \mathsf{L}_n\, bi \,;\, + \,;\, +^{-1} \,;\, [\mathsf{L}_n\, bi, \times 2^n]^{-1} \,;\, \mathsf{snd}\, bi^{-1} \\
&\quad = \quad \{\text{equation 5}\} \\
&\quad \mathsf{fst}\, bi \,;\, \mathsf{row}_n(\mathsf{snd}\, bi \,;\, + \,;\, s{+}^{-1} \,;\, \mathsf{fst}\, bi^{-1}) \,;\, \mathsf{snd}\, bi^{-1} \qquad (9)
\end{aligned}$$

The next step is to push the abstractions at each end of the row ($\mathsf{fst}\, bi$ and $\mathsf{snd}\, bi^{-1}$) onto the internal arcs of the row so that the components of the row are then themselves in the form of representation changers. This is done using a row induction. Using the two lemmas (which are checked by arithmetic)

$$[bi, bi] \,;\, + \vdash \|[0,1,2]\|$$

and

$$\|[0,1,2]\| \,;\, s{+}^{-1} \,;\, \mathsf{fst}\, bi \vdash \mathsf{snd}\, bi{>}$$

it can be shown that

$$\mathsf{fst}\, bi \,;\, (\mathsf{snd}\, bi \,;\, + \,;\, s{+}^{-1} \,;\, \mathsf{fst}\, bi^{-1}) \vdash \mathsf{snd}\, bi{>} \qquad (10)$$

This is then used as the condition for a row induction.

$$\begin{aligned}
&\mathsf{fst}\, bi \,;\, \mathsf{row}_n(\mathsf{snd}\, bi \,;\, + \,;\, s{+}^{-1} \,;\, \mathsf{fst}\, bi^{-1}) \,;\, \mathsf{snd}\, bi^{-1} \\
&\quad = \quad \{\text{equation 10 and row induction}\} \\
&\quad \mathsf{row}_n([bi, bi] \,;\, + \,;\, s{+}^{-1} \,;\, [bi, bi]^{-1}) \,;\, \mathsf{snd} {<} bi \\
&\quad = \quad \{\text{defintions of } bb{+} = [bi, bi] \,;\, + \text{ and } bbs{+} = [bi, bi] \,;\, s{+}\} \\
&\quad \mathsf{row}_n(bb{+} \,;\, bbs{+}^{-1}) \,;\, \mathsf{snd} {<} bi \\
&\quad = \quad \{\text{definition of } halfadd = bb{+} \,;\, bbs{+}^{-1} \text{ and } {<}bi = sig\} \\
&\quad \mathsf{row}_n\, halfadd \,;\, \mathsf{snd}\, sig
\end{aligned}$$

We can stop here because $bb+ \,;\, bbs+^{-1}$ is just a half-adder. The snd *sig* on the right shows that if $n = 0$ the circuits consists of a single concrete wire – a thing that can carry a signal of type *sig*.

The derivation of a circuit to add a bit to a carry-save number to give a binary number and a carry-out is almost identical. This time the component is a full adder.

$$
\begin{array}{ll}
\multicolumn{2}{l}{[bi, \mathsf{L}_n\, bb+] \,;\, + \,;\, +^{-1} \,;\, [\mathsf{L}_n\, bi, (bi \,;\, \times 2^n)]^{-1}} \\
= & \{\text{equation } 5\} \\
& \mathsf{fst}\; bi \,;\, \mathsf{row}_n(\mathsf{snd}\; bb+ \,;\, + \,;\, s+^{-1} \,;\, \mathsf{fst}\; bi^{-1}) \,;\, \mathsf{snd}\; bi^{-1} \\
= & \{\mathsf{fst}\; bi \,;\, (\mathsf{snd}\; bb+ \,;\, + \,;\, s+^{-1} \,;\, \mathsf{fst}\; bi^{-1}) \vdash \mathsf{snd}\; bi\!> \text{ and induction}\} \\
& \mathsf{row}_n([bi, bb+] \,;\, + \,;\, s+^{-1} \,;\, [bi, bi]^{-1}) \,;\, \mathsf{snd} <\!bi \\
= & \{\text{defintion of } bbs+ = [bi, bi] \,;\, s+ \text{ and } <\!bi = sig\} \\
& \mathsf{row}_n([bi, bb+] \,;\, + \,;\, bbs+^{-1}) \,;\, \mathsf{snd}\; sig \\
= & \{\text{defintion of } fulladd = [bi, bb+] \,;\, + \,;\, bbs+^{-1}\} \\
& \mathsf{row}_n\, fulladd \,;\, \mathsf{snd}\; sig
\end{array}
$$

In our third example, we relate a carry-save number and a top-carry that is a nit (rather than a bit) to a nit and a binary number. The process of refinement is the same as in the previous examples. We use one of (5) to (8) to introduce a row or column and then use either row or column induction to push the remaining abstractions inside the array.

$$
\begin{array}{ll}
\multicolumn{2}{l}{[\mathsf{L}_n\, bb+, (ni \,;\, \times 2^n)] \,;\, + \,;\, +^{-1} \,;\, [ni, \mathsf{L}_n\, bi]^{-1}} \\
= & \{\text{equation } 7\} \\
& \mathsf{snd}\; ni \,;\, \mathsf{col}_n(\mathsf{fst}\; bb+ \,;\, s+ \,;\, +^{-1} \,;\, \mathsf{snd}\; bi^{-1}) \,;\, \mathsf{fst}\; ni^{-1} \\
= & \{\mathsf{snd}\; ni\!> \dashv (\mathsf{fst}\; bb+ \,;\, s+ \,;\, +^{-1} \,;\, \mathsf{snd}\; bi^{-1}) \,;\, \mathsf{fst}\; ni^{-1} \text{ and induction}\} \\
& \mathsf{snd} <\!ni \,;\, \mathsf{col}_n([bb+, ni] \,;\, s+ \,;\, +^{-1} \,;\, [ni, bi]^{-1}) \\
= & \{\text{defintion of } ti = [ni, bi] \,;\, + \text{ and } <\!ni = sig\} \\
& \mathsf{snd}\; sig \,;\, \mathsf{col}_n([bb+, ni] \,;\, s+ \,;\, ti^{-1}) \\
= & \{\text{definition of } fulladd' = [bb+, ni] \,;\, s+ \,;\, ti^{-1}\} \\
& \mathsf{snd}\; sig \,;\, \mathsf{col}_n\, fulladd'
\end{array}
$$

The column component is a full adder with a carry that flows from range to domain, so the whole circuit has an upward-flowing carry-chain, as shown in Fig. 5. In the domain, the carry-save number is an input and the nit is an output; in the range, the nit is an input and the binary number an output. This choice of inputs and outputs ensures that the full adder components are driven correctly.

5 Derivation of an Array Multiplier

In each of the above examples, refinement stopped after one iteration. In more complicated examples, one would expect to have to go down through several levels

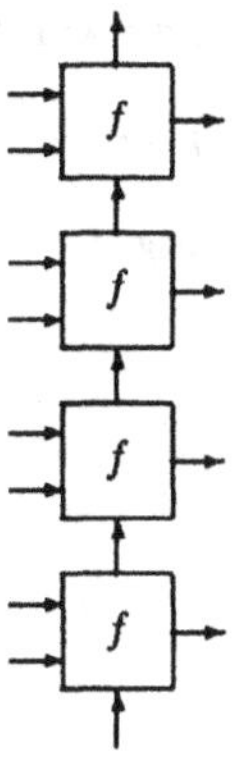

Fig. 5. col_4 *fulladd'* showing constraints on data-flow in an implementation

of hierarchy before reaching primitives. We now demonstrate this hierarchical approach in a substantial calculation, deriving the design of an n-bit by m-bit binary multiplier.

The initial specification of the circuit is

$$[\mathsf{L}_n\, bi, \mathsf{L}_m\, bi] \,;\, \times \,;\, (\mathsf{L}_{n+m}\, bi)^{-1}$$

which is to say, that it should relate a pair of an n-bit and an m-bit binary number to the $(n+m)$-bit number that represents their product. From (2) and (3) about the product of an abstraction and a ladder, we have that

$$\begin{aligned}[\mathsf{L}_n\, bi, \mathsf{L}_m\, bi] \,;\, \times &= dstl_m \,;\, \mathsf{L}_m([\mathsf{L}_n\, bi, bi] \,;\, \times) \\ &= dstr_n \,;\, \mathsf{L}_n([bi, \mathsf{L}_m\, bi] \,;\, \times)\end{aligned}$$

but there seems to be no obvious next move.

Because we have a wealth of laws involving carries, we will introduce a carry-in to the circuit. The carry will be represented in n-bit binary – a design decision that we have made before starting the derivation. We could have chosen a different representation, and in [9] we derive a multiplier design which arises naturally from chosing carry-save representation for the carry at this point.

To take account of the n-bit binary carry, the domain abstraction must be changed to

$$[\mathsf{L}_n\, bi, ([\mathsf{L}_n\, bi, \mathsf{L}_m\, bi] \,;\, \times)] \,;\, +$$

and in order to make an opportunity to apply (5) we change the range representation, so that the answer comes in two parts: the least significant m bits as a binary number, and the n most significant bits as a binary number of weight 2^m. The corresponding abstraction is

$$[\mathsf{L}_m\, bi, (\mathsf{L}_n\, bi \,;\, \times 2^m)] \,;\, +$$

so the specification of the circuit which we will develop is

$$\mathcal{M} = [\mathsf{L}_n\, bi, ([\mathsf{L}_n\, bi, \mathsf{L}_m\, bi] \,;\, \times)] \,;\, + \,;\, +^{-1} \,;\, [\mathsf{L}_m\, bi, (\mathsf{L}_n\, bi \,;\, \times 2^m)]^{-1}$$

an interpretation of which is shown in Fig. 6.

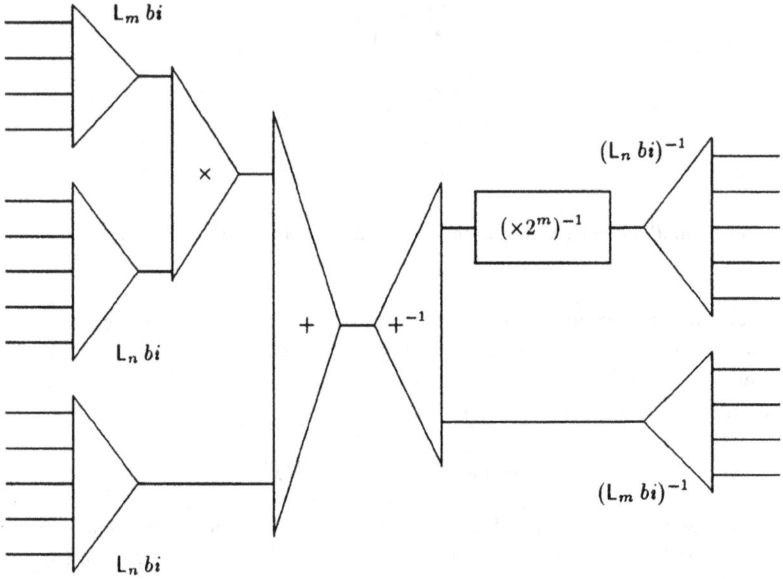

Fig. 6. $\mathcal{M} = [\mathsf{L}_n\, bi, [\mathsf{L}_n\, bi, \mathsf{L}_m\, bi] \,;\, \times] \,;\, + \,;\, +^{-1} \,;\, [\mathsf{L}_m\, bi, (\mathsf{L}_n\, bi \,;\, \times 2^m)]^{-1}$ **with** $m, n = 4, 5$

The first step in the refinement is to rearrange the product of the two ladders of bits into a ladder. We can then use (5) to introduce a row.

$$\begin{aligned}
\mathcal{M} &= [\mathsf{L}_n\, bi, ([\mathsf{L}_n\, bi, \mathsf{L}_m\, bi] \,;\, \times)] \,;\, + \,;\, +^{-1} \,;\, [\mathsf{L}_m\, bi, (\mathsf{L}_n\, bi \,;\, \times 2^m)]^{-1} \\
&= \quad \{\text{product of an abstraction and a ladder (equation 2)}\} \\
&\quad \mathsf{snd}\ dstl_m \,;\, [\mathsf{L}_n\, bi, \mathsf{L}_m([\mathsf{L}_n\, bi, bi] \,;\, \times)] \,;\, + \,;\, +^{-1} \,;\, [\mathsf{L}_m\, bi, (\mathsf{L}_n\, bi \,;\, \times 2^m)]^{-1} \\
&= \quad \{\text{equation 5}\} \\
&\quad [\mathsf{L}_n\, bi, dstl_m] \,;\, \mathsf{row}_m(\mathsf{snd}([\mathsf{L}_n\, bi, bi] \,;\, \times) \,;\, + \,;\, s{+}^{-1} \,;\, \mathsf{fst}\ bi^{-1}) \,;\, \mathsf{snd}(\mathsf{L}_n\, bi)^{-1}
\end{aligned}$$

Next, we push the abstraction $\mathsf{L}_n\, bi$ rightwards along the row; the result is a row of representation changers. By arithmetic we can show that

$$[\mathsf{L}_n\, bi, ([\mathsf{L}_n\, bi, bi] \,;\, \times)] \;\vdash\; [(\mathsf{L}_n\, bi){>}, (\mathsf{L}_n\, bi){>}]$$

and

$$[(\mathsf{L}_n\, bi){>}, (\mathsf{L}_n\, bi){>}] \,;\, + \,;\, s{+}^{-1} \,;\, \mathsf{fst}\ bi^{-1} \;\vdash\; \mathsf{snd}\,(\mathsf{L}_n\, bi){>}$$

Combining these judgements gives

$$\mathsf{fst}(\mathsf{L}_n\, bi)\ ;\ \big(\mathsf{snd}([\mathsf{L}_n\, bi, bi]\ ;\ \times)\ ;\ +\ ;\ s{+}^{-1}\ ;\ \mathsf{fst}\, bi^{-1}\big) \vdash \mathsf{snd}\,(\mathsf{L}_n\, bi)> \tag{11}$$

and, as in the previous examples, this can be used as the condition for a row induction.

$$\begin{aligned}
&[\mathsf{L}_n\, bi, dstl_m]\ ;\ \mathsf{row}_m\ \big(\mathsf{snd}([\mathsf{L}_n\, bi, bi]\ ;\ \times)\ ;\ +\ ;\ s{+}^{-1}\ ;\ \mathsf{fst}\, bi^{-1}\big)\ ;\ \mathsf{snd}(\mathsf{L}_n\, bi)^{-1}\\
&= \quad \{\text{row induction from equation } 11\}\\
&\quad \mathsf{snd}\ dstl_m\ ;\\
&\quad\quad \mathsf{row}_m([\mathsf{L}_n\, bi, ([\mathsf{L}_n\, bi, bi]\ ;\ \times)]\ ;\ +\ ;\ s{+}^{-1}\ ;\ [bi, \mathsf{L}_n\, bi]^{-1})\ ;\\
&\quad\quad\quad \mathsf{snd} <(\mathsf{L}_n\, bi)\\
&= \quad \{\text{naming } c_2 = [\mathsf{L}_n\, bi, ([\mathsf{L}_n\, bi, bi]\ ;\ \times)]\ ;\ +\ ;\ s{+}^{-1}\ ;\ [bi, \mathsf{L}_n\, bi]^{-1}\}\\
&\quad \mathsf{snd}\ dstl_m\ ;\ \mathsf{row}_m\ c_2\ ;\ \mathsf{snd} <(\mathsf{L}_n\, bi)\\
&= \quad \{<(\mathsf{L}_n\, bi) = \mathsf{map}_n\ sig\}\\
&\quad \mathsf{snd}\ dstl_m\ ;\ \mathsf{row}_m\ c_2\ ;\ \mathsf{snd}\ \mathsf{map}_n\ sig
\end{aligned}$$

So $\mathcal{M}$ can be implemented by some wiring and a row of c_2 cells; we are now free to concentrate on the design of c_2.

Because c_2 is the component of a row, it is good idea to aim to implement it as a column (rather than a row) of components. Rows of rows do not make attractive layouts, but rows of columns do. This bias towards a column influences the refinement of c_2, making it like a mirror image of the refinement so far: where we had *dstl*, we now have *dstr*; where we had a row, we now have a column.

The first steps are to rewrite $[\mathsf{L}_n\, bi, bi]\ ;\ \times$ as a ladder and to note that we can implement $[bi, bi]\ ;\ \times$ using an *and* gate. The resulting sum of two ladders can again be rewritten as a ladder.

$$\begin{aligned}
c_2 &= [\mathsf{L}_n\, bi, ([\mathsf{L}_n\, bi, bi]\ ;\ \times)]\ ;\ +\ ;\ s{+}^{-1}\ ;\ [bi, \mathsf{L}_n\, bi]^{-1}\\
&= \quad \{\text{product of a ladder and an abstraction (equation 3)}\}\\
&\quad [\mathsf{L}_n\, bi, (dstr_n\ ;\ \mathsf{L}_n([bi, bi]\ ;\ \times))]\ ;\ +\ ;\ s{+}^{-1}\ ;\ [bi, \mathsf{L}_n\, bi]^{-1}\\
&= \quad \{and\ ;\ bi = [bi, bi]\ ;\ \times\ ;\ bi^{-1}\ ;\ bi = [bi, bi]\ ;\ \times\}\\
&\quad [\mathsf{L}_n\, bi, (dstr_n\ ;\ \mathsf{L}_n(and\ ;\ bi))]\ ;\ +\ ;\ s{+}^{-1}\ ;\ [bi, \mathsf{L}_n\, bi]^{-1}\\
&= \quad \{\text{sum of two ladders (equation 1)}\}\\
&\quad \mathsf{snd}\ dstr_n\ ;\ zip_n\ ;\ \mathsf{L}_n([bi, and\ ;\ bi]\ ;\ +)\ ;\ s{+}^{-1}\ ;\ [bi, \mathsf{L}_n\, bi]^{-1}\\
&= \quad \{\text{defintion of } bb{+} = [bi, bi]\ ;\ +\}\\
&\quad \mathsf{snd}\ dstr_n\ ;\ zip_n\ ;\ \mathsf{L}_n(\mathsf{snd}\ and\ ;\ bb{+})\ ;\ s{+}^{-1}\ ;\ [bi, \mathsf{L}_n\, bi]^{-1}
\end{aligned}$$

Unfortunately, the resulting pattern does not match our rule for introducing columns (8). We would like the abstraction on the left to be of the form $[\mathsf{L}_n\, a, \times 2^n];+$ but we have only a single ladder. Again we introduce a carry, forced to zero, to get the required pattern. Adding zero to an integer makes no difference, and multiplying zero by 2^n gives zero, so if $a \vdash int$, it can be replaced by

$$\begin{aligned}
a &= \pi_1^{-1}\ ;\ [a, \llbracket 0 \rrbracket]\ ;\ +\\
&= \pi_1^{-1}\ ;\ [a, (\llbracket 0 \rrbracket\ ;\ \times 2^n)]\ ;\ +
\end{aligned}$$

In this case, certainly $\mathsf{L}_n(\mathsf{snd}\ and\ ;\ bb+) \vdash int$, so

$$\begin{aligned}
&\mathsf{snd}\ dstr_n\ ;\ zip_n\ ;\ \mathsf{L}_n(\mathsf{snd}\ and\ ;\ bb+)\ ;\ s+^{-1}\ ;\ [bi, \mathsf{L}_n\ bi]^{-1} \\
=&\quad \{\text{zero introduction}\} \\
&\mathsf{snd}\ dstr_n\ ;\ zip_n\ ;\ \pi_1^{-1}\ ; \\
&\quad [\mathsf{L}_n(\mathsf{snd}\ and\ ;\ bb+), (\|[0]\|\ ;\ \times 2^n)]\ ;\ +\ ;\ s+^{-1}\ ;\ [bi, \mathsf{L}_n\ bi]^{-1} \\
=&\quad \{\text{equation } 8\} \\
&\mathsf{snd}\ dstr_n\ ;\ zip_n\ ;\ \pi_1{}^{-1}\ ;\ \mathsf{snd}\ \|[0]\|\ ; \\
&\quad \mathsf{col}_n(\mathsf{fst}(\mathsf{snd}\ and\ ;\ bb+)\ ;\ s+\ ;\ s+^{-1}\ ;\ \mathsf{snd}\ bi^{-1})\ ;\ \mathsf{fst}\ bi^{-1} \\
=&\quad \{\text{naming } c_3 = \mathsf{fst}(\mathsf{snd}\ and\ ;\ bb+)\ ;\ s+\ ;\ s+^{-1}\ ;\ \mathsf{snd}\ bi^{-1}\} \\
&\mathsf{snd}\ dstr_n\ ;\ zip_n\ ;\ \pi_1{}^{-1}\ ;\ \mathsf{snd}\ \|[0]\|\ ;\ \mathsf{col}_n\ c_3\ ;\ \mathsf{fst}\ bi^{-1}
\end{aligned}$$

Now we would like to push abstractions onto the internal arcs of the column. However the most obvious strategy, pushing bi^{-1} from right to left, does not work. It is *not* the case that $\mathsf{snd}\ bi{>} \dashv c_3\ ;\ \mathsf{fst}\ bi^{-1}$. In fact since

$$\langle\langle high, \langle high, high\rangle\rangle, -1\rangle\ \ (c_3\ ;\ \mathsf{fst}\ bi^{-1})\ \ \langle low, low\rangle$$

the relevant abstract wire in the circuit can even take a negative value. However, the internal arcs are of type $ti{>} = \|[-1, 0, 1]\|$, rather than $bi{>}$ as we had hoped.

Before pushing a trit abstraction through the column, we need to introduce $\mathsf{snd}\ ti^{-1}$ at the right. This can be done because trits and bits are related by

$$bi\ =\ \pi_2{}^{-1}\ ;\ [low, sig]\ ;\ ti$$

That is to say, a bit represents the same integer as a trit in which the wire of negative weight is held low and the positive wire is connected to the same level as the given bit.

The best way to proceed is to find a c_4 for which $c_3\ ;\ \mathsf{fst}\ ti^{-1} = \mathsf{snd}\ ti^{-1}\ ;\ c_4$ and then use column induction.

$$\begin{aligned}
c_3\ ;\ \mathsf{fst}\ ti^{-1} &= \mathsf{fst}(\mathsf{snd}\ and\ ;\ bb+)\ ;\ s+\ ;\ s+^{-1}\ ;\ \mathsf{snd}\ bi^{-1}\ ;\ \mathsf{fst}\ ti^{-1} \\
&= \quad \{s+\ ;\ s+^{-1} = \mathsf{snd}\ +^{-1}\ ;\ rsh\ ;\ \mathsf{fst}\ s+\} \\
&\quad \mathsf{fst}(\mathsf{snd}\ and\ ;\ bb+)\ ;\ \mathsf{snd}\ +^{-1}\ ;\ rsh\ ;\ [(s+\ ;\ ti^{-1}), bi^{-1}] \\
&= \quad \{\text{right-shifting}\} \\
&\quad \mathsf{snd}(+^{-1}\ ;\ \mathsf{snd}\ bi^{-1})\ ;\ rsh\ ;\ \mathsf{fst}(\mathsf{fst}(\mathsf{snd}\ and\ ;\ bb+)\ ;\ s+\ ;\ ti^{-1}) \\
&= \quad \{\mathsf{snd}\ ni{>} \dashv \mathsf{fst}\ bb+\ ;\ s+\ ;\ ti^{-1}\} \\
&\quad \mathsf{snd}(+^{-1}\ ;\ \mathsf{snd}\ bi^{-1})\ ;\ rsh\ ;\ \mathsf{fst}([(\mathsf{snd}\ and\ ;\ bb+), ni{>}]\ ;\ s+\ ;\ ti^{-1}) \\
&= \quad \{\text{right-shifting, and definition of } ni{>} = ni^{-1}\ ;\ ni\} \\
&\quad \mathsf{snd}(+^{-1}\ ;\ [ni^{-1}, bi^{-1}])\ ;\ rsh\ ;\ \mathsf{fst}([(\mathsf{snd}\ and\ ;\ bb+), ni]\ ;\ s+\ ;\ ti^{-1}) \\
&= \quad \{\text{definition of } ti = [ni, bi]\ ;\ +\} \\
&\quad \mathsf{snd}\ ti^{-1}\ ;\ rsh\ ;\ \mathsf{fst}([(\mathsf{snd}\ and\ ;\ bb+), ni]\ ;\ s+\ ;\ ti^{-1}) \\
&= \quad \{ti^{-1} \vdash \mathsf{snd}\ sig \text{ and right-shifting}\} \\
&\quad \mathsf{snd}\ ti^{-1}\ ;\ rsh\ ;\ [([(\mathsf{snd}\ and\ ;\ bb+), ni]\ ;\ s+\ ;\ ti^{-1}), sig]
\end{aligned}$$

$$
\begin{aligned}
= &\quad \{\text{definition of } \mathit{fulladd}' = [\mathit{bb}+, \mathit{ni}] \mathbin{;} s+ \mathbin{;} \mathit{ti}^{-1}\} \\
&\mathsf{snd}\ \mathit{ti}^{-1} \mathbin{;} \mathit{rsh} \mathbin{;} [(\mathsf{fst\,snd}\ \mathit{and} \mathbin{;} \mathit{fulladd}'), \mathit{sig}] \\
= &\quad \{\text{naming } c_4 = \mathit{rsh} \mathbin{;} [(\mathsf{fst\,snd}\ \mathit{and} \mathbin{;} \mathit{fulladd}'), \mathit{sig}]\} \\
&\mathsf{snd}\ \mathit{ti}^{-1} \mathbin{;} c_4
\end{aligned}
$$

The cell c_4 consists of wiring, an *and* gate and a full adder. It requires skill and experience to be able to recognise opportunities to use existing hardware. We expect that tables of readily-implementable relations would be a useful tool for the practising designer, just as books of data sheets have been in the past.

The introduction of trits and the above calculation both required some ingenuity, but the rest of the derivation is largely mechanical. Because c_4 need not be refined any further, we have almost completed the design of c_2. The ti^{-1} that we need for the column induction is introduced at the bottom of the column, pushed up the column, and finally implemented when it meets a zero at the top.

$$
\begin{aligned}
c_2 &= \mathsf{snd}\ \mathit{dstr}_n \mathbin{;} \mathit{zip}_n \mathbin{;} {\pi_1}^{-1} \mathbin{;} \mathsf{snd}\ \|[0]\| \mathbin{;} \mathsf{col}_n\ c_3 \mathbin{;} \mathsf{fst}\ \mathit{bi}^{-1} \\
&= \quad \{\text{bits as trits, since } \mathit{bi} = {\pi_2}^{-1} \mathbin{;} [\mathit{low}, \mathit{sig}] \mathbin{;} \mathit{ti}\} \\
&\quad \mathsf{snd}\ \mathit{dstr}_n \mathbin{;} \mathit{zip}_n \mathbin{;} {\pi_1}^{-1} \mathbin{;} \mathsf{snd}\ \|[0]\| \mathbin{;} \mathsf{col}_n\ c_3 \mathbin{;} \mathsf{fst}(\mathit{ti}^{-1} \mathbin{;} [\mathit{low}, \mathit{sig}] \mathbin{;} \pi_2) \\
&= \quad \{\text{column induction from } c_3 \mathbin{;} \mathsf{fst}\ \mathit{ti}^{-1} = \mathsf{snd}\ \mathit{ti}^{-1} \mathbin{;} c_4\} \\
&\quad \mathsf{snd}\ \mathit{dstr}_n \mathbin{;} \mathit{zip}_n \mathbin{;} {\pi_1}^{-1} \mathbin{;} \mathsf{snd}(\|[0]\| \mathbin{;} \mathit{ti}^{-1}) \mathbin{;} \mathsf{col}_n\ c_4 \mathbin{;} \mathsf{fst}([\mathit{low}, \mathit{sig}] \mathbin{;} \pi_2)
\end{aligned}
$$

The final step is to eliminate the remaining ti. The representation of zero as a trit is a pair of equal signals, $\|[0]\| \mathbin{;} \mathit{ti}^{-1} = \|[0]\| \mathbin{;} \mathit{any} \mathbin{;} \mathit{sig} \mathbin{;} \mathit{fork}$, so

$$
\begin{aligned}
{\pi_1}^{-1} \mathbin{;} \mathsf{snd}(\|[0]\| \mathbin{;} \mathit{ti}^{-1}) &= {\pi_1}^{-1} \mathbin{;} \mathsf{snd}(\mathit{any} \mathbin{;} \|[0]\| \mathbin{;} \mathit{any} \mathbin{;} \mathit{sig} \mathbin{;} \mathit{fork}) \\
&= {\pi_1}^{-1} \mathbin{;} \mathsf{snd}(\mathit{sig} \mathbin{;} \mathit{fork})
\end{aligned}
$$

This completes the first implementation for the multiplier: gathering all the parts we have that

$$
\begin{aligned}
\mathcal{M} = {} & \mathsf{snd}\ \mathit{dstl}_m \mathbin{;} \mathsf{row}_m\ c_2 \mathbin{;} \mathsf{snd}\ \mathsf{map}_n\ \mathit{sig} \\
& \text{where } c_2 = \mathsf{snd}\ \mathit{dstr}_n \mathbin{;} \mathit{zip}_n \mathbin{;} \pi_1^{-1} \mathbin{;} \mathsf{snd}(\mathit{sig} \mathbin{;} \mathit{fork}) \mathbin{;} \\
& \qquad\qquad \mathsf{col}_n\ c_4 \mathbin{;} \mathsf{fst}([\mathit{low}, \mathit{sig}] \mathbin{;} \pi_2) \\
& \qquad\quad c_4 = \mathit{rsh} \mathbin{;} [\mathsf{fst\,snd}\ \mathit{and} \mathbin{;} \mathit{fulladd}', \mathit{sig}]
\end{aligned}
$$

However, if we try to draw this circuit it turns out to be far from beautiful. The instances of *dstl*, *dstr* and *zip* in particular consist of large bundles of wires which are knitted together in untidy knots.

This would not matter if we only needed to produce a netlist for input to some absolutely general automatic layout generator such as a programmer for configurable logic arrays, or a VLSI layout system. However if we want to interface to a design system using a more structured circuit description, we must rearrange the wiring to give a more satisfactory implementation. Most refinements will naturally result in circuits that need to be improved by moving wires or components about.

6 Rearrangement

A further calculation, about as long as that which has gone before, will allow the wires of the implementation of the multiplier to be 'combed out' into a regular pattern. The strategy for this part of the calculation is to represent as much as possible of the structure of the circuits as adjacent rows and columns of simple components – even the wiring – and then to interleave these rows and columns.

Because the beside and below operators distribute over each other in this way:

$$(A \updownarrow B) \leftrightarrow (C \updownarrow D) = (A \leftrightarrow C) \updownarrow (B \leftrightarrow D)$$

(a property which Richard Bird calls 'abiding' [3]) it follows by induction that adjacent columns can be combined

$$(\mathsf{col}_n\, R) \leftrightarrow (\mathsf{col}_n\, S) = \mathsf{col}_n(R \leftrightarrow S)$$
$$(\mathsf{row}_n\, R) \updownarrow (\mathsf{row}_n\, S) = \mathsf{row}_n(R \updownarrow S)$$

at least when $n > 0$, and similarly rows that are below each other can be combined; and more generally $\mathsf{row}_n\, \mathsf{col}_m\, R = \mathsf{col}_m\, \mathsf{row}_n\, R$, provided $n > 0$ and $m > 0$.

The pattern

$$\begin{aligned} A \between B &= (A \updownarrow \mathit{swp}) \leftrightarrow (\mathit{swp} \updownarrow B) \\ &= (A \leftrightarrow \mathit{swp}) \updownarrow (\mathit{swp} \leftrightarrow B) \end{aligned}$$

which is illustrated in Fig. 7 is one which arises often in the course of these interleavings, and has the pleasant property that

$$\mathsf{col}_n(A \between B) = \mathsf{fst}\ \mathit{zip}_n^{-1}\ ;\ ((\mathsf{col}_n\, A) \between (\mathsf{col}_n\, B))\ ;\ \mathsf{snd}\ \mathit{zip}_n$$
$$\mathsf{row}_n(A \between B) = \mathsf{snd}\ \mathit{zip}_n^{-1}\ ;\ ((\mathsf{row}_n\, A) \between (\mathsf{row}_n\, B))\ ;\ \mathsf{fst}\ \mathit{zip}_n$$

again provided $n > 0$, which can be used right-to-left to rewrite matching rows and columns as single structures of more complex cells, possibly eliminating the large number of wire crossings indicated by zip_n.

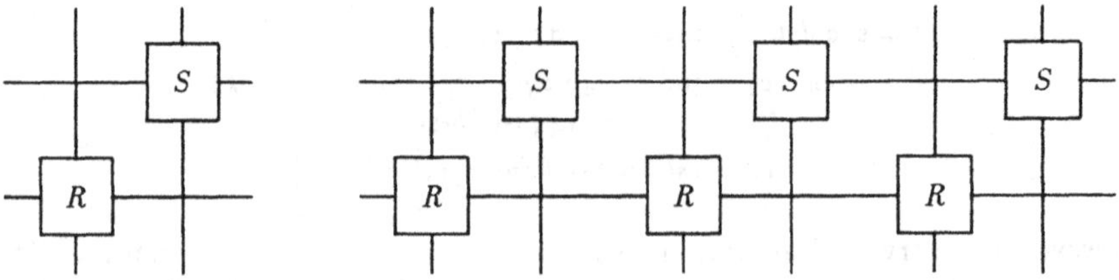

Fig. 7. $R \between S$ and $\mathsf{row}_3(R \between S)$

If R and S are relations for which $R \updownarrow S = \mathsf{fst}\ \mathit{swp}\ ;\ (S \updownarrow R)\ ;\ \mathsf{snd}\ \mathit{swp}$, which is the case if either R or S is itself swp, then

$$\mathsf{col}_n(R \updownarrow S) = \mathsf{fst}\ \mathit{zip}_n\ ;\ (\mathsf{col}_n\, R \updownarrow \mathsf{col}_n\, S)\ ;\ \mathsf{snd}\ \mathit{zip}_n^{-1}$$

which can be used as a right-to-left rule to interleave two columns of largely independent cells. The dual is a statement about rows, that if $R \leftrightarrow S = \mathsf{snd}\ swp\ ; (S \leftrightarrow R)\ ; \mathsf{fst}\ swp$

$$\mathsf{row}_n(R \leftrightarrow S) = \mathsf{snd}\ zip_n\ ; (\mathsf{row}_n\ R \leftrightarrow \mathsf{row}_n\ S)\ ; \mathsf{fst}\ zip_n^{-1}$$

Not all the structure of the implementation of the multiplier yet consists of rows and columns, but we have already seen that the instances of *dstr* and *dstl* have implementations as columns and rows of simpler cells, for

$$\begin{aligned} dstr_n &= \mathsf{col}_n(fork\ ; \mathsf{fst}\ \pi_2)\ ; \pi_2 \\ dstl_n &= \mathsf{row}_n(fork\ ; \mathsf{snd}\ \pi_1)\ ; \pi_1 \end{aligned}$$

and other structures and wiring components can also be expanded into rows and columns in this way, for

$$\begin{aligned} \mathsf{map}_n\ R &= \pi_1^{-1}\ ; \mathsf{col}_n(\pi_1\ ; R\ ; \pi_2^{-1})\ ; \pi_2 \\ &= \pi_2^{-1}\ ; \mathsf{row}_n(\pi_2\ ; R\ ; \pi_1^{-1})\ ; \pi_1 \\ \mathsf{map}_n\ R\ ; fork &= \mathsf{map}_n(R\ ; fork)\ ; zip_n^{-1} \\ \mathsf{fst}\ \mathsf{map}_n\ R\ ; swp &= \mathsf{col}_n(\mathsf{fst}\ R\ ; swp) \\ \mathsf{snd}\ \mathsf{map}_n\ R\ ; swp &= \mathsf{row}_n(\mathsf{snd}\ R\ ; swp) \end{aligned}$$

and so on.

6.1 Tidying the Wiring

Although we performed this part of the calculation in about the same detail as the first part of the derivation, we are not satisfied that our presentation is yet sufficiently lucid, nor that it conveys any new insights, to justify its inclusion in full in the paper. In particular, it has proved very hard to control the size of the expressions and they are not easily read. Accordingly we will only outline the shape of the calculation, confident in our ability if challenged to explain the calculation in more detail than a human reader could want. Indeed we intend that these calculations should be explained to a machine that would check their accuracy, and machine assistance in such a detailed presentation is certainly necessary.

Since we expect that the structure of the circuit will change as we bring in more wiring from the outside, the restructuring will be done from the inside out. The component in the middle of the circuit, if we agree to disregard c_4 which we claim is in essence already implemented, begins

$$\begin{aligned} \mathsf{snd}\ dstr_n\ ; zip_n = \quad & \{\text{expanding } dstr \text{ and interleaving the columns}\} \\ & rsh\ ; [zip_n, \pi_2]\ ; \mathsf{col}_n\ c_5\ ; \pi_2 \\ & \text{where } c_5 = \left(\pi_1\ ; \pi_2^{-1}\right) \between \left(\mathsf{fst}\ \pi_1^{-1}\ ; (swp \updownarrow (\pi_2\ ; fork))\right) \end{aligned}$$

The component c_5 can be simplified, by first expressing it in terms of beside and below, and then using the properties of the projections, the abiding of beside with below, and reassociating operations with left and right shifts. Thus we arrive at

$$c_5 = [\pi_1^{-1}, \pi_2]\ ; (swp \updownarrow (\pi_2\ ; fork))\ ; [\pi_2^{-1}, lsh]$$

which, since c_5 is just a wiring relation, can be checked by showing that each side relates the same tuples in the same ways. Then by a further column induction

$$\begin{aligned} \mathsf{snd}\ dstr_n\ ;\ zip_n &= rsh\ ;\ \mathsf{fst}\ zip_n\ ;\ \mathsf{col}_n\ c_6\ ;\ \pi_2 \\ &\text{where } c_6 = \mathsf{fst}\ \pi_1^{-1}\ ;\ (swp \updownarrow (\pi_2\ ;\ fork))\ ;\ \mathsf{snd}\ lsh \end{aligned}$$

Substituting in the definition of c_2, this column can be merged with the $\mathsf{col}_n\ c_4$ which it is beside, ultimately yielding

$$\begin{aligned} c_2 = {} & \mathsf{snd}\,\mathsf{snd}(\pi_1^{-1}\ ;\ \mathsf{snd}(sig\ ;\ fork))\ ; \\ & ((zip_n\ ;\ \pi_2^{-1}) \leftrightarrow (\mathsf{col}_n(c_6 \leftrightarrow c_4)))\ ; \\ & \mathsf{fst}(\pi_2\ ;\ \pi_2\ ;\ [low, sig]\ ;\ \pi_2) \end{aligned} \tag{12}$$

When this is substituted back into the implementation of $\mathcal{M}$, it appears alongside an instance of $dstl_m$ which can be expanded as a row, and merged with it:

$$\begin{aligned} \mathcal{M} &= \mathsf{snd}\ dstl_m\ ;\ \mathsf{row}_m\ c_2\ ;\ \mathsf{snd}\ \mathsf{map}_n\ sig \\ &= rsh\ ;\ ((\mathsf{row}_m\ c_2) \updownarrow (dstl_m\ ;\ \pi_1^{-1}))\ ;\ \mathsf{snd}(\pi_1\ ;\ \mathsf{map}_n\ sig) \\ &= \quad \{\text{expanding } dstl \text{ and row below row}\} \\ & \quad rsh\ ;\ \mathsf{snd}\ \mathsf{map}_m\ \pi_2^{-1}\ ;\ \mathsf{row}_m(c_2 \updownarrow ((\pi_1\ ;\ fork) \leftrightarrow swp))\ ;\ \mathsf{snd}(\pi_1\ ;\ \mathsf{map}_n\ sig) \end{aligned}$$

Now note that $\mathsf{snd}\,\mathsf{fst}\,\mathsf{map}_n\ id \dashv c_2$, from which it follows that the *fork* and *swp* in this expression are also n signals wide. These can be expanded and interleaved in the same way, and the resulting column combined with the $\mathsf{col}_n(c_6 \leftrightarrow c_4)$ obtained by substituting for c_2 using (12). The result of this calculation is

$$\begin{aligned} \mathcal{M} = {} & rsh\ ;\ [zip_n, \mathsf{map}_m(\pi_1^{-1}\ ;\ \mathsf{snd}(sig\ ;\ fork))]\ ; \\ & \mathsf{row}_m\ \mathsf{col}_n\ cell\ ; \\ & [\mathsf{map}_m(\pi_2\ ;\ [low, sig]\ ;\ \pi_2), \mathsf{map}_n(\pi_1\ ;\ sig)] \\ & \text{where } cell = \mathsf{fst}(\mathsf{snd}\ fork\ ;\ rsh)\ ;\ ((c_6 \leftrightarrow c_4) \updownarrow swp) \end{aligned} \tag{13}$$

which is a grid of cells, surrounded by some simple wiring.

The only part of the wiring which we would prefer not to implement is the instance of zip_n which would involve $n(n-1)/2$ wire crossings. We choose to have the user of the multiplier implement this, and change the specification of our design to $\mathsf{fst}\ zip_n^{-1}\ ;\ lsh\ ;\ \mathcal{M}$. That is to say that we will offer the user a component which must be given one of the factors already interleaved with the carry in. The user may well want the carry in to be zero anyway.

6.2 Tidying the Cell

Now consider the cell of which this grid is composed.

$$\begin{aligned} cell &= \mathsf{fst}(\mathsf{snd}\ fork\ ;\ rsh)\ ;\ ((c_6 \leftrightarrow c_4) \updownarrow swp) \\ &\text{where } c_6 = \mathsf{fst}\ \pi_1^{-1}\ ;\ (swp \updownarrow (\pi_2\ ;\ fork))\ ;\ \mathsf{snd}\ lsh \\ &\qquad\quad c_4 = rsh\ ;\ [\mathsf{fst}\ \mathsf{snd}\ and\ ;\ fulladd', sig] \end{aligned}$$

It is now necessary to simplify the wiring of this cell, and further to make sure that each of the remaining components is in our repertoire of standard components.

This means, for example, that we will not leave *fork* in the final implementation, because *fork* may in general involve forking a large bus of wires, involving many wire crossings. We will only be satisfied when each fork in the circuit divides a single signal wire, that is when each is an instance of $sfork = sig \,;\, fork$. Similar considerations apply to the other primitive wiring tiles.

A calculation about as long as that in the preceding section of the paper, but conceptually simpler in that it involves no rows and columns, shows that

$$\begin{aligned} cell = {} & (\mathsf{fst\,fst}\ \pi_1^{-1} \,;\, ((sswp \updownarrow (\pi_2 \,;\, sfork)) \updownarrow sswp) \,;\, \mathsf{snd}(\mathsf{snd}\ sfork \,;\, srsh)) \leftrightarrow \\ & ((\mathsf{fst}(slsh \,;\, \mathsf{snd}\ and) \,;\, srsh \,;\, [fulladd', sig]) \updownarrow (sswp \leftrightarrow sswp)) \\ \text{where } sfork &= sig \,;\, fork \\ sswp &= [sig, sig] \,;\, swp \\ slsh &= [[sig, sig], sig] \,;\, lsh \qquad = [sig, sig] \updownarrow [sig, sig] \\ srsh &= [[sig, sig], [sig, sig]] \,;\, rsh = [[sig, sig], sig] \leftrightarrow [sig, sig] \end{aligned}$$

Figure 8 illustrates a layout for the cell, suggested by this equation, annotated with indications of which wires are necessarily inputs to or outputs from the cell.

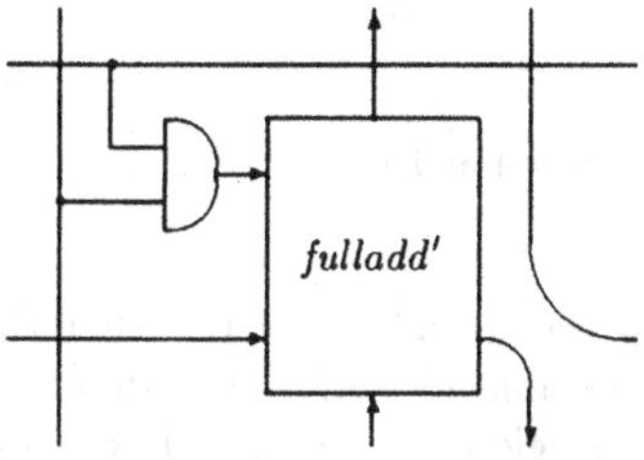

Fig. 8. the *cell* of the multiplier

This completes the development of a bit-level regular array that implements the multiplier, because substituting back into (13),

$$\begin{aligned} \mathsf{fst}\ zip_n^{-1} \,;\, lsh \,;\, \mathcal{M} = \mathsf{snd}\ \mathsf{map}_n(\pi_1^{-1} \,;\, \mathsf{snd}\ sfork) \,; \\ \mathsf{row}_m\ \mathsf{col}_n\ cell \,; \\ [\mathsf{map}_m(\pi_2 \,;\, [low, sig] \,;\, \pi_2), \mathsf{map}_n\ \pi_1] \end{aligned}$$

where *cell* is implemented entirely in terms of bit-operations. The layout which this suggests for $\mathsf{fst}\ zip^{-1} \,;\, lsh \,;\, \mathcal{M}$ is shown in Fig. 9.

This design is essentially the one in [13], where we presented it rather more abruptly. That paper systematically transforms the circuit into a bit-systolic array able to deliver a multiplication once on each cycle of a clock whose period

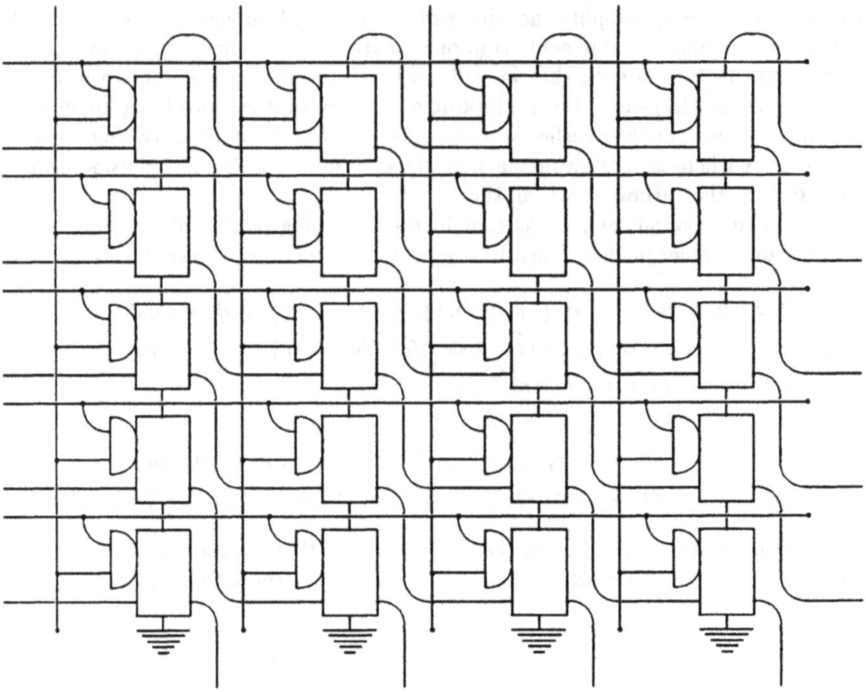

Fig. 9. fst zip^{-1} ; lsh ; $\mathcal{M}$ with $m = 4$ and $n = 5$

is roughly the delay through a full adder. Notice that the circuit is hexagonally-connected, although we were able to derive it without needing to introduce any special constructors, nor to develop an algebra of hex-connected circuits.

7 Conclusion

We have shown in some detail how the design of a complex and subtle circuit is a natural consequence of a relatively abstract specification and some simple design decisions. These decisions are ones that can largely be explained at the level of the abstract specification. The most positive thing about this calculation is that it is an argument for the correctness of the design *before the fact.* That is to say that once the design has been done there is no need to construct a separate proof of correctness.

That a derivation can be driven by high-level decisions is promising because it suggests that there is a systematic way of delineating a design space which could be explored for alternative implementations of the same specification. For example the choice of representation for the carries has driven us to this design whereas another choice would have produced the carry-save multiplier [9]; other designs are suggested by the possibility of choosing some other base (than two) to represent any of the inputs or outputs or carries. Perhaps base four might yield a design with fewer, larger cells? We hope that as more experience is gained in deriving designs in this way, better methods will emerge of aiming more directly at good implementations.

The derivation proceeds by refinement: a process of reducing the complexity of the data on the 'wires' of the circuit until only bits remain, and all of the components are implementable. We found very useful the ability to deal in a uniform way both with the abstractions such as integers and multiplication, and with concrete signal levels and gates. Distinguishing between numbers and signal levels is something which we have not done in the past, and we believe that it makes the step to 'known' hardware components easier to recognise.

As far as was possible we tried to be honest about the design process: we followed the calculation through with no particular target implementation in mind. For example, at the point where we discovered that $\mathsf{snd}\, bi\!> \not\dashv c_3 \,;\, \mathsf{fst}\, bi^{-1}$ and were driven to introduce trits into the design, we did so 'blindly'. One of us was unable to proceed with the design, and the other suggested this invention step with no global knowledge of the design, nor of the ultimate consequences of this decision.

Notice that the two wires representing this trit in the final circuit carry signals in opposite directions. That is to say, one is an input to a component and the other is an output from it. The calculations in Ruby, and the introduction of the trit data type, were performed in complete ignorance of this fact. This abstraction from inputs and outputs is a consequence of the decision to represent circuits by relations, and a valuable separation of concerns. The cost is of course that when claiming that we have implemented a primitive relation by some electronic component we must check that it has an appropriate allocation of inputs and outputs, compatible with those of the rest of the circuit of which it is a part.

We think that the first half of the derivation presented in this paper is a convincing account of the development of the algorithm which is implemented by the final circuit. It is disappointing however that the rearrangement of the circuit remains tedious and difficult; we hope that this can be improved, perhaps by a better understanding of the sort of calculation involved. It is clear, however, that mechanical assistance in checking the details of our calculations would be a great help. The very low levels of circuit design are already heavily automated, and it is one of our aims to expand the range of such mechanical assistance to higher and more abstract levels of the design process.

The techniques used here should be applicable to a range of similar arithmetic circuits, but we see no reason why they should not be extended to other fields. All that is necessary is that solutions can be constructed from components whose combinations have algebraic properties like those of the arithmetic building-blocks which we used here. We are investigating, as an example, the derivation of routing circuits implemented as arrays of binary decisions and arbiters [11].

Acknowledgement. The work reported in this paper has been supported by grants from the UK Science and Engineering Research Council, and was done while Mary Sheeran held a Royal Society of Edinburgh BP Research Fellowship at the Department of Computing Science of the University of Glasgow.

References

1. R. C. Backhouse, P. J. de Bruin, P. Hoogendijk, G. Malcolm, E. Voermans and J. C. S. P. van der Woude, *A relational theory of datatypes*, unpublished.
2. Richard Bird, *Ain Introduction to the Theory of Lists*, in M. Broy (ed.), *Logic of programming and calculi of discrete design*, NATO advanced study institutes, Series F: Computer and systems sciences, Springer-Verlag, 1987. pp. 5–42.
3. Richard Bird, *Lectures on constructive functional programming*, in M. Broy (ed.), *Constructive methods in computing science*, NATO advanced study institutes, Series F: Computer and systems sciences, Springer-Verlag, 1989.
4. D. Borrione and A. Salem, *Proving an on-line multiplier with OBJ and TACHE: a practical experience*, in [6].
5. S-K. Chin, *Verified synthesis functions for negabinary arithmetic hardware*, in [6].
6. L. J. M. Claesen, *Applied formal methods for correct VLSI design*, North-Holland, 1989.
7. Rogardt Heldal, Carsten Kehler Holst, Phil Wadler (eds.), *Functional programming*, Glasgow 1991, Springer-Verlag (Workshops in Computing), 1992.
8. Graham Hutton and Ed Voermans, *A calculational theory of pers as types*, in [7].
9. Geraint Jones and Mary Sheeran, *Relations and Refinement in Circuit Design* in Carroll Morgan and J. C. P. Woodcock (eds.), Proceedings of the 3rd Refinement Workshop, 9–11 January 1990, Springer-Verlag (Workshops in Computing), 1991. pp. 133–152.
10. Geraint Jones and Mary Sheeran, *Circuit design in Ruby*, in Jørgen Staunstrup (ed.), *Formal methods for VLSI design*, North-Holland, 1990. pp. 13–70.
11. Mark B. Josephs, Rudolph H. Mak, Jan Timen Udding, Tom Verhoeff and Jelio T. Yantchev, High-level design of an asynchronous packet routing chip, in [16].
12. L. G. L. T. Meertens, *Constructing a calculus of programs*, in J. L. A. van de Snepscheut (ed.), *Mathematics of program construction*, Springer-Verlag LNCS 375, 1989.
13. Mary Sheeran, *Designing regular array architectures using higher-order functions*, in J.-P. Jouannaud (ed.), *Functional programming languages and computer architecture*, Springer-Verlag LNCS 201, 1985. pp. 220–237.
14. Mary Sheeran, *A note on abstraction in Ruby*, in [7].
15. M. Simonis, *Formal verification of multipliers* in [6].
16. Jørgen Staunstrup and R. Sharp (eds.), *Designing Correct Circuits*, Lyngby 1992, IFIP Transactions A, Volume 5, North-Holland, 1992.
17. D. Verkest, L. Claesen and H. De Man, *A proof of the non-restoring division algorithm and its implementation on the Cathedral-II ALU*, in [16].

An Operational Semantics for the Guarded Command Language *

Johan J. Lukkien

Eindhoven University of Technology
Computer Science
P.O. Box 513
5600 MB Eindhoven, The Netherlands

Abstract. In [6], Dijkstra and Scholten present an axiomatic semantics for Dijkstra's guarded command language through the notions of weakest precondition and weakest liberal precondition. The informal notion of a computation is used as a justification for the various definitions. In this paper we present an operational semantics in which the notion of a computation is made explicit. The novel contribution is a generalization of the notion of weakest precondition. This generalization supports reasoning about general properties of programs (i.e, not just termination in a certain state).

1 Introduction

In [6], Dijkstra and Scholten present an axiomatic semantics for Dijkstra's guarded command language through the notions of weakest precondition and weakest liberal precondition. The informal notion of a computation is used as a justification for the various definitions. In this paper we present an operational semantics in which a computation is defined explicitly as a sequence of states through which execution of the program may evolve, starting from a certain initial state. This way of characterizing program executions is fairly conventional. Earlier work on the topic may be found in [8]; in [1] such a characterization is used to define liveness and safety formally. These state sequences play an important rôle in temporal logic (e.g., [10]).

The contributions of this paper are twofold. The way repetition is dealt with differs a little from the usual approach. Repetition has been defined both as the solution of a recursive equation and as the limit of its unfoldings. In this paper we also use a definition in terms of the limit of unfoldings but we explicitly locate limit series in these unfoldings such as to include unbounded choice in an easy way. Then we show that the repetition is also a solution of a recursive equation. This equation is used to prove that the weakest precondition of the repetition is the strongest solution of a recursive equation and that the weakest liberal precondition is the weakest solution of a recursive equation. The main contribution of this paper is a generalization of the notion of weakest precondition. This generalization supports reasoning about general properties of programs (i.e, not just termination in a certain state), including temporal properties. The alternative definition of the repetition is

* The research described in this paper was conducted while the author was on leave from Groningen University at the California Institute of Technology.

especially helpful in determining which extreme solution of a recursive equation to choose as the weakest precondition of the repetition, for such a property.

We use the term *operational* for this semantics because it determines the states that some abstract mechanism executing the program goes through. Sometimes this term is reserved for a labelled transition system in which the atomic actions of the program are the labels for transitions between states. A semantics like ours is then usually called *denotational.*

This paper proceeds as follows. In the next section we introduce the language and its operational semantics. In the third section we introduce properties of programs and the generalized notion of weakest precondition; we discuss Dijkstra's *wp* and *wlp* in detail. We also generalize the notion of (non)determinate programs. We show how to introduce the temporal property *ever* (or *eventually*, the operator "$\diamond$" from temporal logic). We end with some conclusions.

2 Operational Semantics

In defining the operational semantics of our language we define what happens if a program written in that language is executed by a machine. The machine has an internal state that is changed according to the program's instructions. This state consists of the values of the variables declared in the program. The semantics of a program is the set of all state sequences (finite or infinite) that may result from executing the program. We call these sequences *computations.* For our purposes it is sufficient to restrict our attention to this state, the space in which it assumes its values and the sequences over this space.

We first introduce some notation.

V = The list of all program variables.
X = A cartesian product of domains, one for each program variable, in the order in which they occur in V. X is called the state space and its elements are called states. X is nonempty.
$Bool = \{true, false\}$.
P = The functions from X to *Bool.* Elements of P are called predicates.
T = The set of all nonempty finite or infinite sequences of states.

For $t \in T$, we use t^∞ to denote an infinite sequence of t's. For strings, we use juxtaposition to denote catenation. This is extended to sets in the obvious way. The catenation of a sequence to the tail of an infinite sequence is that infinite sequence. For a string s, $|s|$ denotes its length, $s.i$ element i if $0 \leq i < |s|$ and $last.s$ its last element if $|s| < \infty$. For strings s and t, $s \sqsubseteq t$ denotes that s is a prefix of t. $\sqsubseteq$ is a partial order on strings. Sometimes it is convenient to distinguish between finite and infinite strings. For A a set of strings, $fin.A$ is the subset of A consisting of the finite strings in A and $inf.A$ the remaining part of A.

P, equipped with the pointwise implication order is a boolean lattice (see, for instance, [5]). We use $[p]$ as a shorthand for $\forall(x : x \in X : p.x)$. Hence, the fact that p precedes q in the order is denoted by $[p \Rightarrow q]$. We use the name *true* for the function that yields *true* everywhere and *false* for the function that yields *false* everywhere.

With these preliminaries, we define programs. A program describes which sequences are possible computations, hence a program may be viewed as a subset of T. However, not all these subsets can be programs. If a subset of T is to be viewed as a program there must be, for every $x \in X$, a sequence in that subset that starts with x, since execution of a program can commence in an arbitrary initial state.

$$\mathit{Prog} = \{S : S \subseteq T \wedge \forall(x : x \in X : \exists(s : s \in S : s.0 = x)) : S\},$$
the set of programs.

So, we do not bother whether a program is a computable set and indeed, many elements of *Prog* are not computable. We only have the restriction with respect to initial states. Notice that since X is nonempty, a program can never be the empty set.

We now look more closely at our programming language. We list the constructs and their intended meaning.

abort	- loop forever
skip	- do nothing
$y := e$	- assign value e to program variable y
$S; U$	- sequential composition
if $[](i :: B_i \rightarrow S_i)$ **fi**	- execute an S_i for which B_i holds
do $B \rightarrow S$ **od**	- repeat S as long as B holds

All these constructs are defined as elements of *Prog*. For every definition we have to verify whether we defined a proper element of *Prog*, i.e., we have to verify the restriction in the definition of *Prog*.

We start with the definitions of the three basic constructs.

$$\begin{aligned} \mathit{abort} &= \{x : x \in X : x^{\infty}\} \\ \mathit{skip} &= X \end{aligned}$$

abort repeats the state in which it is started forever. *skip* does not do anything: initial state and final state coincide.

In order to define the assignment statement, we introduce substitution. Let d be a function from X to the component of the state space in which some program variable y assumes its values.

$$x[y/d.x].z = \begin{cases} x.z & \text{if } y \neq z \\ d.x & \text{if } y = z \end{cases}$$

Using this, the assignment statement is defined by

$$y := e = \{x : x \in X : x(x[y/e.x])\}.$$

The assignment changes the state in which it is started in that at position $y, e.x$ is stored. Clearly, all three constructs satisfy the constraint imposed in the definition of *Prog*.

We define sequential composition in a slightly more general context than only for members of *Prog*. We define it for general subsets of T. For $A, B \subseteq T$,

$$A; B = \{a, x, b : ax \in A \wedge xb \in B : axb\} \cup \mathit{inf}.A.$$

$A; B$ contains the infinite sequences in A and all sequences formed from a finite sequence $a \in A$ and a sequence $b \in B$ such that the last element of a is the first element of b, by catenating a without its last element, and b. In the case that B is nonempty, we can omit the term $\mathit{inf}.A$ because of our convention with respect to the catenation of infinite sequences.

Property 1. *Let* $A, B \subseteq T$.

$$(i)\ \forall(s : s \in A; B : \exists(a : a \in A : a \sqsubseteq s))$$
$$(ii)\ B \in \mathit{Prog} \Rightarrow \forall(a : a \in A : \exists(s : s \in A; B : a \sqsubseteq s))$$

Proof. (i) follows directly from the definition of ";". (ii) Choose $a \in A$. If $|a| = \infty$, $a \in A; B$. If $|a| < \infty$, there exists a $xb \in B, x \in X$ such that $a = cx$ since $B \in \mathit{Prog}$. From the definition of ";" it follows that $cxb \in A; B$. Obviously, $a \sqsubseteq cxb$.

From property 1, it follows that $A, B \in \mathit{Prog}$ implies $A; B \in \mathit{Prog}$.

In both the alternative construct and the repetition we have the notion of a guard. A guard is a predicate. Operationally, evaluation of a guard is the restriction to the states for which it yields *true*. We do want to record the evaluation of a guard such that an infinite number of evaluations yields an infinite sequence. So, for a predicate p we define

$$p^{guard} = \{x : x \in X \wedge p.x : xx\}.$$

Now we can define the alternative construct. We abbreviate **if** $[\!](i :: B_i \rightarrow S_i)$ **fi** by *IF*.

$$\mathit{IF} = \bigcup(i :: B_i^{guard}; S_i) \cup (\forall(i :: \neg B_i))^{guard}; \mathit{abort}$$

We have to show again that we defined a proper member of *Prog*. Notice that, for a predicate p, $p^{guard}; S$ is the restriction of S to those sequences that start with a state for which p holds, with the initial state repeated. Since for every $x \in X$ we have $\forall(i :: \neg B_i.x) \vee \exists(i :: B_i.x)$, the fact that *IF* is an element of *Prog* follows from the fact that S_i and *abort* are elements of *Prog*.

Finally, we define repetition. Executing the repetitive construct **do** $B \rightarrow S$ **od**, corresponds to the repeated, sequential execution of the body, S, possibly an infinite number of times. We therefore first study some properties of the sequential composition.

Property 2. *Let* $A, B \subseteq T$.

$$(i)\ \mathit{inf}.(A; B) = \mathit{inf}.A \cup (\mathit{fin}.A); \mathit{inf}.B$$
$$(ii)\ \mathit{fin}.(A; B) = (\mathit{fin}.A); \mathit{fin}.B$$

Proof. Immediate from the definition.

Property 3. *For $A, B, C \subseteq T$ such that A and B contain no infinite sequences,*

$$(A; B); C = A; (B; C).$$

Proof. Notice that, according to property 2(*i*), *inf.*$(A; B)$ is also empty. Choose $s \in (A; B); C$. By definition, s can be written as uxc such that $ux \in A; B$ and $xc \in C$. ux may be written in turn as $ayb, ay \in A, yb \in B$. If b is the empty string, we have $x = y$ and $x \in B$. Hence, $xc \in B; C$ and $s = axc \in A; (B; C)$. If b is not the empty string, we can write yb as $yb'x$ hence $yb'xc \in B; C$ and $s = ayb'xc \in A; (B; C)$. In a similar way we can prove $A; (B; C) \subseteq (A; B); C$.

Property 4. *";" distributes over arbitrary union in both arguments.*

Proof. Let I be a collection of subsets of T and $B \subseteq T$.

$$\begin{aligned}
& \bigcup(A : A \in I : A); B \\
= \;& \{\text{definition ";"}\} \\
& \{a, x, b : ax \in \bigcup(A : A \in I : A) \wedge xb \in B : axb\} \cup \mathit{inf}.(\bigcup(A : A \in I : A)) \\
= \;& \{\text{interchange unions, definition } \mathit{inf}\} \\
& \bigcup(A : A \in I : \{a, x, b : ax \in A \wedge xb \in B : axb\}) \cup \bigcup(A : A \in I : \mathit{inf}.A) \\
= \;& \{\text{calculus}\} \\
& \bigcup(A : A \in I : \{a, x, b : ax \in A \wedge xb \in B : axb\} \cup \mathit{inf}.A) \\
= \;& \{\text{definition}\} \\
& \bigcup(A : A \in I : A; B)
\end{aligned}$$

A similar proof can be given for the distributive property in the other argument.

Property 5. *";" is associative.*

Proof. Let $A, B, C \subseteq T$. We prove this property by showing *inf.*$((A; B); C) =$ *inf.*$(A; (B; C))$ and *fin.*$((A; B); C) =$ *fin.*$(A; (B; C))$.

$$\begin{aligned}
& \mathit{inf}.((A; B); C) \\
= \;& \{\text{property } 2(i)\} \\
& \mathit{inf}.(A; B) \cup \mathit{fin}.(A; B); \mathit{inf}.C \\
= \;& \{\text{property } 2(i), (ii)\} \\
& \mathit{inf}.A \cup (\mathit{fin}.A); \mathit{inf}.B \cup ((\mathit{fin}.A); \mathit{fin}.B); \mathit{fin}.C \\
= \;& \{\text{property } 3\} \\
& \mathit{inf}.A \cup (\mathit{fin}.A); \mathit{inf}.B \cup (\mathit{fin}.A); ((\mathit{fin}.B); \mathit{fin}.C) \\
= \;& \{\text{property } 4\} \\
& \mathit{inf}.A \cup (\mathit{fin}.A); (\mathit{inf}.B \cup (\mathit{fin}.B); \mathit{fin}.C)
\end{aligned}$$

$$
\begin{aligned}
&= \quad \{\text{property } 2(i)\} \\
&\quad \mathit{inf}.A \cup (\mathit{fin}.A); \mathit{inf}.(B;C) \\
&= \quad \{\text{property } 2(i)\} \\
&\quad \mathit{inf}.(A;(B;C))
\end{aligned}
$$

$$
\begin{aligned}
&\quad \mathit{fin}.((A;B);C) \\
&= \quad \{\text{property } 2(ii)\} \\
&\quad (\mathit{fin}.(A;B)); \mathit{fin}.C \\
&= \quad \{\text{property } 2(ii)\} \\
&\quad ((\mathit{fin}.A); \mathit{fin}.B); \mathit{fin}.C \\
&= \quad \{\text{property } 3\} \\
&\quad (\mathit{fin}.A); ((\mathit{fin}.B); \mathit{fin}.C) \\
&= \quad \{\text{property } 2(ii) \text{ twice (same steps)}\} \\
&\quad \mathit{fin}.(A;(B;C))
\end{aligned}
$$

Property 6. *skip is a neutral element of ";" for both arguments.*

Proof. Immediate, from the definitions of ";" and *skip*.

Definition 7. *For $A \subseteq T$, we define the powers of A as follows.*

$$
\begin{aligned}
&A^0 = \mathit{skip} \\
&A^{n+1} = A; A^n \quad \text{for } n \geq 0.
\end{aligned}
$$

We need this notion in the definition of the repetition but we need more. As mentioned previously, in an execution of a repetition the body may be executed an infinite number of times. The operational interpretation is an infinite sequence as a limit of finite ones. We therefore continue with studying limits of sequences.

The prefix order on T is a partial order. We are going to define the above limit as a least upper bound of some set. Since T is not a complete lattice, not every set needs to have a least upper bound. However, we show that each ascending chain has one. Consider an ascending chain in T: $c_i, 0 \leq i$. There are two possibilities: there exists an $N \in \mathbb{N}$ such that $c_j = c_N$ for $j \geq N$ or no such N exists. In the first case we have $c_N = \sqcup\{i : 0 \leq i : c_i\}$. In the second case we can find a function $f : \mathbb{N} \to \mathbb{N}$ such that $c_{f.i}, 0 \leq i$ is strictly increasing. Define $d \in T$ elementwise as $d.i = c_{f.i}.i, 0 \leq i$. Notice that d is of infinite length. We show that d is the least upper bound of the chain. Suppose $c_k \not\sqsubseteq d$ for some k. Since $|d| = \infty$, this implies that there exists a j such that $c_k.j \neq d.j$. Hence, $c_k.j \neq c_{f.j}.j$. It follows that c_k and $c_{f.j}$ are incomparable which contradicts the assumption that the c_i form a chain. We conclude that d is an upper bound of the chain. Notice that every upper bound of $\{i : 0 \leq i : c_i\}$ is of infinite length. Assume e is another upper bound, different from d. We have that d and e differ at some index i. It follows that $c_{f.i}$ is not a prefix of e which is in contradiction with the assumption that e is an upper bound of the chain. We conclude that d is the least upper bound of the chain. It follows that, for an ascending chain, the least upper bound is well defined. Notice however that T is not a cpo since the empty string is not an element of T.

Definition 8. *A characterizing chain for an infinite sequence $l \in T$ is an ascending chain $l_i, 0 \leq i$ such that*

$$(i)\ l_i \neq l_{i+1},$$
$$(ii)\ l = \sqcup\{i : 0 \leq i : l_i\}.$$

Definition 9. *l is a loop point of $A \subseteq T$ if there exists a characterizing chain $l_i, 0 \leq i$ for l such that $l_i \in A^i$. The set of loop points of A is denoted by loop.A.*

Property 10. *For $A \subseteq T$ we have loop.A $\cup$ inf.A $=$ A; loop.A.*

Proof. The proof is by mutual inclusion. First we prove $\subseteq$. Clearly, *inf.A* $\subseteq$ *A*; *loop.A*. Choose $l \in loop.A$ and $l_i, 0 \leq i$ a characterizing chain for l. From the definition of characterizing chain, it follows that all l_i are finite. We can write $l_i, 1 \leq i$ as $l_1 m_i$ since the l_i form an ascending chain and $l = l_1 m$ since l is the least upper bound of the l_i. We now have $(last.l_1)m_i \in \{last.l_1\}; A^{i-1}$. We construct a new chain, $n_i, 0 \leq i$, as follows: $n_i = (last.l_1)m_{i+1}$. The limit of this chain is $(last.l_1)m \in loop.A$. Since $l_1 \in A$, $\{l_1\}; (last.l_1)m \in A; loop.A$.

For the other inclusion, choose $l \in A; loop.A$. Then either $l = axm$ for $ax \in fin.A$, $xm \in loop.A$ or $l \in inf.A$. In the last case we are done. In the first case, we have a characterizing chain for xm, $xm_i, 0 \leq i$. We construct a new chain, l_i, as follows. $l_0 = (ax).0$, $l_i = axm_{i-1}, 1 \leq i$. Now we have $l_i \in A^i, 0 \leq i$ and $l = \sqcup\{i : 0 \leq i : l_i\}$ hence $l \in loop.A$.

Definition 11. *For $A \subseteq T$ we define A^ω as follows.*

$$A^\omega = \bigcup(n : 0 \leq n : A^n) \cup loop.A$$

Now we are ready to give the definition of the repetition. We abbreviate **do** $B \rightarrow S$ **od** by *DO*.

$$DO = (B^{guard}; S)^\omega; \neg B^{guard}$$

We have to verify that we defined a proper element of *Prog*. Suppose there exists $x \in X$ such that there is no sequence in *DO* starting with x. Since sequences starting with x are present in $(B^{guard}; S)^\omega$, it follows that they are all finite and terminate in a state satisfying B. Define a characterizing chain as follows. $l_0 = x$, choose $l_{i+1} \in \{l_i\}; S$, arbitrarily. The least upper bound of this chain is a loop point of $(B^{guard}; S)$ starting with x. Since it is a loop point, it is in *DO*. This is a contradiction. It follows that *DO* is an element of *Prog*.

The above definition of *DO* is a rather complicated one. It has the advantage however, of defining *DO* uniquely. A more conventional definition of *DO* starts with the first unfolding. A definition of *DO* is then obtained by solving

$$DO = \textbf{if}\ B \rightarrow S; DO\ [\!]\ \neg B \rightarrow skip\ \textbf{fi}$$

viewed as an equation in sets. This equation itself however, cannot serve as a definition because it does not have a unique solution. This means that we must distinguish among the solutions, which may require a topological approach. This is done for instance in [8] where it is proven that only one solution of the above equation exists that is nonempty and closed with respect to a certain topology, based on a metric. This proof is given only for a language with bounded choice. In our definition we could restrict ourselves to the prefix order on strings and we did not need a metric or a topology. Furthermore, we do not need any restrictions on the alternative construct.

The main reason that we deviate from the usual definition in terms of the first unfolding is that this semantics serves as the basis of a generalized notion of weakest precondition. This generalization allows us to introduce weakest preconditions for various properties. We use the first unfolding to obtain a recursive equation for the weakest precondition of a repetition. We use the above definition to distinguish among the solutions of this equation.

Since we are going to need that *DO* is semantically equivalent to its first unfolding, we prove it as a theorem. This is done in the remainder of this section.

Lemma 12. $S^{\omega} = \mathit{skip} \cup S; S^{\omega}$

Proof.

$$\begin{array}{ll} & S^{\omega} \\ = & \{\text{definition}\} \\ & \bigcup(n : 0 \leq n : S^n) \cup \mathit{loop}.S \\ = & \{\text{property 10: } \mathit{inf}.S \subseteq \bigcup(n : 0 \leq n : S^n)\} \\ & \bigcup(n : 0 \leq n : S^n) \cup S; \mathit{loop}.S \\ = & \{\text{definition 7, domain split}\} \\ & \mathit{skip} \cup \bigcup(n : 1 \leq n : S^n) \cup S; \mathit{loop}.S \\ = & \{\text{definition 7}\} \\ & \mathit{skip} \cup \bigcup(n : 0 \leq n : S; S^n) \cup S; \mathit{loop}.S \\ = & \{\text{property 4}\} \\ & \mathit{skip} \cup S; \bigcup(n : 0 \leq n : S^n) \cup S; \mathit{loop}.S \\ = & \{\text{property 4}\} \\ & \mathit{skip} \cup S; (\bigcup(n : 0 \leq n : S^n) \cup \mathit{loop}.S) \\ = & \{\text{definition 11}\} \\ & \mathit{skip} \cup S; S^{\omega} \end{array}$$

Theorem 13. $DO = \textbf{if } B \rightarrow S; DO \;[]\; \neg B \rightarrow \mathit{skip} \textbf{ fi}$

Proof.

$$\begin{array}{ll} & \textbf{if } B \rightarrow S; DO \; [\!] \neg B \rightarrow \textit{skip} \textbf{ fi} \\ = & \{\text{definition}\} \\ & B^{guard}; S; DO \cup \neg B^{guard}; \textit{skip} \cup \textit{false}^{guard}; \textit{abort} \\ = & \{\textit{false}^{guard} = \emptyset, \text{ definition } DO\} \\ & B^{guard}; S; (B^{guard}; S)^{\omega}; \neg B^{guard} \cup \neg B^{guard}; \textit{skip} \\ = & \{\text{property 6, twice}\} \\ & B^{guard}; S; (B^{guard}; S)^{\omega}; \neg B^{guard} \cup \textit{skip}; \neg B^{guard} \\ = & \{\text{property 4}\} \\ & (B^{guard}; S; (B^{guard}; S)^{\omega} \cup \textit{skip}); \neg B^{guard} \\ = & \{\text{lemma 12}\} \\ & (B^{guard}; S)^{\omega}; \neg B^{guard} \\ = & \{\text{definition}\} \\ & DO \end{array}$$

This concludes our definition of the operational semantics of the guarded command language.

3 Properties of Programs

In [6], attention is focused on one particular property of a program, viz., whether or not it terminates in a state that satisfies a certain condition. It has been recognized that, especially in the area of parallel programming, other properties are important as well. For instance, the properties "during execution, a condition will become true" and "a state satisfying a certain condition is always followed by a state satisfying some other condition" are properties defined and used in temporal logic. The fact that a program has a certain property means that all its computations share a common characteristic. In other words, a property may be regarded as a set of computations and a program has a property if all its computations are in the property. We add to our notation

$Prop = \mathcal{P}(T)$, the set of properties.

For a particular program and property it is fairly well possible that only part of the computations of the program are in the property. The function $wp.S.p$ defines the ***weakest precondition*** for a program S such that all its computations (starting from that precondition) have the property: termination in a state satisfying p (and similarly, wlp). This is generalized to other properties as follows.

Definition 14. *The weakest precondition is a function* $w : Prog \times Prop \rightarrow P$.

$$w.S.Q.x \;=\; \{x\}; S \subseteq Q$$

Now we introduce wlp and wp as special cases.

Definition 15. *The termination functions $lt, t : P \to Prop$ are defined by*

$$\begin{aligned} lt.p &= \{s,w : sw \in T \land w \in X \land (|sw| = \infty \lor p.w) : sw\}, \\ t.p &= \{s,w : sw \in T \land w \in X \land |sw| < \infty \land p.w : sw\}. \end{aligned}$$

The functions $wlp, wp : Prog \times P \to P$ are defined by

$$\begin{aligned} wlp.S.p &= w.S.(lt.p), \\ wp.S.p &= w.S.(t.p). \end{aligned}$$

We use the operational semantics to derive *wlp* and *wp* for the constructs in the language. We do not derive them all; some of them are left to the reader. From now on we will often use a somewhat different characterization of sequential composition. From property 4 it follows that for $S, U \subseteq T$,

$$S;U = \bigcup(s,x : sx \in S \land x \in X : \{sx\};U).$$

For arbitrary program S we have

$$\begin{aligned} & wlp.S.p.x \\ = & \quad \{\text{definition } wlp\} \\ & w.S.(lt.p).x \\ = & \quad \{\text{definition } w\} \\ & \{x\};S \subseteq lt.p \end{aligned}$$

and a similar formula for wp. This yields for *skip*

$$\begin{aligned} & wlp.skip.p.x \\ = & \quad \{\text{definition } skip\} \\ & \{x\};X \subseteq lt.p \\ = & \quad \{\text{definition ``;''}\} \\ & \{x\} \subseteq lt.p \\ = & \quad \{\text{definition } lt\} \\ & p.x. \end{aligned}$$

For $y := e$,

$$\begin{aligned} & wlp.(y := e).p.x \\ = & \quad \{\text{definitions assignment and ``;''}\} \\ & \{x(x[y/e.x])\} \subseteq lt.p \\ = & \quad \{\text{definition } lt\} \\ & p.(x[y/e.x]). \end{aligned}$$

For both the assignment and *skip*, *wlp* and *wp* coincide. In a similar way we derive

$$\begin{aligned} wlp.abort.p &= true, \\ wp.abort.p &= false. \end{aligned}$$

Sequential composition is a little more complicated.

$$
\begin{array}{ll}
 & wlp.(S;U).p.x \\
= & \{\text{definition } wlp\} \\
 & \{x\};S;U \subseteq lt.p \\
= & \{\text{definitions ";" and } lt\} \\
 & \forall(t,w : tw \in \bigcup(s,v : sv \in \{x\};S \wedge v \in X : \{sv\};U) \wedge w \in X : \\
 & \qquad |tw| = \infty \vee p.w) \\
= & \{tw \in \bigcup(s,v : sv \in \{x\};S \wedge v \in X : \{sv\};U) \equiv \\
 & \qquad tw = svuw \text{ for } sv \in \{x\};S, vuw \in \{v\};U\} \\
 & \forall(s,v,u,w : sv \in \{x\};S \wedge v \in X \wedge vuw \in \{v\};U \wedge w \in X : \\
 & \qquad |svuw| = \infty \vee p.w) \\
= & \{\text{calculus}\} \\
 & \forall(s,v,u,w : sv \in \{x\};S \wedge v \in X \wedge vuw \in \{v\};U \wedge w \in X : |sv| = \infty \vee \\
 & \qquad |vuw| = \infty \vee p.w) \\
= & \{\text{nesting}\} \\
 & \forall(s,v : sv \in \{x\};S \wedge v \in X : \forall(u,w : vuw \in \{v\};U \wedge w \in X : |sv| = \infty \vee \\
 & \qquad |vuw| = \infty \vee p.w)) \\
= & \{u \text{ and } w \text{ not free in } sv, \text{ renaming the dummy}\} \\
 & \forall(s,v : sv \in \{x\};S \wedge v \in X : |sv| = \infty \vee \\
 & \qquad \forall(u,w : uw \in \{v\};U \wedge w \in X : |uw| = \infty \vee p.w)) \\
= & \{\text{definition } lt\} \\
 & \forall(s,v : sv \in \{x\};S \wedge v \in X : |sv| = \infty \vee \{v\};U \subseteq lt.p) \\
= & \{\text{definition } wlp\} \\
 & \forall(s,v : sv \in \{x\};S \wedge v \in X : |sv| = \infty \vee wlp.U.p.v) \\
= & \{\text{definition } wlp\} \\
 & wlp.S.(wlp.U.p).x
\end{array}
$$

In a similar way we obtain $wp.(S;U).p = wp.S.(wp.U.p)$. The alternative construct is simple again and yields

$$
\begin{array}{l}
wlp.IF.p = \forall(i : B_i : wlp.S_i.p), \\
wp.IF.p = \exists(i :: B_i) \wedge \forall(i : B_i : wp.S_i.p).
\end{array}
$$

The repetition is the most interesting construct. Using theorem 13 we derive

$$
\begin{array}{ll}
 & wlp.DO.p.x \\
= & \{\text{theorem 13}\} \\
 & wlp.(\textbf{if } B \rightarrow S;DO\ [\!]\ \neg B \rightarrow skip\ \textbf{fi}).p.x \\
= & \{wlp \text{ for } IF, \text{";" and } skip\} \\
 & (B \wedge wlp.S.p.(wlp.DO.p)\ \vee\ \neg B \wedge p).x.
\end{array}
$$

We conclude that *wlp.DO.p* is a solution of the equation in unknown predicate Y:

$$[Y \equiv B \wedge wlp.S.Y \ \vee \ \neg B \wedge p] \tag{1}$$

According to the theorem of Knaster-Tarski ([5, 6]) this equation has extreme solutions if the righthandside is a monotonic function of Y, i.e.,

$$[Y \Rightarrow Y'] \Rightarrow [(B \wedge wlp.S.Y \ \vee \ \neg B \wedge p) \Rightarrow (B \wedge wlp.S.Y' \ \vee \ \neg B \wedge p)]$$

which reduces to

$$[Y \Rightarrow Y'] \Rightarrow [wlp.S.Y \Rightarrow wlp.S.Y'].$$

This is a direct consequence of the definition of *wlp*. Hence, equation (1) has a weakest and a strongest solution.

Theorem 16. *wlp.DO.p is the weakest solution of (1).*

Proof. Let y be an arbitrary solution of (1). We have to prove that y implies $wlp.DO.p$. Choose x such that $y.x$ holds and $s \in \{x\}; DO$. It is our obligation to prove that $wlp.S.p.x$ holds, i.e.,

$$|s| < \infty \Rightarrow p.(last.s)$$

We first prove by induction on n:

$$|t| < \infty \wedge t \in \{x\}; (B^{guard}; S)^n \Rightarrow y.(last.t) \tag{2}$$

For $n = 0$ we have the case $t = x$ and $y.x$ is given. Consider a finite sequence $t \in \{x\}; (B^{guard}; S)^{n+1}$. Then $t = avb, av \in \{x\}; (B^{guard}; S)^n, v \in X$. According to the induction hypothesis, $y.v$ holds. Since $vb \in \{x\}; B^{guard}; S, B.v$ holds. From the fact that y solves (1) we conclude $wlp.S.y.v$. This yields $y.(last.b)$, hence $y.(last.t)$ which concludes our proof of (2).

Assuming $|s| < \infty$ we can write s as tww with $tw \in \{x\}; (B^{guard}; S)^n$ for some $n \in \mathbb{N}$ and $w \in X$ (the doubling of the last element stems from the $\neg B^{guard}$ in the definition of DO). Now we have

$$
\begin{array}{ll}
 & tww \in DO \\
\Rightarrow & \{(2), \text{definition } DO\text{: finite sequences end in } \neg B\} \\
 & \neg B.w \wedge y.w \\
\Rightarrow & \{y \text{ is a solution of } (1)\} \\
 & p.w
\end{array}
$$

which is what we had to prove.

Using theorem 13 again we obtain a similar equation for *wp*. It follows that *wp.DO.p* is a solution of the equation in unknown predicate Y:

$$[Y \equiv B \wedge wp.S.Y \ \vee \ \neg B \wedge p] \tag{3}$$

Also *wp.S* is a monotonic function. We have

Theorem 17. *wp.DO.p is the strongest solution of (3).*

Proof. We have to prove $[wp.DO.p \Rightarrow y]$ for every solution y of (3). We prove this by contradiction. Let y be such a solution and suppose there exists $x \in X$ such that $wp.DO.p.x \wedge \neg y.x$. We show the existence of a characterizing chain l_k, $k \geq 0$ starting in x of a loop point l such that $\forall(k : 0 \leq k : \neg y.(last.l_k))$. We have to show that we can find such an l_k in every $\{x\};(B^{guard};S)^k$. We prove this by induction. For $k = 0$ we have $l_0 = x$ and $\neg y.x$ is given. Suppose we have $l_k \in \{x\};(B^{guard};S)^k$ satisfying $l_k < \infty$ (any infinite computation starting in x is already a contradiction). We denote the last element of l_k by w and prove that $B.w$ holds.

$$
\begin{array}{ll}
 & \neg B.w \\
= & \{\text{definition } DO\} \\
 & l_k w \in \{x\}; DO \ \wedge \neg B.w \\
= & \{wp.DO.p.x \text{ is given, } l_k \text{ is finite}\} \\
 & l_k w \in t.p \wedge \neg B.w \\
= & \{\text{definition } t\} \\
 & p.w \wedge \neg B.w \\
= & \{y \text{ solves } (3)\} \\
 & y.w \\
= & \{\text{induction hypothesis: } \neg y.w\} \\
 & \mathit{false}
\end{array}
$$

We now have $B.w \wedge \neg y.w$. In view of y being a solution of (3) this implies $\neg wp.S.y.w$ hence there exists $s \in \{l_k\}; B^{guard}; S$ such that $s \notin t.y$. Because $wp.DO.p.x$ holds, $|s| < \infty$. This s is our l_{k+1}.

The limit of this chain is a loop point starting in x which is in DO. Since a loop point is infinite, this contradicts the assumption $wp.DO.p.x$.

This concludes our discussion of *wlp* and *wp*.

We proceed with some comments on the notion of nondeterminate programs. In [6], the following definition is given: program S is deterministic if

$$[wp.S.p \equiv \neg wlp.S.\neg p], \text{ for all } p. \tag{4}$$

We change the terminology a bit; a program that satisfies (4) is called *determinate* since there may be more than one computation but with respect to the final states these computations are indistinguishable. A program is deterministic if there is only one possible computation for every initial state, i.e., $|\{x\};S| = 1$ for every $x \in X$.

As a definition of determinate programs, (4) does not suffice anymore since we generalized the notion of weakest precondition. As an example, according to this definition the following program is determinate.

if *true* $\rightarrow x := 3; y := 4$
$[]$ *true* $\rightarrow y := 4; x := 3$
fi

As long as we are interested in final states only, the two possibilities in this program are indistinguishable. If we generalize this idea to other properties we can say that a program is determinate if the fact whether a computation is in a property is completely determined by the initial state. In other words, there are no initial states with computations both inside and outside the property.

Definition 18. *Program S is determinate with respect to property $Q \in Prop$ if*

$$\{x\}; S \subseteq Q \ \vee \ \{x\}; S \subseteq (T \setminus Q)$$

or equivalently,

$$\neg(\{x\}; S \subseteq Q) \ \equiv \ \{x\}; S \subseteq (T \setminus Q)$$

for all $x \in X$. S is determinate with respect to a class of properties if it is determinate with respect to all properties in the class.

Using the definition of weakest precondition we derive that program S is determinate with respect to property Q if

$$[\neg w.S.Q \ \equiv \ w.S.(T \setminus Q)].$$

Now we analyze what this means for wp and wlp. If we look at one particular postcondition p we obtain for wp:

$$
\begin{aligned}
& T \setminus t.p \\
= \quad & \{\text{definition } t\} \\
& \{t : t \in T \wedge (|t| = \infty \vee \neg p.(last.t)) : t\} \\
= \quad & \{\text{definition } lt\} \\
& lt.\neg p
\end{aligned}
$$

Hence, program S is determinate with respect to termination in postcondition p if

$$[\neg wp.S.p \ \equiv \ wlp.S.\neg p].$$

By quantification over all postconditions we obtain the original definition.

Notice that we are quite fortunate to be able to characterize the complement of $t.p$ (and, of course, of $lt.p$). For other properties we may not find such a concise characterization.

Finally, we have a look at some other properties of programs. In recent work weakest preconditions were introduced to reason about alternative properties of programs ([11, 7, 12]) in an axiomatic way. The above operational semantics can be used for properties like these as well. (As a matter of fact, the introduction of these properties was the major motivation to start this work in the first place.) Consider for example the property: "ever, during execution, condition q holds" (the operator "$\diamond$" from temporal logic). This property is captured in

$$ever.q = \{s, w, t : swt \in T \wedge |s| < \infty \wedge w \in X \wedge q.w : swt\}.$$

We are interested in determining the weakest precondition for a program such that the computations, starting in a state satisfying this precondition are in *ever.q*. Hence we are interested in the function *weakest ever.*

$$wev.S.q = w.S.(ever.q)$$

Again we can analyse what this yields for the various language constructs. As it turns out, sequential composition is a problem. Informally, this is seen as follows: a computation in $S;U$ that is in *ever.q* may reach a state satisfying q either during S or during U. But

$$wev.(S;U).q = wev.S.q \vee wp.S.(wev.U.q)$$

is not the right formula since it does not allow initial states in which computations of both flavors start. We have to include, therefore, the final state in the definition of the function. As a consequence, we also have a liberal version.

$$wev.S.q.r = w.S.(ever.q \cup t.r)$$
$$wlev.S.q.r = w.S.(ever.q \cup lt.r)$$

By a whole lot of string manipulation we obtain now

$$wev.(S;U).q.r = wev.S.q.(wev.U.q.r)$$
$$wlev.(S;U).q.r = wlev.S.q.(wlev.U.q.r).$$

The fact that the final state of a program is of interest, even if the property that we are interested in does not refer explicitly to the final state comes from the fact that we look at preconditions, combined with the existence of sequential composition in our programming language. As a result, this final state will show up in any function that we introduce. This is quite unfortunate since it increases the number of arguments that our functions have. Consider as a second example the property *leads-to.p.q* defined as follows.

$$leads\text{-}to.p.q \;=\; \{s : \forall(a,b : s = ab \wedge |a| < \infty : b \in ever.p \Rightarrow b \in ever.q) : s\}$$

Informally, a computation is in *leads-to.p.q* if a state satisfying p is always followed by a state satisfying q (where "followed by" is understood to include coincidence). In the function *wto*, we include the final state for the same reason as we did it for *wev*.

$$wto.S.p.q.r = w.S.(leads\text{-}to.p.q \cup t.r)$$

For sequential composition, the following function is obtained.

$$wto.(S;U).p.q.r = wto.S.p.q.(wev.U.q.r) \wedge wlp.S.(wto.U.p.q.r)$$

In [9], these properties are analyzed in detail. Especially interesting are the results for the repetition. Using theorem 13, recursive equations are obtained for both properties. The function $wev.DO.q.r$ turns out to be the strongest solution of its equation while $wto.DO.p.q.r$ is the weakest solution. This is especially remarkable since both of them are true generalizations of wp.

$$wev.S.false.r = wp.S.r$$
$$wto.S.true.false.r = wp.S.r$$

Reasoning about programs in an operational domain is quite cumbersome. Therefore, it is a good idea to isolate some characteristics of the properties of interest and to use these characteristics, together with an axiomatic definition of the properties to prove facts about programs. This is done in [12, 9].

4 Conclusion

In this paper we have presented an operational semantics for the guarded command language in terms of sequences of states. The definition of the repetition was given in a closed form which allowed us to derive that the repetition is equivalent to its first unfolding. In this way we could prove that $wp.DO.p$ is the strongest solution of its defining equation and $wlp.DO.p$ the weakest. A generalization of the notion of weakest precondition allows us to reason about different properties in a similar way as we do about final states. Although we did not exploit that in detail in this paper we gave examples of two temporal properties. The generalization also gave rise to a nice characterization of nondeterminate programs.

Originally, this work started from our interest in parallel programs. We did not address any parallelism at all in this paper. However, we did describe nondeterministic choice. This means that we can describe parallelism in a similar way as, for instance, in Unity ([4], apart from the fairness constraint) or in action systems ([2, 3]), viz., as the interleaving of atomic actions.

5 Acknowledgements

I want to thank the two referees for helpful comments on the first version of the paper. The major part of this paper was part of my thesis and I want to thank my supervisor Jan van de Snepscheut for discussing the subjects presented in this paper and for his support. I also thank Wim Hesselink for his comments on notation and presentation.

References

1. Alpern, B., Schneider, F.B.: Defining liveness. I.P.L. **21** (1985) 181-185
2. Back, R.J.R.: Refinement calculus, part II: parallel and reactive programs. report ser. A, no. 93 (1989) Åbo Akademi Finland
3. Back R.J.R., Sere, K.: Stepwise refinement of action systems. In: Mathematics of program construction (J.L.A. van de Snepscheut (ed)), Springer-Verlag LNCS **375** (1989) 115-138
4. Chandy, K.M., Misra, J.: Parallel programming, a foundation. Addison-Wesley publishing company, Reading 1988

5. Davey, B.A., Priestley, H.A.: Introduction to lattices and order. Cambridge University Press, Cambridge 1990
6. Dijkstra, E.W., Scholten, C.S.: Predicate calculus and program semantics. Springer-Verlag, New York 1990.
7. Knapp, E., A predicate transformer for progress, I.P.L. **33** (1989/1990) 323-330
8. Kuiper, R.: An operational semantics for bounded nondeterminism equivalent to a denotational one. In: Algorithmic Languages, de Bakker/van Vliet (eds) 373-398, IFIP, North Holland 1981
9. Lukkien, J.J.: Parallel program design and generalized weakest preconditions. PhD thesis, Groningen University The Netherlands 1991
10. Manna, Z., Pnueli, A.: How to cook a temporal proof system for your pet language. Proceedings POPL (ACM) 141-153, Austin 1983
11. Morris, J.M.: Temporal predicate transformers and fair termination. Acta Informatica **27** (1990) 287-313
12. van de Snepscheut, J.L.A., Lukkien, J.J.: Weakest preconditions for progress. Formal Aspects of Computing **4** (1992) 195-236

Shorter Paths to Graph Algorithms

Bernhard Möller and Martin Russling

Institut für Mathematik, Universität Augsburg,
Universitätsstr. 2, W-8900 Augsburg, Germany,
e-mail: {moeller,russling}@uni-augsburg.de

Abstract. We illustrate the use of formal languages and relations in compact formal derivations of some graph algorithms.

1 Introduction

The transformational or calculational approach to program development has by now a long tradition (see Burstall, Darlington 1977, Bauer et al. 1985, Bauer et al. 1989, Meertens 1986, Bird 1989). In it, one starts from a (possibly non-executable) specification and transforms it into a (hopefully efficient) program using semantics-preserving rules. Many derivations, however, suffer from the use of lengthy expressions involving formulae from predicate calculus. However, in particular in the case of graph algorithms the calculus of formal languages and relations allows considerable compactification. We use a simplified and straightened version of the framework introduced in Möller 1991 to illustrate this with derivations of algorithms for computing the length of a shortest path between two graph vertices and for cycle detection.

2 The Framework

In connection with graph algorithms we use formal languages to describe sets of paths. The letters of the underlying alphabet are interpreted as graph nodes. As a special case of formal languages we consider relations of arities ≤ 2. Relations of arity 1 represent node sets, whereas binary relations represent the edge sets. The only two nullary relations (the singleton relation consisting just of the empty word and the empty relation) play the role of the Boolean values. This also allows easy definitions of assertions, conditional, and guards.

Essential operations on languages are (besides union, intersection, and difference) concatenation, composition, and join. As special cases of composition we obtain image and inverse image as well as tests for intersection, emptiness, and membership. The join corresponds to path concatenation on directed graphs; special cases yield restriction.

Proofs are either straightforward or given by Möller 1991 and therefore omitted.

2.1 Operations on Sets

Given a set A we denote by $\mathcal{P}(A)$ its **powerset**. The cardinality of A is, as usual, denoted by $|A|$. To save braces, we identify a singleton set with its only element.

Frequently, we will extend set-valued operations

$$f : A_1 \times \cdots \times A_n \rightarrow \mathcal{P}(A_{n+1}) \qquad (n > 0)$$

to the powersets $\mathcal{P}(A_i)$ of the A_i. In these cases we use the same symbol f also for the extended function

$$f : \mathcal{P}(A_1) \times \cdots \times \mathcal{P}(A_n) \rightarrow \mathcal{P}(A_{n+1})$$

defined by

$$f(U_1, \ldots, U_n) \stackrel{\text{def}}{=} \bigcup_{x_1 \in U_1} \cdots \bigcup_{x_n \in U_n} f(x_1, \ldots, x_n) \tag{1}$$

for $U_i \subseteq A_i$. By this definition, the extended operation distributes through union in all arguments:

$$f(U_1, \ldots U_{i-1}, \bigcup_{j \in J} U_{ij}, U_{i+1}, \ldots, U_n) = \bigcup_{j \in J} f(U_1, \ldots U_{i-1}, U_{ij}, U_{i+1}, \ldots, U_n) \ . \tag{2}$$

By taking $J = \emptyset$ we obtain strictness of the extended operation w.r.t. $\emptyset$:

$$f(U_1, \ldots U_{i-1}, \emptyset, U_{i+1}, \ldots, U_n) = \emptyset \ . \tag{3}$$

By taking $J = \{1, 2\}$ and using the equivalence

$$U \subseteq V \Leftrightarrow U \cup V = V$$

we also obtain monotonicity w.r.t. $\subseteq$ in all arguments:

$$\begin{aligned} &U_{i1} \subseteq U_{i2} \Rightarrow \\ &\quad f(U_1, \ldots U_{i-1}, U_{i1}, U_{i+1}, \ldots, U_n) \subseteq f(U_1, \ldots U_{i-1}, U_{i2}, U_{i+1}, \ldots, U_n) \ . \end{aligned} \tag{4}$$

Moreover, bilinear equational laws are preserved (see e.g. Lescanne 1982).

2.2 Languages and Relations

Consider an alphabet A. We denote the empty word over A by ε and concatenation by $\bullet$. It is associative, with ε as the neutral element:

$$u \bullet (v \bullet w) = (u \bullet v) \bullet w \ , \tag{5}$$

$$\varepsilon \bullet u = u = u \bullet \varepsilon \ . \tag{6}$$

As usual, a singleton word is not distinguished from the only letter it contains. The **length** of a word u, i.e., the number of letters from A in u, is denoted by $||u||$.

A **(formal) language** is a set of words over A. Concatenation is extended pointwise to languages. Since the above laws are bilinear, they carry over to languages U, V, W over A:

$$U \bullet (V \bullet W) = (U \bullet V) \bullet W \ , \tag{7}$$

$$\varepsilon \bullet U = U = U \bullet \varepsilon \ . \tag{8}$$

The **diagonal** V^Δ over a subset $V \subseteq A$ is defined by

$$V^\Delta \stackrel{\text{def}}{=} \bigcup_{x \in V} x \bullet x \ . \tag{9}$$

A **relation of arity** n is a language R such that all words in R have length n. Note that $\emptyset$ is a relation of any arity. For $R \neq \emptyset$ we denote the arity of R by $\mathsf{ar}\, R$. There are only two 0-ary relations, viz. $\emptyset$ and ε.

2.3 Composition

For languages V and W over alphabet A we define their **composition** $V\,;W$ by

$$V\,;W \stackrel{\text{def}}{=} \bigcup_{x\in A}\ \bigcup_{v\bullet x\in V}\ \bigcup_{x\bullet w\in W} v\bullet w\ . \tag{10}$$

If V and W are binary relations this coincides with the usual definition of relational composition (see e.g. Tarski 1941, Schmidt, Ströhlein 1989).

Composition is associative:

$$U\,;(V\,;W) = (U\,;V)\,;W \ \Leftarrow\ \forall\, y\in V : \|y\| \geq 2\ . \tag{11}$$

Composition associates with concatenation:

$$U\bullet(V\,;W) = (U\bullet V)\,;W \ \Leftarrow\ \forall\, y\in V : \|y\| \geq 1\ , \tag{12}$$

$$U\,;(V\bullet W) = (U\,;V)\bullet W \ \Leftarrow\ \forall\, y\in V : \|y\| \geq 1\ . \tag{13}$$

We shall omit parentheses whenever one of these laws applies. Moreover, $\bullet$ and ; bind stronger than $\cup$ and $\cap$.

Interesting special cases of relational composition arise when one of the operands has arity 1. Suppose $1 = \mathsf{ar}\,R \leq \mathsf{ar}\,S$. Then

$$R\,;S = \bigcup_{x\in R}\ \bigcup_{x\bullet v\in S} v\ .$$

In other words, $R\,;S$ is the image of R under S. Likewise, if $1 = \mathsf{ar}\,T \leq \mathsf{ar}\,S$, then $S;T$ is the inverse image of T under S. For these reasons we may define domain and codomain of a binary relation R by

$$\mathsf{dom}\,R \stackrel{\text{def}}{=} R\,;A\ , \tag{14}$$

$$\mathsf{cod}\,R \stackrel{\text{def}}{=} A\,;R\ . \tag{15}$$

Suppose now $\mathsf{ar}\,R = 1 = \mathsf{ar}\,S$ and $\|x\| = 1 = \|y\|$. Then

$$R\,;S = \begin{cases} \varepsilon \text{ if } R\cap S \neq \emptyset\ , \\ \emptyset \text{ if } R\cap S = \emptyset\ , \end{cases} \tag{16}$$

$$R\,;R = \begin{cases} \varepsilon \text{ if } R \neq \emptyset\ , \\ \emptyset \text{ if } R = \emptyset\ , \end{cases} \tag{17}$$

$$x\,;R = R\,;x = \begin{cases} \varepsilon \text{ if } x \in R\ , \\ \emptyset \text{ if } x \notin R\ , \end{cases} \tag{18}$$

$$x\,;y = y\,;x = \begin{cases} \varepsilon \text{ if } x = y\ , \\ \emptyset \text{ if } x \neq y\ . \end{cases} \tag{19}$$

Because these "tests" will be used frequently, we introduce more readable notations for them by setting

$$R \neq \emptyset = R\,;R\ , \tag{20}$$

$$x \in R = x\,;R\ , \tag{21}$$

$$(x = y) = x\,;y\ , \tag{22}$$

$$R \subseteq S = (R\cup S = S)\ . \tag{23}$$

For binary R and $x \in \mathsf{dom}\, R, y \in \mathsf{cod}\, R$ we have

$$x\,;R\,;y = \begin{cases} \varepsilon \text{ if } x \bullet y \in R \ , \\ \emptyset \text{ otherwise } . \end{cases} \tag{24}$$

Finally, we note that diagonals are neutral w.r.t. composition. Assume $P \supseteq \mathsf{dom}\, V$ and $Q \supseteq \mathsf{cod}\, V$. Then

$$P^{\Delta}\,; V = V \ , \tag{25}$$

$$V\,; Q^{\Delta} = V \ . \tag{26}$$

2.4 Assertions

As we have just seen, the nullary relations ε and $\emptyset$ characterize the outcomes of certain test operations. More generally, they can be used instead of Boolean values; therefore we call expressions yielding nullary relations **assertions**. Note that in this view "false" and "undefined" both are represented by $\emptyset$. Negation is defined by

$$\overline{\emptyset} \stackrel{\text{def}}{=} \varepsilon \ , \tag{27}$$

$$\overline{\varepsilon} \stackrel{\text{def}}{=} \emptyset \ . \tag{28}$$

Note that this operation is not monotonic.

For assertions B, C we have e.g. the properties

$$B \bullet C = B \cap C \ , \tag{29}$$

$$B \bullet B = B \ , \tag{30}$$

$$B \bullet \overline{B} = \emptyset \ , \tag{31}$$

$$B \cup \overline{B} = \varepsilon \ , \tag{32}$$

$$\overline{B \bullet C} = \overline{B} \cup \overline{C} \ . \tag{33}$$

Conjunction and disjunction of assertions are represented by their intersection and union. To improve readability, we write $B \wedge C$ for $B \cap C = B \bullet C$ and $B \vee C$ for $B \cup C$.

For assertion B and arbitrary language R we have

$$B \bullet R = R \bullet B = \begin{cases} R \text{ if } B = \varepsilon \ , \\ \emptyset \text{ if } B = \emptyset \ . \end{cases} \tag{34}$$

Hence $B \bullet R$ (and $R \bullet B$) behaves like the expression

$$B \ \triangleright\ R = \mathsf{if}\ B\, \mathsf{then}\ R\, \mathsf{else}\ \mathsf{error}\ \mathsf{fi}$$

in Möller 1989. We will use this construct for propagating assertions through recursions.

2.5 Conditional

Using assertions we can also define a conditional by

$$\mathsf{if}\ B\, \mathsf{then}\ R\, \mathsf{else}\ S\ \mathsf{fi} \stackrel{\text{def}}{=} B \bullet R \ \cup\ \overline{B} \bullet S \ , \tag{35}$$

for assertion B and languages R, S. Note that this operation is not monotonic in B.

2.6 Join

A useful derived operation is provided by a special case of the join operation as used in database theory (see e.g. Date 1988). Given two languages R, S, their **join** $R \bowtie S$ consists of all words that arise from "glueing" together words from R and from S along a common intermediate letter. By our previous considerations, the beginnings of words ending with $x \in A$ are obtained as $R\,;x$ whereas the ends of words which start with x are obtained as $x\,;S$. Hence we define

$$R \bowtie S \stackrel{\text{def}}{=} \bigcup_{x \in A} R\,;x \bullet x \bullet x\,;S \ . \tag{36}$$

Again, $\bowtie$ binds stronger than $\cup$ and $\cap$.

Join and composition are closely related. To explain this we consider two binary relations $R, S \subseteq A \bullet A$:

$$R\,;S = \bigcup_{z \in A} \{x \bullet y : x \bullet z \in R \wedge z \bullet y \in S\} \ ,$$

$$R \bowtie S = \bigcup_{z \in A} \{x \bullet z \bullet y : x \bullet z \in R \wedge z \bullet y \in S\} \ .$$

Thus, whereas $R\,;S$ just states whether there is a path from x to y via some point $z \in Q$, the relation $R \bowtie S$ consists of exactly those paths $x \bullet z \bullet y$. In particular, the relations

$$\begin{array}{l} R \ , \\ R \bowtie R \ , \\ R \bowtie (R \bowtie R) \ , \\ \vdots \end{array}$$

consist of the paths of edge numbers $1, 2, 3, \ldots$ in the directed graph associated with R.

Other interesting special cases arise when the join is taken w.r.t. the minimum of the arities involved. Suppose $1 = \mathsf{ar}\,R \leq \mathsf{ar}\,S$. Then

$$\begin{aligned} & R \bowtie S \\ = \ & \bigcup_{x \in A} R\,;x \bullet x \bullet x\,;S \\ = \ & \bigcup_{x \in R} x \bullet x\,;S \ . \end{aligned}$$

In other words, $R \bowtie S$ is the restriction of S to R. Likewise, for T with $1 = \mathsf{ar}\,T \leq \mathsf{ar}\,S$, the language $S \bowtie T$ is the corestriction of S to T.

If even $\mathsf{ar}\,R = \mathsf{ar}\,S = 1$ we have

$$R \bowtie S = R \cap S \ . \tag{37}$$

In particular, if $\mathsf{ar}\,R = 1$ and $\|x\| = 1 = \|y\|$,

$$R \bowtie R = R \ , \tag{38}$$

$$x \bowtie R = R \bowtie x = \begin{cases} x \text{ if } x \in R \ , \\ \emptyset \text{ if } x \notin R \ , \end{cases} \tag{39}$$

$$x \bowtie y = y \bowtie x = \begin{cases} x \text{ if } x = y \ , \\ \emptyset \text{ if } x \neq y \ . \end{cases} \tag{40}$$

For binary R, $x \in \mathsf{dom}\, R$, and $y \in \mathsf{cod}\, R$ this implies

$$x \bowtie R \bowtie y = \begin{cases} x \bullet y \text{ if } x \bullet y \in R \ , \\ \emptyset \quad \text{otherwise} \ . \end{cases} \tag{41}$$

In special cases the join can be expressed by a composition: Assume $\mathsf{ar}\, P = 1 = \mathsf{ar}\, Q$. Then

$$P \bowtie R = P^{\Delta} ; R \ , \tag{42}$$

$$R \bowtie Q = R ; Q^{\Delta} \ . \tag{43}$$

By the associativity of composition (11) also join and composition associate:

$$(R \bowtie S) ; T = R \bowtie (S ; T) \ , \tag{44}$$

$$R ; (S \bowtie T) = (R ; S) \bowtie T \ , \tag{45}$$

provided $\mathsf{ar}\, S \geq 2$.

Moreover, also joins associate:

$$R \bowtie (S \bowtie T) = (R \bowtie S) \bowtie T \ . \tag{46}$$

2.7 Closures

Consider a binary relation $R \subseteq A \bullet A$. We define the **(reflexive and transitive) closure** R^* of R by

$$R^* \stackrel{\text{def}}{=} \bigcup_{i \in \mathbb{N}} R^i \ , \tag{47}$$

where, as usual,

$$R^0 \stackrel{\text{def}}{=} A^{\Delta} \ , \tag{48}$$

$$R^{i+1} \stackrel{\text{def}}{=} R ; R^i \ . \tag{49}$$

It is well-known that R^* is the least fixpoint of the recursion equations

$$R^* \ = \ A^{\Delta} \cup R ; R^* \ = \ A^{\Delta} \cup R^* ; R \ . \tag{50}$$

Let G be the directed graph associated with R, i.e., the graph with vertex set A and arcs between the vertices corresponding to the pairs in R. We have

$$x ; R^i ; y = \begin{cases} \varepsilon \text{ if there is a path with } i \text{ edges from } x \text{ to } y \text{ in } G \ , \\ \emptyset \text{ otherwise} \ . \end{cases} \tag{51}$$

Likewise,

$$x ; R^* ; y = \begin{cases} \varepsilon \text{ if there is a path from } x \text{ to } y \text{ in } G \ , \\ \emptyset \text{ otherwise} \ . \end{cases} \tag{52}$$

For $s \subseteq A$, the set $s ; R^*$ gives all points in A reachable from points in s via paths in G, whereas $R^*; s$ gives all points in A from which some point in s can be reached. Finally,

$$s ; R^* ; t = \begin{cases} \varepsilon \text{ if } s \text{ and } t \text{ are connected by some path in } G \ , \\ \emptyset \text{ otherwise} \ . \end{cases} \tag{53}$$

As usual, we set

$$R^+ \stackrel{\text{def}}{=} R\,;R^* = R^*\,;R\ . \tag{54}$$

Analogously, we define the **path closure** $R^{\Rightarrow}$ of R by

$$R^{\Rightarrow} \stackrel{\text{def}}{=} \bigcup_{i\in\mathbb{N}} {}^{i}R\ , \tag{55}$$

where now

$$^{0}R \stackrel{\text{def}}{=} A\ , \tag{56}$$

$$^{i+1}R \stackrel{\text{def}}{=} R \bowtie {}^{i}R\ . \tag{57}$$

It is the least fixpoint of the recursion equations

$$R^{\Rightarrow} = A \cup R \bowtie R^{\Rightarrow} = A \cup R^{\Rightarrow} \bowtie R\ . \tag{58}$$

The path closure consists of all finite paths in G. Hence

$$x \bowtie R^{\Rightarrow} \bowtie y \tag{59}$$

is the language of all paths between x and y in G. Analogously to R^+ we define the **proper path closure** by

$$R^{\rightarrow} \stackrel{\text{def}}{=} R \bowtie R^{\Rightarrow} = R^{\Rightarrow} \bowtie R = R^{\Rightarrow} \backslash A\ . \tag{60}$$

3 Graph Algorithms

We now want to apply the framework in case studies of some simple graph algorithms.

3.1 Length of a Shortest Connecting Path

Specification and First Recursive Solution. We consider a finite set A of vertices and a binary relation $R \subseteq A \bullet A$. The problem is to find the length of a shortest path from a vertex x to a vertex y. Therefore we define

$$\mathit{shortestpath}(x,y) \stackrel{\text{def}}{=} \mathit{min}(\mathit{edgelengths}(x \bowtie R^{\Rightarrow} \bowtie y))\ , \tag{61}$$

where, for a set S of (non-empty) paths,

$$\mathit{edgelengths}(S) \stackrel{\text{def}}{=} \bigcup_{s\in S} (||s|| - 1) \tag{62}$$

calculates the set of path lengths, i.e., the number of edges in each path, and, for a set N of natural numbers,

$$\mathit{min}(N) \stackrel{\text{def}}{=} \begin{cases} k \text{ if } k \in N \wedge N \subseteq k; \leq\ , \\ \emptyset \text{ if } N = \emptyset\ . \end{cases} \tag{63}$$

It is obvious that *edgelengths* is strict and distributes through union. Moreover, for unary S,

$$edgelengths(S \bowtie T) = 1 + edgelengths(S\,;T) \ , \tag{64}$$

and

$$min(M \cup N) = min(min(M) \cup min(N)) \ , \tag{65}$$

$$min(0 \cup M) = 0 \ . \tag{66}$$

For deriving a recursive version of ***shortestpath*** we generalize this to a function *sp* which calculates the length of a shortest path from a set S of vertices to a vertex y:

$$sp(S, y) \stackrel{\text{def}}{=} min(edgelengths(S \bowtie R^{\Rightarrow} \bowtie y)) \ . \tag{67}$$

The embedding

$$shortestpath(x, y) = sp(x, y) \tag{68}$$

is straightforward.

We calculate

$sp(S, y)$

$=$ ⦃ definition ⦄

$min(edgelengths(S \bowtie R^{\Rightarrow} \bowtie y))$

$=$ ⦃ by (58) ⦄

$min(edgelengths(S \bowtie (A \ \cup \ R \bowtie R^{\Rightarrow}) \bowtie y))$

$=$ ⦃ distributivity ⦄

$min(edgelengths(S \bowtie A \bowtie y) \ \cup \ edgelengths(S \bowtie R \bowtie R^{\Rightarrow} \bowtie y))$

$=$ ⦃ by (37) ⦄

$min(edgelengths(S \bowtie y) \ \cup \ edgelengths(S \bowtie R \bowtie R^{\Rightarrow} \bowtie y))$.

By (39) the subexpression $S \bowtie y$ can be simplified according to whether $y \in S$ or not.

Case 1: $y \in S$.

$min(edgelengths(S \bowtie y) \ \cup \ edgelengths(S \bowtie R \bowtie R^{\Rightarrow} \bowtie y))$

$=$ ⦃ by (39), since $y \in S$ ⦄

$min(edgelengths(y) \ \cup \ edgelengths(S \bowtie R \bowtie R^{\Rightarrow} \bowtie y))$

$=$ ⦃ definition of *edgelengths* ⦄

$min(0 \ \cup \ edgelengths(S \bowtie R \bowtie R^{\Rightarrow} \bowtie y))$

$=$ ⦃ by (66) ⦄

0 .

Case 2: $y \notin S$.

$$
\begin{aligned}
& min(edgelengths(S \bowtie y) \cup edgelengths(S \bowtie R \bowtie R^{\Rightarrow} \bowtie y)) \\
= \ & \{\!| \text{ by (39), since } y \notin S \ |\!\} \\
& min(edgelengths(\emptyset) \cup edgelengths(S \bowtie R \bowtie R^{\Rightarrow} \bowtie y)) \\
= \ & \{\!| \text{ strictness, neutrality } |\!\} \\
& min(edgelengths(S \bowtie R \bowtie R^{\Rightarrow} \bowtie y)) \\
= \ & \{\!| \text{ by (64) } |\!\} \\
& min(1 + edgelengths(S \,;\, R \bowtie R^{\Rightarrow} \bowtie y)) \\
= \ & \{\!| \text{ distributivity } |\!\} \\
& 1 + min(edgelengths(S \,;\, R \bowtie R^{\Rightarrow} \bowtie y)) \\
= \ & \{\!| \text{ definition } |\!\} \\
& 1 + sp(S \,;\, R \,,\, y) \ .
\end{aligned}
$$

Altogether we have derived the recursion equation

$$sp(S, y) \ = \ \text{if } y \in S \text{ then } 0 \text{ else } 1 + sp(S \,;\, R \,,\, y) \text{ fi } . \tag{69}$$

Note, however, that termination cannot be guaranteed for this recursion. To make progress in that direction we show some additional properties of sp.

Lemma 1. $sp(S \cup T \,,\, y) \ = \ min(sp(S, y) \cup sp(T, y))$.

Proof. $sp(S \cup T \,,\, y)$

$$
\begin{aligned}
= \ & \{\!| \text{ definition } |\!\} \\
& min(edgelengths((S \cup T) \bowtie R^{\Rightarrow} \bowtie y)) \\
= \ & \{\!| \text{ distributivity } |\!\} \\
& min(edgelengths(S \bowtie R^{\Rightarrow} \bowtie y) \cup edgelengths(T \bowtie R^{\Rightarrow} \bowtie y)) \\
= \ & \{\!| \text{ by (65) } |\!\} \\
& min(min(edgelengths(S \bowtie R^{\Rightarrow} \bowtie y)) \cup min(edgelengths(T \bowtie R^{\Rightarrow} \bowtie y))) \\
= \ & \{\!| \text{ definition } |\!\} \\
& min(sp(S, y) \cup sp(T, y)) \ .
\end{aligned}
$$

□

We now consider again the case $y \notin S$. From (69) we obtain

$$sp(S \,;\, R \,,\, y) \leq sp(S, y) \ , \tag{70}$$

and hence

$$
\begin{array}{ll}
& sp(S, y) \\
= & \{\!\!\{ \text{ by } y \notin S \text{ and (69) } \}\!\!\} \\
& 1 + sp(S\,;R\,,\ y) \\
= & \{\!\!\{ \text{ by (70) } \}\!\!\} \\
& 1 + min(sp(S, y) \cup sp(S\,;R\,,\ y)) \\
= & \{\!\!\{ \text{ by Lemma 1 } \}\!\!\} \\
& 1 + sp(S \cup S\,;R\,,\ y)\ ,
\end{array}
$$

so that a second recursion equation for *sp* is

$$sp(S, y) \ = \ \mathsf{if}\ y \in S\ \mathbf{then}\ 0\ \mathbf{else}\ 1 + sp(S \cup S\,;R\,,\ y)\ \mathsf{fi}\ . \tag{71}$$

Now, although the first parameter is non-decreasing in each recursive call, still non-termination is guaranteed if there is no path from S to y. However, in that case by finiteness of A the recursive calls of *sp* eventually become stationary, i.e., eventually $S = S \cup S\,;R$ holds, which is equivalent to $S\,;R \subseteq S$. We consider that case in the following lemma:

Lemma 2. *If* $y \notin S$ *and* $S\,;R \subseteq S$ *then* $S \bowtie R^{\Rightarrow} \bowtie y = \emptyset$, *i.e., there is no path from set* S *to vertex* y, *and therefore* $sp(S, y) = \emptyset$.

Proof. Using the least fixpoint property of $R^{\Rightarrow}$ we use computational induction (see e.g. Manna 1974) to show $S \bowtie R^{\Rightarrow} \bowtie y \subseteq \emptyset$. We use the predicate

$$P[X] \overset{\text{def}}{\Leftrightarrow} S \bowtie X \bowtie y \subseteq \emptyset\ .$$

The induction base $P[\emptyset]$ is trivial by strictness. For the induction step we have to show $P[X] \Rightarrow P[A \cup R \bowtie X]$. Assume $P[X]$. We calculate

$$
\begin{array}{ll}
& S \bowtie (A \cup R \bowtie X) \bowtie y \\
= & \{\!\!\{ \text{ distributivity } \}\!\!\} \\
& S \bowtie A \bowtie y \cup S \bowtie R \bowtie X \bowtie y \\
= & \{\!\!\{ \text{ by (37) } \}\!\!\} \\
& S \bowtie y\ \cup\ S \bowtie R \bowtie X \bowtie y \\
= & \{\!\!\{ \text{ by (39), since } y \notin S, \text{ and neutrality } \}\!\!\} \\
& S \bowtie R \bowtie X \bowtie y \\
= & \{\!\!\{ \text{ by (42) } \}\!\!\} \\
& S^{\Delta}\,;R \bowtie X \bowtie y \\
\subseteq & \{\!\!\{ \text{ by } S^{\Delta} \subseteq S \bullet S \text{ and monotonicity } \}\!\!\} \\
& S \bullet S\,;R \bowtie X \bowtie y \\
\subseteq & \{\!\!\{ \text{ by } S\,;R \subseteq S \text{ and monotonicity } \}\!\!\}
\end{array}
$$

$$
\begin{array}{ll}
 & S \bullet S \bowtie X \bowtie y \\
\subseteq & \{\!\![\text{ by } P[X] \text{ and monotonicity }]\!\!\} \\
 & S \bullet \emptyset \\
= & \{\!\![\text{ strictness }]\!\!\} \\
 & \emptyset \;.
\end{array}
$$

Now the claim is immediate from the definition of *sp*. □

Altogether we have:

$$shortestpath(x,y) \;=\; sp(x,y) \;,$$

$$sp(S,y) \;=\; \mathsf{if}\ y \in S\ \mathsf{then}\ 0 \ \mathsf{else\ if}\ S\,;R \subseteq S\ \mathsf{then}\ \emptyset \ \mathsf{else}\ 1 + sp(S \cup S\,;R\;,\ y)\ \mathsf{fi}\ \mathsf{fi}\;. \tag{72}$$

Now termination is guaranteed, since S increases for each recursive call and is bounded by the finite set A of all vertices.

Improving Efficiency. One may argue that in the above version accumulating vertices in the parameter S is not efficient because it makes calculating $S\,;R$ more expensive. So, in an improved version of the algorithm, we shall keep as few vertices as possible in the parameter S and the set of vertices already visited in an additional parameter T, tied to S by an assertion. Let

$$sp2(S,T,y) \stackrel{\text{def}}{=} (S \cap T = \emptyset \wedge y \notin T) \bullet sp(S \cup T\,,\ y)\;, \tag{73}$$

with the embedding

$$shortestpath(x,y) \;=\; sp2(x,\emptyset,y)\;. \tag{74}$$

Now assume $S \cap T = \emptyset \wedge y \notin T$. Again we distinguish two cases:

Case 1: $y \in S$.

$$
\begin{array}{ll}
 & sp2(S,T,y) \\
= & \{\!\![\text{ definition }]\!\!\} \\
 & sp(S \cup T\;,\ y) \\
= & \{\!\![\text{ by } y \in S \subseteq S \cup T \text{ and (72) }]\!\!\} \\
 & 0 \;.
\end{array}
$$

Case 2: $y \notin S$.

$$
\begin{aligned}
& sp2(S,T,y) \\
= \quad & \{\!\!\{ \text{ definition } \}\!\!\} \\
& sp(S \cup T \, , \, y) \\
= \quad & \{\!\!\{ \text{ by } y \notin S \cup T \text{ and (72) } \}\!\!\} \\
& \textsf{if } (S \cup T)\,;R \subseteq S \cup T \textsf{ then } \emptyset \\
& \qquad\qquad\qquad\qquad \textsf{else } 1 + sp(S \cup T \cup (S \cup T)\,;R \, , \, y) \textsf{ fi} \\
= \quad & \{\!\!\{ \text{ set theory } \}\!\!\} \\
& \textsf{if } (S \cup T)\,;R \subseteq S \cup T \textsf{ then } \emptyset \\
& \qquad\qquad\qquad\qquad \textsf{else } 1 + sp(((S \cup T)\,;R)\backslash(S \cup T) \;\cup\; (S \cup T) \, , \, y) \textsf{ fi} \\
= \quad & \{\!\!\{ \text{ definition and } y \notin S \cup T \}\!\!\} \\
& \textsf{if } (S \cup T)\,;R \subseteq S \cup T \textsf{ then } \emptyset \\
& \qquad\qquad\qquad\qquad \textsf{else } 1 + sp2(((S \cup T)\,;R)\backslash(S \cup T) \, , \, S \cup T \, , \, y) \textsf{ fi} \; .
\end{aligned}
$$

Altogether,

$$
shortestpath(x,y) \;=\; sp2(x,\emptyset,y) \; ,
$$

$$
\begin{aligned}
sp2(S,T,y) \;=\; & (S \cap T = \emptyset \wedge y \notin T) \bullet \\
& \textsf{if } y \in S \\
& \textsf{then } 0 \\
& \textsf{else if } (S \cup T)\,;R \subseteq S \cup T \\
& \qquad \textsf{then } \emptyset \\
& \qquad \textsf{else } 1 + sp2(((S \cup T)\,;R)\backslash(S \cup T) \, , \, S \cup T \, , \, y) \textsf{ fi fi} \; .
\end{aligned}
$$

This version still is very inefficient. However, a simple analysis shows that the assertion of *sp2* can be strengthened by the conjunct $T\,;R \subseteq S \cup T$. Thus, one can simplify the program to

$$
shortestpath(x,y) \;=\; sp3(x,\emptyset,y) \; ,
$$

$$
\begin{aligned}
sp3(S,T,y) \;=\; & (S \cap T = \emptyset \wedge y \notin T \wedge T;R \subseteq S \cup T) \bullet \\
& \textsf{if } y \in S \\
& \textsf{then } 0 \\
& \textsf{else if } S\,;R \subseteq S \cup T \\
& \qquad \textsf{then } \emptyset \\
& \qquad \textsf{else } 1 + sp3((S\,;R)\backslash(S \cup T) \, , \, S \cup T \, , \, y) \textsf{ fi fi} \; .
\end{aligned}
$$

The formal derivation steps for this are similar to the ones above and hence we omit them.

Termination is guaranteed, since T increases for each recursive call and is bounded by the finite set A of all vertices.

Note that a tail-recursive variant can easily be derived from *sp3* by introducing an accumulator. A corresponding algorithm in iterative form can be found in the literature, e.g. in Gondran, Minoux 1979 (but there unfortunately not faultless).

Further, our algorithm also solves the problem whether a vertex y is reachable from a vertex x, since

$$\textit{reachable}(x,y) = (\textit{shortestpath}(x,y) \neq \emptyset) \ . \tag{75}$$

3.2 Cycle Detection

Problem Statement and First Solution. Consider again a finite set A of vertices and a binary relation $R \subseteq A \bullet A$. The problem consists in determining whether R contains a **cyclic path**, i.e., a path in which a node occurs twice.

Lemma 3. *The following statements are equivalent:*
(1) *R contains a cyclic path.*
(2) $R^+ \cap A^\Delta \neq \emptyset$.
(3) $R^{|A|} \neq \emptyset$.
(4) $R^{|A|} ; A \neq \emptyset$.
(5) $A ; R^{|A|} \neq \emptyset$.

Proof. (1) $\Rightarrow$ (2) Let $p = u \bullet x \bullet v \bullet x \bullet w$ with $x \in A$ and $u, v, w \in A^{(*)}$ be a cyclic path. Then $x \bullet x \in R^+$ and the claim follows.

(2) $\Rightarrow$ (3) Assume $x \bullet x \in R^+$ and let n be the smallest number such that there are $x_0, \ldots, x_n \in A$ with $\bigcup_{i=0}^{n-1} x_i \bullet x_{i+1} \subseteq R$ and $x_0 = x = x_n$. Then $\bullet_{i=0}^{|A|} x_{i \bmod n}$ is a path as well and hence the claim holds.

(3) $\Rightarrow$ (4) Trivial, since $R^{|A|} ; A$ is the domain of $R^{|A|}$.

(4) $\Rightarrow$ (5) Trivial, since a relation with nonempty domain also has a nonempty codomain.

(5) $\Rightarrow$ (1) We have $y \in A ; R^{|A|}$ iff there is an $x \in A$ and a path from x to y with $|A| + 1$ nodes. By the pigeonhole principle this path must contain at least one node twice and hence is cyclic. □

By (5) we may specify our problem as

$$\textit{hascycle} \stackrel{\text{def}}{=} (A ; R^{|A|} \neq \emptyset) \ .$$

To compute $A ; R^{|A|}$ we define $A_i \stackrel{\text{def}}{=} A ; R^i$ and use the properties of the powers of R:

$$\begin{aligned} A_0 &= A ; R^0 = A ; A^\Delta = A \ , \\ A_{i+1} &= A ; R^{i+1} = A ; (R^i ; R) = (A ; R^i) ; R = A_i ; R \ . \end{aligned}$$

The associated function

$$f : X \mapsto X ; R$$

is monotonic. We now prove a general theorem about monotonic functions on noetherian partial orders. A partial order $(M, \leq)$ is **noetherian** if every descending sequence in it becomes stationary or, equivalently, if each of its nonempty subsets has a minimal element.

Theorem 4. *Let $(M, \leq)$ be a noetherian partial order and let $f : M \to M$ be monotonic.*
(1) If for $x \in M$ we have $f(x) \leq x$, then $x_\infty \stackrel{\text{def}}{=} \text{glb}\ \{f^i(x) : i \in \mathbb{N}\}$ exists and is a fixpoint of f.
(2) If for $x, y \in M$ we have $f(x) \leq x$ and $x_\infty \leq y \leq x$, then also y_∞ exists and $x_\infty = y_\infty$.
(3) If M has a greatest element $\top$ then $\top_\infty$ is the greatest fixpoint of f.

Proof. (1) By assumption we have $f^1(x) = f(x) \leq f^0(x) = x$. Now a straightforward induction using monotonicity shows $f^{i+1}(x) \leq f^i(x)$ for all i so that the $f^i(x)$ form a descending chain. Since M is noetherian, the chain of the $f^i(x)$ has to become stationary, i.e., there is some k such that $f^k(x) = f^{k+1}(x)$. But then $f^j(x) = f^k(x)$ for all $j \geq k$ and hence $f^k(x) = \text{glb}\ X$, where $X \stackrel{\text{def}}{=} \{f^i(x) : i \in \mathbb{N}\}$, so that $x_\infty = f^k(x)$. But then $x_\infty = f^k(x) = f^{k+1}(x) = f(f^k(x)) = f(x_\infty)$, i.e., x_∞ is a fixpoint of f.
(2) A straightforward induction using monotonicity of f shows that $x_\infty \leq f^i(y) \leq f^i(x)$ for all $i \in \mathbb{N}$. Hence x_∞ is a lower bound for $Y \stackrel{\text{def}}{=} \{f^i(y) : i \in \mathbb{N}\}$. Let z be another lower bound for Y. Then z is also a lower bound for X defined in (1) and hence $z \leq x_\infty$. Hence $x_\infty = \text{glb}\ Y = y_\infty$.
(3) Trivially, $f(\top) \leq \top$, and hence $\top_\infty$ exists by (1). Let x be a fixpoint of f. A straightforward induction using monotonicity of f shows that $x \leq f^i(\top)$ for all $i \in \mathbb{N}$, so that x is a lower bound for $\{f^i(\top) : i \in \mathbb{N}\}$. But then $x \leq \top_\infty = \text{glb}\ \{f^i(\top) : i \in \mathbb{N}\}$. □

A similar theorem has been stated by Cai, Paige 1989.

Corollary 5. *If $\top_\infty \leq x$ for some $x \in M$ then $\top_\infty = x_\infty$.*

Proof. By (2) of the above theorem. □

To actually calculate x_∞ we define a function *inf* by

$$inf(y) \stackrel{\text{def}}{=} (x_\infty \leq y \leq x) \bullet x_\infty$$

which determines x_∞ using an upper bound y. We have the embedding $x_\infty = inf(x)$. Now from the proof of the above theorem the following recursion is immediate:

$$inf(y) = (x_\infty \leq y \leq x) \bullet \text{ if } y = f(y) \text{ then } y \text{ else } inf(f(y)) \text{ fi }.$$

This recursion terminates for every y satisfying $f(y) \leq y$, since monotonicity then also shows $f(f(y)) \leq f(y)$, so that in each recursive call the parameter decreases properly. In particular, the call $inf(x)$ terminates. This algorithm is an abstraction of many iteration methods on finite sets.

We now return to the special case of cycle detection. By finiteness of A the partial order $(\mathcal{P}(A), \subseteq)$ is noetherian with greatest element A. Therefore A_∞ exists. Moreover, we have

Corollary 6. $A_{|A|} = A_\infty$.

Proof. The length of any properly descending chain in $\mathcal{P}(A)$ is at most $k+1$. Hence we have $A_{k+1} = A_k$ and thus $A_k = A_\infty$. □

So we have reduced our task to checking whether $A_\infty \neq \emptyset$, i.e., whether $inf(A) \neq \emptyset$. For our special case the recursion for inf reads (omitting the trivial part $W \subseteq A$)

$$inf(W) = (A_\infty \subseteq W) \bullet \text{ if } W = W\,;R \text{ then } W \text{ else } inf(W\,;R) \text{ fi } .$$

We want to improve this by avoiding the computation of $W\,;R$. By the above considerations we may strengthen the assertion of inf by adding the conjunct $W;R \subseteq W$. We define

$$src(W) \stackrel{\text{def}}{=} W\backslash(W\,;R) .$$

This is the set of **sources** of W, i.e., the set of nodes in W which do not have a predecessor in W.

Now, assuming $W\,;R \subseteq W$, we have $W = W\,;R \Leftrightarrow src(W) = \emptyset$ and $W\,;R = W\backslash src(W)$ so that we can rewrite inf into

$$\begin{aligned} inf(W) = (A_\infty \subseteq W \wedge W\,;R \subseteq W) \bullet \\ \text{if } src(W) = \emptyset \text{ then } W \text{ else } inf(W\backslash src(W)) \text{ fi } . \end{aligned}$$

This is an improvement in that $src(W)$ usually will be small compared to W; moreover, the computation of $src(W)$ can be facilitated by a suitable representation of R.

Plugging this into our original problem of cycle recognition we obtain

$$hascycle = hcy(A) , \tag{76}$$

where

$$\begin{aligned} hcy(W) = (A_\infty \subseteq W \wedge W\,;R \subseteq W) \bullet \\ \text{if } src(W) = \emptyset \text{ then } W \neq \emptyset \text{ else } hcy(W\backslash src(W)) \text{ fi } , \end{aligned} \tag{77}$$

which is one of the classical algorithms which works by successive removal of sources (see e.g. Berghammer 1986). Note that Lemma 3(4) suggests a dual specification to the one we have used; replaying our development for it would lead to an algorithm that works by successive removal of sinks.

Improving Efficiency. We want to improve the computation of the sets $src(W)$. We observe that

$$\begin{aligned} & x \in src(W) \\ = \; & x \in W\backslash(W\,;R) \\ = \; & x \in W \wedge x \notin W\,;R \\ = \; & x \in W \wedge R\,;x \cap W = \emptyset \\ = \; & x \in W \wedge |R\,;x \cap W| = 0 . \end{aligned}$$

So we define for $W \subseteq A$ the relation $in(W)$ by

$$x\,;in(W) \stackrel{\text{def}}{=} |R\,;x \cap W| . \tag{78}$$

Hence $x\,;in(W)$ gives the indegree of x w.r.t. W and

$$src(W) = W \cap in(W);0 . \tag{79}$$

In a final implementation, $in(W)$ will, of course, be realized by an array. We aim at an incremental updating of in in the course of our algorithm. We calculate

$$
\begin{aligned}
& x\,;in(W\backslash src(W)) \\
= \quad & \{\!\!\{ \text{ definition } \}\!\!\} \\
& |R\,;x \cap (W\backslash src(W))| \\
= \quad & \{\!\!\{ \text{ set theory } \}\!\!\} \\
& |(R\,;x \cap W)\backslash src(W)| \\
= \quad & \{\!\!\{ \ |A\backslash B| = |A| - |A \cap B| \ \}\!\!\} \\
& |(R\,;x \cap W)| - |R\,;x \cap W \cap src(W)| \\
= \quad & \{\!\!\{ \ src(W) \subseteq W \ \}\!\!\} \\
& |(R\,;x \cap W)| - |R\,;x \cap src(W)| \\
= \quad & \{\!\!\{ \text{ definition } \}\!\!\} \\
& x\,;in(W) - x\,;in(src(W)) \ .
\end{aligned}
$$

For binary relations f, g with the same domain and subsets of $\mathbb{N}$ as codomains and arithmetic operator $\wr$ we define $f \wr g$ by

$$x\,;(f \wr g) \stackrel{\text{def}}{=} (x\,;f) \wr (x\,;g) \ . \tag{80}$$

Then

$$in(W\backslash src(W)) = in(W) - in(src(W)) \ . \tag{81}$$

For the computation of in we observe that

$$in(\emptyset) = \mathbf{0} \ , \tag{82}$$

where

$$x\,;\mathbf{0} \stackrel{\text{def}}{=} 0 \ . \tag{83}$$

If $S \neq \emptyset$ and $q \in S$ is arbitrary we have

$$
\begin{aligned}
& x\,;in(S) \\
= \ & x\,;in(q \cup S\backslash q) \\
= \ & |R\,;x \cap (q \cup S\backslash q)| \\
= \ & |(R\,;x \cap q) \cup (R\,;x \cap S\backslash q)| \\
= \ & |R\,;x \cap q| + |R\,;x \cap S\backslash q| \\
= \ & x\,;in(q) + x\,;in(S\backslash q) \ ,
\end{aligned}
$$

where

$$x\,;in(q) \ = \ \textsf{if } q\,;R\,;x \textsf{ then } 1 \textsf{ else } 0 \textsf{ fi} \ . \tag{84}$$

Then

$$in(S) = in(q) + in(S\backslash q) \ . \tag{85}$$

We forego a transformation of in into tail recursive form, since this is completely standard using associativity of $+$.

Now we can administer the source sets more efficiently: We introduce additional parameters for carrying along $src(W)$ and $in(W)$ and adjust these parameters by the technique of finite differencing (see e.g. Partsch 1990). We set, for $S \subseteq W \subseteq A$ and relation f,

$$hc(W, S, f) \stackrel{\text{def}}{=} (S = src(W) \wedge f = in(W)) \bullet hcy(W) \ , \tag{86}$$

with the embedding

$$hcy(W) = hc(W, src(W), in(W)) \ . \tag{87}$$

Now

$$
\begin{array}{ll}
 & hc(W, S, f) \\
= & \{\!\mid \text{definitions} \mid\!\} \\
 & \textsf{if } src(W) = \emptyset \textsf{ then } W \neq \emptyset \textsf{ else } hcy(W \backslash src(W)) \\
= & \{\!\mid \text{assertion} \mid\!\} \\
 & \textsf{if } S = \emptyset \textsf{ then } W \neq \emptyset \textsf{ else } hcy(W \backslash S) \\
= & \{\!\mid \text{embedding} \mid\!\} \\
 & \textsf{if } S = \emptyset \textsf{ then } W \neq \emptyset \textsf{ else } hc(W \backslash S\ ,\ src(W \backslash S)\ ,\ in(W \backslash S)) \\
= & \{\!\mid \text{introducing auxiliaries} \mid\!\} \\
 & \textsf{if } S = \emptyset \textsf{ then } W \neq \emptyset \\
 & \qquad \textsf{else let } T \stackrel{\text{def}}{=} W \backslash S \\
 & \qquad\qquad \textsf{let } g \stackrel{\text{def}}{=} in(T) \\
 & \qquad\qquad \textsf{in } \ hc(T, src(T), g) \ \textsf{fi} \\
= & \{\!\mid \text{by (81) and (79)} \mid\!\} \\
 & \textsf{if } S = \emptyset \textsf{ then } W \neq \emptyset \\
 & \qquad \textsf{else let } T \stackrel{\text{def}}{=} W \backslash S \\
 & \qquad\qquad \textsf{let } g \stackrel{\text{def}}{=} f - in(S) \\
 & \qquad\qquad \textsf{in } \ hc(T,\ T \cap g\,;0,\ g) \ \textsf{fi} \ .
\end{array}
$$

A final improvement would consist in merging the computation of g with that of $T \cap g\,;0$ using the tupling strategy (see e.g. Partsch 1990).

4 Conclusion

The calculus of formal languages and relations has proved to speed up derivations, in particular the way from "non-operational" specifications involving the closures R^* and $R^{\Rightarrow}$ to first recursive solutions. But also the tuning steps in improving the recursions have benefitted from the quantifier-free notation. If the resulting derivations still appear lengthy, this is to a great deal due to the fact that the assertions have been constructed in a stepwise fashion (for mastering complexity) rather than in one blow. Further case studies which should demonstrate the viability of the approach in more complicated examples are under way. Also, we are working on the definition of a more general program development language based on this approach. While other authors use a purely relational approach employing mostly even only binary relations, we find that relations with their fixed arity are too unflexible and lead to a lot of unnecessary encoding and decoding.

Acknowledgement

We are grateful to H. Partsch and to the anonymous referees for a number of valuable remarks.

References

F.L. Bauer, R. Berghammer, M. Broy, W. Dosch, F. Geiselbrechtinger, R. Gnatz, E. Hangel, W. Hesse, B. Krieg-Brückner, A. Laut, T.A. Matzner, B. Möller, F. Nickl, H. Partsch, P. Pepper, K. Samelson, M. Wirsing, H. Wössner: The Munich project CIP. Volume I: The wide spectrum language CIP-L. Lecture Notes in Computer Science **183**. Berlin: Springer 1985

F.L. Bauer, B. Möller, H. Partsch, P. Pepper: Formal program construction by transformations — Computer-aided, Intuition-guided Programming. IEEE Transactions on Software Engineering **15**, 165–180 (1989)

R. Berghammer: A transformational development of several algorithms for testing the existence of cycles in a directed graph. Institut für Informatik der TU München, TUM-I8615

R. Bird: Lectures on constructive functional programming. In M. Broy (ed.): Constructive methods in computing science. NATO ASI Series. Series F: Computer and systems sciences **55**. Berlin: Springer 1989, 151–216

R.M. Burstall, J. Darlington: A transformation system for developing recursive programs. J. ACM **24**, 44–67 (1977)

J. Cai, R. Paige: Program derivation by fixed point computation. Science of Computer Programming **11**, 197–261 (1989)

C.J. Date: An introduction to database systems. Vol. I, 4th edition. Reading, Mass.: Addison-Wesley 1988

M. Gondran, M. Minoux: Graphes et algorithmes. Paris: Eyrolles 1979

P. Lescanne: Modèles non déterministes de types abstraits. R.A.I.R.O. Informatique théorique **16**, 225–244 (1982)

Z. Manna: Mathematical theory of computation. New York: McGraw-Hill 1974

L.G.L.T. Meertens: Algorithmics — Towards programming as a mathematical activity. In J. W. de Bakker et al. (eds.): Proc CWI Symposium on Mathematics and Computer Science. CWI Monographs Vol **1**. Amsterdam: North-Holland 1986, 289–334

B. Möller: Applicative assertions. In: J.L.A. van de Snepscheut (ed.): Mathematics of Program Construction. Lecture Notes in Computer Science **375**. Berlin: Springer 1989, 348–362

B. Möller: Relations as a program development language. In B. Möller (ed.): Constructing programs from specifications. Proc. IFIP TC2/WG 2.1 Working Conference on Constructing Programs from Specifications, Pacific Grove, CA, USA, 13–16 May 1991. Amsterdam: North-Holland 1991, 373–397

H.A. Partsch: Specification and transformation of programs — A formal approach to software development. Berlin: Springer 1990

G. Schmidt, T. Ströhlein: Relationen und Graphen. Berlin: Springer 1989. English version: Relations and graphs (forthcoming)

A. Tarski: On the calculus of relations. J. Symbolic Logic 6, 73–89 (1941)

Logical Specifications for Functional Programs

Theodore S. Norvell and Eric C.R. Hehner
norvell@cs.utoronto.ca hehner@cs.utoronto.ca

Department of Computer Science
University of Toronto

Abstract. We present a formal method of functional program development based on step-by-step transformation.
In their most abstract form, specifications are essentially predicates that relate the result of the specified program to the free variables of that program. In their most concrete form, specifications are simply programs in a functional programming language. Development from abstract specifications to programs is calculational.
Using logic in the specification language has many advantages. Importantly it allows nondeterministic specifications to be given, and thus does not force overspecification.

0 Introduction

A great deal of research has focused on transforming functional programs into equivalent functional programs. The original program can be considered to be an executable specification.

In this paper we wish to consider not only executable specifications, but also implicit specifications that relate the input and result of a functional program in ways that give no indication of any practical way to compute the result. Such a specification can be more abstract and more declarative than an executable specification.

We take the following point of view, applicable to programming in imperative, functional, or any other kind of language: Specifications describe those *observations* that are acceptable and programs are one sort of specification. A specification x can be refined to another specification y if and only if x describes every observation y describes. Within such a framework, nondeterminism presents no difficulty and the validity of refinement is a very simple relationship. In the case of functional (expression) programming each observation consists of the state in which an expression is evaluated and and a value for the whole expression.

To describe acceptable observations, various notations can be used. Common notations include predicate calculus and set notation. Neither of these is satisfactory for expressions, as they disagree with existing notation for deterministic expressions. Instead we use a calculus of nondeterministic expressions known as bunch theory. Implicit specifications written in predicate calculus fit well into this calculus.

0.0 The Structure of the Paper

The structure of this paper is as follows. Section 1 presents a theory of nondeterministic expressions that is used throughout the rest of the paper. Section 2 introduces a functional programming language and a specification language. The programming language will be a subset of the specification language. Examples of using this specification language are given in Sect. 3. In Sect. 4 the relation of ***refinement*** is introduced. A specification y refines a specification x (written $x \sqsupseteq y$) iff every way that y can be satisfied also satisfies x. A program is a specification that can be executed with acceptable efficiency and so needs no further refinement. Section 5 presents a number of theorems that are of help in proving the refinement relation. Section 6 shows how these theorems can be used to derive programs from specifications by a number of small and formally justified steps. Higher order functions are discussed in Sect. 7. Section 8 presents a method of specifying time bounds. In Sect. 9 we look at pattern matching. Finally Sect. 10 discusses related research.

1 Nondeterministic Expressions

We generalize the notion of expression to allow "don't care" nondeterminism (also known as "erratic" nondeterminism). Our generalized expressions are known as ***bunch expressions*** (Hehner 1984).

Given expressions x and y, the expression x,y called the ***bunch union*** of x and y denotes a value that could be the value of x or could be the value of y. Bunch union is associative, commutative, and idempotent. Ordinary operators distribute over bunch union. For example, the following three expressions are equivalent

$$(1,4)+(5,2)$$
$$(1+5),(1+2),(4+5),(4+2)$$
$$3,6,9$$

The identity of bunch union is written as *null*. It represents the empty bunch. At the opposite end of the spectrum is *all*, representing the union of all expressions.

A bunch x is a ***subbunch*** of a bunch y if and only if there is a bunch z such that x,z is equivalent to y. We write $x : y$. This is a partial order. For all bunch expressions x, we have $null : x$. Equality of bunch expressions will be written as $x \equiv y$, meaning $x : y$ and $y : x$. Ordinary equality is written $x = y$ and differs from $x \equiv y$ in that it distributes over bunch union. Thus

$$(1,2)=2 \quad \equiv \quad \textbf{true}, \textbf{false}$$

whereas

$$1,2 \quad \not\equiv \quad 2$$

Certain bunch expressions will be called *elements*. Each number is an element as are the constants **true** and **false**. A list of elements is an element. Which functional values are elements will be discussed later. We say e is an element of x if $e : x$ and e is an element.

Bunches may be considered as sets but without the nesting (sets of sets), using a simpler notation (no curly braces), and with distribution of operations over the elements. The main reasons for using bunches rather than sets are notational convenience, and that they specialize properly to deterministic values, whereas sets do not.

2 The Specification Language

Our specification language will be an extension to a simple functional programming language. The expressions of the specification language are bunch expressions.

Specifications may contain free variables. These represent the input to the expression, *i.e.* the state in which it is evaluated. Each variable represents an element.

As a simple example of a specification, $n+1$ is the specification of a number one greater than state variable n. It happens that this specification is also a program. By using bunch expressions, we allow for choice in the specification. A specification of a number that is one, two, or three greater than n is $n+(1,2,3)$.

2.0 The Programming Language Subset

For this paper, we will use the simple language illustrated in Fig. 0. Expressions in this language will be called *programs* to distinguish them from more general specifications.

Types		
Naturals	nat	$0, 1, 2, \ldots$
Subranges	$9,..\,12$	$9, 10, 11$
Booleans	$bool$	**true**, **false**
Lists	nat^*	$[\,], [0], [0;0], \ldots$
Expressions		
Numerical Expressions	$1+5$	6
Boolean Expressions	$1=5$	**false**
Conditionals	**if** $1=5$ **then** 3 **else** 4	4
Lists	$[1;2;3;4]$	$[1;2;3;4]$
Lists	$[9;..\,12]$	$[9;10;11]$
List catenation	$[1;2]^+[3;4]$	$[1;2;3;4]$
List indexing	$[1;2;3;4]\,2$	3
List indexing	$[1;2;3;4]\,[3;2;1;0]$	$[4;3;2;1]$
List length	$\#[1;2;3;4]$	4
Functions	$\lambda m : nat^* \succ \lambda i : nat \succ m\,[0;..\,i]$	
Application	$(\lambda m : nat^* \succ \lambda i : nat \succ m\,[0;..\,i])\,[1;2;3;4]\,2$	[1;2]
Let	**let** $i=3 \rightarrow [1;2;3;4]\,[0;..\,i]$	$[1;2;3]$

Fig. 0. A simple functional programming language

Types The types of this language are bunches. The bunch $bool$ has elements **true** and **false**. The bunch nat has elements 0, 1, 2, and so on. A subrange of the naturals is written $i,..\,j$ for naturals i and j. This subrange includes i but excludes j. Given a type T, the type T^* is the bunch of all finite lists with items (list members) in T. An elementary list is one whose items are all elements.

Expressions The usual boolean and numerical operators are provided, as well as a standard if-expression.

Lists are written in square brackets with semicolons separating the items. A useful notation forms a list of contiguous naturals; $[i;..j]$ begins with i and continues up to (but not including) j. Lists may be catenated using $^+$. List indexing is written as juxtaposition and lists are indexed from 0. A list may be indexed by a list, producing a list of results.

Functions and let-expressions introduce new identifiers which may be used within their bodies. Function application is written as juxtaposition.

Notably absent from this language is any form of recursive definition. Recursion is treated in section 4.1.

Semantics The formal semantics of the operators of the programming language can be given axiomatically. A listing of all the axioms would be rather long. We list a few as examples.

$$\begin{aligned}
\textbf{if true then } x \textbf{ else } y &\equiv x \\
\textbf{if false then } x \textbf{ else } y &\equiv y \\
\#[\,] &\equiv 0 \\
\#[x] &\equiv 1 \\
\#(x \mathbin{^+} y) &\equiv \#x + \#y \\
&\vdots
\end{aligned}$$

The treatment of errors (for example division by 0) is a matter of some choice. We can treat errors as equivalent to *all*, or we can omit axioms that allow us to reason about erroneous computations. Either way of treating errors is consistent with the rest of this paper.

2.1 Specification Language Extensions

The programming language presented so far can be used to write executable specifications which can then be transformed to more efficient programs using conventional techniques.

Instead of stopping at an executable specification language, we will allow any bunch expression to be used as a specification. In this section, we present a number of constructs that are of use in writing specifications. They extend the programming language to allow greater ease and range of expression.

In this section, P and Q will stand for first order predicates, e for an element, x, y, and z for specifications, and i for an identifier.

Until Sect. 7 we will only consider elements that are first order, that is numbers, booleans, and lists of first order elements. Functional elements will be discussed in Sect. 7.

Predicates are boolean expressions. However the nondeterminism of the specification language is not extended to the predicates. For example $i \leq (2,3)$ is not an acceptable predicate because, in any state where $i = 3$, it is equivalent to $(\textbf{true}, \textbf{false})$.

In each state, a predicate must be either true or false, never both, never neither (though the logic may not be complete enough to say which).

Programs Any program is also a specification. Furthermore, any way of constructing programs from programs can be used to construct specifications from specifications. Thus

$$[x; y]$$

and

$$\textbf{if } P \textbf{ then } x \textbf{ else } y$$

are both specifications provided that P is a predicate and x and y are specifications, even though P, x, and y may not be programs.

Solutions The expression $\S i \cdot P$ is equivalent to the bunch of all elements i for which P is true. For example, $\S i \cdot i : nat \wedge i < 3$ is the bunch $0, 1, 2$. The axiom for this quantifier is:

$$(e : \S i \cdot P) \quad = \quad (\text{Substitute } e \text{ for } i \text{ everywhere in } P)$$

with the usual caveats for substitution.

Null The specification *null* refines all specifications. This specification is not satisfied by any result. The axiom for *null* is

$$null \quad \equiv \quad \S i \cdot \textbf{false}$$

In imperative programming, the corresponding specification is that which has, as its weakest precondition predicate transformer, $\lambda R \cdot \textbf{true}$.

All The specification *all* is refined by all specifications. It can be used by the specifier to indicate that she doesn't care about the result. The axiom for *all* is

$$all \quad \equiv \quad \S i \cdot \textbf{true}$$

This is the bunch of all elements.

Union and Intersection The specification x,y specifies that at least one of specifications x and y must be met. The specification $x`y$ specifies that both x and y must be met. Their axioms are

$$\begin{aligned} x, y \quad &\equiv \quad \S i \cdot (i : x) \vee (i : y) \\ x`y \quad &\equiv \quad \S i \cdot (i : x) \wedge (i : y) \end{aligned}$$

for i not free in x or y.

Assert The specification $P \succ\!\!- x$ expresses that x must be met when P is true, and otherwise any result will do. Its axiom is

$$P \succ\!\!- x \quad \equiv \quad \textbf{if } P \textbf{ then } x \textbf{ else } \mathit{all}$$

In this usage, P is called an *assertion*.

Guard The specification $P \rightarrow x$ expresses that x must be met when P is true, and is otherwise impossible to meet. Its axiom is

$$P \rightarrow x \quad \equiv \quad \textbf{if } P \textbf{ then } x \textbf{ else } \mathit{null}$$

In this usage, P is called a *guard*.

Seen as unary operators, $P \succ\!\!-$ and $P \rightarrow$ are duals and adjoint.

Try The specification **try** x expresses that x must be met if possible. Its axiom is

$$\textbf{try } x \quad \equiv \quad (x \not\equiv \mathit{null}) \succ\!\!- x \quad \equiv \quad \textbf{if } x \not\equiv \mathit{null} \textbf{ then } x \textbf{ else } \mathit{all}$$

The specification **try** x **else** y expresses that x must be met if possible, and if not, y must be met. Its axiom is

$$\textbf{try } x \textbf{ else } y \quad \equiv \quad \textbf{if } x \not\equiv \mathit{null} \textbf{ then } x \textbf{ else } y$$

This construct expresses a kind of backtracking or dynamic exception handling where failure is expressed by *null*.

Unlike our other specification constructs, **try** and **try else** are not monotonic in all their specification operands, with respect to the subbunch ordering.

Lambda The specification language has a more general abstraction operator than the programming language. For identifier i and expression x, the following is an expression

$$\lambda i \cdot x$$

For any element e

$$(\lambda i \cdot x)\, e \quad \equiv \quad (\text{Substitute } e \text{ for } i \text{ everywhere in } x)$$

Furthermore application distributes over bunch formation, so, for example,

$$f\ \mathit{null} \equiv \mathit{null}$$
$$f\,(y, z) \equiv (f\ y), (f\ z)$$

Thus variables always represent elements.

Lambda abstraction is untyped with respect to the programming language types. However, in order to prevent paradoxical expressions, it is typed with respect to the order of the arguments. Until Sect. 7 all arguments will be first order, that is, nonfunctional.

Let Likewise, the specification language has a more general **let** construct. It is defined by

$$\textbf{let}\ i \cdot x \quad \equiv \quad (\lambda i \cdot x)\ all$$

Typically x is of the form $P \rightarrow y$, in which case it can be seen that $\textbf{let}\ i \cdot (P \rightarrow y)$ is the union over all elements i such that P is true, of y.

2.2 Syntactic Issues

Precedence The precedence of operators used in this paper will be first juxtaposition (application and indexing) and then in order

				$=$							
				$\neq$						Binders	
$\#$			,	$<$					**if then else**		$\equiv$
$*$	$\times$	$+$								$\rightarrow$	
			;..	$>$	$\neg$	$\wedge$	$\vee$	$\Rightarrow$	**try**		$\sqsupseteq$
$¢$	$/$	$-$	‘							$\succ$	
				$\leq$					**try else**		$\underline{\square}$
										$\gg$	
				$\geq$							
				$:$							

Binders (λ, **let** , §, $\forall$, $\exists$), $\rightarrow$, and $\succ$ are right associative, so that we can write, for example,

$$\lambda i \cdot (P \succ (\textbf{let}\ j \cdot (Q \rightarrow R \succ x)))$$

as

$$\lambda i \cdot P \succ \textbf{let}\ j \cdot Q \rightarrow R \succ x$$

Relation operators are *continuing*, so we can write, for example

$$(x \sqsupseteq y) \wedge (y \sqsupseteq z)$$

as

$$x \sqsupseteq y \sqsupseteq z$$

Syntactic Sugar For all binders, we allow the following abbreviation. If the textually first identifier to appear in the body is the bound variable, then the bound variable and the subsequent dot can be omitted. Thus

$$\lambda i \cdot i : T \succ x$$

can be written as

$$\lambda i : T \succ x$$

and

$$\textbf{let}\ i \cdot i = e \rightarrow x$$

can be written as

$$\textbf{let}\ i = e \rightarrow x$$

In programs, we always use the abbreviated notation.

3 Writing Specifications

In this section several examples are given of using the specification language.

We remind the reader that the free variables together represent the state in which the expression is evaluated and thus each free variable represents an element. Restrictions on these variables, i.e. the type of the state, will be stated informally.

An implementation is obliged to give a result described by the specification. Thus *null* is unimplementable. The specification

$$\textbf{if } x = 0 \textbf{ then } \mathit{null} \textbf{ else } 1$$

can be satisfied in states such that $x \neq 0$ but not when $x = 0$. Perhaps the specifier has no intention of providing a state for which the specification is *null*, but to the implementor every input is a possibility. A specification is called *implementable* if there is no state in which it is equivalent to *null*.

3.0 Searching

Suppose that L is a list variable of a type T^* and x is a variable of type T. Informally, we need to find an index of an item x of a list L. A first attempt at formally specifying this is

$$\S i \cdot L\, i = x$$

This says that we want any i such that $L\, i = x$. However, x may not occur in L at all. For such a case, the above specification is *null*, and so the specification is unimplementable. Suppose that we intend to use the specification only when x occurs in the list. Then we don't care what the result would be if x did not occur, and the specification should be

$$(\exists i \cdot L\, i = x) \succ\!\!-\, (\S i \cdot L\, i = x)$$

This is still not entirely satisfactory if it is not guaranteed by the axioms concerning lists that $L\, i = x$ is false for values of i that are not valid indices of L. The next specification covers this situation

$$(\exists i : 0,..\#L \wedge L\, i = x) \succ\!\!-\, (\S i : 0,..\#L \wedge L\, i = x)$$

(Note the use of the syntactic sugar from Sect. 2.2.) The **try** operator can be used to make this more concise:

$$\textbf{try } (\S i : 0,..\#L \wedge L\, i = x)$$

It is noteworthy that this is a nondeterministic problem. When x appears more than once in the list, the result can be any suitable index. A deterministic specification language would necessitate overspecification.

3.1 Fermat's Last Theorem

Quite often an informal search specification will be of the form "if there is an x such that $P\,x$, then $f\,x$, else y". The **if then else** construct can not be used to formalize this as x will not be available in the then-part. A solution is to use the **try else** construct. For example, the following specification is [ ] if Fermat's Last Theorem is true and is some counterexample otherwise.

$$\begin{array}{l}\textbf{try}\;(\textbf{let}\;n : nat+3 \rightarrow \textbf{let}\;i : nat \rightarrow \textbf{let}\;j : nat \rightarrow \textbf{let}\;k : nat \rightarrow \\ \qquad i^n + j^n = k^n \rightarrow [n;i;j;k]\;) \\ \textbf{else}\;[\,]\end{array}$$

3.2 Sorting

Suppose that $\leq$ is a relation, on a type T, that is reflexive, transitive, and total (that is, for all x and y in T, either $x\leq y$ or $y\leq x$). We wish to specify that, given a list, we want a permutation of it that is sorted with respect to this relation. We will present two equivalent specifications to illustrate the range of styles that the specification language permits.

A Logic Oriented Specification The first specification is more logic oriented. It proceeds by defining a predicate describing the desired relationship between the input and output of the program. First we define a function that returns the number of times an item occurs in a list.

$$count\;L\;x \quad \stackrel{\text{def}}{\equiv} \quad ¢\S j : 0,..\#L \wedge x = L\,j$$

This uses the counting operator ¢ that gives the number of elements in a bunch. Now we define what it is for one list to be a permutation of another

$$Perm\;L\;M \quad \stackrel{\text{def}}{\equiv} \quad \forall x \cdot count\;L\;x = count\;M\;x$$

Next is a predicate that indicates a list is monotone

$$Mono\;M \quad \stackrel{\text{def}}{\equiv} \quad \forall j : 1,..\#M \Rightarrow M\,(j-1) \leq M\,j$$

The final predicate states that one list is a sorted permutation of another

$$Sortof\;L\;M \quad \stackrel{\text{def}}{\equiv} \quad Perm\;L\;M \wedge Mono\;M$$

Finally this predicate is used to form the specification:

$$sort \quad \stackrel{\text{def}}{\equiv} \quad \lambda L : T^* \succ \S M : T^* \wedge Sortof\;L\;M$$

An Expression Oriented Specification The second specification is more expression oriented. First we define a permutation function as the smallest function satisfying

$$perm \;\equiv\; \lambda L{:}T^* \succ L, (\,\mathbf{let}\ M : perm\ L \rightarrow \mathbf{let}\ i \cdot \mathbf{let}\ j \cdot 0 \le i < j < \#M \rightarrow M[0;..i] + [M\ j] + M[i+1;..j] + [M\ i] + M[j+1;..\#M]\,)$$

(The meaning of "smallest function" will be explained in section 4.2.) This function nondeterministically returns any permutation of its argument. Next we define the bunch of all ordered lists over T as the smallest bunch satisfying

$$ordered \;\equiv\; [\,], (\mathbf{let}\ M : ordered \rightarrow \mathbf{let}\ t : T \wedge (\forall i : 0,..\#M \Rightarrow t \le M\ i) \rightarrow [t] + M)$$

Finally one can specify *sort* as

$$sort \;\stackrel{\text{def}}{\equiv}\; \lambda L : T^* \succ ordered\ `\ perm\ L$$

4 Refinement

4.0 The Refinement Relation

We define the refinement relation $x \sqsupseteq y$ to mean that $y : x$ universally. By "universally" we mean in all states, that is for all assignments of elements to the free variables of expressions x and y. We say x *is refined by* y. For example, that 1,2 is refined by 1 is written

$$1,2 \;\sqsupseteq\; 1$$

For another example,

$$n : nat \succ n + (1,2) \;\sqsupseteq\; n + 1$$

The refinement relation is a partial order on specifications.

Programming from a specification x is the finding of a program y such that $x \sqsupseteq y$. To simplify this process, we find a sequence of specifications $x_0 \cdots x_n$ where x_0 is x and x_n is y, and where $x_i \sqsupseteq x_{i+1}$ is a fairly trivial theorem, for each i. This is a formalization of the process of stepwise refinement.

Note that some authors write $x \sqsubseteq y$ for refinement where we write $x \sqsupseteq y$. Perhaps they believe that "bigger is better," but we find the analogy with standard set notation ($\supseteq$) too strong to resist.

4.1 Programming with Refined Specifications

At this point we can add one final construct to the programming language. Any specification x can be considered to be a program provided a program y is supplied such that $x \sqsupseteq y$. We can think of x as a subprogram name and of y as its subprogram body.

Recursion and mutual recursion are allowed. Since $\sqsupseteq$ is reflexive, it is always possible to refine x with x itself. This leads to correct programs, but ones that take an infinite amount of time to execute. This will be discussed further in Sect. 8.

A programming notation for recursion could be defined, but we have chosen not to do so.

4.2 Function Refinement

Because we wish to speak of refinement of functions, we must extend the subbunch relation to functions. This is done by defining

$$(\lambda i \cdot y) \quad : \quad (\lambda i \cdot x)$$

if for all elementary i,

$$y \quad : \quad x$$

Thus if $x \sqsupseteq y$, then $\lambda i \cdot x \sqsupseteq \lambda i \cdot y$.

5 Laws of Programming

In this section we will present a number of theorems that can be used to prove refinement relations. Numerous other theorems could be presented; this is a selection of those most useful for developing programs.

Some of the following laws show mutual refinement, that is both $x \sqsupseteq y$ and $y \sqsupseteq x$; we will use $x \underline{\square}\, y$ to show this. Some of the following laws apply to both assertions and guards; we will use $\gg$ to mean one of $\succ\!\!-$ or $\twoheadrightarrow$. That is, the laws where $\gg$ appears (even if more than once) each abbreviate exactly two laws, one for $\succ\!\!-$ and one for $\twoheadrightarrow$.

Union elimination: $x, y \sqsupseteq x$
If introduction/elimination: $x \underline{\square}$ **if** P **then** x **else** x
Case analysis:

$$\textbf{if } P \textbf{ then } x \textbf{ else } y \quad \underline{\square} \quad \textbf{if } P \textbf{ then } (P \gg x) \textbf{ else } y$$

$$\textbf{if } P \textbf{ then } x \textbf{ else } y \quad \underline{\square} \quad \textbf{if } P \textbf{ then } x \textbf{ else } (\neg P \gg y)$$

Let introduction/elimination: *If i is not free in x, then*

$$x \quad \underline{\square} \quad \textbf{let } i \cdot x$$

The example law for let: *If e is an element and* (Substitute e for i everywhere in P), *then*

$$\textbf{let } i \cdot P \twoheadrightarrow x \quad \sqsupseteq \quad \textbf{let } i \cdot P \twoheadrightarrow (\text{Substitute } e \text{ for } i \text{ anywhere in } x)$$

The example law for §: *If e is an element and* (Substitute e for i everywhere in P), *then*

$$\S i \cdot P \quad \sqsupseteq \quad e$$

Guard introduction: $x \sqsupseteq P \twoheadrightarrow x$
Assertion elimination: $P \succ\!\!- x \sqsupseteq x$
Guard strengthening: *If* $(Q \Rightarrow P)$ *universally, then* $P \twoheadrightarrow x \sqsupseteq Q \twoheadrightarrow x$
Assertion weakening: *If* $(Q \Rightarrow P)$ *universally, then* $Q \succ\!\!- x \sqsupseteq P \succ\!\!- x$
Assertion/guard use: *If* $(P \Rightarrow y : x)$ *universally, then* $P \gg x \sqsupseteq P \gg y$
Assertion/guard combining/splitting: $P \gg Q \gg x \quad \underline{\square} \quad P \wedge Q \gg x$
Adjunction: $(P \succ\!\!- x \sqsupseteq y) = (x \sqsupseteq P \twoheadrightarrow y)$

One point: *If e is an element,*

$$i = e \gg x \quad \square \quad i = e \gg \text{(Substitute e for i anywhere in x)}$$

Application introduction/elimination: *If $\lambda i \cdot x$ distributes over bunch union,* then

$$(\lambda i \cdot x)\, y \quad \square \quad \text{(Substitute y for i everywhere in x)}$$

Lambda introduction: *If $x \sqsupseteq y$ then $\lambda i \cdot x \sqsupseteq \lambda i \cdot y$*

There are a great many laws for moving assertions and guards. Inward movement laws say that assertions and guards that apply to a specification apply to any part of the specification. For example,

An example inward movement law: $P \succ x, y \quad \sqsupseteq \quad P \succ (P \succ x), y$

Outward movement laws say that assertions and guards that apply to all parts of a specification apply to the whole specification.

An example outward movement law: *If i is not free in P, then*

$$\text{let } i \cdot P \to x \quad \sqsupseteq \quad P \to \text{let } i \cdot x$$

Except for **try** and **try else**, all the operators we have introduced that form specifications from specification operands are monotonic in those operands, with respect to the refinement relation. This gives rise to a number of monotonicity laws that will be used implicitly. For example,

An example monotonicity law: *If $x \sqsupseteq y$, then*

$$\text{if } P \text{ then } x \text{ else } z \quad \sqsupseteq \quad \text{if } P \text{ then } y \text{ else } z$$

Monotonicity laws allow application of the other laws deep within the structure of a specification.

6 Deriving Programs

In this section, we demonstrate a programming methodology based on the refinement relation.

6.0 Searching

Our searching specification from Sect. 3.0 was

$$(\exists i : 0,.. \#L \wedge L\, i = x) \succ (\S i : 0,.. \#L \wedge L\, i = x)$$

We add a parameter j so we can specify searching in the part of list L preceding index j

$$search\_before \quad \overset{\text{def}}{=} \quad \lambda j : 1,.. 1 + \#L \succ (\exists i : 0,.. j \wedge L\, i = x) \succ (\S i : 0,.. \#L \wedge L\, i = x)$$

The original specification is refined by

$$search\_before\,(\#L)$$

It remains to supply a program that refines *search_before*. Let j represent any element of type $1,..\,1{+}\#L$. We start by refining *search_before* j. (Note that hints appear *between* the two specifications they apply to.)

$$
\begin{array}{ll}
& search\_before\ j \qquad\qquad \textbf{if}\text{ introduction and case analysis}\\
\sqsupseteq & \textbf{if}\ L\,(j-1)=x\ \textbf{then}\ (L\,(j-1)=x \succ search\_before\ j)\\
& \textbf{else}\ (L\,(j-1)\neq x \succ search\_before\ j)\\
\multicolumn{2}{l}{\text{Definition of } search\_before\text{, assertion combining, and assertion weakening}}\\
\sqsupseteq & \textbf{if}\ L\,(j-1)=x\ \textbf{then}\ (L\,(j-1)=x \succ (\S i:0,..\,\#L \wedge L\,i=x))\\
& \textbf{else}\ (L\,(j-1)\neq x \wedge (\exists i:0,..\,j \wedge L\,i=x) \succ (\S i:0,..\,\#L \wedge L\,i=x))\\
\multicolumn{2}{l}{\text{Assertion use, example law, and assertion elimination in the then-part}}\\
\multicolumn{2}{l}{\text{Logic and assertion weakening in the else-part}}\\
\sqsupseteq & \textbf{if}\ L\,(j-1)=x\ \textbf{then}\ j-1\\
& \textbf{else}\ ((\exists i:0,..\,j{-}1 \wedge L\,i=x) \succ (\S i:0,..\,\#L \wedge L\,i=x))\\
\multicolumn{2}{l}{\text{If there exists an } i \text{ in } 0,..\,j{-}1\text{, then } j>1}\\
\sqsupseteq & \textbf{if}\ L\,(j-1)=x\ \textbf{then}\ j-1\\
& \textbf{else}\ ((\exists i:0,..\,j{-}1 \wedge L\,i=x) \wedge j:2,..\,1{+}\#L \succ\\
& \qquad (\S i:0,..\,\#L \wedge L\,i=x)\,)\\
\multicolumn{2}{l}{\text{Assertion weakening and assertion splitting}}\\
\sqsupseteq & \textbf{if}\ L\,(j-1)=x\ \textbf{then}\ j-1\\
& \textbf{else}\ ((j-1):1,..\,1{+}\#L \succ\\
& \qquad (\exists i:0,..\,j{-}1 \wedge L\,i=x) \succ\\
& \qquad (\S i:0,..\,\#L \wedge L\,i=x)\,)\\
\multicolumn{2}{l}{\text{Definition of } search\_before \text{ and application introduction}}\\
\sqsupseteq & \textbf{if}\ L\,(j-1)=x\ \textbf{then}\ j-1\\
& \textbf{else}\ search\_before\,(j-1)
\end{array}
$$

We can add to both sides the range assertion on j and then use the function refinement law of Sect. 4.2 (lambda introduction). This gives us

$$
\begin{array}{ll}
& search\_before\\
\sqsupseteq & \lambda j:1,..\,1{+}\#L \succ \textbf{if}\ L\,(j-1)=x\ \textbf{then}\ j-1\\
& \qquad\qquad\qquad\quad \textbf{else}\ search\_before\,(j-1)
\end{array}
$$

6.1 Sorting

As a second example of deriving programs we derive a merge sort program from the sorting specification given in Sect. 3.2.

$$
\begin{array}{lll}
& sort\ L & \\
\sqsupseteq & \S M \cdot Sortof\ L\ M & \text{Definition}\\
\sqsupseteq & \textbf{if}\ \#L \leq 1 & \textbf{if}\text{ introduction and case analysis}\\
& \textbf{then}\ (\#L \leq 1 \succ \S M \cdot Sortof\ L\ M) &\\
& \textbf{else}\ (\#L > 1 \succ \S M \cdot Sortof\ L\ M) &
\end{array}
$$

The then-branch is refined as follows

$$\#L \leq 1 \succ \S M \cdot Sortof\ L\ M$$ Assertion use and elimination

$$\sqsupseteq\ \S M \cdot \#L \leq 1 \Rightarrow Sortof\ L\ M$$ Example

$$\sqsupseteq\ L$$

We now refine the else-branch. The first idea is to divide and conquer.

$$\#L > 1 \succ \S M \cdot Sortof\ L\ M$$

Let introduction and assertion elimination

$$\sqsupseteq\ \mathbf{let}\ T \cdot \mathbf{let}\ U \cdot \S M \cdot Sortof\ L\ M$$ Guard introduction

$$\sqsupseteq\ \mathbf{let}\ T \cdot \mathbf{let}\ U \cdot L = T^+ U \rightarrow \S M \cdot Sortof\ L\ M$$ One point

$$\sqsupseteq\ \mathbf{let}\ T \cdot \mathbf{let}\ U \cdot L = T^+ U \rightarrow \S M \cdot Sortof\ (T^+ U)\ M$$

We break off the derivation at this point to consider the next move.

Having divided the list, we will sort the two parts. We need to replace the predicate *Sortof* by one in terms of the sorted parts. We call that predicate *Mergeof* and the desired theorem is

$$Mergeof\ (sort\ T)\ (sort\ U)\ M \Rightarrow Sortof\ (T^+ U) M$$

One definition that yields this theorem is

$$Mergeof\ T\ U\ M \quad \stackrel{\text{def}}{\equiv} \quad (Mono\ T \wedge Mono\ U) \Rightarrow Sortof\ (T^+ U)\ M$$

Using the theorem we continue the derivation with

$$\sqsupseteq\ \mathbf{let}\ T \cdot \mathbf{let}\ U \cdot L = T^+ U \rightarrow \S M \cdot Mergeof\ (sort\ T)\ (sort\ U)\ M$$

We defer the implementation of *Mergeof* so for now we just define

$$merge \quad \stackrel{\text{def}}{\equiv} \quad \lambda X : T^* \succ \lambda Y : T^* \succ Mono\ X \wedge Mono\ Y \succ \S M \cdot Mergeof\ X\ Y\ M$$

So the derivation continues

$$\sqsupseteq\ \mathbf{let}\ T \cdot \mathbf{let}\ U \cdot L = T^+ U \rightarrow merge\ (sort\ T)\ (sort\ U)$$

Guard strengthening

$$\sqsupseteq\ \mathbf{let}\ T \cdot \mathbf{let}\ U \cdot T = L[0;..\ \#L\ \mathbf{div}\ 2] \wedge U = L[\#L\ \mathbf{div}\ 2;..\ \#L] \rightarrow$$
$$merge\ (sort\ T)\ (sort\ U)$$

Guard movement

$$\sqsupseteq\ \mathbf{let}\ T = L[0;..\ \#L\ \mathbf{div}\ 2] \rightarrow$$
$$\mathbf{let}\ U = L[\#L\ \mathbf{div}\ 2;..\ \#L] \rightarrow$$
$$merge\ (sort\ T)\ (sort\ U)$$

We now need to refine the *merge* specification. Assuming X and Y of the right types we have

$$merge\ X\ Y$$ Definition

$$\sqsupseteq\ Mono\ X \wedge Mono\ Y \succ \S M \cdot Mergeof\ X\ Y\ M$$

if introduction and case analysis

$$\sqsupseteq\ \mathbf{if}\ X = [\,]\ \mathbf{then}\ (Mono\ Y \wedge X = [\,] \succ \S M \cdot Mergeof\ X\ Y\ M)$$
$$\mathbf{else\ if}\ Y = [\,]\ \mathbf{then}\ (Mono\ X \wedge Y = [\,] \succ \S M \cdot Mergeof\ X\ Y\ M)$$
$$\mathbf{else}\ (\ Mono\ X \wedge Mono\ Y \wedge X \neq [\,] \neq Y \succ$$
$$\S M \cdot Mergeof\ X\ Y\ M\ )$$

Assertion use and example

$$
\begin{array}{ll}
\sqsupseteq & \textbf{if } X = [\,] \textbf{ then } Y \qquad\qquad\qquad\qquad \textbf{if} \text{ introduction} \\
& \textbf{else if } Y = [\,] \textbf{ then } X \\
& \textbf{else} \,(\, \mathit{Mono}\, X \wedge \mathit{Mono}\, Y \wedge X \neq [\,] \neq Y \succ \\
& \quad \S M \cdot \mathit{Mergeof}\; X\; Y\; M\,) \\
\sqsupseteq & \textbf{if } X = [\,] \textbf{ then } Y \\
& \textbf{else if } Y = [\,] \textbf{ then } X \\
& \textbf{else if } X\,0 \leq Y\,0 \textbf{ then} \,(\, \mathit{Mono}\, X \wedge \mathit{Mono}\, Y \wedge X \neq [\,] \neq Y \wedge X\,0 \leq Y\,0 \succ \\
& \qquad \S M \cdot \mathit{Mergeof}\; X\; Y\; M\,) \\
& \textbf{else} \,(\, \mathit{Mono}\, X \wedge \mathit{Mono}\, Y \wedge X \neq [\,] \neq Y \wedge Y\,0 \leq X\,0 \succ \\
& \quad \S M \cdot \mathit{Mergeof}\; X\; Y\; M\,)
\end{array}
$$

Assertion use, example, and definition of *merge*

$$
\begin{array}{ll}
\sqsupseteq & \textbf{if } X = [\,] \textbf{ then } Y \\
& \textbf{else if } Y = [\,] \textbf{ then } X \\
& \textbf{else if } X\,0 \leq Y\,0 \textbf{ then } [X\,0] + \mathit{merge}\, (X[1;..\#X])\; Y \\
& \textbf{else } [Y\,0] + \mathit{merge}\; X\; (Y[1;..\#Y])
\end{array}
$$

Summarizing the above, we have proven

$$
\begin{array}{ll}
& \mathit{sort} \\
\sqsupseteq & \lambda L : T^* \succ \textbf{if } \#L \leq 1 \\
& \qquad \textbf{then } L \\
& \qquad \textbf{else} \,(\, \textbf{let } T = L[0;..\#L \textbf{ div } 2] \rightarrow \\
& \qquad\quad \textbf{let } U = L[\#L \textbf{ div } 2;..\#L] \rightarrow \\
& \qquad\quad \mathit{merge}\, (\mathit{sort}\; T)\, (\mathit{sort}\; U)\,)
\end{array}
$$

and

$$
\begin{array}{ll}
& \mathit{merge} \\
\sqsupseteq & \lambda X : T^* \succ \lambda Y : T^* \succ \\
& \textbf{if } X = [\,] \textbf{ then } Y \\
& \textbf{else if } Y = [\,] \textbf{ then } X \\
& \textbf{else if } X\,0 \leq Y\,0 \textbf{ then } [X\,0] + \mathit{merge}\, (X[1;..\#X])\; Y \\
& \textbf{else } [Y\,0] + \mathit{merge}\; X\; (Y[1;..\#Y])
\end{array}
$$

7 Higher Order Programming

In Sect. 4.2 we extended the subbunch relation to functions. This allows one to develop functions that have functional results. For example:

$$
\begin{array}{ll}
& \lambda i : \mathit{nat} \succ \lambda j : \mathit{nat} \succ i + j + (0, 1, 2) \\
\sqsupseteq & \lambda i : \mathit{nat} \succ \lambda j : \mathit{nat} \succ i + j + 1
\end{array}
$$

We are not yet ready to develop functions that have functional parameters.

Recall that parameters always represent elements. We extend the notion of elementhood to functions before talking about passing functions as arguments.

In order to avoid circularity in the definition of "element" and to preclude paradoxical expressions, we impose a simple type system on bunches (Church 1940). Expressions containing elements of primitive types such as *bool*, *nat*, and lists of such, we say are of type ι. A lambda expression is written $\lambda i_m \cdot x$ where m is a type. If x is of type n, $\lambda i_m \cdot x$ is of type $m \mapsto n$. In determining the type of the body x or any expression within it, it is assumed that i has type m. A function of type $m \mapsto n$, can be applied only to arguments of type m; the type of the application is n.

We say that a lambda expression $\lambda i_m \cdot x$ is an element iff for each element e of type n, $(\lambda i_m \cdot x)e$ is an element. For example, the elements of $\lambda i_\iota \cdot 1, 2$ are $\lambda i_\iota \cdot 1$ and $\lambda i_\iota \cdot 2$.

This definition has the interesting, but not problematic, consequence that there are non-*null* functions that are proper subbunches of elements. For example,

$$(\lambda i_\iota \cdot i = 0 \rightarrow 0) \quad : \quad (\lambda i_\iota \cdot 0)$$

To avoid cluttering specifications, including programs, with subscripts, we adopt the following convention:

$$\lambda i : x \succ\!\!- z$$

abbreviates

$$\lambda i_m \cdot i : x \succ\!\!- z$$

and

$$\lambda i\ x : y \succ\!\!- z$$

abbreviates

$$\lambda i_{m \mapsto n} \cdot i\ x : y \succ\!\!- z$$

where x is of type m and y is of type n. Similarly for functions of more arguments. This makes sense because a function i that maps elements of x to elements of y is accurately described by the predicate $i\ x : y$.

The definition of application is the same for functional parameters as for non-functional parameters. That is, it is the union over all substitutions of elements of the argument for the parameter.

Let us look at how definitions of application and elementhood affect higher order functions. Suppose we have a higher order function *map* defined by

$$\lambda f\ nat : nat \succ\!\!- \lambda L : nat^* \succ\!\!- \S M : nat^* \wedge \#M = \#L \wedge (\forall i : 0,.. \#L \Rightarrow M\ i = f\ (L\ i))$$

then the application

$$map\ (\lambda i : nat \succ\!\!- i + (1, 2))\ [0; 0]$$

is equivalent to

$$[1; 1], [2; 2]$$

This is perhaps a somewhat surprising consequence, but the alternative of allowing parameters to represent nondeterministic functions has serious pitfalls (see (Meertens 1986) and the discussion in Sect. 10 below).

As the *map* example suggests, the formalism presented here can be used to provide formal definitions of, and prove properties of, higher order operators such as those of Bird (1987).

8 Termination and Timing

As noted previously, programs that are correct according to the calculus given so far in this paper may specify nonterminating computations. This is because any specification x may be used as a program provided it is refined by a program, with recursion allowed. For example, we might refine x by **if** b **then** x **else** x or even by just x.

It is possible (and often reasonable) to verify that a program terminates, or to verify a time bound for it, by analysing the program after it has been derived without explicit consideration of time. If the verification fails, it is back to the drawing board. Such analysis is discussed in, for example, (Sands 1989). In this section we explore an alternative idea, that of incorporating timing (and hence termination) requirements into the original specification and refining such specifications to obtain a program.

8.0 Specifications with Time

Rather than deal with termination and nontermination as a duality, we deal with the time required for a computation to complete. First we must expand the idea of an observation to include the time that is required for a computation. *Specifications with time* are written as $P@T$ where P is a specification of a value and T is a number specification. The kind of numbers used in the T part may include an infinity value. Nondeterministic expressions may be used to give a range of acceptable times. (Syntactically @ binds closer than any operator, even juxtaposition.)

Programming and other operators on specifications are lifted to specifications with time according to a *timing policy*. A timing policy reflects implementation decisions (such as whether operands are evaluated in sequence or parallel), language design decisions (such as strictness), and decisions about how much operations should cost. We will exhibit a particular timing policy based on sequential implementation, strict application, and charging at least one unit of time for each recursive call.

Primitives such as multiplication are lifted to specifications with time as

$$x@a \times y@b \quad \equiv \quad (x \times y)@(a + b)$$

The **if** is lifted as

$$\textbf{if } x@a \textbf{ then } y@b \textbf{ else } z@c \quad \equiv \quad (\textbf{if } x \textbf{ then } y \textbf{ else } z)@(a + \textbf{if } x \textbf{ then } b \textbf{ else } c)$$

Specifications of functions with time specify both the time required to produce the functions and the time required to apply it (as a function of its argument). The specification $(\lambda i \cdot x@a)@b$ specifies a function that takes b time units to produce and $(\lambda i \cdot a)y$ time units to apply to y. The following way of lifting application models eager evaluation where the cost of evaluating the argument is assessed at the point of application. Let $(\lambda i \cdot x@a)^{\nu}$ mean $\lambda i \cdot x$ and $(\lambda i \cdot x@a)^{\tau}$ mean $\lambda i \cdot a$. Now

$$f@b\, y@c \quad \equiv \quad (f^{\nu}\, y)@(b + c + f^{\tau}\, y)$$

Reference to a refined specification is allowed as a programming construct (Sect. 4.1), but extra time may be optionally added. For example, if $x@0$ is a refined specification, one may make reference to $x@1$. In any loop of references, by this timing policy, at least 1 time unit must be added in the loop. Thus the observation that $x@a \sqsupseteq x@a$, although true, does not allow us to use $x@a$ in a program. On the other hand, if $x@a \sqsupseteq x@(a+1)$ is true, $x@a$ (or $x@(a+1)$) may be used as a program. For example, the observation that $x@\infty \sqsupseteq x@(\infty+1)$ means that $x@\infty$ may be used as a program; but $x@\infty$ is not a very useful specification. Recursive reference should be a bit clearer with an example.

Since the suffix @0 occurs quite frequently we will take the liberty of not writing it, leaving it implicit.

8.1 An Example

Let ΣL be the sum of the elements of a list of naturals L. Our specification with time of a summation function is

$$sum \overset{\text{def}}{\equiv} \lambda L : nat^* \succ (\Sigma L)@(\#L)$$

The time required to produce the summation function must be 0, that is no recursive calls are allowed, by the convention of not writing @0. This is easily achieved if we write the function as a constant. The $@\#L$ means that the time required to apply the summation function to a list L is $\#L$. We will write sum' for the same specification with the (implicit) @0 replaced by @1. However ΣL is specified in detail, the following should hold

$$\begin{array}{l} L = [\,] \succ \Sigma L \sqsupseteq 0 \\ L \neq [\,] \succ \Sigma L \sqsupseteq L\,0 + \Sigma(L[1;..\,\#L]) \end{array}$$

The following are also true

$$\begin{array}{l} L = [\,] \succ \#L \sqsupseteq 0 \\ L \neq [\,] \succ \#L \sqsupseteq 1 + \#(L[1;..\,\#L]) \end{array}$$

With these theorems we can quickly derive the obvious program

$$\begin{array}{lll} & sum & \textbf{if } \text{introduction; case analysis; first and third theorems} \\ \sqsupseteq & \lambda L : nat^* \succ \textbf{if } L = [\,] \textbf{ then } 0@0 & \text{Second and fourth theorems} \\ & \qquad \textbf{else } (L \neq [\,] \succ (\Sigma L)@(\#L)) & \\ \sqsupseteq & \lambda L : nat^* \succ \textbf{if } L = [\,] \textbf{ then } 0 & \\ & \qquad \textbf{else } (L\,0 + \Sigma(L[1;..\,\#L]))@(1 + \#(L[1;..\,\#L])) & \\ & \text{Application for specifications with time} & \\ \sqsupseteq & \lambda L : nat^* \succ \textbf{if } L = [\,] \textbf{ then } 0 \textbf{ else } L\,0 + sum'\,(L[1;..\,\#L]) & \end{array}$$

8.2 Higher Order Specifications with Time

The time taken to apply a function obtained from an application of a higher-order function may well depend on the time to apply a closure. The τ and ν notation allows specification of such functions. An efficient map function is specified by

$$\begin{array}{l} \lambda f\ nat : nat \succ \lambda L : nat^* \succ \\ (\S M : nat^* \wedge \#M = \#L \wedge (\forall i : 0,..\,\#L \Rightarrow M\,i = f^{\nu}\,(L\,i)))@(\sum_{i:0,..\,\#L} 1 + f^{\tau}\,(L\,i)) \end{array}$$

9 Pattern Matching

In modern function programming languages, functions are generally defined by a sequence of equations with the appropriate definition being picked according to pattern matching. Likewise the **case** construct of, for example, Haskell works by pattern matching. We look here at how this syntactic device can be given a semantics using the notation and theory presented earlier.

Since function definition by pattern matching can be understood in terms of the **case** construct we discuss only that. Consider the **case** expression

$$\mathbf{case}\ x\ \mathbf{of}\ \{f\,i \rightarrow y; \\ g\,j \rightarrow z\,\}$$

Where f and g are functions mapping types T and U respectively to a third type. The case expression can be understood as the specification

$$x : f\,T, g\,U \succ\!\!- \mathbf{let}\ k : x \twoheadrightarrow (\mathbf{let}\ i \cdot k : f\,i \twoheadrightarrow y), \\ (\mathbf{let}\ j \cdot k : g\,j \twoheadrightarrow z)$$

This interpretation of the case statement is nondeterministic when patterns overlap. Sequential pattern matching is modelled somewhat differently. The above case expression can be modelled as

$$\mathbf{let}\ k : x \twoheadrightarrow \mathbf{if}\ k : f\,T\ \mathbf{then}\,(\mathbf{let}\ i \cdot k : f\,i \twoheadrightarrow y) \\ \mathbf{else\,if}\ k : g\,U\ \mathbf{then}\,(\mathbf{let}\ j \cdot k : g\,j \twoheadrightarrow z) \\ \mathbf{else}\ all$$

10 Related Work

The use of logic to express the relationship between input and output dates back to work by Turing (Morris and Jones 1984), and is more recently found in the work of, for example, Hoare (1969) and Dijkstra (1975).

The uniform treatment of abstract specifications and programs is becoming common in imperative programming methodologies. Back (1987), Hehner (1984), Morgan (1988), and Morris (1990), building on the work of Dijkstra (1975), all extend imperative languages to include arbitrary specifications. A new methodology of Hehner (1990) treats the programming language as a subset of logic and uses logic as the full specification language.

Some of the specification constructs presented here are based on constructs that have been used in imperative specification. The $\twoheadrightarrow$, **try**, and **try else** operators, for example, are similar to operators described by Morgan (1988) and/or Nelson (1987).

In the functional programming community nondeterministic specifications have been avoided, perhaps because it is feared that nondeterminism does not mix with referential transparency. An exception is the work of Søndergaard and Sestoft (1988, 1990) which explores several varieties of nondeterminism and their relationships to referential transparency. Redelmeier (1984) used a form of weakest precondition semantics to define a programming language, but did not pursue nondeterminism or program derivation. Three bodies of work in functional program transformation do allow nondeterministic specifications. These are the CIP project (Bauer *et al.* 1987), Meertens's essay (Meertens 1986), and Hoogerwoord's thesis (Hoogerwoord 1989).

The CIP project involves not only functional programming, but also algebraic specification of data types and imperative programming. Only the functional programming aspects of CIP are discussed here. CIP is also a transformational approach based on nondeterministic specification. In CIP each specification is associated with a set of values called its breadth. One specification refines another if its breadth is a subset of the other's. CIP includes a **some** quantifier which closely parallels the § quantifier presented here. The significant differences between CIP and the formalism presented here are mainly in the treatment of errors, and predicates.

Errors in CIP are represented by a bottom value. The presence of the bottom value in the breadth of a specification means that the specification may lead to error. Many transformation rules have special side conditions about errors, especially in predicates. In the present formalism, errors are represented by *all* or by incompleteness with a resulting simplification.

Predicates in CIP are simply boolean specifications. This has a unifying effect, but, as with errors, adds side conditions to transformation rules, for example saying that the predicate must be deterministic and must not be in error. In the present formalism, we do not specify the exact language used for predicates, but we do assume that each predicate is either true or false in each state, although the logic may not be complete enough to say which. For example $0/0 = 5$ is not considered to be in error, nor to be nondeterministic. As in CIP, the side conditions about determinism are there, but are somewhat hidden. We are currently looking at allowing nondeterministic predicates without complicating the laws.

Recently Möller (1989) proposed an "assertion" construct for CIP. His construct, $P \rhd x$ is similar to both our guard and assertion in that it is x when P is true, but differs from both our constructs in that it is the bottom (error) value when P is false. It is faithful to the notion of assertions as safety nets. By contrast, our assertion construct is used to represent context. The difference is illustrated by the assertion elimination law, which does not hold for Möller's assertions.

Meertens, in his excellent paper on transformational programming (Meertens 1986), discusses nondeterministic functional programs as a unified notation for specifications and programs. Unfortunately, Meertens confuses *null* (in his notation $\llbracket/0$) with the undefined value (the error value). This leads him to choose between rejecting the validity of $x \sqsupseteq \mathit{null}$ and rejecting that $\sqsupseteq$ means "may (as a task) be replaced by." The solution is to accept both, regard *null* as the over-determined value, and use the undetermined value *all* to represent errors.

Meertens uses direct substitution for application. He also adopts the rule $(f, g)x \equiv f\,x, g\,x$. He correctly notes that these seemingly reasonable choices lead to contradictions. The following example is given

$$
\begin{aligned}
f &\overset{\text{def}}{\equiv} \lambda x \cdot x \\
g &\overset{\text{def}}{\equiv} \lambda x \cdot 3 \\
F &\overset{\text{def}}{\equiv} \lambda \phi \cdot \phi\,1 + \phi\,2
\end{aligned}
$$

then

$$3,6 \equiv (1{+}2),(3{+}3) \equiv Ff, Fg \equiv F(f,g) \equiv (f,g)1{+}(f,g)2 \equiv (1,3){+}(2,3) \equiv 3,4,5,6$$

Our formalism avoids this paradox by carefully defining elementhood and allowing only elements as the values of parameters.

One outgrowth of Meertens's paper is the so-called Bird-Meertens formalism. Initially, nondeterministic specification was ignored (see e.g. (Bird 1987)). In (Bird 1990), Bird discusses nondeterministic specifications, but not the refinement order on them.

Hoogerwoord in his thesis (Hoogerwoord 1989) develops a calculational method of functional program development based on logical specifications. In contrast to the present paper, he does not treat specifications and expressions as objects of the same sort, and thus does not have a refinement calculus; rather, specifications are predicates that describe the desired expressions. Nondeterminism is not allowed in expressions themselves, but a specification may, of course, underdetermine the meaning of the desired expression.

11 Conclusions

We have presented a simple refinement calculus for functional programming and an attendant programming methodology. The key aspect of this calculus is that it treats specifications and executable expressions uniformly. This allows the programmer to formally express and to verify each step of a stepwise refinement. The calculus includes timing, not just for analysis after program development, but as a guide to development.

Several of the specification operators presented and used herein are new or new to functional programming, as far as we know. These include $\succ$, $\twoheadrightarrow$, **let** , **try**, and **try else**.

The specification language is a small extension to a functional programming language. The extension allows the specifier to state the relationship between the free variables and the result of an expression. Because logic can be used to state this relationship, the language is expressive and natural to anyone familiar with logic. The specifier needs to state exactly the desired relationship and nothing more; there is no requirement that the relationship be functional. Furthermore, the relationship can be expressed in ways that are completely nonalgorithmic.

Acknowledgements We would like to thank Ray Blaak, Andrew Malton, and the referees for helpful comments on earlier drafts. We gratefully acknowledge the support of the Department of Computer Science at the University of Toronto and the Natural Sciences and Engineering Research Council.

References

R.J.R. Back. A calculus of refinement for program derivations. Technical Report 54, Department of Computer Science, Åbo Akademi, Finland, 1987.

F.L. Bauer, H. Ehler, A. Horsch, B. Möller, H. Partsch, O. Puakner, and P. Pepper. *The Munich Project CIP: Volume II: The Program Transformation System CIP-S.* Number 292 in Lecture Notes in Computer Science. Springer-Verlag, 1987.

R.S. Bird. Introduction to the theory of lists. In M. Broy, editor, *Logic of Programming and Calculi of Discrete Design,* number 36 in NATO ASI Series F. Springer, 1987.

R.S. Bird. A calculus of functions for program derivation. In David A. Turner, editor, *Research Topics in Functional Programming,* The UT Year of Programming Series. Addison-Wesley, 1990.

Alonzo Church. A formulation of the simple theory of types. *J. Symbolic Logic,* 5:56–68, 1940.

E.W. Dijkstra. Guarded commands, nondeterminacy, and formal derivation of programs. *Communications of the ACM,* 18(8):453–457, 1975.

Eric C.R. Hehner. *The Logic of Programming.* Prentice-Hall International, 1984.

Eric C.R. Hehner. A practical theory of programming. *Science of Computer Programming,* 14:133–158, 1990.

C.A.R. Hoare. An axiomatic basis for computer programming. *Communications of the ACM,* 12(10):576–580, 583, 1969.

Rob Hoogerwoord. The design of functional programs: a calculational approach. PhD Thesis, Technische Universiteit Eindhoven, 1989.

Lambert Meertens. Algorithmics. In J.W. de Bakker, M. Hazewinkel, and J.K. Lenstra, editors, *Mathematics and Computer Science,* number 1 in CWI Monographs. North-Holland, 1986.

Bernhard Möller. Applicative assertions. In J.L.A. van de Snepscheut, editor, *Mathematics of Program Construction,* number 375 in Lecture Notes in Computer Science. Springer-Verlag, 1989.

Carroll Morgan. The specification statement. *Trans. on Programming Languages and Systems,* 10(3):403–419, 1988.

F.L. Morris and C.B. Jones. An early program proof by Alan Turing. *Annals of the History of Computing,* 6(2):139–143, 1984.

Joseph M. Morris. Programs from specifications. In E. W. Dijkstra, editor, *Formal Development of Programs and Proofs,* pages 81–115. Addison-Wesley, 1990.

Greg Nelson. A generalization of Dijkstra's calculus. Technical Report 16, Digital Systems Research Center, Palo Alto, CA, U.S.A., April 1987. Also published in *Trans. on Programming Languages and Systems,* 11(4):517–561, 1989.

D. Hugh Redelmeier. *Towards Practical Functional Programming.* PhD thesis, University of Toronto, 1984.

David Sands. Complexity analysis for a lazy higher-order language. In *Proceedings of the 1989 Glasgow Functional Programming Workshop,* Workshops in Computing. Springer-Verlag, 1989.

Harald Søndergaard and Peter Sestoft. Nondeterminism in functional languages. Technical Report 88/18, Department of Computer Science, University of Melbourne, Australia, 1988.

Harald Søndergaard and Peter Sestoft. Referential transparency, definiteness and unfoldability. *Acta Informatica,* 27(6):505–518, 1990.

Inorder Traversal of a Binary Heap and its Inversion in Optimal Time and Space

Berry Schoenmakers

Department of Mathematics and Computing Science
Eindhoven University of Technology
P.O. Box 513, 5600 MB Eindhoven, The Netherlands
wsinbs@win.tue.nl

Abstract. In this paper we derive a linear-time, constant-space algorithm to construct a binary heap whose inorder traversal equals a given sequence. We do so in two steps. First, we invert a program that computes the inorder traversal of a binary heap, using the proof rules for program inversion by W. Chen and J.T. Udding. This results in a linear-time solution in terms of binary trees. Subsequently, we data-refine this program to a constant-space solution in terms of linked structures.

1 Introduction

In [7] an elegant sorting algorithm is presented which exploits the presortedness of the input sequence. The first step of this variant of Heapsort comprises the conversion of the input sequence in a "mintree," i.e., a binary heap whose inorder traversal equals the input sequence.[1] For this conversion, the authors of [7] provide a complicated, yet linear algorithm, consisting of no fewer than four repetitions. In this paper we show that the practical significance of the sorting algorithm can be increased considerably by deriving a conversion algorithm that consists of a single repetition only.

The derivation proceeds in two steps. In the first step we derive an algorithm in terms of binary trees. We do so by inverting a program that solves the "inverse problem," i.e., it computes the inorder traversal of a binary heap in linear time. To guarantee the correctness of this inversion, we apply the proof rules given by W. Chen and J.T. Udding in [3]. These proof rules support stepwise program inversion.

Subsequently, in the second step, we refine this algorithm to a program that operates on linked structures instead of binary trees. Just as in [7], our object is to minimize space utilization. It turns out that we can implement the construction—in a simple way—such that only $O(1)$ additional space is required. This contrasts favourably with the complicated method the authors of [7] seem to have in mind to achieve this for their algorithm—as far as we can conclude from their hint in the footnote, where they remark that the construction can be done "without wasting space." We also present a refinement using an array representation for binary trees, since this is advantageous when the input sequence is also represented by an array, as is often the case.

[1] Actually, they use a "maxtree," but we have taken the liberty to change this into "mintree" so as to comply with [2, pp. 55-67].

2 Problem Statement

The problem is specified in terms of lists of type [Int] and binary trees of type ⟨Int⟩. Later, these types will be refined to pointer and array types.

For lists and trees we use the following notations. [] stands for the empty list, $[a]$ stands for the singleton list with element a, and catenation is denoted by $+\!\!+$. Furthermore, the head, tail, front, and last element of a nonempty list s are denoted by $hd.s$, $tl.s$, $ft.s$, and $lt.s$, respectively. Hence, s may be written both as $[hd.s] +\!\!+ tl.s$ and as $ft.s +\!\!+ [lt.s]$. The length of s is denoted by $\#s$. Similarly, $\langle\,\rangle$ stands for the empty tree, and $\langle t, a, u\rangle$ for a nonempty tree with left subtree t, root a, and right subtree u. The left subtree, root, and right subtree of nonempty tree t are denoted by $l.t$, $m.t$, and $r.t$, respectively.

In terms of these types, the problem is stated as follows. Let $\bar{t}$ denote the *inorder traversal* of tree t. That is,

$$\begin{aligned} \overline{\langle\,\rangle} &= [] \\ \overline{\langle t, a, u\rangle} &= \bar{t} +\!\!+ [a] +\!\!+ \bar{u} \ . \end{aligned}$$

Furthermore, let $H.t$ denote that tree t satisfies the *heap condition*, defined by

$$\begin{aligned} H.\langle\,\rangle &\equiv \text{true} \\ H.\langle t, a, u\rangle &\equiv H.t \ \wedge\ a \leq {\downarrow}t \ \wedge\ a \leq {\downarrow}u \ \wedge\ H.u \ , \end{aligned}$$

in which $\downarrow$ denotes the *minimum* of a tree (with ${\downarrow}\langle\,\rangle = \infty$). Then, given list s satisfying $s = S$, the problem is to design an $O(\#S)$ program with postcondition

$$H.t \ \wedge\ \bar{t} = S \ .$$

Here, S denotes a specification variable that may not be used in the program.

3 Proof Rules for Program Inversion

Assuming some familiarity with program inversion, we confine ourselves to a brief summary of the results of [3]. There, the inverse of a program is defined as follows.

> Program T is said to be an inverse of program S under precondition P when $\{P \wedge Q\}\ S\ ;\ T\ \{Q\}$ for any predicate Q.

Obviously, **skip** is then its own inverse. For each of the other constructs of Dijkstra's guarded command language, Chen and Udding provide proof rules to support stepwise program inversion. Below, the rules for assignments and sequential compositions are simply copied from [3]. The rules for the other two constructs are instantiations of the more general rules presented in [3].

Proof rule for assignments

$$\frac{P \Rightarrow def(E0) \ \wedge\ def((E1)^{x}_{E0}) \ \wedge\ x = (E1)^{x}_{E0}}{\{P \wedge Q\}\ x := E0\ ;\ x := E1\ \{Q\} \text{ for any } Q}$$

Here, $def(E)$ means that expression E is well-defined.

Proof rule for sequential compositions

$$\begin{array}{l} \{P\}\ S0\ \{R\} \\ \{P \wedge Q\}\ S0\ ;\ T0\ \{Q\} \text{ for any } Q \\ \{R \wedge Q\}\ S1\ ;\ T1\ \{Q\} \text{ for any } Q \\ \hline \{P \wedge Q\}\ S0\ ;\ S1\ ;\ T1\ ;\ T0\ \{Q\} \text{ for any } Q \end{array}$$

Proof rule for selections

$$\begin{array}{l} \{P \wedge B0\}\ S0\ \{C0 \wedge \neg C1\} \\ \{P \wedge B1\}\ S1\ \{C1 \wedge \neg C0\} \\ P \Rightarrow B0 \vee B1 \\ \{P \wedge B0 \wedge Q\}\ S0\ ;\ T0\ \{Q\} \text{ for any } Q \\ \{P \wedge B1 \wedge Q\}\ S1\ ;\ T1\ \{Q\} \text{ for any } Q \\ \hline \{P \wedge Q\}\ \mathbf{if}\ B0 \rightarrow S0\ [\!]\ B1 \rightarrow S1\ \mathbf{fi}\ ;\ \mathbf{if}\ C0 \rightarrow T0\ [\!]\ C1 \rightarrow T1\ \mathbf{fi}\ \{Q\} \text{ for any } Q \end{array}$$

Proof rule for repetitions

$$\begin{array}{l} \{P \wedge \neg C\}\ \mathbf{do}\ B \rightarrow S\ \mathbf{od}\ \{\text{true}\} \\ \{P \wedge B\}\ S\ \{P \wedge C\} \\ \{P \wedge B \wedge Q\}\ S\ ;\ T\ \{Q\} \text{ for any } Q \\ \hline \{P \wedge \neg C \wedge Q\}\ \mathbf{do}\ B \rightarrow S\ \mathbf{od}\ ;\ \mathbf{do}\ C \rightarrow T\ \mathbf{od}\ \{Q\} \text{ for any } Q \end{array}$$

Note that an inverse constructed according to these rules is such that it *exactly* retraces the steps of the program inverted. Also notice that such an inverse is *deterministic* by construction.

4 The Program to Be Inverted

As outlined in Section 1, we will first solve the "inverse problem:" given t, satisfying $H.t \wedge \bar{t} = S$, construct an $O(\#S)$ program with postcondition $s = S$. Since this problem has been solved in a neat way already many times (e.g., in [5, 6, 3]), we merely present a solution without derivation. In each of the programs in [5, 6, 3], a list of trees, which we name q, is used, and the loop invariant is something like

$$s \mathbin{+\!\!+} \bar{t} \mathbin{+\!\!+} \bar{\bar{q}} = S \ , \tag{P0}$$

where $\bar{\bar{\cdot}}$ denotes the inorder traversal of a *list of trees*:

$$\begin{array}{ll} \bar{\bar{[]}} & = [] \\ \overline{\overline{[t] \mathbin{+\!\!+} q}} & = \bar{t} \mathbin{+\!\!+} \bar{\bar{q}} \ . \end{array}$$

Starting from this invariant, the following program is easily calculated:

$\{\ \bar{t} = S\ \}$
$\quad s, q := [\,], [\,]$
$\{$ invariant: $\mathrm{P0} \wedge \mathrm{P1}$; bound: $2\#S - (2\#s + \#q)\ \}$
$;\ \mathbf{do}\ t \neq \langle\,\rangle \vee q \neq [\,] \rightarrow$
$\quad\quad \mathbf{if}\ t \neq \langle\,\rangle \rightarrow t, q := l.t,\ [\langle\langle\,\rangle, m.t, r.t\rangle] \mathbin{+\!\!+} q$
$\quad\quad [\!]\ t = \langle\,\rangle \rightarrow s, t, q := s \mathbin{+\!\!+} [m.(hd.q)],\ r.(hd.q),\ tl.q$
$\quad\quad$ **fi**
$\quad$ **od**
$\{\ s = S\ \}$,

where P1 is required to prove the invariance of P0:

$$(\forall i : 0 \leq i < \#q : q.i \neq \langle\,\rangle\ \wedge\ l.(q.i) = \langle\,\rangle)\ . \qquad \text{(P1)}$$

Having convinced ourselves of the correctness of this program, we can now ignore the above invariant and bound function, and annotate the program so as to facilitate its inversion. In finding the appropriate annotations, we are guided by the proof rules for program inversion. Of course, the precondition that t satisfies the heap condition must be exploited:

$\{\ H.t\ \wedge\ \bar{t} = S\ \}$
$\quad s, q := [\,], [\,]$
$\{\ \mathrm{P}\ \wedge\ \neg(s \neq [\,] \vee q \neq [\,])\ \wedge\ (H.t\ \wedge\ \bar{t} = S)\ \}$
$;\ \mathbf{do}\ t \neq \langle\,\rangle \vee q \neq [\,] \rightarrow$
$\quad \{\ \mathrm{P}\ \wedge\ (t \neq \langle\,\rangle \vee q \neq [\,])\ \}$
$\quad\quad \mathbf{if}\ t \neq \langle\,\rangle \rightarrow t, q := l.t,\ [\langle\langle\,\rangle, m.t, r.t\rangle] \mathbin{+\!\!+} q \qquad \{\ \mathrm{A}\ \wedge\ \neg\mathrm{B}\ \}$
$\quad\quad [\!]\ t = \langle\,\rangle \rightarrow s, t, q := s \mathbin{+\!\!+} [m.(hd.q)],\ r.(hd.q),\ tl.q \qquad \{\ \mathrm{B}\ \wedge\ \neg\mathrm{A}\ \}$
$\quad\quad$ **fi**
$\quad \{\ \mathrm{P}\ \wedge\ (s \neq [\,] \vee q \neq [\,])\ \}$
$\quad$ **od**
$\{\ s = S\ \wedge\ t = \langle\,\rangle\ \wedge\ q = [\,]\ \}$.

In these annotations, P denotes the *strongest* invariant for the above repetition that holds initially—hence, P implies both P0 and P1. Conditions A and B correspond to $C0$ and $C1$ in the proof rule for selections, and will be defined in the next section.

5 The Inversion

First of all, let us explain how the problem stated in Section 2 can be solved using an inverse of the repetition of the above program. To that end, let **DO** denote the repetition of this program, and let **OD** denote an inverse of **DO** under precondition $\mathrm{P} \wedge \neg(s \neq [\,] \vee q \neq [\,])$. According to the definition of a program's inverse (Section 3), this means that

$$\{\ \mathrm{P}\ \wedge\ \neg(s \neq [\,] \vee q \neq [\,])\ \wedge\ Q\ \}\ \mathbf{DO}\ ;\ \mathbf{OD}\ \{\ Q\ \}$$

for any predicate Q. The annotations in the above program show that $H.t \wedge \bar{t} = S$ is a precondition of **DO**, hence we may take this for Q. Moreover, we see that $s = S \wedge t = \langle\,\rangle \wedge q = [\,]$ is a postcondition of **DO**. Since this postcondition is a one-point predicate, it is the strongest postcondition of **DO**, and we thus conclude that

$$\begin{array}{l}\{\ \mathrm{P} \wedge \neg(s \neq [\,] \vee q \neq [\,]) \wedge (H.t \wedge \bar{t} = S)\ \} \\ \quad \mathbf{DO} \\ \{\ s = S \wedge t = \langle\,\rangle \wedge q = [\,]\ \} \\ ;\mathbf{OD} \\ \{\ H.t \wedge \bar{t} = S\ \}\ .\end{array}$$

As a consequence, we can use **OD** to solve the problem as follows:

$$\begin{array}{l}\{\ s = S\ \} \\ \quad t, q := \langle\,\rangle, [\,] \\ \{\ s = S \wedge t = \langle\,\rangle \wedge q = [\,]\ \} \\ ;\mathbf{OD} \\ \{\ H.t \wedge \bar{t} = S\ \}\ .\end{array}$$

Thus, we have to invert **DO**. Inspection of the proof rule for repetitions reveals that this gives rise to inversion of the body of **DO**. So, let **IF** denote the body of **DO**. Then we have to determine an inverse of **IF** under precondition $\mathrm{P} \wedge (t \neq \langle\,\rangle \vee q \neq [\,])$. To this end, we have to find conditions A and B such that the above annotation is valid (cf. the proof rule for selections, first two antecedents). Moreover, A and B have to be boolean expressions that may be used as guards in our inverse of **IF**.

Before we derive A and B, however, we first invert the assignments in **IF**, since this is also required by the proof rule for selections (last two antecedents). For the assignment

$$t, q := l.t,\ [\langle\langle\,\rangle, m.t, r.t\rangle] +\!\!\!+\, q\ , \tag{a}$$

the following assignment is an inverse under precondition $t \neq \langle\,\rangle$:

$$t, q := \langle t, m.(hd.q), r.(hd.q)\rangle,\ tl.q\ .$$

This is easily verified by applying "substitution" (a) to $\langle t, m.(hd.q), r.(hd.q)\rangle,\ tl.q$, as prescribed by the proof rule for assignments.

The other assignment to be inverted is

$$s, t, q := s +\!\!\!+ [m.(hd.q)],\ r.(hd.q),\ tl.q\ , \tag{b}$$

for which we find

$$s, t, q := ft.s,\ \langle\,\rangle,\ [\langle\langle\,\rangle, lt.s, t\rangle] +\!\!\!+\, q$$

as an inverse under precondition

$$t = \langle\,\rangle \wedge q \neq [\,] \wedge hd.q \neq \langle\,\rangle \wedge l.(hd.q) = \langle\,\rangle\ . \tag{$\star$}$$

Since we only have to invert assignment (b) under precondition $P \wedge t = \langle\,\rangle \wedge q \neq []$, this suffices, because this precondition implies ($\star$) on account of invariant P1. (Recall that, by definition, P implies any invariant of **DO**.)

Our final problem is now to determine A and B. To this end we closely examine the effects of the assignments in **IF**. In order to get useful results, we use that—as may be expected—t is initially a heap, and that, consequently, P implies P2 as well, with

$$H.t \;\wedge\; (\forall i : 0 \leq i < \#q : H.(q.i)) \;. \tag{P2}$$

Examination of assignment (a) reveals that it has $m.t \geq m.(hd.q)$ as a postcondition; consequently, this condition is a candidate for A and its negation is a candidate for B. Unfortunately, however, $m.t < m.(hd.q)$ does not hold after assignment (b), since it turns out that P also implies

$$\boldsymbol{m}.t \geq \boldsymbol{mhd}.q \;\wedge\; (\forall i : 0 \leq i < \#q - 1 : m.(q.i) \geq m.(q.(i{+}1))) \;, \tag{P3}$$

with

$$\boldsymbol{m}.t = \begin{cases} m.t & , t \neq \langle\,\rangle \\ \infty & , t = \langle\,\rangle \end{cases} ,$$

and

$$\boldsymbol{mhd}.q = \begin{cases} m.(hd.q) & , q \neq [] \\ -\infty & , q = [] \end{cases} .$$

Here, functions $\boldsymbol{m}$ and $\boldsymbol{mhd}$ are defined such that $\boldsymbol{m}.\langle\,\rangle \geq \boldsymbol{mhd}.q$ for all q and $\boldsymbol{m}.t \geq \boldsymbol{mhd}.[]$ for all t. Note that $\boldsymbol{m}$ is not needed in the universal quantification in P3, since the trees in q are nonempty due to invariant P1.

Since the above examination of the first alternative does not give us a solution for A and B, we now examine the second alternative. We observe that after its assignment $lt.s \leq \boldsymbol{m}.t$, since $m.(hd.q) \leq \boldsymbol{m}.(r.(hd.q))$ is a precondition in this case (on account of P2). But again—unfortunately—we have a similar postcondition for the other alternative. More precisely, P implies

$$\boldsymbol{lt}.s \leq \boldsymbol{m}.t \;, \tag{P4}$$

with

$$\boldsymbol{lt}.s = \begin{cases} lt.s & , s \neq [] \\ -\infty & , s = [] \end{cases} .$$

Thus, also our examination of the second alternative does not give the desired result. However, the above examinations resulted in two additional invariants, P3 and P4, which we can exploit as follows. Firstly, we have $\boldsymbol{lt}.s \geq \boldsymbol{mhd}.q$ after assignment (b), by the axiom of assignment and the invariance of P3. And, secondly, we have after assignment (a) that $\boldsymbol{lt}.s \leq \boldsymbol{mhd}.q$, by the axiom of assignment and the invariance of P4. So, we are done if we can conclude that $\boldsymbol{lt}.s \neq \boldsymbol{mhd}.q$ after one of the two guarded commands. For this purpose we introduce a *skewed* version of H, defined by

$$\begin{aligned} \widehat{H}.\langle\,\rangle &\equiv \text{true} \\ \widehat{H}.\langle t, a, u\rangle &\equiv \widehat{H}.t \;\wedge\; a \leq \downarrow t \;\wedge\; a < \downarrow u \;\wedge\; \widehat{H}.u \;, \end{aligned}$$

and we replace H by $\widehat{H}$ in the annotations of the preceding programs. Consequently, we have that P also implies

$$\widehat{H}.t \;\wedge\; (\forall i : 0 \le i < \#q : \widehat{H}.(q.i)) \quad , \text{ and} \tag{P2a}$$

$$\boldsymbol{lt}.s < \boldsymbol{m}.t \quad . \tag{P4a}$$

Now it follows that $\boldsymbol{lt}.s < \boldsymbol{mhd}.q$ holds after assignment (a).

As a result, we obtain the following linear-time program for the original problem:

```
{ s = S }
  t,q := ⟨ ⟩, []
; do s ≠ [] ∨ q ≠ [] →
    if lt.s < mhd.q  →  t,q := ⟨t, m.(hd.q), r.(hd.q)⟩, tl.q
    [] lt.s ≥ mhd.q  →  s,t,q := ft.s, ⟨ ⟩, [⟨⟨ ⟩, lt.s, t⟩] ++ q
    fi
  od
{ Ĥ.t ∧ t̄ = S, hence H.t ∧ t̄ = S }  .
```

Notice that the replacement of $H.t$ by $\widehat{H}.t$ is harmless in the sense that we are still able to solve the problem for any sequence S.

6 A Nondeterministic Solution

In the previous section we have arbitrarily chosen to replace H by $\widehat{H}$. Of course, we may also decide to replace H by the symmetrical counterpart of $\widehat{H}$. This leads to a solution with $\le$ instead of $<$ in the first alternative, and $>$ instead of $\ge$ in the second alternative. Since this implies that P0–P4 is an invariant for both solutions, we infer that P0–P4 is invariant under both assignments in **OD**, if $\boldsymbol{lt}.s = \boldsymbol{mhd}.q$. Therefore we also have the following nondeterministic solution, in which we have made the types explicit to facilitate the data refinements in the next section.

```
proc C (in s:[Int] ; out t:⟨Int⟩)
 { s = S }
  |[var q:[⟨Int⟩] ;
    t,q := ⟨ ⟩, []
   ; do s ≠ [] ∨ q ≠ [] →
       if lt.s ≤ mhd.q  →  t,q := ⟨t, m.(hd.q), r.(hd.q)⟩, tl.q
       [] lt.s ≥ mhd.q  →  s,t,q := ft.s, ⟨ ⟩, [⟨⟨ ⟩, lt.s, t⟩] ++ q
       fi
     od
  ]|
 { H.t ∧ t̄ = S }
corp .
```

The correctness of this nondeterministic procedure has to be established in the conventional way, because the proof rules of [3] can only be used to derive deterministic programs.

7 Two Data Refinements

We shall refine procedure C in two steps. In the first step, we do away with type $[\langle \text{Int} \rangle]$. Subsequently, we replace values of type $\langle \text{Int} \rangle$ by linked structures. Since the use of pointers in the resulting refinement is limited, we can also use arrays instead of pointers, as will be shown in Section 7.3.

7.1 Elimination of type $[\langle \text{Int} \rangle]$

The important observation is that the trees in list q all have empty left subtrees on account of invariant P1:

$$(\forall i : 0 \le i < \#q : q.i \ne \langle\,\rangle \;\wedge\; l.(q.i) = \langle\,\rangle) \;.$$

Such lists of trees may be represented by binary trees as follows: [] is represented by $\langle\,\rangle$, and a nonempty list $[\langle\langle\,\rangle, a, t\rangle] \mathbin{+\!\!+} q$ is represented by tree $\langle u, a, t\rangle$, where u is the representation of q. Thus, under this representation, we can replace type $[\langle \text{Int} \rangle]$ by $\langle \text{Int} \rangle$, resulting in procedure $C1$:

$$
\begin{array}{l}
\textbf{proc } C1 \; (\textbf{in } s{:}[\text{Int}] \;;\; \textbf{out } t{:}\langle\text{Int}\rangle) \\
\quad |[\textbf{var } q{:}\langle\text{Int}\rangle \;; \\
\qquad t, q := \langle\,\rangle, \langle\,\rangle \\
\quad\; ;\, \textbf{do } s \ne [\,] \;\vee\; q \ne \langle\,\rangle \;\rightarrow \\
\qquad\quad \textbf{if } \mathit{lt}.s \le m.q \;\rightarrow\; t, q := \langle t, m.q, r.q\rangle,\; l.q \\
\qquad\quad [\!]\; \mathit{lt}.s \ge m.q \;\rightarrow\; s, t, q := \mathit{ft}.s,\; \langle\,\rangle,\; \langle q, \mathit{lt}.s, t\rangle \\
\qquad\quad \textbf{fi} \\
\qquad \textbf{od} \\
\quad]| \\
\quad \textbf{corp} \;,
\end{array}
$$

with $m.q = \begin{cases} m.q & , q \ne \langle\,\rangle \\ -\infty & , q = \langle\,\rangle \;. \end{cases}$

Now, only type $\langle \text{Int} \rangle$ remains to be refined.

7.2 Implementation in terms of pointers

We adopt a Pascal-like notation for pointers and tuples ("records"). Binary trees are represented in the common way by values of type B, which is defined as

$$\text{B} = \uparrow\langle l{:}\text{B}, m{:}\text{Int}, r{:}\text{B}\rangle \;.$$

Hence, **nil** represents $\langle\,\rangle$ and for the assignment $t, q := \langle t, m.q, r.q\rangle,\; l.q$ we have as obvious refinement, in which q is now of type B:

$$|[\textbf{var } b{:}\text{B};\; \textbf{new}(b)\,;\; b\!\uparrow\; := \langle t, q\!\uparrow\!.m, q\!\uparrow\!.r\rangle \,;\; t, q, b := b, q\!\uparrow\!.l, q \,;\; \textbf{dispose}(b) \;]| \;.$$

Assuming that disposed cells are recycled, we see that this block does not change the number of cells in use. However, we can avoid the calls to **new** and **dispose** altogether by recycling the cell, to which q points initially, in-line: the latter cell can be used instead of the cell returned by **new**(b). The required assignment is now a simple *rotation* of three pointers: $t, q, q\!\uparrow\!.l := q, q\!\uparrow\!.l, t$.

Without further comment, we thus obtain:

```
proc C2 (in s:[Int] ; out t:B)
  |[var q:B ;
    t,q := nil, nil
    ; do s ≠ [] ∨ q ≠ nil →
        if lt.s ≤ q↑.m  →  t,q,q↑.l := q,q↑.l,t
        [] lt.s ≥ q↑.m  →  |[var b:B; new(b) ; b↑ := ⟨q,lt.s,t⟩ ; s,t,q := ft.s,nil,b ]|
        fi
      od
  ]|
corp ,
```

with $\textbf{nil}\uparrow.m = -\infty$. Execution of this program for an input sequence of length N gives rise to N calls to **new**, from which we infer that exactly the number of cells required for the representation of the output tree is used. Hence, this program uses only constant extra space.

7.3 Implementation in terms of arrays

In case type [Int] is refined by an array type, it is advantageous to refine ⟨Int⟩ by an array type as well. To that end we introduce a global array a (again, in Pascal-like notation):

$$a : \textbf{array}\ \text{Nat}\ \textbf{of}\ \langle l{:}\text{Nat}\,, m{:}\text{Int}\,, r{:}\text{Nat}\rangle\ .$$

In this context, we can associate a binary tree with each natural number as follows: 0 represents ⟨ ⟩, and n, $n{>}0$, represents a nonempty tree with root $a[n].m$, and with $a[n].l$ and $a[n].r$ representing its left and right subtree, respectively. (We assume that a is such that this definition defines a finite tree for all naturals—as we did, without mentioning, for values of type B.)

The desired refinement of $C1$ can now be obtained from $C2$ by replacing B by Nat, **nil** by 0, $q\uparrow$ by $a[q]$, and $b\uparrow$ by $a[b]$. Furthermore, we turn b into a global variable (initially $b = 0$), and replace **new**(b) by $b := b{+}1$. As a consequence, procedure call $C3(s,t)$ establishes that t and segment $a[0..\#s]$ represent the tree corresponding to s, where

```
proc C3 (in s:[Int] ; out t:Nat)
  |[var q:Nat;
    t,q := 0, 0
    ; do s ≠ [] ∨ q ≠ 0 →
        if lt.s ≤ a[q].m  →  t,q,a[q].l := q,a[q].l,t
        [] lt.s ≥ a[q].m  →  b := b+1 ; a[b] := ⟨q,lt.s,t⟩ ; s,t,q := ft.s,0,b
        fi
      od
  ]|
corp ,
```

with $a[0].m = -\infty$.

Now, note that a call to $C3$ establishes $(\forall i : 0 \le i < \#s : s.i = a[\#s{-}i].m)$. Hence, in case type [Int] is refined by an array type, there is the opportunity to omit component m from the elements of a without loss of efficiency.

8 Concluding Remarks

The problem of constructing a heap from its inorder traversal had already been posed and solved by R.S. Bird [2, pp. 55–67]. Not being satisfied with Bird's derivation, we first solved the problem—from scratch—using elementary techniques from functional programming. Having digested [3], however, in which an algorithm to construct a tree from its preorder and inorder traversal is derived by means of program inversion, it appeared to us that the same approach should be applicable to Bird's problem. Surprisingly so, the technique of program inversion led rather straightforwardly to a nice conversion algorithm.

In fact, the only problem we encountered was to find suitable conditions A and B, and, in retrospect, it turns out that an investigation of the relations between ***lt***.*s*, ***m***.*t*, and ***mhd***.*q* does the job. In order to arrive at a solution for A and B, however, we had to replace H by $\widehat{H}$. This has to do with the fact that t is in general not uniquely determined by $H.t \wedge \bar{t} = S$, but it is, for instance, by $\widehat{H}.t \wedge \bar{t} = S$.

We have confined the data refinement to the essential steps, viz. the elimination of type [⟨Int⟩] and the representation of type ⟨Int⟩ either by a pointer type or by an array type. We remark that it is crucial that the first step makes q and t of the same type, so that the **new** and **dispose** operations cancel out in the second step—leading to a constant-space solution.

The incentive to record the data refinement of the conversion algorithm has been its application in the sorting algorithm. As for the efficiency of procedures $C2$ and $C3$ we see that they are optimal with respect to time as well as space. We have found a similar result for the reconstruction of a binary tree from its preorder and inorder traversals in [1]. We remark that the latter result may also be achieved by refining the algorithm derived—by program inversion—in [3]. Compared to the way this result is achieved in [1], we observe that such an approach gives a much better separation of concerns; e.g., we do not have to discuss a tricky implementation of the "recursion stack," a discussion in which algorithmic details and the representation of trees play a role at the same time.

Apart from the adaptive sorting algorithm [7], the conversion algorithm has many other applications. For example, the "largest rectangle under a histogram" can easily be computed once the histogram—which is just a list of natural numbers—has been converted into the corresponding heap (see [2]). Other applications of these heaps can be found in [4], which also contains a description of a linear-time conversion algorithm. In [4] the heaps are called "Cartesian trees" after Vuillemin, who introduced these structures in [8].

References

1. Andersson, A., Carlsson, S.: Construction of a Tree from its Traversals in Optimal Time and Space. Information Processing Letters **34** (1990) 21–25
2. Bird, R.S.: Lectures on Constructive Functional Programming. Technical monograph PRG **69**, Oxford University Computing Laboratory (1988)
3. Chen, W., Udding, J.T.: Program Inversion: More Than Fun! Science of Computer Programming **15** (1990) 1–13

4. Gabow, H.N., Bentley, J.L., Tarjan, R.E.: Scaling and related techniques for geometry problems. Proc. 16th Annual ACM Symposium on Theory of Computing (1984) 135–143
5. Gries, D.: Inorder Traversal of a Binary Tree. In: E.W. Dijkstra (ed.), The Formal Development of Programs and Proofs, Addison-Wesley, Amsterdam (1990)
6. Gries, D., v.d. Snepscheut, J.L.A.: Inorder Traversal of a Binary Tree and its Inversion. In: E.W. Dijkstra (ed.), The Formal Development of Programs and Proofs, Addison-Wesley, Amsterdam (1990)
7. Levcopoulos, Ch., Petersson, O.: Heapsort—Adapted for Presorted Files. In: F. Dehne, J.-R. Sack, N. Santoro (eds.), Algorithms and Data Structures, LNCS **382** (1989) 499–509
8. Vuillemin, J.: A unifying look at data structures. Communications of the ACM **23** (1980) 229–239

Acknowledgement

One of the referees is acknowledged for pointing out references [8] and [4].

A Calculus for Predicative Programming

Emil Sekerinski

Forschungszentrum Informatik Karlsruhe, Haid-und-Neu Strasse 10-14, 7500 Karlsruhe, Germany, sekerinski@fzi.de

Abstract. A calculus for developing programs from specifications written as predicates that describe the relationship between the initial and final state is proposed. Such specifications are well known from the specification language Z. All elements of a simple sequential programming notation are defined in terms of predicates. Hence programs form a subset of specifications. In particular, sequential composition is defined by 'demonic composition', nondeterministic choice by 'demonic disjunction', and iteration by fixed points. Laws are derived which allow proving equivalence and refinement of specifications and programs by a series of steps. The weakest precondition calculus is also included. The approach is compared to the predicative programming approach of E. Hehner and to other refinement calculi.

1 Introduction

We view a specification as a predicate which describes the admissible final state of a computing machine with respect to some initial state. A program is a predicate restricted to operators which can be implemented efficiently. Hence the task of a programmer is to transform a specification written in the rich mathematical notation into a corresponding one expressed in the restricted programming notation, perhaps by a series of transformation steps. In this report, the programming notation consists of assignment, sequential composition, conditional, nondeterministic choice, variable declaration, and iteration.

The predicative programming approach was originally proposed by Eric Hehner in [5] for both a sequential and concurrent programming notation, and later refined in [6] and [7]. The benefit of this approach is that there are no separated worlds for specification and programming with cumbersome proof rules for the transition from specification to programs; rather the laws can be used for development of programs from specifications, for the transformation of programs to equivalent, perhaps more efficient ones, and for deriving properties of programs. By using programming operators like the conditional for specifying as well, we can also get clearer specifications. Another benefit is that when presenting the calculus, we can start with the specification notation and gradually introduce the programming operators by their definition.

For specifications we use a style which is similar to the specification language Z [15]. Specifications are basically predicates relating the initial to the final state, given by the values of primed and unprimed symbols respectively. We will use the Z notation whenever appropriate, for example for the predicates and for the refinement relation. However, as our aim are compact calculations, we refrain form using the graphical conventions for the layout. We also ignore the typing problems and rather concentrate on the definitions and properties of the 'control structures'. It should be also noted that we prefer to view predicates as boolean valued functions and will make extensive use of quantifications over all predicates over a range.

The calculus presented here allows stating the equivalence of two specifications (or programs), not just refinement. When developing a program from a specification by a series of transformations, we like to state for each intermediate result whether it is equivalent to or a refinement of previous one, even though only refinement is required (similarly as when proving that one real expression is less than another real expression). Stating equivalence expresses that no premature design decisions have been made.

However, it should be clarified what the equivalence of an executable program with a specification does mean: For some initial state, a specification is either defined, in which case it does relate to some final state, or is undefined. A program, when executed, does either terminates with some valid result, performs some undefined operation (like indexing out of range), or does not terminate at all. ¿From our point of view nontermination is as undesirable as an undefined operation. Hence we do not distinguish them and represent both by ***undefinedness.***

This allows us to define sequential composition by 'demonic composition' and nondeterministic choice by 'demonic disjunction'. When comparing relational composition and disjunction with sequential composition and nondeterministic choice, it turns out that the former two 'deliver a result' whenever a sensible result exists. They are therefore called angelic operators. However, they cannot be implemented effectively. In contrast, the latter two are undefined whenever the possibility of failure exists. They are called demonic operators, as if several possibilities for execution exist, the implementor is free to choose one arbitrarily (and we have to be prepared that always the worst one is chosen).

This approach leads to a nice way for expressing Dijkstras [2] weakest preconditions $wp(P, b)$: Let b be a condition, i.e. a specification over the initial state only. P generalized to the case that it is a specification, not necessarily a program. The meaning of $wp(P, b)$ is given by sequential composition $P; b$. This is made possible as the condition b is just a specification and sequential composition is defined for any specification. As a consequence, the definition of the sequential composition and the conditional by weakest preconditions

$$wp(P; Q, b) = wp(P, wp(Q, b))$$

$$wp(\text{if } c \text{ then } P \text{ else } Q \text{ end}, b) = c \wedge wp(P, b) \vee \neg c \wedge wp(Q, b)$$

correspond to the associativity of sequential composition and distributivity of sequential composition over the conditional.

$$(P; Q); b = P; (Q; b)$$

$$\text{if } c \text{ then } P \text{ else } Q \text{ end}; b = \text{if } c \text{ then } P; b \text{ else } Q; b \text{ end}$$

An obvious advantage is that we saved introducing a new function. However, a deeper advantage is that by mere notation we save applications of theorems like associativity and make theorems look simpler and easier to memorize.

The following section introduces all programming operators except iteration. Each definition is followed by a number of properties of the operator in question. The third section introduces the refinement ordering and states relationships between the programming operators and the refinement ordering. The fourth section discusses weakest preconditions. Iteration is treated in the fifth section: First the definition in terms of fixed points is given, its soundness verified, and the main iteration theorem derived. The main iteration theorem makes use of weakest preconditions. The last section presents a small example of the use of the calculus, in particular of the main iteration theorem.

2 Straight Line Programming Operators

Basic Notation ("Basics")

The boolean values are written as $\top$ and $\bot$, the boolean operators are written as $\neg, \wedge, \vee, \Rightarrow, \Leftrightarrow$ with their usual meaning, $\neg$ binding strongest, $\Rightarrow$ weakest, and $\Leftrightarrow$ binding as weak as $\Rightarrow$. Let P, Q stand for predicates. For the universal and existential quantifiers, we will also use the "restricted" forms:

$$(\forall d \mid P \cdot Q) = (\forall d \cdot P \Rightarrow Q)$$

$$(\exists d \mid P \cdot Q) = (\exists d \cdot P \wedge Q)$$

where d is the dummy over which the quantification ranges, with the possibility of quantifying over list of dummies. The substitution of a free symbol s in P by some expression e is written as $P[s := e]$, and the simultaneous substitution of symbols of the list S by the corresponding ones of the list of expressions E as $P[S := E]$. The *everywhere* operator $[P]$ stands for the universal quantification of the symbols of interest in P. We write $=$ and $\preceq$ over predicates for the universal equivalence and universal implication:

$$(P = Q) \Leftrightarrow [P \Leftrightarrow Q]$$

$$(P \preceq Q) \Leftrightarrow [P \Rightarrow Q]$$

They bind weaker than all other boolean operators. The *context* of a specification is a list of symbols, called the *variables*. For a variable v we allow priming by v', and similarly for lists of variables. A *specification* describes how the initial values of the variables relate to their final values. Hence in the context V, a specification is written as a predicate over V, V', and possibly some (unprimed) constants. Note that, although a specification can only be understood with respect to a context, we will leave the context in this paper usually implicit.

A *condition* is a specification in which no primed symbols occur, i. e. which does not say anything about the final values of the variables. For a condition b, b' stands for that condition with all the free (unprimed) variables of the context primed.

Throughout the paper, P, Q, R, X, Y stand for specifications, b, c for conditions, V for the current context, v for a variable (element of the context V), and x, y for symbols not in V. An example of a condition is the domain ΔP of a specification P:

$$\Delta P = (\exists V' \cdot P)$$

The domain of a condition is the condition itself. The domain operator distributes over disjunctions but not over conjunctions. However, if one of the conjuncts is a condition, it can be moved out of the domain:

(1) $\Delta(b \wedge P) = b \wedge \Delta P$

The void specification II (also called ***skip***) leaves all variables of the context unmodified:

$$\text{II} \Leftrightarrow (V' = V)$$

Relational composition of specifications is defined by identifying the final state of the first component with the initial state of the second component:

$$(P \circ Q) \Leftrightarrow (\exists V'' \cdot P[V' := V''] \wedge Q[V := V''])$$

By convention, Δ binds as strong as $\neg$ and $\circ$ weaker than $\Rightarrow$ and $\Leftrightarrow$ but stronger than $=$ and $\preceq$. Relational composition has zero $\perp$, identity II, is associative, and distributes through disjunction in both directions. The domain of the relational composition can be "pushed" into the second operand:

(2) $\Delta(P \circ Q) = P \circ \Delta Q$

For conjunctions with conditions, following laws hold:

(3) $b \wedge P \circ Q = b \wedge (P \circ Q)$

(4) $P \circ b \wedge Q = P \wedge b' \circ Q$

For the purpose of defining sequential composition, we introduce the condition $P \triangleright b$. Informally, we can think of $P \triangleright b$ as characterizing those initial states which only relate to final states satisfying b', if they relate to any state at all.

$$P \triangleright b = ((\forall V' \cdot P \Rightarrow b')$$

The subsequent laws about $\triangleright$ can be proved by predicate calculus:

(5) $(P \triangleright b) \wedge (P \circ b) = (P \triangleright b) \wedge \Delta P$

(6) $P \triangleright (b \wedge c) = (P \triangleright b) \wedge (P \triangleright c)$

In the sequel, we will visit each programming operator in turn and uncover many laws holding for each operator.

Assignment (":=")

The assignment $v := e$ changes variable v to e and leaves all other variables of the context unmodified. v must be an element of the context.

$$v := e = \text{II}[v := e]$$

We assume that the expression e is everywhere defined, therefore:

(1) $\Delta(v := e) = \top$

Sequential Composition (";")

The sequential composition $P;Q$ means that first P is executed, then Q. If P is nondeterministic, Q must be defined whatever choice in P is taken. Hence $P;Q$ behaves like the relational composition restricted to those initial states which only lead to intermediate states for which Q is defined.

$$P;Q = (P \triangleright \Delta Q) \wedge (P \circ Q)$$

Sequential composition is assigned the same binding power as "$\circ$". $P;Q$ is defined if P is defined and P leads to a state for which Q is defined.

(1) $\Delta(P;Q) = \Delta P \wedge (P \triangleright \Delta Q)$

Proof L.H.S. $= \Delta((P \triangleright \Delta Q) \wedge (P \circ Q))$ — by def. ";"
$= (P \triangleright \Delta Q) \wedge \Delta(P \circ Q)$ — by (1) under "Basics"
$= (P \triangleright \Delta Q) \wedge (P \circ \Delta Q)$ — by (2) under "Basics"
$=$ R. H. S. — by (5) under "Basics"

Sequential composition has zero $\perp$, identity II, and is associative.

(2) $P;\perp = \perp;P = \perp$

(3) $P;\mathrm{II} = \mathrm{II};P = P$

(4) $P;(Q;R)q = (P;Q);R$

Associativity is proved in the appendix. In general, sequential composition does neither distribute through disjunction nor conjunction. However, if one of the conjuncts is a condition, following theorems hold:

(5) $b \wedge P;Q = b \wedge (P;Q)$

Proof L.H.S. $= ((b \wedge P) \triangleright \Delta Q) \wedge (b \wedge P \circ Q)$ — by def. ";"
$= ((b \wedge P) \triangleright \Delta Q) \wedge b \wedge (P \circ Q)$ — by (3) under "Basics"
$= (P \triangleright \Delta Q) \wedge b \wedge (P \circ Q)$ — as for any $c : (b \wedge P) \triangleright c = b \Rightarrow (P \triangleright c)$
$=$ R.H.S — by def. ";"

(6) $P;b \wedge Q = (P \triangleright b) \wedge (P;Q)$

Proof L.H.S. $= (P \triangleright \Delta(b \wedge Q)) \wedge (P \circ b \wedge Q)$ — by def. ";"
$= (P \triangleright (b \wedge \Delta Q)) \wedge (P \wedge b' \circ Q)$ — by (1) and (4) under "Basics"
$= (P \triangleright \Delta Q) \wedge ((P \triangleright b) \wedge P \wedge b' \circ Q)$ — by (6) and (3) under "Basics"
$= (P \triangleright \Delta Q) \wedge ((P \triangleright b) \wedge P \circ Q)$ — as $(P \triangleright b) \wedge P \preceq b'$
$=$ R.H.S. — by (3) under "Basics", def. ";"

Sequential compositions with assignments can be simplified as follows:

(7) $v := e; P = P[v := e]$

Conditional ("if")

The conditional "if b then P else Q end" behaves like P if b initially holds, and as Q if b does not initially hold. The choice between P and Q depends only on b.

$$\text{if } b \text{ then } P \text{ else } Q \text{ end} = b \wedge P \vee \neg b \wedge Q$$

(1) $\Delta \text{if } b \text{ then } P \text{ else } Q \text{ end} = \text{if } b \text{ then } \Delta P \text{ else } \Delta Q \text{ end}$

There are many simple laws for manipulating conditionals. We give some which will be needed later on.

(2) $b \wedge \text{if } b \text{ then } P \text{ else } Q \text{ end} = b \wedge P$

(3) $\neg b \wedge \text{if } b \text{ then } P \text{ else } Q \text{ end} = \neg b \wedge Q$

(4) $b \wedge \text{if } c \text{ then } P \text{ else } Q \text{ end} = \text{if } c \text{ then } b \wedge P \text{ else } b \wedge Q \text{ end}$

(5) $\text{if } b \text{ then } P \text{ else } Q \text{ end} = \text{if } b \text{ then } b \wedge P \text{ else } \neg b \wedge Q \text{ end}$

(6) $\text{if } b \text{ then } P \text{ else } P \text{ end} = P$

(7) $\text{if } b \text{ then } P \text{ else } Q \text{ end}; R = \text{if } b \text{ then } P; R \text{ else } Q; R \text{ end}$

Laws (2) to (6) are best proved by direct manipulations in the predicate calculus, law (7) by casewise reasoning, which is expressed as follows:

(8) "Principle of casewise reasoning"
$(P = Q) \Leftrightarrow (R \wedge P = R \wedge Q) \wedge (\neg R \wedge P = \neg R \wedge Q)$

For the proof of law (7), we distinguish the cases "b" and "$\neg b$". We consider the first case only, as the second follows the same pattern.

Proof $b \wedge (\text{if } b \text{ then } P \text{ else } Q \text{ end}; R) = b \wedge \text{if } b \text{ then } P; R \text{ else } Q; R \text{ end}$
$\Leftrightarrow b \wedge \text{if } b \text{ then } P \text{ else } Q \text{ end}; R = b \wedge \text{if } b \text{ then } P; R \text{ else } Q; R \text{ end}$ — by (5) under ";"
$\Leftrightarrow b \wedge P; R = b \wedge P; R$ — by (2)

We will also use the shorthand "if b then P end":

$$\text{if } b \text{ then } P \text{ end} = \text{if } b \text{ then } P \text{ else II end}$$

Choice ("$\sqcap$")

The *binary choice* $P \sqcap Q$ means that initially either P or Q is chosen for execution. As we have no control over the choice, we must be prepared for the worst case in the following sense: $P \sqcap Q$ is only defined if both P and Q are defined, and if defined, the result is either that of P or Q.

$$P \sqcap Q = \Delta P \wedge \Delta Q \wedge (P \vee Q)$$

(1) $\Delta(P \sqcap Q) = \Delta P \wedge \Delta Q$

Proof L.H.S.$= \Delta(\Delta P \wedge \Delta Q \wedge (P \vee Q))$ by def. "$\sqcap$"
$= \Delta P \wedge \Delta Q \wedge \Delta(P \vee Q)$ by (1) under "Basics"
$= \Delta P \wedge \Delta Q \wedge (\Delta P \vee \Delta Q)$ Δ distributes over $\vee$
$=$ R.H.S. law of absorption

Binary choice has zero $\perp$, is idempotent and symmetric (which all follows immediately from the definition) and associative.

(2) $\quad P \sqcap \perp = \perp \sqcap P = \perp$

(3) $\quad P \sqcap P = P$

(4) $\quad P \sqcap Q = Q \sqcap P$

(5) $\quad P \sqcap (Q \sqcap R) = (P \sqcap Q) \sqcap R$

Proof L.H.S.$= \Delta P \wedge \Delta(Q \sqcap R) \wedge (P \vee (Q \sqcap R))$ by def. "$\sqcap$"
$= \Delta P \wedge \Delta Q \wedge \Delta R \wedge (P \vee (\Delta Q \wedge \Delta R \wedge (Q \vee R)))$ by (1), def. "$\sqcap$"
$= \Delta P \wedge \Delta Q \wedge \Delta R \wedge (P \vee Q \vee R)$ by pred. calc.
$=$ R.H.S. by repeating the argument

Furthermore, restricting one of the operands by a condition is the same as restricting the whole choice by the condition.

(6) $\quad (b \wedge P) \sqcap Q = b \wedge (P \sqcap Q)$

Proof L.H.S.$= \Delta(b \wedge P) \wedge \Delta Q \wedge ((b \wedge P) \vee Q)$ by def. "$\sqcap$"
$= b \wedge \Delta P \wedge \Delta Q \wedge (b \wedge P \vee Q)$ by (1) under "Basics"
$=$ R.H.S. by pred. calc., def. "$\sqcap$"

Sequential composition distributes through binary choice in both directions. The proofs are given in the appendix.

(7) $\quad P; (Q \sqcap R) = (P; Q) \sqcap (P; R)$

(8) $\quad (P \sqcap Q); R = (P; R) \sqcap (Q; R)$

The following two laws express that choosing between P and Q under a condition is the same as first choosing and then evaluating the condition.

(9) if b then $P \sqcap Q$ else R end $=$ if b then P else R end $\sqcap$ if b then Q else R end

(10) if b then P else $Q \sqcap R$ end $=$ if b then P else Q end $\sqcap$ if b then P else R end

Finally, binary choice distributes through the conditional.

(11) if b then P else Q end $\sqcap R =$ if b then $P \sqcap R$ else $Q \sqcap R$ end

The last three laws are most easily proved by casewise reasoning.

The *unrestricted choice* $(\sqcap x \cdot P)$ choses the initial value of x in P arbitrarily. We assume that x is not in the current context.

$$(\sqcap x \cdot P) = (\forall x \cdot \Delta P) \wedge (\exists x \cdot P)$$

The symbol x is *bound* by the choice. The rules for renaming bound symbols of quantifiers apply here as well. The choice $(\sqcap x \cdot P)$ is defined only in those initial states for which P will accept any value of x.

(12) $\Delta(\sqcap x \cdot P) = (\forall x \cdot \Delta P)$

We give some laws about unrestricted choice without proof. The order by which the value of two symbols are chosen does not matter.

(13) $(\sqcap x \cdot \sqcap y \cdot P) = (\sqcap y \cdot \sqcap x \cdot P)$

Choosing the value of x and then choosing between P and Q is the same as first choosing between P and Q and then the value of x.

(14) $(\sqcap x \cdot P \sqcap Q) = (\sqcap x \cdot P) \sqcap (\sqcap x \cdot Q)$

Choosing a value for x and accepting any (final) value of y' in P is equivalent to accepting any value of y' and choosing a value for x.

(15) $(\sqcap x \cdot \exists y' \cdot P) = (\exists y' \cdot \sqcap x \cdot P)$

Finally if x is not free in P, following laws for sequential composition hold:

(15) $(\sqcap x \cdot P; Q) = P; (\sqcap x \cdot Q)$

(16) $(\sqcap x \cdot Q; P) = (\sqcap x \cdot Q); P$

Variable Declaration ("var")

The variable declaration $(\text{var } x \cdot P)$ extends the current context V by x. We assume that x is not in the current context. The initial value of x is arbitrary and any final value is acceptable.

$$(\text{var } x \cdot P) = (\sqcap x \cdot \exists x' \cdot P)$$

Both x and x' are bound by the variable declaration. $(\text{var } x \cdot P)$ is defined if, for any initial value of x, P is defined.

(1) $\Delta(\text{var } x \cdot P) = (\forall x \cdot \Delta P)$

The order, in which variables are declared, does not matter:

(2) $(\text{var } x \cdot \text{var } y \cdot P) = (\text{var } y \cdot \text{var } x \cdot P)$

Proof L.H.S. $= (\sqcap x \cdot \exists x' \cdot \sqcap y \cdot \exists y' \cdot P)$ by def. "var"
$= (\sqcap x \cdot \sqcap y \cdot \exists x' \cdot \exists y' \cdot P)$ by (15) under "$\sqcap$"
$= (\sqcap y \cdot \sqcap x \cdot \exists y' \cdot \exists x' \cdot P)$ by (14) under "$\sqcap$", pred. calc.
$= (\sqcap y \cdot \exists y' \cdot \sqcap x \cdot \exists x' \cdot P)$ by (15) under "$\sqcap$"
$=$ R.H.S. by def. "var"

We give some laws about unrestricted law without proof. Let R be a specification in which (in the current context) neither x nor x' are free.

(3) $(\text{var } x \cdot P \sqcap Q) = (\text{var } x \cdot P) \sqcap (\text{var } x \cdot P)$

(4) $(\text{var } x \cdot R; P) = R; (\text{var } x \cdot P)$

(5) $(\text{var } x \cdot P; R) = (\text{var } x \cdot P); R$

(6) $(\text{var } x \cdot x := e) = \text{II}$

3 The Refinement Relation ("$\sqsubseteq$")

We define an ordering relation $\sqsubseteq$ on specifications with the intention that $P \sqsubseteq Q$ holds if Q is an acceptable replacement for P: When expecting P, we cannot tell whether Q has been executed in place of P.

$$(P \sqsubseteq Q) \Leftrightarrow [\Delta P \Rightarrow \Delta Q \wedge (Q \Rightarrow P)]$$

$[P \Rightarrow Q]$ states that Q has to be defined where P is and $[\Delta P \Rightarrow (Q \Rightarrow P)]$ states that within the domain of P, Q is at least as deterministic as P. Outside the domain of P, Q may behave arbitrarily. We assign $\sqsubseteq$ the same lowest binding power as $=$ and $\preceq$. Our confidence in the above definition is reassured by the following basic law about $\sqsubseteq$, which we could have taken as the definition as well.

(1) $(P \sqsubseteq Q) \Leftrightarrow (P \sqcap Q = P)$

This means that Q serves all the purposes P does, but it may serve more. Hence we call Q *better* than P and P *worse* than Q. The refinement relation is a partial order in that it is reflexive, antisymmetric, transitive, and it has $\perp$ as bottom element.

(2) $P \sqsubseteq P$

(3) $(P \sqsubseteq Q) \wedge (Q \sqsubseteq P) \Leftrightarrow (P = Q)$

(4) $(P \sqsubseteq Q) \wedge (Q \sqsubseteq R) \Rightarrow (P \sqsubseteq R)$

(5) $\perp \sqsubseteq P$

The proofs are straightforward using theorem (1) and the laws given under $\sqcap$. There are many laws about $\sqsubseteq$, some of which are given below. Their use in program development is for restricting the nondeterminism and for enlarging the domains of specifications.

(6) $b \wedge P \sqsubseteq P$

(7) $P \sqcap Q \sqsubseteq P$

(8) $(P \sqsubseteq Q \sqcap R) \Leftrightarrow (P \sqsubseteq Q) \wedge (P \sqsubseteq R)$

Again, they are most easily proved using (1). All the programming operators introduced in the last section are monotonic with respect to $\sqsubseteq$ in all their specification operands:

$$(9) \quad (P \sqsubseteq Q) \Rightarrow$$
$$\begin{aligned}
&(P;R \sqsubseteq Q;R) \wedge \\
&(R;P \sqsubseteq R;Q) \wedge \\
&(\text{if } b \text{ then } P \text{ else } R \text{ end} \sqsubseteq \text{if } b \text{ then } Q \text{ else } R \text{ end}) \wedge \\
&(\text{if } b \text{ then } R \text{ else } P \text{ end} \sqsubseteq \text{if } b \text{ then } R \text{ else } Q \text{ end}) \wedge \\
&(P \sqcap R \sqsubseteq Q \sqcap R) \wedge \\
&(R \sqcap P \sqsubseteq R \sqcap Q) \wedge \\
&((\sqcap x \cdot P) \sqsubseteq (\sqcap x \cdot Q)) \wedge \\
&((\text{var } x \cdot P) \sqsubseteq (\text{var } x \cdot Q))
\end{aligned}$$

The proof is most easily carried out using following theorem. Let $R(X)$ be a specification over X.

(10) $R(P \sqcap Q) = R(P) \sqcap R(Q) \Rightarrow (R(X)$ monotonic in $X)$

Proof Assuming $R(P \sqcap Q) = R(P) \sqcap R(Q)$ holds for any P, Q, we have to show that $R(X) \sqsubseteq R(Y)$ holds for any X, Y with $X \sqsubseteq Y$.

$$\begin{aligned}
& R(X) \sqsubseteq R(Y) \\
\Leftrightarrow\ & R(X) \sqcap R(Y) = R(Y) && \text{by (1)} \\
\Leftrightarrow\ & R(X) \sqcap R(Y) = R(X \sqcap Y) && \text{by assumption and (1)} \\
\Leftrightarrow\ & \top && \text{by assumption}
\end{aligned}$$

Theorem (9) follows from the distributivity of $\sqcap$ over all the programming operators as shown in the last section.

4 Weakest Precondition ("wp")

Informally, the weakest precondition of a specification P with respect to a condition b characterizes those initial states, for which P is defined and which only relate to states within b. The following theorem supports the informal claim that $P;b$ is the weakest precondition of P with respect to b.

(1) $P;b = (P \triangleright b) \wedge \Delta P$

$$\begin{aligned}
\textit{Proof}\ \text{L.H.S.} &= (P \triangleright \Delta b) \wedge (P \circ b) && \text{by def. ";"} \\
&= (P \triangleright b) \wedge (P \circ b) && \text{as } \Delta b = b \\
&= \text{R.H.S.} && \text{by (5) under "Basics"}
\end{aligned}$$

Before further elaborating the properties of weakest preconditions, we give two small theorems about conditions, which will be useful in proofs.

(2) $(b \sqsubseteq c) \Leftrightarrow (b \preceq c)$

(3) $b \sqcap c = b \wedge c$

We further underpin the correspondence of $P;b$ and $wp(P,b)$ by investigating Dijkstras "healthiness" properties. The "law of the excluded miracle: $wp(P, \bot) = \bot$" translates to $P;\bot = \bot$, which follows from the fact that $\bot$ is zero of ";". The monotonicity of $wp(P,b)$ in b with respect to implication is expressed as follows:

(4) $(b \preceq c) \Rightarrow (P; b \preceq P; c)$

Proof $\text{L.H.S.} \Leftrightarrow b \sqsubseteq c$ by (2)
$\Rightarrow P; b \sqsubseteq P; c$ as ";" $\sqsubseteq$-monotonic
$\Leftrightarrow \text{R.H.S.}$ by (2)

Now consider the property "$wp(P, b \wedge c) = wp(P, b) \wedge wp(P, c)$", which translates to

(5) $P; b \wedge c = (P; b) \wedge (P; c)$

Rather than proving this theorem we prove the following generalization, which expresses that $P; b \wedge Q$ behaves as $P; Q$ restricted to the weakest predcondition of P with respect to b.

(6) $P; b \wedge Q = (P; b) \wedge (P; Q)$

Proof $\text{L.H.S.} = (P \triangleright b) \wedge (P; Q)$ by (6) under ";"
$= (P \triangleright b) \wedge \Delta P \wedge (P; Q)$ as $P; Q \preceq \Delta P$
$= \text{R.H.S.}$ by (1)

For practical program development, it is important to be able to calculate the weakest precondition of a given specification with respect to some postcondition. Here we give the laws for weakest preconditions of the programs. We assume that x is not in the current context and x is not free in b.

(7) $\bot; b = \bot$

(8) $\text{II}; b = b$

(9) $x := e; b = b[x := e]$

(10) $(P; Q); b = P; (Q; b)$

(11) $\text{if } c \text{ then } P \text{ else } Q \text{ end}; b = \text{if } c \text{ then } P; c \text{ else } Q; b \text{ end}$

(12) $(P \sqcap Q); b = (P; b) \wedge (Q; b)$

(13) $(\sqcap x \cdot P); b = (\forall x \cdot P; b)$

(14) $(\text{var } x \cdot P); b = (\forall x \cdot P; b)$

Laws (7) to (11) follow directly from laws given in earlier sections. Law (12) follows from the distributivity of ; over $\sqcap$ and property (3). For (13) we observe:

Proof $\text{L.H.S.} = (\sqcap x \cdot P; b)$ by () under $\sqcap$
$= \Delta(\sqcap x \cdot P; b)$ for any $b : \Delta b = b$
$= \text{R.H.S.}$ by (1) under $\sqcap$

The proof of (14) follows the same pattern. Finally we state that the weakest precondition with respect to $\top$ is the domain:

(15) $\Delta P = P; \top$

Our use of weakest preconditions will be for proving properties of iterations.

5 Iteration

In this section we define the iteration "while b do P end", show that it is well defined, and give the main iteration theorem. The next section applies the main iteration theorem to the linear search. For the definition of the iteration we observe, that the body P may be any specification, and hence unbounded nondeterministic. It is known that unbounded nondeterminism leads to ***noncontinuity***, which rules out the definition of iteration as a limit of a countable sequence of specifications. Hence, we define the iteration as a fixed point based on the refinement relation as a partial order with bottom element. We give a self-contained account of the theory.

Let W stand for "while b do P end". As "$W =$ if b then $P; W$ end" is an essential property of W, we like to define the iteration W as the solution of the recursive equation

(*) $\quad Y =$ if b then $P; Y$ end

. However, in case the iteration does not terminate for some states, there may be many solutions. Consider for example "while $\perp$ do II end": any Y will solve (*). Having decided that nontermination is represented by undefinedness, we have to choose the *least defined* solution, if there are several. Now consider the nondeterministic iteration "while $0 \leq x < 5$ do $x' > x$ end". Again there are several solutions of (*), for example $x' < 0 \vee x' = 7$ and $x' < 0 \vee x' \geq 5$. The effect of the iteration is given by the *most nondeterministic* one. Hence, for the definition of the iteration we have to take the least defined and most nondeterministic solution, i.e. the *worst* (least with respect to the refinement ordering) one:

$$(Q = \text{worst } Y \text{ such that } r(Y)) \Leftrightarrow r(Q) \wedge (\forall X \mid r(X) \cdot Q \sqsubseteq X)$$

$$\text{while } b \text{ do } P \text{ end} = \text{worst } Y \text{ such that } (Y = \text{if } b \text{ then } P; Y \text{ end})$$

where $r(X)$ is for any specification X either $\top$ or $\perp$. We are obviously faced with the question of existence and uniqueness of the solutions, i.e. with the well-definedness of iteration. We do so by giving an equivalent, but explicit definition of iteration. To this end, we introduce the choice over a range of specifications. Let r be for any specification X either $\top$ or $\perp$:

$$(\sqcap X \mid r \cdot P) = (\forall X \mid r \cdot \Delta P) \wedge (\exists X \mid r \cdot P)$$

We state some consequences of this definition. The next theorem states a relationship between universal quantification and choice over ranges. It can be seen as a generalization of (8) under "$\sqsubseteq$" and is used in the proof of the subsequent theorem. We assume that X is not free in Q.

(1) $\quad Q \sqsubseteq (\sqcap X \mid r \cdot P) \Leftrightarrow (\forall X \mid r \cdot Q \sqsubseteq P)$

Proof L.H.S. $\Leftrightarrow [\Delta Q \Rightarrow \Delta(\sqcap X \mid r \cdot P) \wedge ((\sqcap X \mid r \cdot P) \Rightarrow Q)]$ — by def. "$\sqsubseteq$"
$\Leftrightarrow [\Delta Q \Rightarrow (\forall X \mid r \cdot \Delta P) \wedge ((\forall X \mid r \cdot \Delta P) \wedge (\exists X \mid r \cdot P) \Rightarrow Q)]$ — by def. "$\sqcap$"
$\Leftrightarrow [\Delta Q \Rightarrow (\forall X \mid r \cdot \Delta P) \wedge ((\exists X \mid r \cdot P) \Rightarrow Q)]$ — by pred. calc.
$\Leftrightarrow [\Delta Q \Rightarrow (\forall X \mid r \cdot \Delta P \wedge (P \Rightarrow Q))]$ — by pred. calc.
$\Leftrightarrow (\forall X \mid r \cdot [Q \Rightarrow \Delta P \wedge (P \Rightarrow Q)])$ — by pred. calc.
$\Leftrightarrow$ R.H.S. — by def. "$\sqsubseteq$"

(2) $(Q = \text{worst } Y \text{ such that } r(Y)) \Leftrightarrow r(Q) \wedge (Q = (\sqcap X \mid r(X) \cdot X))$

The implication from left to right states that if Q is the worst solution of $r(Q)$, then $r(Q)$ holds (trivially), and Q is given by some unique expression. Hence, if a worst solution exists, it is unique. However, there does not necessarily exist a solution Y for $r(Y)$, but if one Q is a solution of $r(Q)$ and Q is $(\sqcap X \mid r(X) \cdot X)$, then the converse implication states that it is the worst one. Here is the proof:

Proof L.H.S. $\Leftrightarrow r(Q) \wedge (\forall X \mid r(X) \cdot Q \sqsubseteq X)$ by def. "worst"
$\Leftrightarrow r(Q) \wedge (Q \sqsubseteq (\sqcap X \mid r(X) \cdot X))$ by (1)
$\Leftrightarrow r(Q) \wedge (Q = (\sqcap X \mid r(X) \cdot X))$ as $r(Q) \Rightarrow ((\sqcap X \mid r(X) \cdot X) \sqsubseteq Q)$
$\Leftrightarrow$ R.H.S.

This established the uniqueness of worst solutions of $r(Y)$. Next we turn our attention to the existence of solutions. We show the existence of solutions only for $r(y)$ of the form $P(Y) = Y$, which suffices our purposes. We do so in two steps, first establishing the existence of solutions for $P(Y) \sqsubseteq Y$, then for $P(Y) = Y$. We need some properties of restricted choices:

(3) $(\forall X \mid r \cdot P = Q) \Rightarrow ((\sqcap X \mid r \cdot P) = (\sqcap X \mid r \cdot Q))$

The proof of this theorem is done by simple manipulations in the predicate calculus. It is used for the justification of the following theorem.

(4) $(\forall X \mid r \cdot P \sqsubseteq Q) \Rightarrow ((\sqcap X \mid r \cdot P) \sqsubseteq (\sqcap X \mid r \cdot Q))$

Proof Assuming $(\forall X \mid r \cdot P \sqsubseteq Q)$, we calculate
$(\sqcap X \mid r \cdot P) \sqsubseteq (\sqcap X \mid r \cdot Q)$
$\Leftrightarrow (\sqcap X \mid r \cdot P) \sqcap (\sqcap X \mid r \cdot Q) = (\sqcap X \mid r \cdot Q)$ by (1) under "$\sqsubseteq$"
$\Leftrightarrow (\sqcap X \mid r \cdot P \sqcap Q) = (\sqcap X \mid r \cdot Q)$ property of "$\sqcap$"
$\Leftrightarrow (\sqcap X \mid r \cdot Q) = (\sqcap X \mid r \cdot Q)$ by assumption and (3) with $[P := P \sqcap Q]$
$\Leftrightarrow \top$

(5) $(P(X) \text{ monotonic in } X) \Rightarrow P(\sqcap X \mid r \cdot X) \sqsubseteq (\sqcap X \mid r \cdot P(X))$

Proof Assuming $P(X)$ monotonic in X, we calculate
$P(\sqcap X \mid r \cdot X) \sqsubseteq (\sqcap X \mid r \cdot P(X))$
$\Leftrightarrow (\forall X \mid r \cdot P(\sqcap X \mid r \cdot X) \sqsubseteq P(X))$ by (1)
$\Leftarrow (\forall X \mid r \cdot (\sqcap X \mid r \cdot X) \sqsubseteq X)$ $P(X)$ monotonic in X
$\Leftrightarrow \top$ property of "$\sqcap$"

(6) $(P(X) \text{ monotonic in } X) \Rightarrow (\text{there exists worst } Y \text{ such that } P(Y) \sqsubseteq Y)$

Proof According to (5) it suffices to show that $(\sqcap X \mid P(X) \sqsubseteq X \cdot X)$ is a solution.
$P(\sqcap X \mid P(X) \sqsubseteq X \cdot X)$
$\sqsubseteq (\sqcap X \mid P(X) \sqsubseteq X \cdot P(X))$ by assumption and (5)
$\sqsubseteq (\sqcap X \mid P(X) \sqsubseteq X \cdot X)$ by (4) with $[P, Q := X, P(X)]$

This result establishes the existence and uniqueness of the worst solution for $P(Y) \sqsubseteq Y$, but iteration is defined by the worst solution of $P(Y) = Y$. The theorem of Knaster and Tarski does the rest:

(7) "Theorem of Knaster-Tarski" If $P(X)$ monotonic in X, then

(a) worst Y such that $P(Y) = Y$, and

(b) worst Y such that $P(Y) \sqsubseteq Y$

have the same solution.

Proof According to (6), (b) has a solution. Let it be denoted by Q. Now we show that Q is also the solution of (a), i.e.

(c) $Q = P(Q)$, and

(d) $(\forall X \mid X = P(X) \cdot Q \sqsubseteq X)$

For (c) we observe:

$$\begin{array}{lr} Q=P(Q) & \\ \Leftrightarrow Q \sqsubseteq P(Q) \wedge P(Q) \sqsubseteq Q & \text{as "}\sqsubseteq\text{" antisymmetric} \\ \Leftrightarrow Q \sqsubseteq P(Q) & \text{as Q solution of (b)} \\ \Leftarrow P(P(Q)) \sqsubseteq P(Q) & \text{as } (\forall X \mid P(X) \sqsubseteq X \cdot Q \sqsubseteq X)) \text{ with } [X := P(Q)] \\ \Leftarrow P(Q) \sqsubseteq Q & P(X) \text{ monotonic in } X \\ \Leftarrow \mathsf{T} & \text{as Q solution of (b)} \end{array}$$

For (d) we observe for any X:

$$\begin{array}{lr} X=P(X) & \\ \Rightarrow P(X) \sqsubseteq X & \text{property of "}\sqsubseteq\text{"} \\ \Rightarrow Q \sqsubseteq X & \text{as } (\forall X \mid P(X) \sqsubseteq X \cdot X \sqsubseteq Q) \end{array}$$

hence $(\forall X \mid X = P(X) \cdot Q \sqsubseteq X)$.

For iteration we observe, that both sequential composition and conditional are monotonic, hence "if b then $P; X$ end" is monotonic in X, which establishes the well-definedness of iterations. Next we turn to the question of how to prove properties about iterations. To this end, we need induction over well-founded sets.

(set C partially ordered by $<$ is well-founded) $\Leftrightarrow$
(all decreasing chains in C are finite)

(mathematical induction over C is valid) $\Leftrightarrow$
(for any predicate $p(c)$ over $c \in C$:

$$(\forall c \mid c \in C \cdot p(c)) \Leftrightarrow (\forall c \mid c \in C \cdot (\forall d \mid d \in C \wedge d < c \cdot p(d)) \Rightarrow p(c)))$$

(8) "Principle of mathematical induction"
(set C is well-founded) $\Leftrightarrow$ (mathematical induction over C is valid)

The definitions above are taken from [4], where also a proof of the principle of mathematical induction is given. This gives finally all the prerequisites for the main theorem about iterations

(9) "*Main Iteration Theorem*" Let D be a set partially ordered by $<$, C a well founded subset of D, t a function from the context V to D, P and Q specification and b a condition. Then

(a) $(\Delta Q \wedge t \notin C \preceq \neg b) \wedge$

(b) $((\forall c \mid c \in C \cdot \Delta Q \wedge (t = c) \wedge b \preceq P; t < c)) \wedge$

(c) (if b then $P; Q$ end $= Q$) $\Rightarrow$

(d) while b do P end $= Q$

The function t is known as the *termination function*. Informally, if t is not in C (but in D), then according to requirement (a) the iteration is sure to terminate as $\neg b$, the condition for termination holds. Otherwise, if t is in C, then the iteration might or might not terminate. However, in case it does not, requirement (b) ensures that t will be decreased by the body of the iteration, and hence eventually terminate: as C is well founded, all decreasing chains in C are finite. Finally, requirement (c) states that the iteration does indeed compute the desired result: Q is a solution of equation defining the iteration.

Proof For the purpose of the proof we define W by

$W =$ while b do P end

and conclude by "unfolding" the iteration:

(e) $W =$ if b then $P; W$ end by (e)

The proof is carried out by showing separately

(f) $\Delta Q \wedge t \notin C \wedge W = t \notin C \wedge Q$

(g) $\Delta Q \wedge t \in C \wedge W = t \in C \wedge Q$

(h) $\neg \Delta Q \wedge W = \perp$

which together, by the principle of casewise reasoning, establish (d). (f) caters for the case that the iteration is defined and terminates immediately, (g) for the case that it will terminate sometime, and (g) for the case that it is undefined. For (f) we observe:

$t \notin C \wedge \Delta Q \wedge W = t \notin C \wedge Q$
$\Leftrightarrow t \notin C \wedge \Delta Q \wedge \neg b \wedge W = t \notin C \wedge \Delta Q \wedge \neg b \wedge Q$ by (a)
$\Leftrightarrow t \notin C \wedge \Delta Q \wedge \neg b \wedge \mathrm{II} = t \notin C \wedge \Delta Q \wedge \neg b \wedge \mathrm{II}$ by (c), (e), and (3) under "if"
$\Leftrightarrow \top$

For the proof of (g) and (h) we need a couple of lemmas:

(i) $b \Rightarrow (P; \Delta Q) = \Delta Q$

(j) $P; \Delta Q \preceq Q$

(k) $\neg b \preceq \Delta Q$

They are all consequences of requirement (c):

if b then $P; Q$ end $= Q$
$\Rightarrow \Delta$if b then $P; Q$ end $= \Delta Q$ rule of Leibnitz
$\Leftrightarrow (b \wedge \Delta(P; Q)) \vee (\neg b \wedge \Delta \mathrm{II}) = \Delta Q$ by (1) under "if", def. "if"
$\Leftrightarrow (b \wedge P; \Delta Q) \vee \neg b = \Delta Q$ by (15) under "wp"
$\Leftrightarrow (P; \Delta Q) \vee \neg b = \Delta Q$ law of absorption
$\Rightarrow (P; \Delta Q \preceq \Delta Q) \wedge (\neg b \preceq \Delta Q)$ pred. calc.

(i) is equivalent to the second last line. Now, for (g) we observe

$t \in C \wedge \Delta Q \wedge W = t \in C \wedge Q$
$\Leftarrow (\forall c \mid c \in C \cdot (t = c) \wedge \Delta Q \wedge W = (t = c) \wedge Q)$ by pred. calc.

According to the principle of mathematical induction, this is proved by deriving

(l) $(t = c) \wedge \Delta Q \wedge W = (t = c) \wedge Q$

from

(m) $(\forall d \mid d \in C \wedge d < c \cdot (t = d) \wedge \Delta Q \wedge W = (t = d) \wedge Q)$

for any $c \in C$. To this end, we calculate:

$(\forall d \mid d \in C \wedge d < c \cdot (t = d) \wedge \Delta Q \wedge W = (t = d) \wedge Q)$
$\Leftrightarrow (t < c) \wedge t \in C \wedge \Delta Q \wedge W = (t < c) \wedge t \in C \wedge Q$ by pred. calc.
$\Leftrightarrow (t < c) \wedge \Delta Q \wedge W = (t < c) \wedge Q$ by (f)
$\Rightarrow (t = c) \wedge \Delta Q \wedge \text{if } b \text{ then } P; ((t < c) \wedge \Delta Q \wedge W) \text{ end} =$
$(t = c) \wedge \Delta Q \wedge \text{if } b \text{ then } P; ((t < c) \wedge Q) \text{ end}$ by rule of Leibnitz
$\Rightarrow (t = c) \wedge \Delta Q \wedge \text{if } b \text{ then } (P; t < c) \wedge (P; \Delta Q) \wedge (P; W) \text{ end} =$
$(t = c) \wedge \Delta Q \wedge \text{if } b \text{ then } (P; t < c) \wedge (P; Q) \text{ end}$ by (6) under "wp"
$\Leftrightarrow (t = c) \wedge \Delta Q \wedge \text{if } b \text{ then } P; W \text{ end} =$
$(t = c) \wedge \Delta Q \wedge \text{if } b \text{ then } P; Q \text{ end}$ by (b), (i), and (4),(5) under "if"
$\Leftrightarrow (t = c) \wedge \Delta Q \wedge W = (t = c) \wedge Q$ by (e) and (c)

For (h) we calculate:

$\neg \Delta Q \wedge W = \bot$
$\Leftrightarrow W \preceq \Delta Q$
$\Leftrightarrow \Delta W \preceq \Delta Q$ as $(P \preceq b) \Leftrightarrow (\Delta P \preceq b)$
$\Leftrightarrow \Delta(\sqcap X \mid X = \text{if } b \text{ then } P; X \text{ end} \cdot X) \preceq \Delta Q$ property of "while"
$\Leftrightarrow (\forall X \mid X = \text{if } b \text{ then } P; X \text{ end} \cdot \Delta X) \preceq \Delta Q$ property of "$\sqcap$"
$\Leftrightarrow (\exists X \mid X = \text{if } b \text{ then } P; X \text{ end} \cdot \Delta X \preceq \Delta Q)$ by pred. calc.
$\Leftarrow \neg b \wedge \text{II} = \text{if } b \text{ then } P; (\neg b \wedge \text{II}) \text{ end}$ as $(\Delta(\neg b \wedge \text{II}) \preceq Q) \Leftrightarrow$ (k)
$\Leftrightarrow \neg b \wedge \text{II} = b \wedge (P; (\neg b \wedge \text{II})) \vee \neg b \wedge \text{II}$ by def. "if"
$\Leftrightarrow \neg b \wedge \text{II} = b \wedge (P; \neg b) \wedge (P; \text{II}) \vee \neg b \wedge \text{II}$ by (6) under "wp"
$\Leftrightarrow \neg b \wedge \text{II} = b \wedge c \wedge (P; \neg b) \wedge P \vee \neg b \wedge \text{II}$ as $P; \neg b \preceq P; \Delta Q \preceq \Delta Q$ by (k),(j)
$\Leftrightarrow \top$ as $b \wedge \Delta Q = \bot$ by (k)

This concludes the proof of the main iteration theorem.

Finally we claim that iteration is monotonic with respect to the refinement relation.

(10) $(P \sqsubseteq Q) \Rightarrow (\text{while } b \text{ do } P \text{ end} \sqsubseteq \text{while } b \text{ do } Q \text{ end})$

Proof For the purpose of the proof we define

$R(Y) = \text{while } b \text{ do } Y \text{ end}$

and conclude by the definition of iteration and Knaster-Tarski:

(a) $R(Y) = \text{worst} Y \text{such that} \text{if } b \text{ then } Y; X \text{ end} \sqsubseteq X$

Recalling the defining properties of "worst", we can express this equivalently by

(b) $\text{if } b \text{ then } Y; R(Y) \text{ end} \sqsubseteq R(Y) \quad \wedge$

(c) $(\text{if } b \text{ then } Y; R(Y) \text{ end} \sqsubseteq X) \Rightarrow (R(Y) \sqsubseteq X)$

for any X, Y. (b) can be rephrased as

(d) $(X \sqsubseteq \text{if } b \text{ then } Y; R(Y) \text{ end}) \Rightarrow (R(Y) \sqsubseteq X)$

for any X. The proof obligation can now be discharged as follows:

$R(P) \sqsubseteq R(Q)$
$\Leftarrow \text{if } b \text{ then } P; R(Q) \text{ end} \sqsubseteq R(Q)$ by (c) with $[X, Y := R(Q), P]$
$\Leftarrow \text{if } b \text{ then } P; R(Q) \text{ end} \sqsubseteq \text{if } b \text{ then } Q; R(Q) \text{ end}$
by (d) with $[X, Y := \text{if } b \text{ then } P; R(Q) \text{ end}, Q]$
$\Leftarrow P \sqsubseteq Q$ by monotonicity of "if" and ";"

6 Example

We demonstrate the usefulness of the main iteration theorem by a small example, the linear search. In the proof, we concentrate rather on the programming calculus than on the problem domain used for this example. We also allow the expression e in $x := e$ to be undefined, with the meaning that $x := e = \Delta e \wedge x := e$, as done in [10], for simplifying the formulation of the theorem.

(1) "Linear Search Theorem". Let x be an integer variable, and $b(x)$ a boolean valued function. Then

$$x := (\min i \mid x \leq i \wedge b(i)) = \text{while } b(x) \text{ do } x := x + 1 \text{ end}$$

Proof For the proof we define $M = (\min i \mid x \leq i \wedge b(i))$ and postulate

$$\Delta(M) = (\exists i \cdot x \leq i \wedge b(i))$$

For the main iteration theorem, we take D :=Nat, C :=Nat$\setminus\{0\}$, $t := M - x$, which implies

$$(t \in C = x < M) \quad \wedge \quad (t \notin C = b(x))$$

and leads to following proof obligations:

(a) $\Delta(x := M) \wedge b(x) \preceq b(x)$

(b) $(\forall c \mid c > 0 \cdot \Delta(x := M) \wedge (M - x = c) \wedge \neg b(x) \preceq x := x - 1; (M - x < c))$

(c) if $\neg b(x)$ then $x := x + 1; x := (\min i \mid x \leq i \wedge b(i))$ end $=$
$x := (\min i \mid x \leq i \wedge b(i))$

Requirement (a) obviously holds. The consequent in (b) can be simplified to $M - x \leq c$ accoding to (9) under "wp", which then indeed follows from the antecedent. For (c) we calculate:

L.H.S. = if $\neg b(x)$ then $x := (\min i \mid x < i \wedge b(i))$ end — by (7) under ";"
= if $\neg b(x)$ then $x := (\min i \mid x < i \wedge b(i))$ else $x := x$ end — by def.'s
= if $\neg b(x)$ then $x := (\min i \mid x \leq i \wedge b(i))$
else $x := (\min i \mid x \leq i \wedge b(i))$ end — by (5) under "if"
= R.H.S. — by (6) under "if"

7 Discussion

In this paper, we have presented a calculus based on demonic composition and demonic disjunction. Although this concepts are known, this choice needs some justification with respect to other approaches.

A common way of defining the meaning of nondeterministic programs is by a total relation over the state space extended by a final fictitious state $\perp$ standing for nontermination. Hence, if an initial state relates to $\perp$, nontermination for that state is possible. For simplifying the definition of sequential composition, the initial state space is extended by $\perp$ as well, which invariantly relates to the final $\perp$. Then nondeterministic choice is given by union and sequential composition by relational composition. In this model, we can express that a program, when started in an initial state, will either terminate in some final states or not terminate at all. However, we might question how useful this distinction is, at least for our purposes: A mechanism which may fail to work (i. e. not terminate) or may work is as unreliable as a mechanism which always fails to work; there is no point in distinguishing between them. Hence, we may impose the restriction that if an initial state relates to $\perp$, it relates to everything else as well. This is, for example, the approach taken by Hoare et al. in [9] and [8]. The correspondence between that model and the model taken here is that an initial state which relates to $\perp$ corresponds to an undefined initial state and vice versa, leaving all other initial states unmodified. Sequential composition and nondeterministic choice are defined here just in such a way to achieve this correspondence. Hence, all laws given in [9] for programs also hold in the calculus presented here.

There exist other equivalent semantic models: For example, nondeterministic programs can also be represented by total functions from the state space X to $\mathbf{P}(X) \cup \{X \cup \{\perp\}\}$ or by a set of functions. A survey of semantic models, which include those for partial correctness and those based on partial relations with $\perp$, can be found in [14]. We use the model of 'demonic composition' and 'demonic disjunction' in a predicative setting (rather than in a relational one) in order to have both the (demonic) programming operators and the (angelic) specification operators available in the simplest possible way: a programming calculus can be explained by starting with the specification notation and gradually introducing the programming operators by giving them meaning through specifications.

In the calculus presented several aspects have been omitted: The concept of the context has been treated rather informally. Although this seems to be appropriate here, things get more complicated in presence of recursion, as each 'recursive call' may lead to an extension of the context. Furthermore, the typing of the variables of the context has been omitted. Recursion can be defined easily using the fixed point approach taken here for iteration. However, it is not easy to give a recursion theorem similar to the iteration theorem. Finally, we have only treated operational refinement and left out data refinement, which is the subject of our further research.

Closely related to our work is the predicative programming approach of E. Hehner [5, 6, 7]. A difference is that in Hehners approach specifications have to be *total* in order to be *implementable*. With this restriction, sequential composition is defined by relational composition and nondeterministic choice by disjunction. Refinement is given by (converse) implication. Termination is treated separately as a special case of timing. The simplicity of the approach makes it very attractive. However, as specifications have to be total, they have often to be written in the form $\Delta P \Rightarrow P$, e.g. $x \geq 0 \Rightarrow y'^2 \leq x < (y'+1)^2$. Moreover, the conjunction of two implementable specifications is not necessarily implementable. By contrast in our approach, like in the specification language Z, any logical expression is an admissible specification. In Z, the conjunction of specifications (i.e. schemas) plays an important role in composing large specifications from smaller ones. On the other hand, we have much more complicated definitions of the programming operators and of the refinement relation. These complexities are caused as nontermination (represented by undefinedness) is always included in our definitions and reasoning. As a consequence one might better not use the definitions directly but rather derive theorems first and use them in program development.

More recently, A. Gravell has also used in an independent work from ours demonic composition for the definition of the programming operators. As his aim was to introduce the programming operators as an extension to standard Z-F set theory, iteration is not defined by fixed points. Sequential composition and conditional are defined such that all unmentioned variables in each component remain unchanged. Although this seems to be natural, it has some surprising consequences. For example, $\top$ acts as the identity of composition, rather than II, and equality and refinement of operations have to take the alphabet (set of all free variables) into account.

We arrive at a similar set of laws as the refinement calculi of R. J. Back [1], C. Morgan [11, 12], and J. Morris [13] do. Apart from the notational difference, these calculi include *miracles*. For example, Morgan writes specifications of the form $x : [pre, post]$, with *pre* a predicate over the initial state and *post* a predicate over initial and final state. The *frame* x is a list of variables which can be changed by the specification. If *pre* implies the domain of *post*, this can be expressed in our approach as $pre \wedge post$. Otherwise, the specification is miraculous, which cannot be expressed by our demonic model. Models which include miracles are discussed by Nelson [14].

Another difference is that refinement calculi define the meaning of both programs and specifications by weakest preconditions. We define programs by specifications and consider weakest preconditions later on, which seems to be more natural. We should also note that the refinement calculi stress, as indicated by the name, refinement rather than equality. Here we have tried to give rules which allow for stating equality whenever it holds.

Our work is much in the same direction as that of King [10]. King proposes a program development which starts with a Z specification and uses the refinement calculus for program development. This requires making a transitions from the Z specification to the refinement notation during the development, which we can avoid in our approach.

Acknowledgements. My thanks go the reviews for their valuable comments on a previous version and to C. Lewerentz and T. Lindner for pointing out mistakes in the final draft. The paper has profited much from discussions with F. Weber and A. Rueping. This work is supported by the German Ministry of Research and Technology (BMFT) under grant number 01 IS 203 M, project "KORSO".

References

1. Back, R. J. R. A Calculus of Refinement for Program Derivations. *Acta Informatica* **23** (1988).
2. Dijkstra, E. W. *A Discipline of Programming.* Prentice Hall, 1976.
3. Dijkstra, E. W. and C. S. Scholten. *Predicate Calculus and Program Semantics.* Springer- Verlag, 1990.
4. Gravell, A. Constructive Refinement of First Order Specifications. to appear in *Proceedings of the 5th Refinement Workshop*, Springer-Verlag, 1992.
5. Hehner, E. C. R. Predicative Programming – Parts I and II. *Communications of the ACM* **27, 2** (1984).
6. Hehner, E. C. R., L. E. Gupta, and A. J. Malton. Predicative Methodology. *Acta Informatica* **23** (1986).

7. Hehner, E. C. R. Termination is Timing. In J. L. A. van de Snepscheut (Ed.) *Mathematics of Program Construction*, Springer-Verlag, 1989.
8. Hoare, C. A. R. and He Jifeng. The weakest prespecification. *Fundamenta Informaticae* **9** (1986).
9. Hoare, C. A. R., I. J. Hayes, He Jifeng, C. C. Morgan, A. W. Roscoe, J. W. Sanders, I. H. Sorensen, J. M. Spivey, and B. A. Sufrin. Laws of Programming. *Communications of the ACM* **30, 8** (1987).
10. King, S. Z and the Refinement Calculus. In D. Bjorner, C. A. R. Hoare, J. Langmaack (Eds.) *VDM '90: VDM and Z - Formal Methods in Software Development* Springer-Verlag, 1990.
11. Morgan, C. The Specification Statement. *ACM Transactions on Programming Languages and Systems* **10,3** (1988).
12. Morgan, C. *Programming from Specifications.* Prentice Hall, 1990.
13. Morris, J. M. Programs from Specifications. In Dijkstra, E. W. (Ed.) *Formal Development of Programs and Proofs.* Addison Wesley, 1990.
14. Nelson, G. A Generalization of Dijkstra's Calculus. *ACM Transactions on Programming Languages and Systems* **11,4** (1989).
15. Spivey, J. M. *The Z Reference Manual.* Prentice Hall, 1989.

Appendix

In this appendix, we prove the associativity of sequential composition and the distributivity of binary choice through sequential composition. Two small lemmas are needed:

(1) $(P \vee Q) \triangleright b = (P \triangleright b) \wedge (Q \triangleright b)$

(2) $(P \circ Q) \triangleright b = P \triangleright Q \triangleright b$

The proofs follow the pattern of manipulating both sides until they are of the form $b \wedge A$, where A is a specification consisting only of angelic operators.

(3) $P; (Q; R) = (P; Q); R$

Proof L.H.S.$= P; (Q \triangleright \Delta R) \wedge (Q \circ R)$ by def. ";"
$= (P \triangleright Q \triangleright \Delta R) \wedge (P; (Q \circ R)$ by (6) under ";"
$= (P \triangleright Q \triangleright \Delta R) \wedge ((P \triangleright \Delta(Q \circ R)) \wedge (P \circ (Q \circ R)))$ by def. ";"
$= P \triangleright ((Q \triangleright \Delta R) \wedge \Delta(Q \circ R)) \wedge (P \circ Q \circ R)$ by (6) under "Basics"
$= P \triangleright ((Q \triangleright \Delta R) \wedge \Delta Q) \wedge (P \circ Q \circ R)$ by (2) and (5) under "Basics"
R.H.S.$= (P \triangleright \Delta Q) \wedge (P \circ R); R$ by def ";"
$= (P \triangleright \Delta Q) \wedge ((P \circ Q); R)$ by (5) under ";"
$= P \triangleright \Delta Q \wedge (P \circ Q) \triangleright \Delta R \wedge (P \circ Q) \circ R$ by def. ";"
$= P \triangleright (Q \triangleright \Delta R \wedge \Delta Q) \wedge (P \circ Q \circ R)$ by (2), and (6) under "Basics"

(4) $P; (Q \sqcap R) = (P; Q) \sqcap (P; R)$

Proof L.H.S.$= P; (\Delta Q \wedge \Delta R \wedge (Q \vee R))$ by def. "$\sqcap$"
$= (P \triangleright (\Delta Q \wedge \Delta R)) \wedge (P; Q \vee R)$ by (6) under ";"
$= (P \triangleright (\Delta Q \wedge \Delta R)) \wedge (P \triangleright \Delta(Q \wedge R)) \wedge (P \circ Q \vee R)$ by def. ";"

$= (P \triangleright (\Delta Q \wedge \Delta R)) \wedge (P \circ Q \vee R)$

by (6) under "Basics", properties of "Δ"

R.H.S.$= ((P \triangleright \Delta Q) \wedge (P \circ Q)) \sqcap ((P \triangleright \Delta R) \wedge (P \circ R))$ by def. ";" twice

$= (P \triangleright \Delta Q) \wedge (P \triangleright \Delta R) \wedge ((P \circ Q) \sqcap (P \circ R))$ by (6) under "$\sqcap$"

$= (P \triangleright \Delta Q) \wedge (P \triangleright \Delta R) \wedge \Delta(P \circ Q) \wedge \Delta(P \circ R) \wedge ((P \circ Q) \vee (P \circ R))$

by def. "$\sqcap$"

$= (P \triangleright \Delta Q) \wedge (P \triangleright \Delta R) \wedge (P \circ \Delta Q) \wedge (P \circ \Delta R) \wedge (P \circ Q \vee R)$

properties of "$\circ$"

$= (P \triangleright \Delta Q) \wedge (P \triangleright \Delta R) \wedge \Delta P \wedge (P \circ Q \vee R)$ by (5) under "Basics"

$= (P \triangleright \Delta Q) \wedge (P \triangleright \Delta R) \wedge (P \circ Q \vee R)$ as $P \circ X \preceq P$

$= P \triangleright (\Delta Q \wedge \Delta R) \wedge P \circ (Q \vee R)$ by (5) under "Basics"

(5) $(P \sqcap Q); R = (P; R) \sqcap (Q; R)$

Proof L.H.S.$= \Delta P \wedge \Delta Q \wedge (P \vee Q); R$ by def. "$\sqcap$"

$= \Delta P \wedge \Delta Q \wedge (P \vee Q; R)$ by (5) under ";"

$= \Delta P \wedge \Delta Q \wedge ((P \vee Q) \triangleright \Delta R) \wedge (P \vee Q) \circ R$ by def. ";"

$= \Delta P \wedge \Delta Q \wedge (P \triangleright \Delta R) \wedge (Q \triangleright \Delta R) \wedge (P \vee Q) \circ R$ by (1)

R.H.S.$= ((P \triangleright \Delta R) \wedge (P \circ R)) \sqcap ((Q \triangleright \Delta R) \wedge (Q \circ R))$ by def. ";" twice

$= (P \triangleright \Delta R) \wedge (Q \triangleright \Delta R) \wedge ((P \circ R) \sqcap (Q \circ R))$ by (6) under "$\sqcap$"

$= (P \triangleright \Delta R) \wedge (Q \triangleright \Delta R) \wedge \Delta(P \circ R) \wedge \Delta(Q \circ R) \wedge ((P \circ R) \vee (Q \circ R))$

by def. "$\sqcap$"

$= (P \triangleright \Delta R) \wedge (Q \triangleright \Delta R) \wedge \Delta P \wedge \Delta Q \wedge (P \vee Q \circ R)$

by (2) and (5) under "Basics"

Derivation of a Parallel Matching Algorithm

Yellamraju V. Srinivas

Kestrel Institute, 3260 Hillview Avenue,
Palo Alto, CA 94304, USA

Abstract. We present a derivation of a parallel version of the Knuth-Morris-Pratt algorithm for finding occurrences of a pattern string in a target string. We show that the failure function, the source of efficiency of the sequential algorithm, is a form of search in an ordered domain. This view enables the generalization of the algorithm both beyond sequential execution and the string data structure. Our derivation systematically uses a divide-and-conquer strategy. The computation tree so generated can be mapped onto time, yielding a naive sequential algorithm, onto a processor tree, yielding a parallel algorithm, or onto a data structure, yielding the failure function.

1 Parallel Pattern Matching

The pattern matching problem for strings consists of finding *occurrences* of a *pattern* string in a *target* string. An occurrence is a piece of the target which is identical to the pattern, together with a correspondence with the pattern—for strings, it is sufficient to indicate the position of the first element of the pattern.

Several parallel algorithms for string matching have been described in the literature [15, 4, 16]. These algorithms are specifically designed to work in parallel for strings, and as such, bear little resemblance to the well known sequential algorithms of Knuth-Morris-Pratt [5] and Boyer-Moore [2]. Indeed, as Galil observes [4, page 145], it is not apparent how to parallelize these sequential algorithms:

> The [Knuth-Morris-Pratt and Boyer-Moore algorithms] do not seem to be parallelizable, because they sequentially construct tables which are used sequentially. So, even giving the tables for free does not seem to help much.

In this paper, we derive a parallel version of the Knuth-Morris-Pratt (KMP) algorithm, thus showing that there is nothing inherently sequential about the failure function. This generalization is made possible by adopting a problem-reduction view of pattern matching. The failure function can be seen as a form of search in an ordered collection of partial occurrences. This view allows our parallel algorithm to be applied to other data structures, and the derivation to be applied to related algorithms. In contrast, the parallel string matching algorithms mentioned above cannot be easily modified to work for other data structures, since these algorithms crucially depend on properties of periodicity in strings. Similarly, at the meta-level, there are several derivations of the sequential KMP algorithm [3, 1, 14, 6, 7]; however, none seems easily modifiable to produce a parallel algorithm. In our derivation, the difference between a sequential and a parallel algorithm is the difference between a front-last split and an equal split of the target string. The derivation is based on the abstract description of pattern matching using categories and sheaves given in [12].

1.1 The Knuth-Morris-Pratt Algorithm

The KMP algorithm [5] is a fast, sequential, pattern matching algorithm for finding occurrences of a constant pattern in a target string. It is linear in the sum of the sizes of the pattern and the target strings. KMP reduces the complexity of the naive algorithm for string matching (check for a match at every position in the target string) by avoiding comparisons whose results are already known (from previous comparisons). In particular, given a character mismatch after the pattern is partially matched, the next possible position in the target where the pattern can match can be computed by using the knowledge of the partial match. This "sliding" of the pattern on a mismatch is the most well known aspect of KMP. We show below an example, where there is a mismatch at the eighth character of the pattern, and the pattern can be slid three positions to the right.

```
slide                 ⟶ a b c a b c a c a b
pattern             a b c a b c a c a b
matches             | | | | | | | ?
target      b a b c b a b c a b c a a b c a b c a b c a c a b c
```

The amounts by which to slide the pattern on possible mismatches can be precomputed in time proportional to the size of the pattern. Thus all occurrences can be enumerated in a single left-ro-right scan of the target string without backing up.

The table assigning the amount of sliding to each mismatch is called the *failure function*. The failure function seems to depend on the sequential traversal of the target, and it is not apparent how to generalize it to a parallel traversal. We follow a more general approach of describing a match in terms of sub-matches; this description only depends on the topological properties of the underlying data structures. Such a description removes the apparent dependency of the failure function on a sequential traversal.

1.2 Methodology

We adopt the derivation style used in the KIDS program synthesis/transformation system [11]. Problems are described using first-order algebraic specifications with subsorts and partial functions. General implementation strategies such as divide-and-conquer, search, etc., are described by algorithm theories; these theories describe the essential computation without committing to any form of control strategy [8]. Algorithm theories are used in actual problems by providing a specification morphism from the algorithm theory to the problem specification. This morphism allows any realization of the algorithm theory to be specialized to the problem at hand. Sometimes, we only have a partial view from the algorithm theory to the problem; e.g., when using a divide-and-conquer theory, we may know how to decompose the problem but not how to compose partial solutions. In such a case, it is possible to complete the specification morphism by inference using the axioms of the algorithm theory. For example, the divide-and-conquer theory described in [9] contains a soundness axiom:

$\langle x_1, x_2 \rangle = \text{decompose}(x) \wedge$ —*decomposition of input*
$\text{io}(x_1, z_1) \wedge \text{io}(x_2, z_2) \wedge$ —*inductive hypothesis*
$x > x_1 \wedge x > x_2 \wedge$ —*well-founded order on input domain*
$z = \text{compose}(z_1, z_2)$ —*composition of outputs*
$\Rightarrow \text{io}(x, z)$ —*input output condition*

By instantiating this axiom using the partial morphism, and by inferencing in the problem domain, we can determine necessary or sufficient conditions on the unknown operators. This inference depends on having appropriate distributive laws relating the known operators and the decomposition of the input. In this paper, we will use this technique several times to determine composition operators.

2 A Specification for String Matching

We use a specification of strings which is different from those normally used, so that our derivation can be generalized to other data structures. Our description of strings only specifies their essential property: each string is a set of elements together with a total order on them, and a labeling function which assigns a label to each element of the string from a fixed set of labels. To simplify the presentation, we only consider non-empty strings.

type String =
sorts Elem
operations
$\_ < \_$: Elem, Elem → Bool
label : Elem → Label —*the set of labels is global*
axioms
—*we only consider non-empty strings*
Elem $\neq \emptyset$
—"<" *is a total order*
$a \not< a$ —*irreflexive*
$a < b \wedge b < c \Rightarrow a < c$ —*transitive*
$\forall a, b \cdot a < b \vee a = b \vee b < a$ —*trichotomy*
end

Normally, a string is represented by listing the elements in order. This is captured in the adjacency relation, which uniquely generates the total order. The adjacency relation can be obviously extended to substrings of a string, and we will overload the symbol for adjacency for this case.

extend String **by**
operations
$\_ \multimap \_$: Elem, Elem → Bool
axioms
$a \multimap b$ **iff** $a < b \wedge \nexists c \cdot a < c \wedge c < b$

With this theory of strings, we can now define an occurrence. Occurrences form a subsort of the sort of functions between the underlying sets of two strings. To be an occurrence, such a function has to be injective, and preserve labels and the adjacency relation.

Notation. We use subscripts to select components of a string, e.g., the elements a string p will be denoted by Elem_p. We use a λ-notation to describe dependent product types, e.g., $\lambda A \cdot A \times f(A)$ denotes a product in which the second component depends on the first via the function f.

sorts $\text{Proto-Arrow} = \lambda p, q\colon \text{String} \cdot p \times (\mathrm{Elem}_p \rightarrow \mathrm{Elem}_q) \times q$
operations

$$\begin{aligned} \text{injective} &: \text{Proto-Arrow} \rightarrow \text{Bool} \\ \text{label-preserving} &: \text{Proto-Arrow} \rightarrow \text{Bool} \\ \text{adjacency-preserving} &: \text{Proto-Arrow} \rightarrow \text{Bool} \end{aligned}$$

axioms

$$\text{injective}(p \xrightarrow{f} q) \textbf{ iff } \forall a, b \in \mathrm{Elem}_p \cdot a \neq b \Rightarrow f(a) \neq f(b)$$
$$\text{label-preserving}(p \xrightarrow{f} q) \textbf{ iff } \forall a \in \mathrm{Elem}_p \cdot \mathrm{label}_p(a) = \mathrm{label}_q(f(a))$$
$$\text{adjacency-preserving}(p \xrightarrow{f} q) \textbf{ iff } \forall a, b \in \mathrm{Elem}_p \cdot a \multimap_p b \Rightarrow f(a) \multimap_q f(b)$$

If a function preserves the adjacency relation, then it is also injective. Hence the test for injectivity is omitted in the definition below.

sorts $\text{Occurrence} \subseteq \text{Proto-Arrow}$ —*the subsort of occurrences*
operations

occurrences : String, String $\rightarrow$ Set(Occurrence)

axioms

$$\text{occurrences}(p, q) = \{ p \xrightarrow{f} q \mid f\colon \mathrm{Elem}_p \rightarrow \mathrm{Elem}_q \wedge \text{label-preserving}(p \xrightarrow{f} q) \wedge \text{adjacency-preserving}(p \xrightarrow{f} q) \}$$

3 Derivation

We systematically derive an efficient parallel algorithm for the string matching problem. Since the matching problem consists of enumerating a set of occurrences, an immediately available realization is a search strategy in which all functions between two strings are enumerated and this set is filtered using the predicates characterizing occurrences. Closer examination of this strategy reveals that it is a combination of problem reduction and search: functions are built from sub-functions (functions on a subset of the domain) and indecomposable sub-functions are produced by searching in the codomain. A straightforward realization of this strategy yields the naive matching algorithm of looking for a matching at every position in the target. In a more sophisticated algorithm, partial occurrences already known are not recomputed.

Although any solution to a matching problem involves both problem reduction and search, we emphasize the former so as to arrive at a general description of an implementation strategy which permits a smooth generalization not only from sequential KMP to parallel KMP, but also from strings to other data structures.

3.1 Decomposing the Target

We will use a divide-and-conquer strategy [9] to compute occurrences. The string constructors "singleton" and "concatenate" provide a way to decompose strings into parts. The functions "first," "last," "prefix," and "suffix," used below have the obvious definition.

Notation. For any function f and for any subset x of its domain, we will use $f|_x$ to denote the restriction of f to x, and $f[x]$ to denote the image of x via f.

operations
  —*constructors for non-empty strings*
  "_:_" : Elem, Label → String
  _ _ : String, String → String
axioms
$p = \text{"}a{:}l\text{"} \Rightarrow \mathrm{Elem}_p = \{a\} \wedge \mathrm{label}_p(a) = l \wedge \neg(a \multimap_p a)$
$p = xy \Rightarrow$
  $\mathrm{Elem}_p = \mathrm{Elem}_x \cup \mathrm{Elem}_y \wedge$
  $\mathrm{label}_p = \mathrm{label}_x \cup \mathrm{label}_y \wedge$
  $a \multimap_p b$ **iff** $a \multimap_x b \vee a \multimap_y b \vee (a = \mathrm{last}(x) \wedge b = \mathrm{first}(y))$

The concatenation of strings can be extended to concatenation of two occurrences (provided they are adjacent).

operations
  —*concatenation of occurrences*
  _ _ : Occurrence, Occurrence → Occurrence —*partial*
axioms
let $u \xrightarrow{f} x, v \xrightarrow{g} y$ **in**
Defined(fg) **iff** $\mathrm{suffix}(f[u], x) \wedge \mathrm{prefix}(g[v], y)$
Defined$(fg) \Rightarrow uv \xrightarrow{fg} xy$
  $\wedge\, fg(a) = f(a)$ **if** $a \in \mathrm{Elem}_u$
  $\wedge\, fg(a) = g(a)$ **if** $a \in \mathrm{Elem}_v$

We now distribute the definition of occurrences over a decomposition of the target.

$$\mathrm{occurrences}(p, \text{"}a{:}l\text{"}) = \textbf{if } p = \text{"}b{:}l\text{"} \textbf{ then } \{\text{"}b{:}l\text{"} \xrightarrow{\{b \mapsto a\}} \text{"}a{:}l\text{"}\} \textbf{ else } \emptyset$$
$$\mathrm{occurrences}(p, xy) = \mathrm{occurrences}(p, x) \cup \mathrm{occurrences}(p, y) \cup \{p \xrightarrow{fg} xy \mid g \in \mathrm{occurrences}(u, x) \wedge h \in \mathrm{occurrences}(v, y) \wedge p = uv \wedge \mathrm{Defined}(fg)\}$$

3.2 The Necessity of Partial Occurrences

¿From the axioms above, we see that to determine the occurrences of p in xy, it is not sufficient to know just about occurrences of p in x and y. We also need to know occurrences of substrings of p. Thus, to distribute the computation of occurrences over a decomposition of the target, we have to first modify the computation so that it produces as results not only occurrences of the given string but also all of its substrings. We are thus forced to consider the *extension* of the occurrence relation, i.e., occurrences of all patterns in all targets.

3.3 Generalizing the Specification

A partial occurrence of a string p in a string q is an occurrence of a substring of p in q. We first modify the function "occurrences" so that it also returns all partial occurrences; the modified function will be called "p-occs."

Notation. The arrow $x \hookrightarrow p$ indicates that x is a (non-empty) substring of p.

operations
p-occs : String, String $\rightarrow$ Set(Occurrence)
axioms
$\text{p-occs}(p, q) = \bigcup_{x \hookrightarrow p} \text{occurrences}(x, q)$

Now that we have generalized the specification into a form which returns all partial occurrences, we can distribute it over a decomposition of the target; this time, the partial occurrences in the string xy are completely determined by those in the strings x and y.

$$\begin{aligned}
&\text{p-occs}(p, \text{“}a{:}l\text{”}) = \{\, \text{“}b{:}l\text{”} \xrightarrow{\{b \mapsto a\}} \text{“}a{:}l\text{”} \mid \text{“}b{:}l\text{”} \hookrightarrow p \,\} \\
&\text{p-occs}(p, xy) = \text{p-occs}(p, x) \cup \text{p-occs}(p, y) \cup \\
&\qquad \{\, uv \xrightarrow{fg} xy \mid u \xrightarrow{f} x \in \text{p-occs}(p, x) \wedge v \xrightarrow{g} y \in \text{p-occs}(p, y) \\
&\qquad\qquad \wedge\, uv \hookrightarrow p \wedge \text{Defined}(fg) \,\}
\end{aligned}$$

3.4 The Naive Algorithm

The axioms for partial occurrences given above can be directly implemented by search. The first set-former is a search for singleton substrings of the pattern which match a piece of the target. The second set-former is a search for compatible partial occurrences which can be concatenated to give larger partial occurrences. The result is a naive string matching algorithm which builds all partial occurrences; hence the algorithm can perform $O(|p|^2\,|q|)$ operations in the worst case, where $|s|$ denotes the length of the string s.

A search procedure is one of several choices for realizing a divide-and-conquer description of a computation: it corresponds to mapping the computation tree directly onto time.

4 Optimizations

We now consider several optimizations which reduce the complexity of the naive algorithm. Two simple optimizations which reduce the size of the set of partial occurrences at each stage are the following:

1. Discard all those partial occurrences which have no potential of being expanded.
2. As a special case of the optimization above, a full occurrence can be discarded after outputting it, since it cannot be further expanded.

4.1 Expandable Occurrences

We formalize below the notion of expandability of a partial occurrence. From now on, we will assume that the relations $u \hookrightarrow p$, prefix(u,p), and suffix(u,p), are all strict, i.e., $u \neq p$.

operations
exp : String, Occurrence $\rightarrow$ Bool
axioms
—*a partial occurrence is expandable*
—*if it can be embedded into a full occurrence*
—*as shown in the commuting diagram below*

$$\exp(p, u \xrightarrow{f} x) \text{ iff } \exists y, g, i, j \cdot \begin{array}{ccc} u & \overset{i}{\hookrightarrow} & p \\ f \downarrow & & \downarrow g \\ x & \overset{j}{\hookrightarrow} & y \end{array}$$

Notation. Expandability of a partial occurrence comprises four cases, and we will use the notation shown on the right below to indicate arrows which satisfy the predicates shown on the left. We assume that f is an occurrence of u in x.

$\text{suffix}(u,p) \wedge \text{prefix}(f[u],x)$	$\overset{f}{\lhd\!\!-\!\!\bullet}$
$\text{prefix}(u,p) \wedge \text{suffix}(f[u],x)$	$\overset{f}{\bullet\!\!-\!\!\rhd}$
$u \hookrightarrow p \wedge f[u] = x$	$\overset{f}{\lhd\!\!-\!\!\rhd}$
$u = p$	$\overset{f}{\bullet\!\!-\!\!\bullet}$

To incorporate expandability into the matching algorithm, we need distributive laws relating the expandability of partial occurrences to the string constructors.

Notation. In the sequel, we will frequently represent the target string as the concatenation xy. The inclusions $i_x\colon x \hookrightarrow xy$ and $i_y\colon y \hookrightarrow xy$ will be used without definition. Thus, if $f\colon u \rightarrow x$ is a partial occurrence with codomain x, then $i_x \circ f\colon u \rightarrow xy$ is a partial occurrence with codomain xy. The situation is similar for $g\colon v \rightarrow y$ and $i_y \circ g\colon v \rightarrow xy$.

—*distributive laws for expandability of partial occurrences*

$\exp(p, i_x \circ f) = \overset{f}{\lhd\!\!-\!\!\bullet}$
$\exp(p, i_y \circ g) = \overset{g}{\bullet\!\!-\!\!\rhd}$
$\text{Defined}(fg) \Rightarrow \exp(p, fg) = \overset{f}{\lhd\!\!-\!\!\rhd} \vee \overset{g}{\lhd\!\!-\!\!\rhd}$

4.2 Enumerating Expandable Occurrences

We now modify the divide-and-conquer specification of Sect. 3.3 to incorporate the optimization of retaining only expandable occurrences. Since full occurrences are not expandable, and they are what we are interested in, we modify the algorithm so that it returns a pair of sets: a set of expandable partial occurrences, and a set of full occurrences.

Notation. For any string q, the collection of expandable occurrences of a pattern p is denoted by e_q (the pattern is implicit), and the collection of full (or total) occurrences by t_q. The collection of partial occurrences, p-occs(p, q), will be abbreviated as o_q.

—*modified specification for matching*
operations
$\text{occs}' : \text{String}, \text{String} \rightarrow \text{Set}(\text{Occurrence}), \text{Set}(\text{Occurrence})$
axioms
let $\text{occs}'(p, q) = \langle e_q, t_q \rangle$ **in**
$e_q = \{\, f \mid f \in o_q \wedge \exp(p, f)\,\}$
$t_q = \{\, f \mid f \in o_q \wedge \bullet\!\overset{f}{\longrightarrow}\!\bullet \,\}$

The base case of singleton target strings is straightforward, and we will not consider it further. The optimization of retaining only expandable occurrences can be systematically propagated into the composition operator as shown in the following diagram:

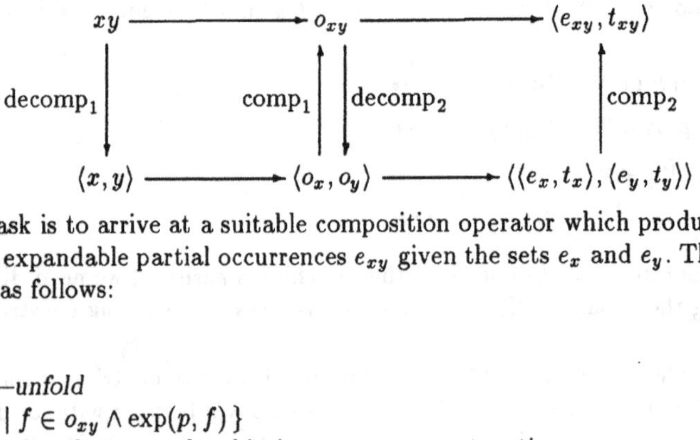

Our task is to arrive at a suitable composition operator which produces the collection of expandable partial occurrences e_{xy} given the sets e_x and e_y. The inference proceeds as follows:

e_{xy}
—*unfold*
$= \{\, f \mid f \in o_{xy} \wedge \exp(p, f)\,\}$
—*distribute membership in* o_{xy} *over concatenation*
$= \{\, i_x \circ f \mid f \in o_x \wedge \exp(p, i_x \circ f)\,\}$
$\cup \{\, i_y \circ g \mid g \in o_y \wedge \exp(p, i_y \circ g)\,\}$
$\cup \{\, fg \mid f \in o_x \wedge g \in o_y \wedge \text{Defined}(fg) \wedge \exp(p, fg)\,\}$
—*apply distributive laws for expandability and fold*
$= \{\, i_x \circ f \mid f \in e_x \;\wedge \triangleleft\!\overset{f}{\longrightarrow}\!\bullet \,\}$
$\cup \{\, i_y \circ g \mid g \in e_y \;\wedge \bullet\!\overset{g}{\longrightarrow}\!\triangleright \,\}$
$\cup \{\, fg \mid f \in e_x \wedge g \in e_y \wedge \text{Defined}(fg) \wedge (\triangleleft\!\overset{f}{\longrightarrow}\!\triangleright \vee \triangleleft\!\overset{g}{\longrightarrow}\!\triangleright)\,\}$

The expression for t_{xy} can be similarly obtained:

$$
\begin{aligned}
&t_{xy}\\
&\qquad \textit{—unfold}\\
&= \{\, f \mid f \in o_{xy} \wedge \overset{p}{\bullet\!\!\longrightarrow\!\!\bullet} \,\}\\
&\qquad \textit{—distribute membership in } o_{xy} \textit{ over concatenation}\\
&= \{\, i_x \circ f \mid f \in o_x \wedge \overset{i_x \circ f}{\bullet\!\!\longrightarrow\!\!\bullet} \,\}\\
&\cup \{\, i_y \circ g \mid g \in o_y \wedge \overset{i_y \circ g}{\bullet\!\!\longrightarrow\!\!\bullet} \,\}\\
&\cup \{\, fg \mid f \in o_x \wedge g \in o_y \wedge \mathrm{Defined}(fg) \wedge \overset{fg}{\bullet\!\!\longrightarrow\!\!\bullet} \,\}\\
&\qquad \textit{—apply distributive laws for } \bullet\!\!\longrightarrow\!\!\bullet \textit{ and fold}\\
&= t_x \cup t_y \cup \{\, fg \mid f \in e_x \wedge g \in e_y \wedge \mathrm{Defined}(fg) \wedge \overset{fg}{\bullet\!\!\longrightarrow\!\!\bullet} \,\}
\end{aligned}
$$

The relationships of the partial occurrences saved is illustrated in Fig. 1.

pattern *a b c a b c a c a b*
target *b a b c b a b c a b c a a b c a b c a b c a c a b c*

		a	subsumed occurrence
		a b c a	subsumed occurrence
partial occurrences	*c a b*	*a b c a b c a*	unsubsumed occurrences
	| | |	| | | | | | |	
piece of target	*c a b*	*c a a b c a b c a*	

Fig. 1. Saving of only potentially expandable partial occurrences

Unfortunately, the optimizations above only improve the worst case complexity of the algorithm to $O(|p|\,|q|)$. The sequential KMP algorithm achieves its linear complexity by using the notion of a failure function, which we explain below.

5 The Subsumption Relation

In an effort to reduce the number of partial occurrences saved, we investigate the relationships between partial occurrences of a pattern in a given target string. We note that given an occurrence $p \rightarrow q$ and any other occurrence $x \rightarrow p$, we get an occurrence of x via the composition $x \rightarrow p \rightarrow q$. The occurrence $x \rightarrow q$ is said to be *subsumed* by the occurrence $p \rightarrow q$. The collection of partial occurrences of a pattern p in a target q forms a preorder under the subsumption relation; if we take the quotient under the equivalence relation of mutual subsumption, we obtain a poset.

operations

$\_ \geq \_ :$ Occurrence, Occurrence $\rightarrow$ Bool
$\_ \cong \_ :$ Occurrence, Occurrence $\rightarrow$ Bool
$\_ > \_ :$ Occurrence, Occurrence $\rightarrow$ Bool
$\_ \succ \_ :$ Occurrence, Occurrence $\rightarrow$ Bool

axioms

—*definition of subsumption*

$y \xrightarrow{g} z \geq x \xrightarrow{f} z$ **iff** $\exists x \xrightarrow{i} y \cdot f = g \circ i$

—*derived consequence:* $\geq$ *is a pre-order*

$f \geq f$

$h \geq g \wedge g \geq f \Rightarrow h \geq f$

—*equivalence induced by subsumption*

$f \cong g$ **iff** $f \geq g \wedge g \geq f$

—*strict subsumption*

$g > f$ **iff** $g \geq f \wedge g \not\cong f$

—*immediate subsumption*

$g \succ f$ **iff** $g > f \wedge \nexists h \cdot g > h > f$

We can use the subsumption relation to reduce the number of partial occurrences saved at each stage (this number is what contributes to the non-linear complexity of the algorithm in Sect. 3.4). Since any subsumed occurrence can be generated via composition with a substring of the pattern, we can adopt the lazy approach of saving only unsubsumed occurrences (generating subsumed ones later if the need arises).

5.1 Equivalence Classes of Partial Occurrences

A set of partial occurrences which mutually subsume one another forms an equivalence class under the relation of mutual subsumption. Such equivalence classes are crucial for reducing the complexity of the naive algorithm. Together with the optimization of discarding all non-expandable partial occurrences, this leaves at most two partial occurrences at each stage: the largest partial occurrences touching either end of the string (see Figs. 1 and 2).

pattern $a_1\, b_2\, c_3\, a_4\, b_5\, c_6\, a_7\, c_8\, a_9\, b_0$
target $b\,a\,b\,c\,b\,a\,b\,c\,a\,b\,c\,a\,a\,b\,c\,a\,b\,c\,a\,b\,c\,a\,c\,a\,b\,c$

a partial occurrence	$a_5\, b_5\, c_6\, a_7$
a partial occurrence	$a_1\, b_2\, c_3\, a_4$
	| | | |
piece of target	$a\;\; b\;\; c\;\; a$

Fig. 2. Equivalence classes of partial occurrences. (Characters in the pattern are subscripted for disambiguation.)

¿From now on, we will consider equivalence classes of partial occurrences. However, to avoid clutter, rather than denoting an equivalence class by $[f]$, we will use the representative f. To justify this usage, we have to ensure that the operations on equivalence classes do no depend upon the representative chosen. The only non-trivial operation is concatenation of occurrences, which we elaborate below:

—equivalence classes of occurrences can be concatenated
—if at least one pair can be concatenated
$\text{Defined}([f][g]) \text{ iff } \exists f' \in [f], g' \in [g] \cdot \text{Defined}(f'g')$
—concatenating equivalence classes eliminates incompatible pairs
—concatenation preserves subsumption and, hence, equivalence
$\text{Defined}([f][g]) \Rightarrow [f][g] = \{\, f'g' \mid f' \in [f] \wedge g' \in [g] \wedge \text{Defined}(f'g') \,\}$

5.2 Efficient Representation of Sets of Partial Occurrences

We formalize the representation of sets of partial occurrences by maximal elements and record the fact that there are at most two maximal elements.

—represent a set of partial occurrences by its maximal elements
$\max(F) = \{\, f \mid f \in F \wedge \nexists f' \in F \cdot f' > f \,\}$

—derived consequence; see Figs. 1 and 2
$\max(F) = \emptyset$
$\vee \max(F) = \{h\} \wedge \overset{h}{\lhd\!\!-\!\!\rhd}$
$\vee \max(F) = \{f\} \wedge \overset{f}{\lhd\!\!-\!\!\bullet}$
$\vee \max(F) = \{g\} \wedge \overset{g}{\bullet\!\!-\!\!\rhd}$
$\vee \max(F) = \{f, g\} \wedge \overset{f}{\lhd\!\!-\!\!\bullet} \wedge \overset{g}{\bullet\!\!-\!\!\rhd}$

We now have to modify the composition operator of Sect. 4.2 to account for this representation of sets of partial occurrences. We again systematically propagate the representation into the composition operator as shown in the following diagram (m_q denotes the maximal elements in the poset of partial occurrences of pattern p in string q):

$$\begin{array}{ccccc}
o_{xy} & \longrightarrow & \langle e_{xy}, t_{xy}\rangle & \longrightarrow & \langle m_{xy}, t_{xy}\rangle \\
\text{decomp}_2 \downarrow & & \text{comp}_2 \uparrow \ \downarrow \text{decomp}_3 & & \uparrow \text{comp}_3 \\
\langle o_x, o_y\rangle & \longrightarrow & \langle\langle e_x, t_x\rangle, \langle e_y, t_y\rangle\rangle & \longrightarrow & \langle\langle m_x, t_x\rangle, \langle m_y, t_y\rangle\rangle
\end{array}$$

We first need some auxiliary axioms stating the monotonicity (with respect to subsumption) of the assembly of occurrences.

—monotonicity of subsumption
$\forall f, g \in o_x, h, k \in o_y \cdot$
$i_x \circ g > i_x \circ f \text{ iff } g > f$
$i_y \circ k > i_y \circ h \text{ iff } k > h$
$gk > i_x \circ f \text{ iff } g \geq f$
$gk > i_y \circ h \text{ iff } k \geq h$
$gk > fh \text{ iff } (g \geq f \wedge k > h) \vee (g > f \wedge k \geq h)$

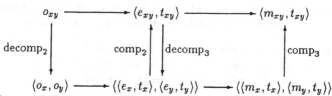

To arrive at a suitable composition operator which produces the representation m_{xy} given the representations m_x and m_y, we rewrite the definition of m_{xy} as follows:

$$
\begin{aligned}
&\quad\textit{—unfold; by definition of maximal elements}\\
m_{xy} &= \{\, f \mid f \in e_{xy} \wedge \nexists f' \in e_{xy} \cdot f' > f \,\}\\
&\quad\textit{—distribute membership in } e_{xy} \textit{ over concatenation}\\
&= \{\, i_x \circ f \mid f \in e_x \wedge \overset{f}{\triangleleft\!\!-\!\!\bullet} \wedge \nexists f' \in e_{xy} \cdot f' > i_x \circ f \,\}\\
&\cup \{\, i_y \circ g \mid g \in e_y \wedge \overset{g}{\bullet\!\!-\!\!\triangleright} \wedge \nexists g' \in e_{xy} \cdot g' > i_y \circ g \,\}\\
&\cup \{\, fg \mid f \in e_x \wedge g \in e_y \wedge \mathrm{Defined}(fg) \wedge (\overset{f}{\triangleleft\!\!-\!\!\triangleright} \vee \overset{g}{\triangleleft\!\!-\!\!\triangleright}) \wedge \nexists h \in e_{xy} \cdot h > fg \,\}
\end{aligned}
$$

We simplify the first component of the union above (the other components can be simplified similarly):

$$
\begin{aligned}
&\{\, i_x \circ f \mid f \in e_x \wedge \overset{f}{\triangleleft\!\!-\!\!\bullet} \wedge \nexists f' \in e_{xy} \cdot f' > i_x \circ f \,\}\\
&\quad\textit{—distribute membership in } e_{xy} \textit{ over concatenation}\\
&= \{\, i_x \circ f \mid f \in e_x \wedge \overset{f}{\triangleleft\!\!-\!\!\bullet}\\
&\qquad\qquad \wedge \nexists f' \in e_x \cdot (i_x \circ f' > i_x \circ f \wedge \overset{f'}{\triangleleft\!\!-\!\!\bullet})\\
&\qquad\qquad \wedge \nexists g' \in e_y \cdot (i_y \circ g' > i_x \circ f \wedge \overset{g'}{\bullet\!\!-\!\!\triangleright}) \qquad \textit{—simplifies to true}\\
&\qquad\qquad \wedge \nexists f' \in e_x, g' \in e_y \cdot (\mathrm{Defined}(f'g') \wedge (\overset{f'}{\triangleleft\!\!-\!\!\triangleright} \vee \overset{g'}{\triangleleft\!\!-\!\!\triangleright}) \wedge f'g' > i_x \circ f) \,\}\\
&\quad\textit{—distribute > over concatenation and simplify}\\
&= \{\, i_x \circ f \mid f \in e_x \wedge \overset{f}{\triangleleft\!\!-\!\!\bullet}\\
&\qquad\qquad \wedge \nexists f' \in e_x \cdot (f' > f \wedge \overset{f'}{\triangleleft\!\!-\!\!\bullet})\\
&\qquad\qquad \wedge \nexists f' \in e_x, g' \in e_y \cdot (\mathrm{Defined}(f'g') \wedge \overset{f'}{\triangleleft\!\!-\!\!\triangleright} \wedge f' \geq f) \,\}
\end{aligned}
$$

The set-former above enumerates elements of e_x; thus, we cannot derive a suitable composition operator which constructs m_{xy} solely from m_x and m_y. This shows that the representation by maximal elements is too weak: it loses essential information about subsumed occurrences. To cure this problem, we change the representation to include a generator which produces subsumed occurrences on demand:

$$
\begin{aligned}
&\mathrm{rep}(F) = \langle \max(F), \mathrm{gen}\colon \mathrm{Occurrence} \rightarrow \mathrm{Set}(\mathrm{Occurrence}) \rangle\\
&\quad\textit{—the generator } \mathrm{gen} \textit{ may depend upon the pattern } p\\
&\textbf{iff } F = \mathrm{gen}{*}(\max(F)) \textbf{ where } \mathrm{gen}{*}(S) = \textstyle\bigcup_{s \in S} \mathrm{gen}(s)
\end{aligned}
$$

A naive realization of the generator above would yield e_x back, thereby not providing any gain in complexity over enumerating e_x directly. To determine properties of this generator, we consider the set-former above and cast it as a search problem. Our goal is exploit the order on partial occurrences.

5.3 Searching in an Ordered Domain

We now describe an algorithm theory for searching for the largest elements in a poset which satisfy a given predicate. This theory is a formalization (and generalization to an arbitrary poset) of Dijkstra's heuristic principle for searching in an array [3, Linear Search Theorem].

The problem of finding can be formalized as follows:

spec Ordered-Search-Spec =
sorts D, R —*domain and range*
operations
$\_ \geq \_ : \mathrm{D}, \mathrm{D} \rightarrow \mathrm{Bool}$ —*partial order on domain*
$P : \mathrm{D} \rightarrow \mathrm{Bool}$ —*predicate for feasible solutions*
$T : \mathrm{D} \rightarrow \mathrm{R}$ —*transformation function on feasible solutions*
$O : \mathrm{D}, \mathrm{R} \rightarrow \mathrm{Bool}$ —*input-output relation*
axioms
—"$\geq$" *is a partial order; axioms omitted*
—*input-output condition: largest elements satisfying P*
—*feasible solutions are transformed by the function T before output*
$O(x, z)$ **iff** $z = T(x) \wedge P(x) \wedge \nexists y \in \mathrm{D} \cdot (P(y) \wedge y > x)$
end

Such a problem can be solved by search. The search theory below (based on [10]) enumerates the elements of the domain in decreasing order, starting from the maximal elements. This order of enumeration maps the partial order onto time: it guarantees that the *first* feasible solution is the *largest* feasible solution [3].

spec Ordered-Search-Imp =
extend Ordered-Search-Spec
sorts D, R
—*the sort of states; each element of the domain is a possible state*
sorts Sd = D
operations
Initial : $\rightarrow$ Sd —*predicate characterizing initial states*
Split : Sd, Sd $\rightarrow$ Bool —*the next state generator*
—Split(s, t) = *t is one possible next state for s*
Extract : R, Sd $\rightarrow$ Bool —*feasible solution extractor*
—Extract(z, s) = *state s yields solution z*
axioms
—*initial states; start with the maximal elements*
Initial(m) **iff** $m \in \mathrm{D} \wedge \nexists x \in \mathrm{D} \cdot x > m$
—*a next state is an immediately subsumed element*
Split(s, t) **iff** $s \succ t \wedge \neg P(s)$
—*feasible solutions*
Extract(z, s) **iff** $P(s) \wedge z = T(s)$
—*the search implements the required input-output relation*
—Split^k *is the k-fold composition of* Split
$O(x, z)$ **iff** $\exists k, m, s \cdot (\mathrm{Initial}(m) \wedge \mathrm{Split}^k(m, s) \wedge \mathrm{Extract}(z, s))$
end

Observe that the only control decision we have made is that $x > y$ implies x is visited before y. The importance of extensional descriptions of computation and minimal commitment of control cannot be overemphasized in algorithm design.

6 The Generalized Knuth-Morris-Pratt Algorithm

The control strategy of searching downwards from the maximal elements can be applied to our set-former,

$$
\begin{aligned}
\{\, i_x \circ f \mid {} & f \in e_x \wedge \overset{f}{\lhd\!\!-\!\!\bullet} \\
& \wedge \nexists f' \in e_x \cdot (f' > f \wedge \overset{f'}{\lhd\!\!-\!\!\bullet}) \\
& \wedge \nexists f' \in e_x, g' \in e_y \cdot (\mathrm{Defined}(f'g') \wedge \overset{f'}{\lhd\!\!-\!\!\rhd} \wedge f' \geq f) \,\},
\end{aligned}
$$

using the following view:

$$
\begin{aligned}
& \mathrm{D} \mapsto e_x \\
& \mathrm{R} \mapsto \mathrm{Occurrence} \\
& > \mapsto > \\
& P \mapsto \lambda f \cdot (\overset{f}{\lhd\!\!-\!\!\bullet} \wedge \nexists f' \in e_x, g' \in e_y \cdot (\mathrm{Defined}(f'g') \wedge \overset{f'}{\lhd\!\!-\!\!\rhd} \wedge f' \geq f)) \\
& T \mapsto \lambda f \cdot i_x \circ f
\end{aligned}
$$

We start the search with m_x because it is the set of maximal elements of e_x. Since m_x can contain at most two elements (see Sect. 5.2), the search stops after one ply, and we get the following set of feasible solutions:

$$
\begin{aligned}
\{\, i_x \circ f \mid {} & f \in m_x \wedge \overset{f}{\lhd\!\!-\!\!\bullet} \\
& \vee f' \in m_x \wedge \overset{f'}{\lhd\!\!-\!\!\rhd} \wedge f' \succ f \wedge \overset{f}{\lhd\!\!-\!\!\bullet} \wedge \nexists g' \in e_y \cdot \mathrm{Defined}(f'g') \,\}
\end{aligned}
$$

The expression above requires a search in e_y, which can be implemented by a similar control strategy. However, we first combine the three components of the union defining m_{xy} so as to eliminate redundant searches:

$$
\begin{aligned}
& m_{xy} \\
& = \{\, i_x \circ f \mid f \in m_x \wedge \overset{f}{\lhd\!\!-\!\!\bullet} \\
& \qquad\quad \vee f' \in m_x \wedge \overset{f'}{\lhd\!\!-\!\!\rhd} \wedge f' \succ f \wedge \overset{f}{\lhd\!\!-\!\!\bullet} \wedge \nexists g' \in e_y \cdot \mathrm{Defined}(f'g') \,\} \\
& \cup \{\, i_y \circ g \mid g \in m_y \wedge \overset{g}{\bullet\!\!-\!\!\rhd} \\
& \qquad\quad \vee g' \in m_y \wedge \overset{g'}{\lhd\!\!-\!\!\rhd} \wedge g' \succ g \wedge \overset{g}{\bullet\!\!-\!\!\rhd} \wedge \nexists f' \in e_x \cdot \mathrm{Defined}(f'g') \,\} \\
& \cup \{\, fg \mid f \in m_x \wedge g \in m_y \wedge \overset{f}{\lhd\!\!-\!\!\rhd} \wedge \overset{g}{\lhd\!\!-\!\!\rhd} \wedge \mathrm{Defined}(fg) \\
& \qquad\quad \vee f \in m_x \wedge g \in e_y \wedge \overset{f}{\lhd\!\!-\!\!\rhd} \wedge \overset{g}{\lhd\!\!-\!\!\bullet} \wedge \mathrm{Defined}(fg) \wedge \\
& \qquad\qquad\qquad \nexists g' \in e_y \cdot (\mathrm{Defined}(fg') \wedge g' > g) \\
& \qquad\quad \vee f \in e_x \wedge g \in m_y \wedge \overset{f}{\bullet\!\!-\!\!\rhd} \wedge \overset{g}{\lhd\!\!-\!\!\rhd} \wedge \mathrm{Defined}(fg) \wedge \\
& \qquad\qquad\qquad \nexists f' \in e_x \cdot (\mathrm{Defined}(f'g) \wedge f' > f) \,\}
\end{aligned}
$$

The last component of the union contains two searches; when one of these searches fails to return a feasible solution, the membership condition of one of the other two sets is satisfied. Thus the expression for m_{xy} can be rewritten in the following simplified form:

case

$f \in m_x \wedge \overset{f}{\triangleleft\!\!-\!\!\bullet} \Rightarrow i_x \circ f \in m_{xy}$

$g \in m_y \wedge \overset{g}{\bullet\!\!-\!\!\triangleright} \Rightarrow i_y \circ g \in m_{xy}$

$f \in m_x \wedge g \in m_y \wedge \overset{f}{\triangleleft\!\!-\!\!\triangleright} \wedge \overset{g}{\triangleleft\!\!-\!\!\triangleright} \wedge \text{Defined}(fg) \Rightarrow fg \in m_{xy}$

$f' \in m_x \wedge g' \in m_y \wedge \overset{f'}{\triangleleft\!\!-\!\!\triangleright} \wedge \overset{g'}{\triangleleft\!\!-\!\!\bullet} \Rightarrow$

  search $\{ g \mid g \in e_y \wedge \overset{g}{\triangleleft\!\!-\!\!\bullet} \}$ for largest g such that Defined$(f'g)$
  if search succeeds **then** $f'g \in m_{xy}$
  else $f' \succ f \wedge \overset{f}{\triangleleft\!\!-\!\!\bullet} \wedge f \in m_{xy}$

$f' \in m_x \wedge g' \in m_y \wedge \overset{f'}{\bullet\!\!-\!\!\triangleright} \wedge \overset{g'}{\triangleleft\!\!-\!\!\triangleright} \Rightarrow$

  search $\{ f \mid f \in e_x \wedge \overset{f}{\bullet\!\!-\!\!\triangleright} \}$ for largest f such that Defined(fg')
  if search succeeds **then** $fg' \in m_{xy}$
  else $g' \succ g \wedge \overset{g}{\bullet\!\!-\!\!\triangleright} \wedge g \in m_{xy}$

end

The expression above captures the essence of the Knuth-Morris-Pratt algorithm: save only the largest partial occurrences, and when they cannot be expanded, fail back to immediately subsumed occurrences and try again. Generating subsumed occurrences corresponds to sliding the pattern on a mismatch. Equivalently, and more profitably, generating subsumed occurrences can be seen as searching in a space of partial occurrences. The latter view is what enables the generalization of KMP beyond the sequential string case; it shows that there is nothing inherently sequential about the failure function.

We have been neglecting the generation of full occurrences until now. In Sect. 4.2 we obtained the following expression for the collection of full occurrences t_{xy}:

$$t_{xy} = t_x \cup t_y \cup \{ fg \mid f \in e_x \wedge g \in e_y \wedge \text{Defined}(fg) \wedge \overset{fg}{\bullet\!\!-\!\!\bullet} \}$$

In terms of the search strategy above, the third component of the union can be implemented as:

case

$f' \in m_x \wedge g' \in m_y \wedge \overset{f'}{\bullet\!\!-\!\!\triangleright} \wedge \overset{g'}{\triangleleft\!\!-\!\!\bullet} \Rightarrow$

  search $\{ g \mid g \in e_y \wedge \overset{g}{\triangleleft\!\!-\!\!\bullet} \}$ for largest g such that Defined$(f'g) \wedge \overset{f'g}{\bullet\!\!-\!\!\bullet}$
  if search succeeds **then** $f'g \in t_{xy}$

  search $\{ f \mid f \in e_x \wedge \overset{f}{\bullet\!\!-\!\!\triangleright} \}$ for largest f such that Defined$(fg') \wedge \overset{fg'}{\bullet\!\!-\!\!\bullet}$
  if search succeeds **then** $fg' \in t_{xy}$

end

6.1 Parallel Traversal: Binary Decomposition of the Target

To obtain a parallel KMP algorithm, we map the decomposition tree generated by the inductive description above onto a tree of processors. To this end, we treat a parallel tree machine as a data type: that of a labeled tree, with the labels denoting the computation that is associated with each node in the tree. A theory interpretation (in the sense of first-order logic; see, for example [13]) from the divide-and-conquer specification of the algorithm to the theory of trees yields the assignment of computations to processors.

The processing of the target by multiple processors may be viewed as a kind of traversal: information about the adjacency relation is obtained incrementally. Each processor in the decomposition tree obtains adjacency information about the two strings handled by the two processors which form its children in the decomposition tree. Although there are several loci of control, the key aspect here is communication: the adjacency relation on the image of each occurrence has to ultimately reach one processor.

The parallel algorithm amounts to sweeping the decomposition tree bottom-up, accumulating partial occurrences for larger and larger pieces of the target (a trace of the algorithm is shown in Fig. 3). The leaves in the processor tree perform the primitive computation—the base case of the divide-and-conquer. Interior nodes compose results produced by their children and pass the result to their parents—this is the composition operator described in the previous section. The algorithm is a combination of problem reduction and search: the processor tree realizes the problem reduction, and each processor performs a local search for partial occurrences which can be concatenated.

Complexity. To obtain the optimum performance on a tree of processors, the target has be decomposed into two roughly equal parts. We can ensure this by assigning elements of the target string to the bottom level of a complete binary tree of processors. With such an assignment, our algorithm terminates after $\log |p|$ stages and performs $O(2\,|t|)$ operations.

7 The Failure Function

We recall that we have chosen to represent a set of partial occurrences by the maximal elements together with a generator. The algorithm in the last section has shown that it is sufficient if this generator produces immediately subsumed occurrences for any given occurrence. We now consider the problem of implementing such a generator. We first start with a specification:

$$\begin{aligned}&\mathrm{rep}(F) = \langle \max(F), \mathrm{gen}\colon \mathrm{Occurrence} \rightarrow \mathrm{Set}(\mathrm{Occurrence})\rangle\\ &\textbf{where } \mathrm{gen}(f) \stackrel{\mathrm{def}}{=} \{\, f' \mid f \succ f' \,\}\end{aligned}$$

We can continue to use a divide-and-conquer strategy (just as for m_{xy} and t_{xy}) and define the generator g_{xy} for the string xy in terms of the generators g_x and g_y. However, this does not result in efficiency gains (which was the purpose of representing sets of partial occurrences by maximal elements). To improve the efficiency, we have to eliminate redundant computations. To this end, we examine the definition of subsumption:

$$y \xrightarrow{g} z \geq x \xrightarrow{f} z \textbf{ iff } \exists x \xrightarrow{i} y \cdot f = g \circ i.$$

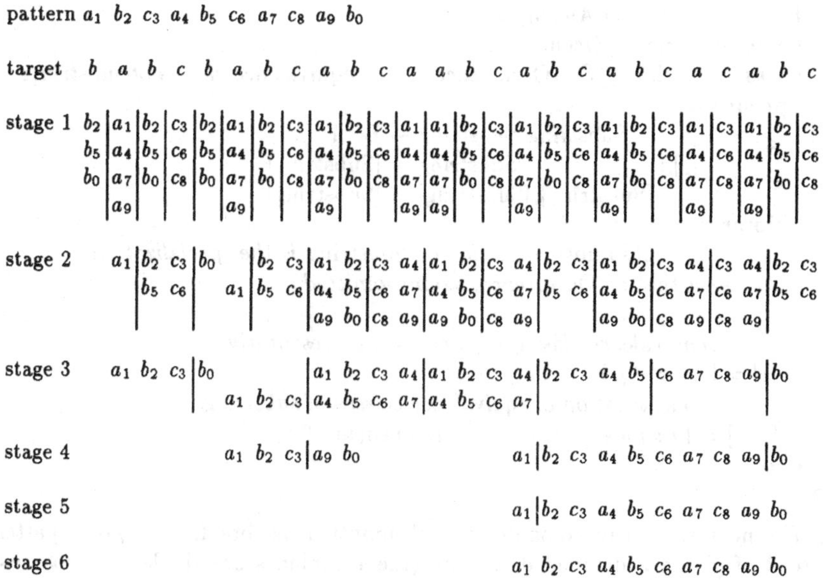
pattern a1 b2 c3 a4 b5 c6 a7 c8 a9 b0
target b a b c b a b c a b c a a b c a b c a b c a c a b c
stage 1
b2 a1 b2 c3 b2 a1 b2 c3 a1 b2 c3 a1 a1 b2 c3 a1 b2 c3 a1 b2 c3 a1 c3 a1 b2 c3
b5 a4 b5 c6 b5 a4 b5 c6 a4 b5 c6 a4 a4 b5 c6 a4 b5 c6 a4 b5 c6 a4 c6 a4 b5 c6
b0 a7 b0 c8 b0 a7 b0 c8 a7 b0 c8 a7 a7 b0 c8 a7 b0 c8 a7 b0 c8 a7 c8 a7 b0 c8
a9 a9 a9 a9 a9 a9 a9 a9 a9
stage 2
a1 b2 c3 b0 b2 c3 a1 b2 c3 a4 a1 b2 c3 a4 b2 c3 a1 b2 c3 a4 c3 a4 b2 c3
b5 c6 a1 b5 c6 a4 b5 c6 a7 a4 b5 c6 a7 b5 c6 a4 b5 c6 a7 c6 a7 b5 c6
a9 b0 c8 a9 a9 b0 c8 a9 a9 b0 c8 a9 c8 a9
stage 3
a1 b2 c3 b0 a1 b2 c3 a4 a1 b2 c3 a4 b2 c3 a4 b5 c6 a7 c8 a9 b0
a1 b2 c3 a4 b5 c6 a7 a4 b5 c6 a7
stage 4
a1 b2 c3 a9 b0 a1 b2 c3 a4 b5 c6 a7 c8 a9 b0
stage 5
a1 b2 c3 a4 b5 c6 a7 c8 a9 b0
stage 6
a1 b2 c3 a4 b5 c6 a7 c8 a9 b0

Fig. 3. Trace of parallel KMP. Characters in the pattern are subscripted for disambiguation. Elements of equivalence classes of partial occurrences are explicitly shown

Since we are only interested in collections of partial occurrences of a given pattern, the domain of an occurrence will always be a substring of the pattern. Thus, the subsumption relation between partial occurrences of a pattern is completely determined by the partial order on pieces of the pattern, as shown in the diagram below.

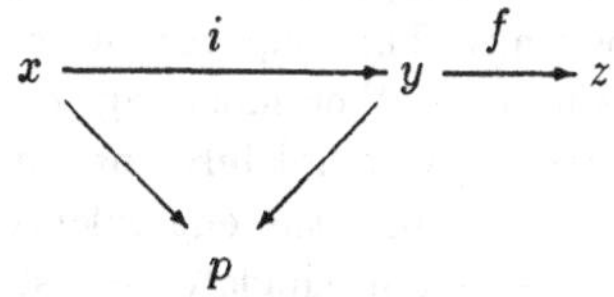

We now associate a subsumption algebra with each string; the algebra is the collection of substrings together with the subsumption relation among them.

Notation. Categorically, a substring is an inclusion arrow $x \hookrightarrow p$, and therefore, the subsumption relation is well-defined on substrings. Rather than be pedantic, we will write the substring $x \hookrightarrow p$ as just x.

type Subsumption-Algebra =
sorts Substring $\subseteq$ Occurrence
sorts [Substring] $\subseteq$ Set(Occurrence) —*equivalence classes of substrings*
operations

$$\begin{aligned} \geq, >, \cong, \succ : &\quad \text{Substring}, \text{Substring} \to \text{Bool} \\ [\_] : &\quad \text{Substring} \to [\text{Substring}] \\ \_\,\_ : &\quad [\text{Substring}], [\text{Substring}] \to [\text{Substring}] \end{aligned}$$

axioms
—*the subsumption relation on* Substring *is the specialization*
—*of that on* Occurrence; *axioms omitted*

—*equivalence class generated by a representative*

$$[x] = \{\, y \mid x \cong y \,\}$$

—*concatenation of equivalence classes of substrings*

$$[x][y] = \{\, uv \mid u \in [x] \wedge v \in [y] \wedge \text{Defined}(uv) \,\}$$

end

The next step is to compute the subsumption algebra for the given pattern. Instead of giving a detailed derivation (the techniques are similar to those used in the derivation of the main part of the algorithm), we briefly indicate the steps. The subsumption algebra for the pattern can be computed by a divide-and-conquer strategy. For example, here is the distributive law for the set of substrings:

$$\begin{aligned} \text{Substring}_{\text{“}a:l\text{”}} &= \{\text{“}a{:}l\text{”}\} \\ \text{Substring}_{xy} &= \text{Substring}_x \cup \text{Substring}_y \cup \\ &\qquad \{\, uv \mid u \in \text{Substring}_x \wedge v \in \text{Substring}_y \wedge \text{Defined}(uv) \,\} \end{aligned}$$

The subsumption relation can be distributed over concatenation as in Sect. 5.2. Computing the immediate subsumption relation requires a search (from the maximum downwards) similar to that in the main part of the KMP algorithm (Sect. 6).

The most complex part of computing the subsumption algebra is the equivalence relation among substrings. This requires the computation of occurrences of all substrings of the pattern in all other substrings of the pattern: a multi-pattern, multi-target matching problem. By completely unrolling the divide-and-conquer description of the subsumption algebra, the equivalence relation can be determined from the equivalence relation among singleton substrings of the pattern. Such a computation is similar in spirit to dynamic programming. The inductive decomposition generated by the divide-and-conquer can be mapped onto a tree of processors (see Sect. 6.1), thus enabling the subsumption algebra to be computed in parallel.

Complexity. For a fixed alphabet, the equivalence relation between singleton substrings can be computed using $O(|p|)$ operations. This relation can be propagated to the rest of the substrings using $O(|p|^2)$ operations; or, if we observe that we only need the subsumption relation for substrings whose length is a power of 2 (see Sect. 6.1), the number of operations reduces to $O(|p| \log |p|)$.

7.1 Precomputation

The carrier of the subsumption algebra for a finite pattern is a finite set. Hence we have the choice of precomputing all the functions in the algebra and recording their extensions in tables. Such a choice is consistent with our overall goal of reducing complexity.

Recapitulating the development so far, we started with an extensional description of the occurrence relation. In an effort to reduce the size of sets of partial occurrences, we first eliminated un-expandable occurrences, and then defined a subsumption relation on occurrences. The subsumption relation enabled us to represent a set of partial occurrences by its maximal elements, together with a generator for the rest. Incorporating this representation in an algorithm for enumerating occurrences determined the ***pattern of invocation*** of this generator, namely, "generate immediately subsumed occurrences." The net effect is that we have mapped an extension—a carrier domain—into an intension—a carrier domain together with operations, i.e., an algebraic description of a data structure.

8 Concluding Remarks

We have adopted a problem reduction approach to pattern matching: occurrences are obtained by gluing together smaller occurrences. Once the essential computation (data flow) has been laid out, the derivation proceeds by eliminating redundant computations and selectively introducing control strategies. The common belief that the failure function of KMP is inherently sequential arises from a lack of separation between data flow and control. We have shown that this belief is false by separating the control—sequential traversal—from the computation encoded in the failure function—search in an ordered domain.

We have also separated the ***structure*** of the computation from its ***realization.*** This separation allows both the introduction of additional structure in the problem description (e.g., the subsumption relation between occurrences) and a rational description of realization decisions. The essential structure of pattern matching is captured in the decomposition tree (or graph) generated by problem reduction. This structure can be realized in several ways:

- in time (the naive algorithm),
- on a processor array (the parallel algorithm), and
- as a data structure (the failure function).

These realizations are not exclusive; we have combined the latter two. The choice between these different realizations is determined by complexity considerations. In the absence of complexity measures, we can only chart the design space, with no basis for choice. In our derivation, we have informally used complexity to motivate the decisions. Much work needs to be done on incorporating such information into derivations formally.

8.1 Rabbitcount → 0

Two steps in our derivation which seem to unmotivated are: (1) the introduction of the subsumption relation between occurrences, and (2) the use of downward search in an ordered domain.

The subsumption relation can be justified by the goal of trying to reduce the size of the set of partial occurrences which is maintained by the algorithm. Two standard ways to reduce the size of a set are to eliminate useless members (see the next section) and to represent the set by a generator. Given the latter goal, the subsumption relation between occurrences arises by lifting the partial order on strings given by the inclusion relation. The latter is already present in the problem: it is the well-founded order required for a divide-and-conquer strategy to terminate.

The technique of searching in an ordered domain is outside the pattern matching domain. However, it is a sufficiently abstract and sufficiently general strategy which is useful in several domains. Our use of an algorithm theory describing this kind of search is consistent with the trend of encoding general implementation strategies (e.g., divide-and-conquer, search, finite differencing) as formal theories which can be instantiated using the specific features of a problem.

8.2 Optimality

Our parallel KMP algorithm is not optimal, in the sense that the product of the number of processors and the execution time is not linear, whereas the sequential KMP algorithm is linear. The parallel algorithms described in [4, 15] are optimal. They achieve optimality by a careful assignment of processors to computations, and by eliminating positions in the target where the pattern cannot occur. Whereas the KMP algorithm works by searching for *positive* occurrences of the pattern, these algorithms work by eliminating *negative* occurrences of the pattern. Thus, they are akin to the Boyer-Moore algorithm for sequential string matching; it is not clear whether the parallel algorithms are optimal with respect to the sub-linear complexity of the Boyer-Moore algorithm.

Our goal in deriving a parallel version of KMP has not been to arrive at the most efficient algorithm possible; such an endeavor usually tends to transfer programming ingenuity to the meta-level, derivation ingenuity. Rather, our goal has been to reduce the design of a reasonably efficient algorithm to the application of a collection of general implementation strategies.

References

1. Bird, R. S., Gibbons, J., Jones, G.: Formal derivation of a pattern matching algorithm. Sci. Comput. Programming **12** (1989) 93–104
2. Boyer, R. S., Moore, J. S.: A fast string searching algorithm. *Commun. ACM 20*, 10 (Oct. 1977) 762–772
3. Dijkstra, E. W. *A Discipline of Programming.* Prentice Hall, Englewood Cliffs, NJ, 1976
4. Galil, Z. Optimal parallel algorithms for string matching. Inf. and Comput. **67** (1985) 144–157
5. Knuth, D. E., Morris, Jr., J. H., Pratt, V. R.: Fast pattern matching in strings. *SIAM J. Comput. 6*, 2 (June 1977) 323–350

6. Morris, J. M.: Programming by expression refinement: The KMP algorithm. In *Beauty is Our Business: A Birthday Salute to Edsger W. Dijkstra*. Springer-Verlag, New York (1990) 327–338
7. Partsch, H. A., Völker, N.: Another case study on reusability of transformational developments: Pattern matching according to Knuth, Morris, and Pratt. Tech. rep., KU Nijmegen, 1990
8. Smith, D. R., Lowry, M. R.: Algorithm theories and design tactics. Sci. Comput. Programming **14** (1990) 305–321
9. Smith, D. R.: Top-down synthesis of divide-and-conquer algorithms. Artificial Intelligence **27** (1985) 43–96
10. Smith, D. R.: The structure and design of global search algorithms. Tech. Rep. KES.U.87.12, Kestrel Institute, Palo Alto, California, July 1988. To appear in Acta Informatica.
11. Smith, D. R.: KIDS: A semiautomatic program development system. IEEE Trans. Softw. Eng. **16** (1990) 1024–1043
12. Srinivas, Y. V.: A sheaf-theoretic approach to pattern matching and related problems. Tech. Rep. KES.U.91.10, Kestrel Institute, Palo Alto, California, Oct. 1991. To appear in Theoretical Computer Science.
13. Turski, W. M., Maibaum, T. S. E.: *The Specification of Computer Programs*. Addison-Wesley, 1987
14. van der Woude, J.: Playing with patterns, searching for strings. Sci. Comput. Programming **12** (1989) 177–190
15. Vishkin, U.: Optimal parallel pattern matching in strings. Inf. and Comput. **67** (1985) 91–113
16. Vishkin, U.: Deterministic sampling: A new technique for fast pattern matching. In *22nd Symposium on Theory of Computing* Baltimore, MD (1990) 170–180

Modular Reasoning in an Object-Oriented Refinement Calculus

Mark Utting[1] and Ken Robinson[2]

[1] Software Verification Research Centre, University of Queensland, St. Lucia, QLD 4072, Australia. Email: marku@cs.uq.oz.au
[2] School of Computer Science and Engineering, University of New South Wales, PO Box 1, Kensington, NSW 2033, Australia. Email: kenr@cs.unsw.oz.au

Abstract. Object-oriented languages typically use late binding for procedure calls on objects. This raises a potential problem for programmers who wish to reason about their programs, because the effects of a procedure call cannot always be determined statically. In this paper we develop a simple model of procedure invocation for object-oriented languages based on the refinement calculus [Morgan and Robinson 87] and define the minimum requirements for a system to support *modular reasoning*. In such systems, reasoning about procedure calls is easier, because the behaviour of a procedure call with arguments of type T can be used as an approximation to its behaviour on more specialised arguments.

1 Introduction

A fundamental feature of object-oriented languages is a form of polymorphism, called *inclusion polymorphism* or *subtyping* [Cardelli and Wegner 85], that allows programs to abstract over a set of objects that have similar behaviour. For example, in a banking system with various types of accounts, a polymorphic procedure that transfers money between two given accounts could be used on any combination of account types (e.g., cheque and savings accounts, or investment and savings accounts), provided they have a common basic functionality. Furthermore, object-oriented languages use late binding to invoke procedures associated with objects, so each account type could have specialised actions associated with withdrawing or depositing money.

Late binding supports the writing of extensible, polymorphic programs, but they can be difficult to reason about, because the effect of a procedure call depends upon the type of object involved. One approach to reasoning about procedure calls that use late binding is to perform an exhaustive case analysis by considering all possible objects that the call could involve. However, this is impractical in large systems and has the severe disadvantage that adding a new type of object to a system can require additional case analysis of existing procedure calls. In this paper, we develop techniques for *modular* reasoning about late binding that overcomes these difficulties. Informally, modular reasoning means that when new types of objects are added to a system, unchanged program modules should not have to be respecified or reverified [Leavens and Weihl 90].

2 Notation

In this section we briefly describe the refinement calculus and the related notation that is used in this paper. We use the same lattice theoretic model of the refinement calculus that is often used in the literature [Back and Wright 89] [Wright 90] [Morris 87] [Gardiner and Morgan 91]. In the next section, we will extend this model to support the late binding aspect of object-oriented programming.

We leave the structure of objects undetermined except to assume that the set of objects (Obj) is a subset of the value space (Val) of programs. Program states (State) are modelled as total functions from a countable set of variables (Var) to values. Predicates (Pred) are modelled as total functions from State to the two point boolean lattice ($\mathsf{Bool} = \{\bot, \top\}$ such that $\bot \Rightarrow \top$). For example, the false predicate maps all states to $\bot$. Programs are modelled as monotonic[3] functions from postconditions to weakest preconditions. Val is ordered by equality and the State and Pred function spaces are ordered by the pointwise extension of their range orderings.[4]

$$\mathsf{State} \mathrel{\hat{=}} \mathsf{Var} \rightarrow \mathsf{Val}$$
$$\mathsf{Pred} \mathrel{\hat{=}} \mathsf{State} \rightarrow \mathsf{Bool}$$
$$\mathsf{Prog} \mathrel{\hat{=}} \mathsf{Pred} \xrightarrow{\mathsf{M}} \mathsf{Pred}$$

With these definitions, Pred is a complete boolean lattice and first order predicate calculus can be embedded in it. The least upper bound, greatest lower bound and complement operators of Pred correspond to disjunction, conjunction and negation, respectively. For a predicate $P \in \mathsf{Pred}$, we write $\lceil P \rceil$ to mean that P is true in all states. Thus, $P \sqsubseteq_{\mathsf{Pred}} Q$ can be written as $\lceil P \Rightarrow Q \rceil$.[5]

Prog is also ordered by the pointwise extension of Pred, but using a slightly different notation to the usual refinement calculus notation. For $S, T \in \mathsf{Prog}$, we define $S \sqsubseteq T$ to be a *predicate*.

$$S \sqsubseteq_{\mathsf{Prog}} T \;\; \hat{=} \;\; (\forall P \in \mathsf{Pred} \bullet S\,P \Rightarrow T\,P)$$

This allows $S \sqsubseteq T$ to be true in some initial states and false in others. The pointwise ordering over Prog is therefore the universal closure of the $\sqsubseteq$ operator (e.g., $\lceil S \sqsubseteq T \rceil$). This pointwise ordering is the usual *refinement* relation of the refinement calculus.[6] Informally, $\left\lceil S \sqsubseteq_{\mathsf{Prog}} T \right\rceil$ means that T is an acceptable replacement for S in any context. Where no ambiguity arises, we often abbreviate $\sqsubseteq_{\mathsf{Pred}}$ or $\sqsubseteq_{\mathsf{Prog}}$ to just $\sqsubseteq$.

Prog is a complete distributive (non-boolean) lattice and its least upper bound and greatest lower bound operators correspond to angelic ($\Diamond$) and demonic ($[]$) choice

---

[3] Recall that a function f is monotonic if $x \sqsubseteq y \Rightarrow f\,x \sqsubseteq f\,y$ for all x and y. We write $A \xrightarrow{\mathsf{M}} B$ for the set of all monotonic functions from A and B.

[4] That is, if $f, g \in A \rightarrow B$, then $f \sqsubseteq g \mathrel{\hat{=}} (\forall a \in A \bullet f\,a \sqsubseteq_B g\,a)$.

[5] Note that all the usual predicate calculus connectives are simply the pointwise extension

of programs [Wright 90]. All the programming constructs of Dijkstra's guarded command language can be represented as members of Prog [Dijkstra 76] (we write his $wp(S, P)$ as functional application: $S\ P$).

An important (though non-executable) statement in the refinement calculus is the specification statement [Morgan 90]. This is written as *frame* : [*pre* / *post*] where *pre* and *post* are predicates that specify the allowable states before and after execution and *frame* is a set of variables whose values may be changed during execution—all other variables remain constant. A specification statement can be refined by strengthening its postcondition, weakening its precondition or removing variables from its frame. The refinement calculus also contains rules for transforming between specification statements and executable code.

We sometimes use predicates and programs over state spaces that contain extra variables in addition to Var, in which case we use explicit subscripts to indicate the additional variables (e.g., $\mathsf{Prog}_{a,b}$ and $\mathsf{Pred}_{a,b}$ where a and b are vectors of variables). To indicate quantification over states containing additional variables, we write $\lceil P \Rightarrow Q \rceil_{a,b}$. Similarly, if $S, S' \in \mathsf{Prog}_v$, we sometimes write $S \sqsubseteq_v S'$ instead of $S \sqsubseteq S'$, to indicate that the postconditions are taken from Pred_v.

Finally, note that in our formulae, function application and postfix substitution operators[7] have the highest precedence and are left associative. As usual, the predicate calculus connectives ($\Rightarrow$, $\wedge$, etc.) have lower precedence than ordering operators such as $\sqsubseteq$ and equality. Quantifiers ($\exists, \forall$), distributive operators ($\bigwedge_{i \in K}$, $\bigsqcup_{i \in K}$, etc.) and lambda expressions have the lowest priority of all and are always explicitly bracketed to indicate their scope.

3 An Object-Oriented Refinement Calculus

To extend the above model of objects and programs into an object-oriented language, we need to model late binding in some way. In a non-object-oriented language, the procedure name in a procedure call statement uniquely determines which procedure is invoked.[8] In contrast, in object-oriented languages the choice of which procedure is invoked by a call statement may depend upon the values of the argument expressions as well as depending upon the procedure name. It is this late binding that makes reasoning about calls more complex in object-oriented languages.

We model this late binding relationship by defining an object-oriented system to be a mapping from the name and argument values of a procedure call to a program. Although the majority of object-oriented languages use only one object to resolve late binding, in this paper we take the more general ***multiple dispatch*** approach, which allows all the arguments in a call statement to be used in resolving late binding.

[7] As in [Morgan 90], we write $P[v \backslash E]$ for the predicate formed from P by replacing all free occurences of v by E.

[8] Some languages generalise this slightly by allowing procedure names to be *overloaded*. This means that a single name is used for several procedures and the static types of the arguments are used in selecting one of those procedures. However, the mapping from a call statement to a unique procedure is still statically determinable, so standard verification techniques can be used.

Definition 1. *The set of all* multiple dispatch object-oriented systems, MOO, *is defined as*

$$\mathsf{MOO} \mathrel{\hat{=}} \mathcal{P} \to \mathsf{Val}^* \to \mathsf{Prog}$$

where $\mathcal{P}$ is a set of procedure names and Val^* *is the set of all finite sequences of values. Throughout this paper we use Ψ as a typical member of* MOO.

$$\Psi \in \mathsf{MOO}.$$

For simplicity, we define MOO as the set of *total* functions. In practice, some input value sequences may be invalid or uninteresting. Such sequences are typically mapped to the **abort** program or perhaps to some error handling procedure like the doesNotUnderstand: method in Smalltalk [LaLonde and Pugh 90].

The multiple dispatch approach is used by several object-oriented languages, though the majority only provide single dispatch. The CLOS language [Keene 89] [Bobrow et al. 88] (an object-oriented extension of Common LISP), supports multiple dispatch and calls each $\Psi\, p$ a *generic function.* A more recent multiple dispatch language, Cecil [Chambers 92], has illustrated that it is possible to provide multiple dispatch facilities and still support encapsulation of data abstractions, the lack of which was often considered to be a weakness of the multiple dispatch approach. Other terminology used in the literature for this approach includes *multi-methods* [Agrawal et al. 91] and *double dispatch* [Hebel and Johnson 90] (for the binary case).

So far, our set of programs, Prog, can inspect and manipulate objects, but they have no access to the Ψ function or to the programs that Ψ returns. We remedy this by defining a multiple dispatch object-oriented call statement, defined in terms of Ψ.[9] We assume a function $(eval\ E\ s)$ that evaluates an expression (or sequence of expressions) E in a state s. For simplicity, we assume that the expressions are restricted so that *eval* always terminates with a well defined result.

Definition 2. *Let p be a procedure name ($p \in \mathcal{P}$), E be a finite sequence of expressions and s be a state. A* multiple dispatch call statement *is written as* $\mathbf{call}\, p(E)$ *and is defined by the following weakest precondition on predicate P*

$$[\![\mathbf{call}\, p(E)]\!]\ P \mathrel{\hat{=}} (\lambda s : \mathsf{State} \bullet \Psi\, p\, (eval\ E\ s)\ P\ s)\ .$$

The rationale for this definition is that the right hand side is equivalent to $(\Psi\, p\, d)\ P$, where d is the result of evaluating the arguments E in the state s. This shows that the essence of a call statement is to look up the Ψ function and return the procedure associated with the given name and argument values. If call statements always contained a sequence of constant values (d), we could define $\mathbf{call}\, p(d) \mathrel{\hat{=}} \Psi\, p\, d$. However, in order to allow late binding, call statements contain a list of argument *expressions* rather than *values*, so it is necessary to evaluate the argument expressions in the initial state before the lookup can be performed. To access the initial

[9] This introduces recursion between Prog and MOO, so the meaning of a system *Sys* that contains multiple dispatch call statements is actually the least fixed point of $\Psi = Sys(\Psi)$. By Tarski's fixpoint theorem, this least fixed point is well defined, since MOO is a complete lattice and *Sys* is monotonic.

state, we rewrite the predicate $\Psi\, p\, d\, P$ as $(\lambda s \bullet \Psi\, p\, d\, P\, s)$ to gain explicit access to the initial state s. Having done this we can then replace d by $eval\, E\, s$, thus using the initial state to evaluate the argument expressions. This shows that, unlike other weakest precondition definitions, the call statement uses the initial state (s) in two ways: to evaluate the argument expressions, and subsequently to evaluate the weakest precondition of the program returned by Ψ. Thus the semantics of the call statement (i.e., its weakest precondition predicate) cannot generally be determined statically, but only in the context of an initial state.

Definition 2 does not define how the parameters of a call statement are passed to the procedure that is executed. There are many different ways in which this can be done, so to avoid committing to a particular solution we treat late binding and parameterisation as separate issues. Essentially, we view all procedures as predicate transformers that act upon a global set of variables. The various substitution operators[10] of the refinement calculus can then be used to parameterise procedures in whatever way is desired. For example, we could write

$$\mathbf{call}\; p(E)[\mathbf{value}\; args \backslash E]$$

to indicate that the E values should be passed by value to the program returned by the call statement. Note that E is used twice in this program; once to resolve the late binding and once for parameterisation.

The semantics of call statements in practical languages usually link the two issues, since leaving them separate introduces the possibility of inconsistencies. For example, one object could be used to resolve the late binding, but a different object passed to the procedure. In fact, in untyped object-oriented languages such as Smalltalk it is the linking of parameterisation with the late binding mechanism that enforces encapsulation. The only way of applying a method to an object a is to use the late binding (message passing) mechanism, and that automatically selects a method from the class of a and also passes a to that method as self. For similar reasons, in the next section, we shall see that it is advantageous to combine some parameterisation with the multiple dispatch call statement.

4 Modular Reasoning

To develop the notion of modular reasoning, we need a ***subtyping*** relationship $\leq$ between objects. It is sufficient to take $\leq$ as being any reflexive and transitive relation over Val, since distinguishing between objects and other kinds of values is not

[10] Three substitution operators are commonly used in the refinement calculus to parameterise procedures [Morgan 90]: call by value, call by result and call by value-result. If u and v are lists of disjoint variables, E is a list of expressions and P is a predicate not involving u (the formal parameters), they are defined as

$$S[\mathbf{value}\; u \backslash E]\, P \mathrel{\widehat{=}} S\, P[u \backslash E]$$
$$S[\mathbf{result}\; u \backslash v]\, P \mathrel{\widehat{=}} (\forall u \bullet S\,(P[v \backslash u]))$$
$$S[\mathbf{valres}\; u \backslash v]\, P \mathrel{\widehat{=}} S\,(P[v \backslash u])\,[u \backslash v]$$

essential to our theory. When $a, b \in \mathsf{Val}$ and $a \leq b$, we say that b is a *subobject* of a, or that a is a *superobject* of b. We extend $\leq$ to finite sequences of values in the obvious way: two sequences $c, d \in \mathsf{Val}^*$ satisfy $c \leq d$ iff c and d have the same length and each member of c is related by $\leq$ to the corresponding member of d.

Intuitively, $c \leq d$ means that d can be substituted for c in any program. That is, any procedure call that has c as an argument can also be given d as an argument, though the effect of the two calls may be quite different because of late binding. However, if we wish to reason about a call statement **call** $p(E)$ in a modular fashion, then the effects of $\Psi\ p\ c$ and $\Psi\ p\ d$ must be related in some way, since we want to be able to reason about the effect of the call statement without knowing exactly which subobjects have been passed as arguments. This is our primary concern in the remainder of the paper: investigating ways of restricting Ψ so that it supports modular reasoning.

For simplicity, we assume that whenever two objects a and b are related by $\leq$, then predicates or programs involving a can be interpreted directly as predicates or programs on b. For example, we want to be able to compare $\Psi\ p\ c$ with $\Psi\ p\ d$ using the usual refinement relation between programs, $\sqsubseteq_{\mathsf{Prog}}$. Informally, this assumption is equivalent to requiring the abstract state of an object and its subobjects to be represented in the same way. In other words, we do not directly support *data refinement* [Gardiner and Morgan 91] between the abstract states of an object and its subobjects (except that the subobjects may possibly define additional state information). In practice, this requirement for the abstract states of a and b to be related is not overly restrictive, since in object-oriented languages that support information hiding, thus allowing the implementation of an object to be hidden from its callers, an *implementation* of a or b may still involve data refinement. However, in this paper we do not address information hiding or data refinement issues.

One obvious way of allowing modular reasoning about call statements is to require Ψ to be monotonic with respect to $\leq$. This means that if $c \leq d$, then passing d to a call statement instead of c can only refine the call statement, since the monotonicity of Ψ ensures that

$$\left[\Psi\ p\ c \sqsubseteq_{\mathsf{Prog}} \Psi\ p\ d\right] .$$

As an example of modular reasoning, consider an addition procedure in a system that contains two and three dimensional point objects. For this example, we choose to represent objects as partial functions from a set of names Var to values, ordered in a way that allows subtype objects to increase the domain of the function. We assume that the value space also contains integers, which are distinct from objects and ordered discretely.

$$\mathsf{Obj} \mathrel{\hat{=}} \mathsf{Var} \nrightarrow \mathsf{Val}$$

$$\mathsf{Val} \mathrel{\hat{=}} \mathsf{Obj} \cup \mathbb{Z}$$

$$a \leq b \mathrel{\hat{=}} \begin{pmatrix} a \in \mathsf{Obj} \land b \in \mathsf{Obj} \\ \mathrm{dom}\, a \subseteq \mathrm{dom}\, b \\ (\forall i : \mathrm{dom}\, a \bullet a(i) \leq b(i)) \end{pmatrix} \lor \begin{pmatrix} a \in \mathbb{Z} \\ b \in \mathbb{Z} \\ a = b \end{pmatrix}$$

Let X, Y and Z be names in Var and *add* be a procedure name. We represent two dimensional (2D) point objects as partial functions with a domain of $\{X, Y\}$, whereas 3D points have a domain of $\{X, Y, Z\}$. The following predicates, $P2$ and $P3$, capture the essential properties of 2D and 3D points respectively.

$$P2(a) \mathrel{\widehat=} \begin{pmatrix} a \in \mathsf{Obj} \\ \operatorname{dom} a = \{X, Y\} \\ a(X) \in \mathbb{Z} \wedge a(Y) \in \mathbb{Z} \end{pmatrix}$$

$$P3(a) \mathrel{\widehat=} \begin{pmatrix} a \in \mathsf{Obj} \\ \operatorname{dom} a = \{X, Y, Z\} \\ a(X) \in \mathbb{Z} \wedge a(Y) \in \mathbb{Z} \wedge a(Z) \in \mathbb{Z} \end{pmatrix}$$

If a and b are both two dimensional point values, we might naively attempt to define the value of $\Psi\, add\, [a, b]$ as

$$S \mathrel{\widehat=} res : \left[\begin{matrix} P2(arg_1) \\ P2(arg_2) \end{matrix} \middle/ \begin{matrix} P2(res) \\ res(X) = arg_1(X) + arg_2(X) \\ res(Y) = arg_1(Y) + arg_2(Y) \end{matrix} \right] [\mathbf{value}\; arg_1, arg_2 \backslash a, b].$$

However, this specification is too strong. The predicates $P2$ and $P3$ are mutually exclusive, so we will not be able to refine S to return 3D points without it becoming infeasible. Even though the object ordering means that 3D points are subobjects of 2D points, predicates such as $P2$ that are true of 2D points may not necessarily be true of 3D points. That is, predicates are not necessarily monotonic with respect to $\leq$. When specifying programs like S it is usually desirable to use monotonic predicates, so we define an operator that weakens any predicate so that it is monotonic.

Definition 3. *If P is a predicate and u is a sequence of (distinct) variables, the* monotonic closure *of P with respect to u ($\overline{P}^u$) is defined as:*

$$\overline{P}^u \mathrel{\widehat=} (\exists u' \bullet P[u \backslash u'] \wedge u' \leq u) \ .$$

We write the monotonic closure of P with respect to all of Var *as $\overline{P}^*$.*

This monotonic closure operator satisfies the following properties, for any sequence of variables u. Proofs of these properties can be found in [Utting 92].

$$\left[P \Rightarrow \overline{P}^u \right]$$

$$\left[P \Rightarrow \overline{Q}^u \right] \Leftrightarrow \left[\overline{P}^u \Rightarrow \overline{Q}^u \right]$$

$$[P \Rightarrow Q] \Rightarrow \left[\overline{P}^u \Rightarrow \overline{Q}^u \right]$$

To illustrate the monotonic closure operator, note that $\overline{P2(a)}^*$ is the same as $P2(a)$ except that $\mathrm{dom}\, a = \{X, Y\}$ becomes $\mathrm{dom}\, a \supseteq \{X, Y\}$. Similarly for $\overline{P3(a)}^*$.

$$\overline{P2(a)}^* \Leftrightarrow \begin{pmatrix} a \in \mathsf{Obj} \\ \mathrm{dom}\, a \supseteq \{X, Y\} \\ a(X) \in \mathbb{Z} \wedge a(Y) \in \mathbb{Z} \end{pmatrix}$$

$$\overline{P3(a)}^* \Leftrightarrow \begin{pmatrix} a \in \mathsf{Obj} \\ \mathrm{dom}\, a \supseteq \{X, Y, Z\} \\ a(X) \in \mathbb{Z} \wedge a(Y) \in \mathbb{Z} \wedge a(Z) \in \mathbb{Z} \end{pmatrix}$$

If we replace all occurences of $P2$ in the specification S by its monotonic closure, we get a specification that still handles 2D points but also allows the inputs and outputs to be subobjects of 2D points.

$$S_2 \mathrel{\widehat{=}} res : \left[\begin{array}{c} \overline{P2(arg_1)}^* \\ \overline{P2(arg_2)}^* \end{array} \middle/ \begin{array}{c} \overline{P2(res)}^* \\ res(X) = arg_1(X) + arg_2(X) \\ res(Y) = arg_1(Y) + arg_2(Y) \end{array} \right] [\mathrm{value}\; arg_1, arg_2 \backslash a, b].$$

This illustrates a methodological principle: for modular reasoning to be practical, specifications should allow for input and output values that refine the expected values.

Now consider how the add procedure should be extended to 3D points. If a and b are both 3D point values, we would like to define the value of $\Psi\; add\; [a, b]$ as

$$S_3 \mathrel{\widehat{=}} res : \left[\begin{array}{c} \overline{P3(arg_1)}^* \\ \overline{P3(arg_2)}^* \end{array} \middle/ \begin{array}{c} \overline{P3(res)}^* \\ res(X) = arg_1(X) + arg_2(X) \\ res(Y) = arg_1(Y) + arg_2(Y) \\ res(Z) = arg_1(Z) + arg_2(Z) \end{array} \right] [\mathrm{value}\; arg_1, arg_2 \backslash a, b].$$

However, this is not a refinement of S_2, because it has a stronger precondition. Yet intuitively, if the only way of calling S_2 and S_3 is via Ψ, then S_3 *does* refine S_2 in some sense, because S_3 is only executed when Ψ is given two 3D point arguments, so its precondition is always satisfied. This indicates that requiring Ψ to be monotonic is a stronger condition than is necessary for modular reasoning.

We can weaken the requirements on Ψ if we change the call statement to always substitute the argument values into the program that it returns. This allows the program to rely upon properties of the arguments that are implicit in the definition of Ψ. For example, if the only circumstances under which Ψ returns a program S is when it is given an argument sequence of $[1, 2]$, then a precondition of $arg_1 = 1 \wedge arg_2 = 2$ in S will be automatically satisfied (assuming that the call statement passes the arguments via $[\mathrm{value}\; arg_1, arg_2 \backslash 1, 2]$). These considerations lead us to define modular reasoning as a weaker form of monotonicity. First, we define a *value passing* call statement that passes its arguments to the program that Ψ returns. We use call by value-result parameterisation, because it encompasses call by value and call by result as special cases and it is equivalent to call by reference if aliasing is banned in some way. The use of value-result parameterisation means that parameters to call statements must be variables, but this is not an important semantic restriction.

To simplify the definition of the value passing call statement, we assume that each procedure name p has an associated *arity* given by $\#p$ and that all call statements involving p have exactly $\#p$ arguments.[11] We also assume that for each procedure name p, programs in the range of $\Psi\, p$ use a standard sequence of variables $arg_1, \ldots, arg_{\#p}$ for the formal parameters. For conciseness, we call this sequence of variables $args$ when the p involved is obvious from context.

Definition 4. *Let p be a procedure name ($p \in \mathcal{P}$), u be a sequence of distinct variables of length $\#p$, P be any predicate and s be any state. A* multiple dispatch value passing call statement *is written as* $\mathbf{callv}\; p(u)$ *and is defined as*

$$[\![\mathbf{callv}\; p(u)]\!]\; P \mathrel{\hat{=}} (\lambda s \bullet \Psi\, p\, (eval\; u\; s)\, [\mathbf{valres}\; args \backslash u]\; P\; s) \;.$$

For convenience, we shall in future leave the evaluation of u implicit, abbreviating the right hand side to $\Psi\, p\, u\, [\mathbf{valres}\; args \backslash u]\; P$.

This value passing call statement defines a relationship between the variables of the caller and callee, so to avoid unnecessary aliasing problems we shall always assume that the caller does not involve $args$ and the callee does not involve any of the argument variables in the call statement. That is, we assume that u and $args$ do not contain any variables in common and that $\mathbf{callv}\; p(u) \in \mathsf{Prog}_u$ and $\Psi\, p\, u \in \mathsf{Prog}_{args}$. We call this the *call disjointness* assumption.

The original call statement (Defn. 2) did not relate the values passed to $\Psi\, p$ to the variables used by the resulting program, so to reason about it in a modular fashion meant that $\Psi\, p$ had to be restricted to be monotonic. However, the value passing call statement (Defn. 4) always passes the argument values to the resulting program. This additional restriction on the call statement means that the following weaker constraint on Ψ is sufficient for modular reasoning.

Definition 5. *Let p be any procedure name and u and v be sequences of distinct variables of length $\#p$. We say that $\Psi\, p$* supports modular reasoning *iff*

$$\left[\, u \le v \Rightarrow \Psi\, p\, u\, [\mathbf{valres}\; args \backslash v] \sqsubseteq_v \Psi\, p\, v\, [\mathbf{valres}\; args \backslash v]\, \right]_{u,v} \;.$$

An object-oriented system $\Psi \in \mathsf{MOO}$ supports modular reasoning iff $\Psi\, p$ supports modular reasoning for all $p \in \mathcal{P}$.

The rationale for this definition is that for some known value in u and some unknown value in v such that $u \le v$, we must be able to use the known behaviour of $\Psi\, p\, u$ as an approximation to the unknown behaviour of $\Psi\, p\, v$. However, there are two additional insights that lead to the precise definition:

[11] An alternative approach would be to define MOO as $(\mathcal{P} \times \mathbb{N}) \to \mathsf{Val}^* \to \mathsf{Prog}$, thus allowing calls to a procedure called p to have different numbers of parameters. This

- Firstly, this approximation does not have to hold in all possible states—only those in which the arguments equal v. This is because for argument values d that are not equal to the value of v we are not interested in the behaviour of $\Psi\ p\ v$, since some other program $\Psi\ p\ d$ will be executed. For this reason, we apply a value result substitution to both $\Psi\ p\ u$ and $\Psi\ p\ v$.
- Secondly, we pass the *same* object values to both programs. We want to make the behaviour of $\Psi\ p\ u$ approximate that of $\Psi\ p\ v$ as closely as possible, and the best way of doing this is to pass it the exact inputs that $\Psi\ p\ v$ will get when invoked via **callv** $p(v)$. Furthermore, passing the same values to both programs means that we can safely allow the programs to make arbitrary observations of the values, including testing their types or making other observations that may distinguish between subobjects and superobjects. Since both programs are observing the same values, the results of any kind of observation will be the same in both programs, so $\Psi\ p\ u$ will be the best approximation to $\Psi\ p\ v$ that is possible.

Definition 5 clearly shows that our modular reasoning requirement is similar to requiring Ψ to be monotonic, except that the right hand side is weaker because the refinement is only required to hold when the arguments to both programs equal v. The following lemma expresses this in an alternative form where $v = args$ appears on the left hand side of the implication.

Lemma 6.

$$\lceil u \leq v \Rightarrow \Psi\ p\ u\,[\mathbf{valres}\ args\backslash v] \sqsubseteq_v \Psi\ p\ v\,[\mathbf{valres}\ args\backslash v]\,\rceil_{u,v}$$
$$\Leftrightarrow \lceil u \leq v \wedge v = args \Rightarrow \Psi\ p\ u \sqsubseteq_{args} \Psi\ p\ v\,\rceil_{u,v,args} \ .$$

Proof.

$\lceil u \leq v \Rightarrow \Psi\ p\ u\,[\mathbf{valres}\ args\backslash v] \sqsubseteq_v \Psi\ p\ v\,[\mathbf{valres}\ args\backslash v]\,\rceil_{u,v}$

$\Leftrightarrow$ {*By definition of refinement*}

$\lceil u \leq v \Rightarrow (\forall P \in \mathsf{Pred}_v \bullet \Psi\ p\ u\,[\mathbf{valres}\ args\backslash v]\ P \Rightarrow \Psi\ p\ v\,[\mathbf{valres}\ args\backslash v]\ P)\,\rceil_{u,v}$

$\Leftrightarrow$ {*By definition of value-result substitution, abbreviating* $P[v\backslash args]$ *to* P'}

$\lceil u \leq v \Rightarrow (\forall P \in \mathsf{Pred}_v \bullet \Psi\ p\ u\ P'\,[args\backslash v] \Rightarrow \Psi\ p\ v\ P'\,[args\backslash v])\,\rceil_{u,v}$

$\Leftrightarrow$ {*Since* P' *is an arbitrary member of* Pred_{args}}

$\lceil u \leq v \Rightarrow (\forall Q \in \mathsf{Pred}_{args} \bullet \Psi\ p\ u\ Q\,[args\backslash v] \Rightarrow \Psi\ p\ v\ Q\,[args\backslash v])\,\rceil_{u,v}$

$\Leftrightarrow$ {*Predicate Calculus:* $\lceil E[args\backslash v] \Leftrightarrow (\forall args \bullet v = args \Rightarrow E)\rceil$ *for any* E.}

$\lceil u \leq v \Rightarrow (\forall Q, args \bullet v = args \Rightarrow \Psi\ p\ u\ Q \Rightarrow \Psi\ p\ v\ Q)\,\rceil_{u,v}$

$\Leftrightarrow$ {*Moving* $\forall Q$ *inwards and* $\forall args$ *outwards*}

$\lceil u \leq v \wedge v = args \Rightarrow \Psi\ p\ u \sqsubseteq_{args} \Psi\ p\ v\,\rceil_{u,v,args}$

□

Returning to the 2D and 3D point example, can we now demonstrate support for modular reasoning about value passing call statements such as **callv** $add(u, v)$?

Since the value passing call statement handles argument substitution, we drop the value substitutions from S_2 and S_3 to give

$$S_{22} \mathrel{\hat=} \mathit{res}: \left[\begin{array}{c} \overline{P2(arg_1)}^* \\ \overline{P2(arg_2)}^* \end{array} \Big/ \begin{array}{c} \overline{P2(res)}^* \\ res(X) = arg_1(X) + arg_2(X) \\ res(Y) = arg_1(Y) + arg_2(Y) \end{array} \right]$$

$$S_{33} \mathrel{\hat=} \mathit{res}: \left[\begin{array}{c} \overline{P3(arg_1)}^* \\ \overline{P3(arg_2)}^* \end{array} \Big/ \begin{array}{c} \overline{P3(res)}^* \\ res(X) = arg_1(X) + arg_2(X) \\ res(Y) = arg_1(Y) + arg_2(Y) \\ res(Z) = arg_1(Z) + arg_2(Z) \end{array} \right]$$

then define Ψ *add* as

$$\Psi\, add\, [a, b] \mathrel{\hat=} \begin{cases} S_{33}, & \textbf{if } \overline{P3(a)}^* \wedge \overline{P3(b)}^* \\ S_{22}, & \textbf{otherwise} \end{cases} .$$

Let $[a, b]$ and $[a', b']$ be value sequences such that $[a, b] \leq [a', b']$. When the two programs $\Psi\, add\, [a, b]$ and $\Psi\, add\, [a', b']$ are equal, the refinement is trivial, so the only interesting case is:

$\Psi\, add\, [a, b] \neq \Psi\, add\, [a', b']$

$\Leftrightarrow \left\{ \textit{Since } \left[x \leq x' \Rightarrow \overline{P(x)}^* \sqsubseteq \overline{P(x')}^* \right], \textit{ we have } \overline{P3(a)}^* \wedge \overline{P3(b)}^* \Rightarrow \overline{P3(a')}^* \wedge \overline{P3(b')}^* \right\}$

$\quad \Psi\, add\, [a, b] = S_{22} \wedge \Psi\, add\, [a', b'] = S_{33}$

$\Leftrightarrow \overline{P3(a')}^* \wedge \overline{P3(b')}^* \wedge \neg \left(\overline{P3(a)}^* \wedge \overline{P3(b)}^*\right)$.

So, it is sufficient to show that

$$\left[\left(\overline{P3(a')}^* \wedge \overline{P3(b')}^*\right) \Rightarrow \left(\begin{array}{l} S_{22}[\textbf{valres } arg_1, arg_2 \backslash a', b'] \\ \sqsubseteq S_{33}[\textbf{valres } arg_1, arg_2 \backslash a', b'] \end{array} \right) \right] .$$

Since the assumption satisfies the preconditions of both programs and the postcondition of S_{33} implies that of S_{22} (since $\overline{P3(res)}^* \Rightarrow \overline{P2(res)}^*$), the refinement is valid.

We conclude that this definition of Ψ *add* does support modular reasoning about statements such as **callv** $add(x, y)$. However, proving that it does is somewhat tedious using Defn. 5, since we must consider all possible pairs of value sequences. In the next section we investigate an alternative way of describing Ψ that is more concise and simplifies the proof requirement slightly.

5 Alternative Representations of Ψ

The definition of object-oriented systems allows Ψ to map each sequence of argument values to a different program. However, it is often the case that Ψ partitions the argument values into sets and maps each such set to a single program. The 2D-3D point example illustrates this: Ψ *add* maps one set of inputs to S_{33} and all other inputs to S_{22}. In this situation, a convenient way of describing $\Psi\, p$ is as a set of predicate-program pairs, $(Q_i, S_i)_{i \in K}$, indexed by some set K. If d is a value sequence of length $\#p$, this allows us to define $\Psi\, p$ as

$$\Psi\, p\, d \mathrel{\widehat{=}} \{\, S_i, \quad \textbf{if } Q_i[args \backslash d] \quad .$$

In other words, $\Psi\, p$ returns the program S_i whose corresponding *guard* (Q_i) is true. For this to define a function, we require the set of guards to partition $\mathsf{Val}^{\#p}$ (i.e., the set of all value sequences of length $\#p$) and the free variables of each Q_i to be a subset of $args$ $(= arg_1, \ldots, arg_{\#p})$. Also, we assume that none of the Q_i predicates are universally false.

The reason for introducing this new representation of $\Psi\, p$ is that it allows us to concisely describe systems that contain only a small number of programs in the range of $\Psi\, p$. Proving that a system supports modular reasoning using Defn. 5 involves considering all possible pairs of value sequences, which is quite impractical. However, we shall shortly show that this new $(Q_i, S_i)_{i \in K}$ representation allows us to prove modular reasoning by only considering all pairs of (S_i, S_j), for $i, j \in K$. This is better, but is still not practical for large systems, as the number of cases grows quadratically with K. Additional techniques for reducing the number of cases are discussed in Sect. 7 and in [Utting 92, Chap. 5].

When $\Psi\, p$ is described in this way, the semantics of **call** $p(u)$ can be expressed directly using the predicate-program pairs. A proof of this theorem is given in [Utting 92].

Theorem 7. *If $\Psi\, p$ is defined as a set of predicate-program pairs, $(Q_i, S_i)_{i \in K}$ then*

$$\left[\, \textbf{callv}\ p(u) \;=\; \left(\mathop{[\!]}_{i} Q_i \rightarrow S_i\right)[\textbf{valres}\ args \backslash u] \,\right] \quad .$$

Since the guards in the demonic choice statement are mutually exclusive and one of them must be true, we could equally well have chosen to embed the choice within an **if** statement, or used angelic choice rather than demonic (see [Wright 90, page 53]).

$$\left[\, \textbf{callv}\ p(u) \;=\; (\textbf{if } \mathop{[\!]}_{i} Q_i \rightarrow S_i \textbf{ fi})[\textbf{valres}\ args \backslash u] \,\right]$$

$$\left[\, \textbf{callv}\ p(u) \;=\; (\Diamond\{Q_i\}\,;\, S_i)[\textbf{valres}\ args \backslash u] \,\right]$$

Theorem 7 and the two equivalences above show that any $\Psi\, p$ can also be described as a single program. That is, we could have defined MOO as $\mathcal{P} \rightarrow \mathsf{Prog}$; a simple model of named procedures that is used by most non-object-oriented languages. However, each of these procedures would have to handle all possible types of argument values, so would typically contain a case analysis, as illustrated by theorem 7. This captures the essential difference between object-oriented languages (which describe $\Psi\, p$ as a collection of separate procedures and use late binding to invoke the appropriate one) and non-object-oriented languages (which typically describe $\Psi\, p$ as a single procedure containing a case analysis to handle the different types of arguments).

We now have three alternative representations of $\Psi\, p$:

$\mathsf{Val}^{\#p} \rightarrow \mathsf{Prog}$: This is the original function representation from Defn. 1. Here the domain is written as $\mathsf{Val}^{\#p}$ to reflect our simplifying assumption that all calls involving p have $\#p$ arguments.

$K \rightarrow \mathsf{Pred} \times \mathsf{Prog}$: As mentioned above, we require systems $(Q_i, S_i)_{i \in K}$ described using this representation to fulfill some additional restrictions. The set of guards $\{i \in K \bullet Q_i\}$ must partition $\mathsf{Var}^{\#p}$, the free variables of each Q_i must be a subset of $args$ ($= arg_1, \ldots, arg_{\#p}$) and none of the guards can be identically **false**.

Prog: This *universal-procedure* representation is included for comparison purposes, because it relates the polymorphic procedure calls and late binding of object-oriented languages to procedures in other languages. It is not used in the remainder of the paper because we are primarily interested in using one procedure $\Psi\ p\ u$ to approximate another, so representing all of $\Psi\ p$ as a single program is not an appropriate model for this purpose.

All three of these models are equivalent in expressive power. Any system described as a function from $\mathsf{Val}^{\#p}$ to Prog can be described in $(Q_i, S_i)_{i \in K}$ form by choosing very specific guards such as $[arg_1, arg_2] = [1, 5]$ (as the guards become more specific, K tends towards $\mathsf{Val}^{\#p}$). Theorem 7 shows that any system described in $(Q_i, S_i)_{i \in K}$ form can also be described as a single program. Finally, any system that describes $\Psi\ p$ as a single program S can be represented as a function from $\mathsf{Val}^{\#p}$ to Prog that maps all inputs to S.

Describing the late binding function as sets of predicate-program pairs is similar to the way that generic functions are specified in CLOS [Keene 89]. However, the predicates in CLOS are limited to testing for membership within a primitive LISP data type or user-defined class or testing for identity (using the LISP *eql* function) with a fixed object. CLOS does not require the predicates to be disjoint; instead it uses various precedence rules and a global ordering over user-defined classes to determine which method should be chosen when several predicates are valid.

When $\Psi\ p$ is described as a set of predicate-program pairs, we also get a more convenient criteria for determining whether or not the definition supports modular reasoning.

Theorem 8. *If $\Psi\ p$ is defined as a set of predicate-program pairs, $(Q_i, S_i)_{i \in K}$, then it supports modular reasoning iff*

$$\left[\forall i, j \in K \bullet \{\overrightarrow{Q_i}^{*} \wedge Q_j\} ; S_i \sqsubseteq_{args} S_j \right]_{args} .$$

Proof. From Lemma 6, the modular reasoning requirement is:

$$\left[u \leq v \wedge v = args \Rightarrow \Psi\ p\ u \sqsubseteq \Psi\ p\ v \right]_{u,v,args}$$

$$\Leftrightarrow \{\textit{Since } \{k \in K \bullet Q_k\} \textit{ partitions } \mathsf{Val}^*\}$$

$$\left[\forall i, j \in K \bullet \begin{pmatrix} Q_i[args \backslash u] \wedge Q_j[args \backslash v] \\ u \leq v \wedge v = args \end{pmatrix} \Rightarrow S_i \sqsubseteq S_j \right]_{u,v,args}$$

$$\Leftrightarrow \{\textit{Replacing } v \textit{ by } args \textit{ throughout}\}$$

$$\left[\forall i, j \in K \bullet \begin{pmatrix} Q_i[args \backslash u] \wedge Q_j \\ u \leq args \end{pmatrix} \Rightarrow S_i \sqsubseteq S_j \right]_{u,args}$$

$\Leftrightarrow$ {*Since* $S_i \sqsubseteq S_j$ *does not involve* u}

$$\left[\forall i,j \in K \bullet (\exists u \bullet Q_i[args\backslash u] \wedge Q_j \wedge u \leq args) \Rightarrow S_i \sqsubseteq S_j \right]_{args}$$

$\Leftrightarrow$ {*Defn. of* $\overrightarrow{Q_i}^*$, *since* Q_j *does not involve* u}

$$\left[\forall i,j \in K \bullet \overrightarrow{Q_i}^* \wedge Q_j \Rightarrow S_i \sqsubseteq S_j \right]_{args}$$

$\Leftrightarrow$ {*Refinement Calculus*}

$$\left[\forall i,j \in K \bullet \{\overrightarrow{Q_i}^* \wedge Q_j\} ; S_i \sqsubseteq S_j \right]_{args}$$

□

This theorem gives significant insight into the requirements that modular reasoning imposes upon each procedure p of a multiple dispatch system Ψ. Let $a_1 < a_2 < a_3 \cdots$ be a chain of acceptable argument sequences for a call statement **callv** $p(a_i)$ and for each a_i define S_i to be the program that the call statement executes when the arguments equal a_i (i.e., $S_i \widehat{=} \Psi\ p\ a_i$). Then, for $i < j$, modular reasoning means that S_j must be a refinement of S_i, except that it may strengthen the precondition of S_i in ways that are consistent with its own guard (Q_j) and the monotonic closure of the guard of S_i ($\overrightarrow{Q_i}^*$).

On the other hand, if S_i and S_j are two programs that work on incomparable sets of inputs (i.e., no values that satisfy Q_i have subtypes that satisfy Q_j, and vice versa) then the assertion $\{\overrightarrow{Q_i}^* \wedge Q_j\}$ is false and the left hand side is equivalent to **abort**. Since **abort** is refined by any program at all, this means that S_i and S_j can be quite unrelated programs, as one would expect.

Theorem 8 also suggests how we might *calculate* a specification for one of the S_i programs. We have

$$\left[\forall i,j \in K \bullet \{\overrightarrow{Q_i}^* \wedge Q_j\} ; S_i \sqsubseteq_{args} S_j \right]_{args}$$

$\Leftrightarrow$ {*Since* S_j *does not involve* i}

$$\left[\forall j \in K \bullet (\exists i \bullet \{\overrightarrow{Q_i}^* \wedge Q_j\} ; S_i) \sqsubseteq_{args} S_j \right]_{args}$$

$\Leftrightarrow$ {*By definition of angelic choice and since* $[S_j \sqsubseteq S_j]$}

$$\left[\forall j \in K \bullet (\Diamond_{i \in K - \{j\}} \{\overrightarrow{Q_i}^* \wedge Q_j\} ; S_i) \sqsubseteq_{args} S_j \right]_{args} .$$

Thus, the weakest specification that S_j must satisfy for the system to support modular reasoning is

$$(\Diamond_{i \in K - \{j\}} \{\overrightarrow{Q_i}^* \wedge Q_j\} ; S_i) .$$

Although this specifies S_j in terms of all other programs in the system, in practice most of the $\{\overrightarrow{Q_i}^* \wedge Q_j\}$ assertions will be false. Since **abort** $\Diamond\ S = S$, this means that we only need to quantify over $i \in K - \{j\}$ such that Q_i is true for some superobjects of the inputs to S_j (i.e., $(\exists args \bullet \overrightarrow{Q_i}^* \wedge Q_j)$). If we start defining a system by defining all the guard predicates, then specify the programs that act upon all the superobjects in the system, this formula allows us to calculate the weakest specification of programs that act upon subobjects.

6 Modular Reasoning in Practice

Having defined precisely what it means for a system to support modular reasoning, it is now worth reviewing how we can reason about call statements. At each point in a program, various static properties are known to hold. That is, each call statement in a program is embedded in a particular context

$$\{Props\}\,;\mathbf{callv}\ p(u)\ ,$$

where *Props* is a predicate that captures all the properties known to hold immediately before the call statement. Although the exact values of the argument variables cannot be determined until the program is executed, the information in *Props* does constrain the possible values of the arguments. If *Props* implies one of the Q_i guards, the following simple theorem shows that the call statement is equivalent to S_i parameterised in the usual way.

Theorem 9. *If $\Psi\, p$ is described in $(Q_i, S_i)_{i\in K}$ form, then for any $i \in K$*

$$\left[\, Q_i[args\backslash u] \Rightarrow S_i[\mathbf{valres}\ args\backslash u] = \mathbf{callv}\ p(u)\,\right]\ .$$

Proof. Since $\{i \in K \bullet Q_i\}$ partitions $\mathsf{Val}^{\#p}$, all Q_j predicates other than Q_i are false. Therefore, Theorem 7 simplifies to $\mathbf{callv}\ p(u) = S_i[\mathbf{valres}\ args\backslash u]$. □

However, with subtyping in object-oriented languages, static properties such as *Props* are usually not precise enough to uniquely identify which guard Q_i is true. For example, *Props* might state that an argument is some type of point object, but would typically not specify whether it was a 2D point, a 3D point, or some other more specialised type of point. In object-oriented systems, such static properties are usually monotonic (i.e., true of all subobjects whenever they are true of u), in order to allow objects to be replaced by subobjects. This means that Theorem 9 is often not useful, since the Q_i guards are often not monotonic, so we cannot determine statically which Q_i is true of the argument expressions. However, in systems that support modular reasoning, the following more useful result holds.

Theorem 10. *If $\Psi\, p$ supports modular reasoning and is described in $(Q_i, S_i)_{i\in K}$ form, then for any $i \in K$*

$$\left[\, \overrightarrow{Q_i}^{*}[args\backslash u] \Rightarrow S_i[\mathbf{valres}\ args\backslash u] \sqsubseteq \mathbf{callv}\ p(u)\,\right]\ .$$

Proof. Let s be any state that satisfies $\overrightarrow{Q_i}^{*}[args\backslash u]$ and let $j \in K$ be the (unique) index such that $Q_j[args\backslash u]$ is true in s. Then, for any postcondition P, we have

$S_i[\mathbf{valres}\ args\backslash u]\ P\ s$

$\Leftrightarrow$ {*Since $\overrightarrow{Q_i}^{*}[args\backslash u]$ and $Q_j[args\backslash u]$ are true in s*}

$(\overrightarrow{Q_i}^{*}[args\backslash u] \wedge Q_j[args\backslash u] \wedge S_i[\mathbf{valres}\ args\backslash u]\ P)\ s$

$\Rightarrow$ {*By Theorem 8 and the monotonicity of value-result substitution [Morgan 88]*}

$S_j[\mathbf{valres}\ args\backslash u]\ P\ s$

$\Leftrightarrow$ {*By Theorem 9, since $Q_j[args\backslash u]$ is true in s*}

$\mathbf{callv}\ p(u)\ P\ s$

□

Theorem 10 gives us considerably more flexibility in reasoning about call statements than Theorem 9, because it gives a number of approximations to their actual behaviour. More importantly, the applicability condition ($\overrightarrow{Q_i}^*[args\backslash u]$) of Theorem 10 is far more useful, since it is weaker than $Q_i[args\backslash u]$ and it is monotonic just as the known static properties of u (e.g., *Props*) usually are.

For example, consider **callv** $add(x, y)$. We may know that x and y contain point objects that have positive X and Y values, but be unable to determine statically whether they are 2D or 3D points or some other more refined kind of point. Therefore we cannot statically determine whether the call statement will execute S_{22} or S_{33}. However, the monotonic closure of the guard for S_{22} is $\overrightarrow{\neg\,(\overline{P3(x)}^* \wedge \overline{P3(y)}^*)}^*$, which simplifies to true, so by Theorem 10 we can always use S_{22} as an approximation of the call statement. Using S_{22}, it is possible to prove that after the call, *res* satisfies

$$\begin{pmatrix} \overline{P2(res)}^* \\ res(X) = x(X) + y(X) \\ res(Y) = x(Y) + y(Y) \\ res(X) > 0 \wedge res(Y) > 0 \end{pmatrix} .$$

If the arguments x and y are actually 3D points, some information about *res* is lost due to using S_{22} as an approximation. Of course, if we can show statically that $\overline{P3(x)}^* \wedge \overline{P3(y)}^*$ then we can also use S_{33} as an approximation of the call statement and it allows us to deduce the additional facts

$$\begin{pmatrix} \overline{P3(res)}^* \\ res(Z) = x(Z) + y(Z) \end{pmatrix} .$$

Theorem 10 means that if a call statement such as **callv** $p(u)$ appears in a context $\{Props\}$; **callv** $p(u)$, where the *Props* predicate describes the set of possible program states before the call, then the set of valid approximations to **callv** $p(u)$ includes any S_i for $i \in K'$ where

$$K' \mathrel{\widehat{=}} \left\{ i \in K \mid Props \Rightarrow \overrightarrow{Q_i}^*[args\backslash u] \right\} .$$

Some of these approximations will give more information or different information than the others. If individual approximations do not give enough information for the situation at hand, they can be combined with the least upper bound operator to give the better approximation, $(\Diamond_{i \in K'} S_i)$.

An alternative way of deducing more about the call statement is to split it into different cases and analyse each case separately. For example, we can rewrite **callv** $add(x, y)$ as

$$\begin{array}{ll} \textbf{if} & (\overline{P3(x)}^* \wedge \overline{P3(y)}^*) \rightarrow \textbf{callv}\ add(x, y) \\ [] & \neg\,(\overline{P3(x)}^* \wedge \overline{P3(y)}^*) \rightarrow \textbf{callv}\ add(x, y) \\ \textbf{fi} & \end{array}$$

which allows us to use the stronger S_{33} as an approximation of the first call statement, because it is in a context where x and y are known to be 3D points (or subtypes thereof).

Case analysis is a general purpose technique for deducing more information about a call statement than could be deduced from considering only a single case. Each case may make additional assumptions about the argument values and those additional assumptions result in a bigger set of approximations to the call statement, thus allowing additional information to be deduced about its behaviour. This is because if T and T' are two sets of approximations to the call statement, then

$$T \subseteq T' \Rightarrow \left[(\Diamond\, T) \sqsubseteq_{\mathsf{Prog}} (\Diamond\, T') \right] .$$

As the number of cases considered increases, we get better approximations to the complete behaviour of the call statement. This technique can obviously become somewhat unwieldy when large numbers of cases are analysed. Fortunately, in many object-oriented systems, subtyping is mainly used to allow multiple implementations of a common abstraction (e.g., a *graphical figure* abstraction might have circles, polygons etc. as subtypes [Utting and Robinson 91]), and in this case a single approximation to a call statement is usually sufficient.

In conclusion, in systems that do not support modular reasoning, the *only* way of reasoning about a call statement is via case analysis of all programs that can possibly be executed (i.e., those whose guards may be satisfied by the argument values). In contrast, in systems that do support modular reasoning, Theorem 10 means that a single program can be used as an approximation to how the call statement will behave for all subobjects of the arguments. This is usually more practical than an exhaustive case analysis, and has the major advantage that the approximations remain unchanged when new procedures on subtype objects are added to a system, provided that the resulting system still supports modular reasoning.

Adding new subtype procedures to a system that does not support modular reasoning would potentially require all call statements to be reverified. Modular reasoning removes the need for this, but means that systems must be restricted to satisfy Defn. 5 (or equivalently, Theorem 8). Thus, we have a tradeoff between

- an unrestricted system in which reasoning about call statements requires exhaustive case analysis and changes to the system can require a major reverification of unchanged programs, and
- a system that is restricted to support modular reasoning. In this case we can use various approximations to reason about call statements and some kinds of changes to the system require no reverification of existing programs.

7 Transitive Modular Reasoning

Although the modular reasoning requirement given in Defn. 5 is a weak form of monotonicity, it does not enjoy an important transitivity property of monotonicity. To illustrate this, let p be any procedure name and u and v be variable sequences such that $u \leq v$, and define relations $M(u,v)$ and $MR(u,v)$ as

$$M(u,v) \mathrel{\widehat{=}} \left[\Psi\, p\, u \sqsubseteq_{\mathsf{Prog}} \Psi\, p\, v \right]$$

$$MR(u,v) \mathrel{\widehat{=}} \left[\Psi\, p\, u\, [\text{valres } args \backslash v] \sqsubseteq_{\mathsf{Prog}} \Psi\, p\, v\, [\text{valres } args \backslash v] \right] .$$

The M (MR) relation is true whenever $\Psi\, p$ satisfies the monotonicity (modular reasoning) requirement at u and v. It is easy to see that M is a transitive relation, because $\sqsubseteq_{\mathsf{Prog}}$ is. However, for MR to be transitive requires

$$[\, MR(u, v) \wedge MR(v, w) \Rightarrow MR(u, w)\,]$$

and this is not necessarily true.

The loss of this transitivity property means that proving that $\Psi\, M$ supports modular reasoning is generally more difficult. For a chain of argument values, $c_1, c_2, \ldots$, it is not sufficient to prove $MR(c_i, c_{i+1})$ for each i, instead we must consider all possible pairings $MR(c_i, c_j)$ such that $i < j$.

To remedy this loss of transitivity, we define another requirement, called the *transitive modular reasoning* requirement, that is stronger than the modular reasoning requirement and is transitive. Essentially, we replace the $v = args$ in Lemma 6 by $v \leq args$.

Definition 11. *Let p be a procedure name and u, v and $args$ be equal length sequences of variables. Then $\Psi\, p$* supports transitive modular reasoning *(with respect to value passing call statements) iff*

$$[\, u \leq v \wedge v \leq args \Rightarrow \Psi\, p\, u \sqsubseteq_{args} \Psi\, p\, v\,]_{u,v,args} \;.$$

Any definition that supports transitive modular reasoning also supports modular reasoning, since $v = args \Rightarrow v \leq args$. The next theorem shows that this stronger relation between programs is indeed transitive.

Theorem 12. *If p is a procedure name, and u, v and w are equal length sequences of variables such that $u \leq v \leq w$, then*

$$\left[\begin{pmatrix} v \leq args \Rightarrow (\Psi\, p\, u \sqsubseteq \Psi\, p\, v) \\ w \leq args \Rightarrow (\Psi\, p\, v \sqsubseteq \Psi\, p\, w) \end{pmatrix} \;\Rightarrow\; (w \leq args \Rightarrow \Psi\, p\, u \sqsubseteq \Psi\, p\, w) \right] \;.$$

Proof. Follows easily from $w \leq args \Rightarrow v \leq args$ and the transitivity of $\sqsubseteq$. □

Next we need an equivalent formulation of the transitive modular reasoning property for when $\Psi\, p$ is described using the predicate-program pair notation.

Theorem 13. *If $\Psi\, p$ is defined in $(Q_i, S_i)_{i \in K}$ form, it supports transitive modular reasoning iff*

$$\left[\forall i, j \in K \bullet \{\overrightarrow{\overline{Q_i}^{*} \wedge Q_j}\}\,; S_i \sqsubseteq_{args} S_j \right]_{args} \;.$$

Proof. Similar to that of Theorem 8. □

The only difference between this requirement and the modular reasoning requirement (Theorem 8) is that this one takes the monotonic closure of the whole assertion $\{\overline{Q_i}^{*} \wedge Q_j\}$ as well as of Q_i. This difference limits the way in which S_j can strengthen the precondition of S_i. It is still possible to strengthen it to some monotonic condition, but it is generally not possible to strengthen it to a non-monotonic precondition, unless the precondition of S_i is also non-monotonic. In practice, this additional restriction is probably quite acceptable, since preconditions are usually written in a monotonic style to allow for subtyping.

The next theorem shows how the transitivity of this stronger relation between S_i and S_j can be used to reduce the number of proofs required to show that a system supports (transitive) modular reasoning. Rather than considering transitivity over chains of individual value sequences, we use subsets of possible argument values to each of the programs. Note that $\overrightarrow{\overline{Q_j}^*} \sqsubseteq_{\mathsf{Pred}} \overrightarrow{\overline{Q_i}^*}$ means that any acceptable arguments to S_j (or subtypes thereof) are also acceptable arguments to S_i (or subtypes thereof).

Theorem 14. *If Ψp is described in $(Q_i, S_i)_{i \in K}$ form and supports transitive modular reasoning and for $i, j, k \in K$, we know that $\overrightarrow{\overline{Q_k}^*} \sqsubseteq \overrightarrow{\overline{Q_j}^*} \sqsubseteq \overrightarrow{\overline{Q_i}^*}$, then*

$$\left[\begin{pmatrix} \{\overrightarrow{\overline{Q_i}^* \wedge Q_j}\}; S_i \sqsubseteq S_j \\ \{\overrightarrow{\overline{Q_j}^* \wedge Q_k}\}; S_j \sqsubseteq S_k \end{pmatrix} \Rightarrow \{\overrightarrow{\overline{Q_i}^* \wedge Q_k}\}; S_i \sqsubseteq S_k \right] .$$

Proof. Since for any predicate P, it is true that $P \sqsubseteq \overline{P}^*$, we have $Q_k \sqsubseteq \overrightarrow{\overline{Q_k}^*} \sqsubseteq \overrightarrow{\overline{Q_j}^*} \sqsubseteq \overrightarrow{\overline{Q_i}^*}$ and $Q_j \sqsubseteq \overrightarrow{\overline{Q_j}^*} \sqsubseteq \overrightarrow{\overline{Q_i}^*}$, so the theorem simplifies to

$$\left[\begin{pmatrix} \{\overrightarrow{\overline{Q_j}^*}\}; S_i \sqsubseteq S_j \\ \{\overrightarrow{\overline{Q_k}^*}\}; S_j \sqsubseteq S_k \end{pmatrix} \Rightarrow \{\overrightarrow{\overline{Q_k}^*}\}; S_i \sqsubseteq S_k \right] .$$

Assuming the left hand side, we have

$$\{\overrightarrow{\overline{Q_k}^*}\}; S_i = \{\overrightarrow{\overline{Q_k}^*}\};\{\overrightarrow{\overline{Q_j}^*}\}; S_i \sqsubseteq \{\overrightarrow{\overline{Q_k}^*}\}; S_j \sqsubseteq S_k$$

□

One nice property of systems that support transitive modular reasoning is that they obey a simple rule that sometimes helps in finding the best approximation of a call statement when several are valid. The rule is only applicable to approximations with related sets of inputs, but this is a common case in practice.

Theorem 15. *If Ψp is described in $(Q_i, S_i)_{i \in K}$ form and supports transitive modular reasoning and and we know that $\overrightarrow{\overline{Q_j}^*} \sqsubseteq_{\mathsf{Pred}} \overrightarrow{\overline{Q_i}^*}$, then S_j is a better approximation of a value passing call statement* **callv** $p(u)$ *than S_i. That is,*

$$\left[\overrightarrow{\overline{Q_j}^*}[args\backslash u] \Rightarrow S_i[\textbf{valres } args\backslash u] \sqsubseteq S_j[\textbf{valres } args\backslash u] \sqsubseteq \textbf{callv } p(u) \right] .$$

Proof. The second refinement follows from Theorem 10. To prove the first refinement, note that $Q_j \sqsubseteq \overrightarrow{\overline{Q_j}^*} \sqsubseteq \overrightarrow{\overline{Q_i}^*}$, so transitive modular reasoning implies:

$$
\begin{aligned}
& \{\overrightarrow{\overline{Q_j}^*}\}; S_i \sqsubseteq S_j \\
\Rightarrow\ & \{\textit{Since value-result substitution is monotonic [Morgan 88]}\} \\
& (\{\overrightarrow{\overline{Q_j}^*}\}; S_i)[\textbf{valres } args\backslash u] \sqsubseteq S_j[\textbf{valres } args\backslash u] \\
\Leftrightarrow\ & \{\overrightarrow{\overline{Q_j}^*}[args\backslash u]\}; S_i[\textbf{valres } args\backslash u] \sqsubseteq S_j[\textbf{valres } args\backslash u] \\
\Leftrightarrow\ & \{\textit{Under the assumption } \overrightarrow{\overline{Q_j}^*}[args\backslash u]\} \\
& S_i[\textbf{valres } args\backslash u] \sqsubseteq S_j[\textbf{valres } args\backslash u]
\end{aligned}
$$

□

Intuitively, this theorem states that procedures defined on subtype values are better approximations than their supertype procedures. This is a useful property, but does not necessarily hold in systems that only support ordinary modular reasoning. For example, if we define Ψ ***add*** as

$$\Psi\ add\,[a,b] \mathrel{\widehat{=}} \begin{cases} S_{33only}, & \textbf{if } P3(a) \wedge P3(b) \\ S_{33}, & \textbf{if } \begin{pmatrix} \overline{P3(a)}^{*} \wedge \neg P3(a) \\ \overline{P3(b)}^{*} \wedge \neg P3(b) \end{pmatrix} \\ S_{22}, & \textbf{otherwise} \end{cases}$$

$$\text{where} \quad S_{33only} \mathrel{\widehat{=}} \{P3(arg_1) \wedge P3(arg_2)\}\,;\,S_{33}$$

then S_{33only} is *not* a better approximation to the call statement than S_{22}, since it aborts for proper subtypes of 3D point values, whereas S_{22} does not abort, but gives some useful information about the X and Y components. This definition of Ψ ***add*** does support ordinary modular reasoning, but not transitive modular reasoning (the non-monotonic strengthening of the precondition of S_{33only} is not allowable in a transitive modular reasoning system).

In conclusion, for practical object-oriented systems, the transitive modular reasoning requirement is probably preferable to the ordinary modular reasoning requirement. The additional restriction does not seem too arduous, and it has two major advantages:

- the required relationship between programs is transitive, which reduces the number of proofs required to show that a system supports transitive modular reasoning, and
- it satisfies Theorem 15, which gives a useful rule for choosing between different approximations to a call statement when several are valid. The importance of this rule is that for any chain of argument values $c_1 < c_2 < \ldots$ we know that $\Psi\ p\ c_1, \Psi\ p\ c_2, \ldots$ is a sequence of increasingly better approximations.

8 The 2D-3D point example

To illustrate how transitive modular reasoning makes it easier to prove that a given Ψ ***p*** supports modular reasoning and to illustrate some of the limitations of modular reasoning, we consider the 2D-3D point example again. This time we shall add two more procedures that specify what happens when a 2D point and a 3D point are added. We define Ψ ***add*** as

$$\Psi\ add\,[a,b] \mathrel{\widehat{=}} \begin{cases} S_{22}, & \textbf{if } \neg\,\overline{P3(a)}^{*} \wedge \neg\,\overline{P3(b)}^{*} \\ S_{23}, & \textbf{if } P2(a) \wedge \overline{P3(b)}^{*} \\ S_{32}, & \textbf{if } \overline{P3(a)}^{*} \wedge P2(b) \\ S_{33}, & \textbf{if } \overline{P3(a)}^{*} \wedge \overline{P3(b)}^{*} \end{cases}$$

where S_{33} and S_{22} are defined as above; S_{23} is defined as

$$S_{23} \mathrel{\widehat{=}} res : \left[\begin{array}{c} \overline{P2(arg_1)}^* \\ \overline{P3(arg_2)}^* \end{array} \Big/ \begin{array}{c} \overline{P3(res)}^* \\ res(X) = arg_1(X) + arg_2(X) \\ res(Y) = arg_1(Y) + arg_2(Y) \\ P2(arg_1) \Rightarrow res(Z) = arg_2(Z) \end{array} \right]$$

and S_{32} is the same as S_{23} but with arg_1 and arg_2 interchanged. For convenience, we name the guard predicates, Q_{33}, Q_{23}, Q_{32} and Q_{22} and let K be the set containing the four subscript names. The monotonic closures of each of these guards, expressed as predicates over arg_1 and arg_2 are

$$\begin{array}{l} \overline{Q_{22}}^* \Leftrightarrow \mathsf{true} \\ \overline{Q_{23}}^* \Leftrightarrow \overline{P2(arg_1)}^* \wedge \overline{P3(arg_2)}^* \\ \overline{Q_{32}}^* \Leftrightarrow \overline{P3(arg_1)}^* \wedge \overline{P2(arg_2)}^* \\ \overline{Q_{33}}^* \Leftrightarrow \overline{P3(arg_1)}^* \wedge \overline{P3(arg_2)}^* \end{array}$$

Although there are 16 possible pairings of these procedures to consider, for the majority of them $\overline{Q_i}^* \wedge Q_j$ is false. That is, we really only need to consider distinct $i, j \in K$ such that Q_j accepts an argument sequence v that has a sequence of superobjects u (i.e., $u \leq v$) that is accepted by Q_i. There are only five such pairs, so we need to show

$$\begin{array}{l} \left[\{\overline{P2(arg_1)}^* \wedge \overline{P3(arg_2)}^*\} ; S_{22} \sqsubseteq S_{23} \right] \\ \left[\{\overline{P3(arg_1)}^* \wedge \overline{P3(arg_2)}^*\} ; S_{23} \sqsubseteq S_{33} \right] \\ \left[\{\overline{P3(arg_1)}^* \wedge \overline{P2(arg_2)}^*\} ; S_{22} \sqsubseteq S_{32} \right] \\ \left[\{\overline{P3(arg_1)}^* \wedge \overline{P3(arg_2)}^*\} ; S_{32} \sqsubseteq S_{33} \right] \\ \left[\{\overline{P3(arg_1)}^* \wedge \overline{P3(arg_2)}^*\} ; S_{22} \sqsubseteq S_{33} \right] \end{array}$$

The last of these refinements follows from the others by transitivity. The remaining four are simple to prove; we consider the second one as an example.

$$\{\overline{P3(arg_1)}^* \wedge \overline{P3(arg_2)}^*\} ; S_{23} \sqsubseteq S_{33}$$

$$\Leftrightarrow res : \left[\begin{array}{c} \overline{P3(arg_1)}^* \\ \overline{P3(arg_2)}^* \end{array} \Big/ \begin{array}{c} \overline{P3(res)}^* \\ res(X) = arg_1(X) + arg_2(X) \\ res(Y) = arg_1(Y) + arg_2(Y) \\ P2(arg_1) \Rightarrow res(Z) = arg_2(Z) \end{array} \right] \sqsubseteq S_{33}$$

$$\Leftrightarrow \{Since\ \overline{P3(arg_1)}^* \Rightarrow \neg\, P2(arg_1)\}$$

$$res : \left[\begin{array}{c} \overline{P3(arg_1)}^* \\ \overline{P3(arg_2)}^* \end{array} \Big/ \begin{array}{c} \overline{P3(res)}^* \\ res(X) = arg_1(X) + arg_2(X) \\ res(Y) = arg_1(Y) + arg_2(Y) \end{array} \right] \sqsubseteq S_{33}$$

$\Leftrightarrow$ {*Since the postcondition of S_{33} is stronger*}

true .

It is interesting to note that if the $P2(arg_1)$ guard in the postcondition was removed, then this refinement would not hold because S_{23} would effectively be treating all arg_1 values as if they had Z components equal to zero, which would contradict S_{33}. On the other hand, if we strengthen S_{23} by adding

$$P3(arg_1) \Rightarrow res(Z) = arg_1(Z) + arg_2(Z)$$

to the postcondition, the system still supports transitive modular reasoning. These scenarios illustrate the basic restriction imposed by (either kind of) modular reasoning; when specifying a program S_i, it is not sufficient to merely specify its behaviour on arguments that satisfy Q_i, instead its behaviour on subtypes of those arguments must also be considered.

If the specification of a type is not sufficently precise about the effects of its procedures, then users of that type may not be able to deduce the properties that they need. On the other hand, if the specification is more precise than necessary, then subtype procedures will be undesirably constrained by the modular reasoning requirement. Finding a useful balance between these two alternatives is not always easy. This is not surprising, since designing type specifications that allow subtyping is similar to the well known problem of designing good interfaces for reusable software components, which is also difficult [Tracz 90].

9 Related Work and Conclusions

We have shown how the refinement calculus can be extended to support object-oriented programming with a generalised procedure call that allows multiple dispatch late binding, in the style of CLOS [Keene 89] or Cecil [Chambers 92]. The concept of modular reasoning has been defined informally in the literature, both for non-object-oriented systems [Ernst et al. 85] and for object-oriented systems [Leavens and Weihl 90]. This paper gives the first formal definition (Defn. 5), which intuitively seems to capture the minimum possible requirements. In addition, we have defined a stronger condition that has practical advantages such as transitivity.

In Utting's thesis [Utting 92], it is shown how aliasing and various forms of incremental modification can easily be incorporated into our model. Restricting our model to *single dispatch* late binding leads to models that are more closely related to languages such as Eiffel [Meyer 88] and POOL [America 91] [America 89], where an encapsulated set of procedures (i.e., a *class* or a *type*) can be associated with each object. These single dispatch models enjoy stronger encapsulation properties than multiple dispatch systems. Also, in the thesis, a more general subtyping relation than $\leq$ is discussed, which directly supports data refinement between objects and their subobjects [Utting 92, Chap. 8]. However, this more general subtyping relation is more complex than the one used here and further research is needed to determine whether all the same results hold.

A number of formal methods that support modular reasoning about subtyping in single dispatch languages have been proposed previously. The Eiffel and POOL languages associate preconditions and postconditions with procedures and informally require subtypes to weaken preconditions and strengthen postconditions. They also associate an *invariant* predicate with each type and allow subtypes to strengthen the invariants of their supertypes. These requirements are one possible instantiation of our more general approach. In particular, strengthening invariants in subtypes is similar to how our model allows the precondition of a program S_j to be strengthened by assuming $\{\overline{Q_i}^{*} \wedge Q_j\}$.[12]

The closest work to ours is that of Leavens [Leavens 89] [Leavens and Weihl 90] [Leavens 91]. This is an algebraic model of subtyping in an object-oriented functional language with multiple dispatching. Essentially, it supports modular reasoning by requiring each subtype object to *simulate* (i.e., be a data refinement of) some supertype object. A Hoare logic is then defined for verifying programs by using *nominal* supertypes as approximations of their subtypes.

One advantage of our approach is that the use of predicate transformer semantics handles assignment easily, whereas Leavens' results do not handle assignment. Our approach may be better suited to handling mutable objects, whereas it seems harder to adapt Leavens' algebraic approach to deal with mutation. However, Leavens' subtyping relation supports data refinement directly, whereas our simple $\leq$ subtyping relation does not, although data refinement may still be used to implement objects in languages that support information hiding. Another difference is that, in his approach, specifications and code are distinct, whereas our refinement calculus extension has the usual advantages of being a wide spectrum language that combines specifications and code into a single framework and emphasises refinement by calculation rather than post-hoc verification.

References

[Agrawal et al. 91] Rakesh Agrawal, Linda G. DeMichiel, and Bruce G. Lindsay. Static type checking of multi-methods. *Sigplan Notices*, 26(11):113–128, November 1991.

[America 89] Pierre America. A behavioural approach to subtyping in object-oriented programming languages. Technical Report 443, Philips Research Laboratories, P.O. Box 80000, 5600 JA Eindhoven, The Netherlands, 1989.

[America 91] Pierre America. Designing an object-oriented programming language with behavioural subtyping. In J. W. de Bakker, W. P. de Roever, and G. Rozenberg, editors, *Foundations of Object-Oriented Languages*, pages 60–90. Springer-Verlag, 1991. Proceedings of REX School/Workshop, May/June 1990, LNCS 489.

[Back and Wright 89] R. J. R Back and J. von Wright. A lattice-theoretic basis for a specification language. In J. L. A. van de Snepscheut, editor, *Mathematics of Program Construction*, pages 139–156. Springer-Verlag, June 1989. LNCS 375.

[12] Note that in a single dispatch language, the Q_i and Q_j predicates depend upon only one input value, corresponding to the self (*Current*) variable of POOL (Eiffel).

[Bobrow et al. 88] Daniel G. Bobrow, Linda G. DeMichiel, Richard P. Gabriel, Sonya E. Keene, Gregor Kiczales, and David A. Moon. Common Lisp object system specification. *SIGPLAN Notices*, 23, September 1988. X3J13 Document 88-002R, June 1988.

[Cardelli and Wegner 85] Luca Cardelli and Peter Wegner. On understanding types, data abstraction, and polymorphism. *Computing Surveys*, 17(4), December 1985.

[Chambers 92] Craig Chambers. Object-oriented multi-methods in Cecil. In O. Lehrmann Madsen, editor, *ECOOP '92*, pages 33–56. Springer-Verlag, 1992. LNCS 615.

[Dijkstra 76] Edsger W. Dijkstra. *A Discipline of Programming*. Prentice-Hall, 1976.

[Ernst et al. 85] George W. Ernst, James A. Menegay, Raymond J. Hookway, and William F. Ogden. Semantics of programming languages for modular verification. Technical Report CES-85-4, Case Western Reserve University, October 1985.

[Gardiner and Morgan 91] Paul Gardiner and Carroll Morgan. Data refinement of predicate transformers. *Theoretical Computer Science*, 87:143–162, 1991.

[Hebel and Johnson 90] Kurt J. Hebel and Ralph E. Johnson. Arithmetic and double dispatching in Smalltalk. *Journal of Object-Oriented Programming*, 2(6), March 1990.

[Keene 89] Sonya E. Keene. Extended vision of methods in CLOS. *Journal of Object-Oriented Programming*, 2(4):76–78, November 1989.

[LaLonde and Pugh 90] Wilf R. LaLonde and John R. Pugh. *Inside Smalltalk: Volume 1*. Prentice-Hall, 1990.

[Leavens and Weihl 90] Gary T. Leavens and William E. Weihl. Reasoning about object-oriented programs that use subtypes. *SIGPLAN Notices*, 25(10):212–223, October 1990. Proceedings of OOPSLA '90.

[Leavens 89] Gary Todd Leavens. *Verifying Object-Oriented Programs that use Subtypes*. PhD thesis, MIT Laboratory for Computer Science, 1989. Technical Report 439.

[Leavens 91] Gary Leavens. Modular specification and verification of object-oriented programs. *IEEE Software*, 8(4):72–80, July 1991.

[Meyer 88] Bertrand Meyer. *Object-Oriented Software Construction*. Prentice-Hall, 1988.

[Morgan and Robinson 87] Carroll Morgan and Ken Robinson. Specification statements and refinement. *IBM Journal of Research and Development*, 31(5), September 1987.

[Morgan 88] Carroll Morgan. Procedures, parameters and abstraction. *Science of Computer Programming*, 11, 1988.

[Morgan 90] Carroll Morgan. *Programming from Specifications*. Prentice Hall, 1990.

[Morris 87] Joseph M. Morris. A theoretical basis for stepwise refinement and the programming calculus. *Science of Computer Programming*, 9:287–306, 1987.

[Tracz 90] Will Tracz. Modularization: Approaches to reuse in Ada. *Journal of Pascal, Ada and Modula-2*, 9(5):10–25, September 1990.

[Utting and Robinson 91] Mark Utting and Ken Robinson. The object-oriented lollipop: An example of subtyping. In *Proceedings of the 1991 Australian Software Engineering Conference*, 1991.

[Utting 92] Mark Utting. *An Object-Oriented Refinement Calculus with Modular Reasoning*. PhD thesis, University of New South Wales, Australia, 1992.

[Wright 90] Joakim von Wright. *A Lattice-theoretical Basis for Program Refinement*. PhD thesis, Åbo Akademi University, Lemminkäisenkatu 14, SF-20500 Turku, Finland, September 1990.

An Alternative Derivation of a Binary Heap Construction Function

Lex Augusteijn

Philips Research Labs, Eindhoven, the Netherlands

1 Introduction

In [Sch92] a derivation of a binary heap construction algorithm is presented, making use of program inversion techniques. In this paper, we present an alternative derivation, using algebraic program transformation techniques in a purely functional setting. The main technique used is that of recursion elimination.

2 Problem specification

We define a binary heap by:

$\langle\rangle$ is a heap,

$\langle l, m, r\rangle$ is a heap when m is an integer, l, r are heaps and $l \geq m$ and $r \geq m$, with

$$\langle\rangle \geq m \;\equiv\; \text{true}$$
$$\langle l, n, r\rangle \geq m \;\equiv\; n \geq m$$

The in-order of a heap is defined as the following function in from heaps to sequences of integers.

$$\text{in}\ \langle\rangle \;=\; \epsilon$$
$$\text{in}\ \langle l, m, r\rangle \;=\; \text{in}\ l +\!\!+ \{m\} +\!\!+ \text{in}\ r$$

For notational convenience, we define the following relation over $\mathbf{N}^* \times \mathbf{N}$.

$$\epsilon \leq m \;\equiv\; \text{true}$$
$$\{n\} +\!\!+ s \leq m \;\equiv\; n \leq m$$

The problem is to find an efficient implementation of the function heaps (from $\mathbf{N}^*$ onto a set of heaps), defined by

$$\text{heaps}\ s \;=\; \{h \mid s = \text{in}\ h\}$$

We will derive an implementation of this function by means of recursion elimination. That implementation can be modified into a version that delivers one heap, instead of a set of possible heaps. This latter, deterministic, function shows linear time behavior, as the function presented in [Sch92].

3 Problem generalization

Instead of rewriting the function heaps, we rewrite a more general version f of this function. As a left to right traversal of the input string (the argument of heaps) should not make use of look-ahead, a more general function should not inspect the whole input string, but only part of it. This part is bounded by a top element m of a heap under construction. The generalized function f takes this m as an argument and delivers not just a set of heaps, but a set of (heap, remainder of input) pairs.

$$\mathsf{f\ m\ s} \ = \ \{(\mathsf{h},\mathsf{t}) \mid \mathsf{s} \ = \ \mathsf{in\ h} +\!\!\!+ \mathsf{t},\ \mathsf{h} \geq \mathsf{m}\}$$

We can express the function heaps by means of f in the following way:

$$\begin{array}{ll} & \mathsf{heaps\ s} \\ = & \text{(definition of heaps and f)} \\ & \{\mathsf{h} \mid (\mathsf{h},\mathsf{t}) \in \mathsf{f}\ (-\infty)\ \mathsf{s},\ \mathsf{t} \ = \ \epsilon\} \\ = & \text{(definition of } \leq\text{)} \\ & \{\mathsf{h} \mid (\mathsf{h},\mathsf{t}) \in \mathsf{f}\ (-\infty)\ \mathsf{s},\ \mathsf{t} \leq -\infty\} \end{array}$$

4 A recursive definition

We can rewrite the definition of f into a recursive form by algebraic program transformations:

$$\begin{array}{ll} & \mathsf{f\ m\ s} \\ = & \text{(definition of f)} \\ & \{(\mathsf{h},\mathsf{t}) \mid \mathsf{s} \ = \ \mathsf{in\ h} +\!\!\!+ \mathsf{t},\ \mathsf{h} \geq \mathsf{m}\} \\ = & \text{(h is either empty or not)} \\ & \{(\langle\rangle,\mathsf{t}) \mid \mathsf{s} \ = \ \mathsf{in}\ \langle\rangle +\!\!\!+ \mathsf{t},\ \langle\rangle \geq \mathsf{m}\} +\!\!\!+ \\ & \{(\langle\mathsf{l},\mathsf{a},\mathsf{r}\rangle,\mathsf{w}) \mid \mathsf{s} = \mathsf{in}\ \langle\mathsf{l},\mathsf{a},\mathsf{r}\rangle +\!\!\!+ \mathsf{w},\ \mathsf{l} \geq \mathsf{a},\ \mathsf{r} \geq \mathsf{a},\ \langle\mathsf{l},\mathsf{a},\mathsf{r}\rangle \geq \mathsf{m}\} \\ = & \text{(definition of in, heap and } \geq\text{)} \\ & \{(\langle\rangle,\mathsf{s})\} +\!\!\!+ \\ & \{(\langle\mathsf{l},\mathsf{a},\mathsf{r}\rangle,\mathsf{w}) \mid \mathsf{s} = \mathsf{in}\ \langle\mathsf{l},\mathsf{a},\mathsf{r}\rangle +\!\!\!+ \mathsf{w},\ \mathsf{l} \geq \mathsf{a},\ \mathsf{r} \geq \mathsf{a},\ \mathsf{a} \geq \mathsf{m}\} \\ = & \text{(definition of in)} \\ & \{(\langle\rangle,\mathsf{s})\} +\!\!\!+ \\ & \{(\langle\mathsf{l},\mathsf{a},\mathsf{r}\rangle,\mathsf{w}) \mid \mathsf{s} = \mathsf{t} +\!\!\!+ \{\mathsf{a}\} +\!\!\!+ \mathsf{u} +\!\!\!+ \mathsf{w},\ \mathsf{t} = \mathsf{in\ l},\ \mathsf{u} = \mathsf{in\ r},\ \mathsf{l} \geq \mathsf{a},\ \mathsf{r} \geq \mathsf{a},\ \mathsf{a} \geq \mathsf{m}\} \\ = & \text{(definition of f)} \\ & \{(\langle\rangle,\mathsf{s})\} +\!\!\!+ \\ & \{(\langle\mathsf{l},\mathsf{a},\mathsf{r}\rangle,\mathsf{w}) \mid \mathsf{s} \ = \ \mathsf{t} +\!\!\!+ \{\mathsf{a}\} +\!\!\!+ \mathsf{u} +\!\!\!+ \mathsf{w},\ \mathsf{t} \ = \ \mathsf{in\ l},\ \mathsf{l} \geq \mathsf{a},\ \mathsf{a} \geq \mathsf{m},\ (\mathsf{r},\mathsf{w}) \in \\ & \mathsf{f\ a}\ (\mathsf{u} +\!\!\!+ \mathsf{w})\} \\ = & \text{(definition of f again, } \mathsf{v} = \mathsf{u} +\!\!\!+ \mathsf{w}\text{)} \\ & \{(\langle\rangle,\mathsf{s})\} +\!\!\!+ \\ & \{(\langle\mathsf{l},\mathsf{a},\mathsf{r}\rangle,\mathsf{w}) \mid (\mathsf{l}, \{\mathsf{a}\} +\!\!\!+ \mathsf{v}) \in \mathsf{f\ m\ s},\ \mathsf{l} \geq \mathsf{a},\ \mathsf{a} \geq \mathsf{m},\ (\mathsf{r},\mathsf{w}) \in \mathsf{f\ a\ v}\} \end{array}$$

5 Intermezzo: recursive descent and ascent

In [Aug93] a form of recursion elimination is presented. It maps a so called recursive *descent* function onto a recursive *ascent* function. Both descent and ascent functions are defined as higher order functions that can be parameterized to obtain an instance.

Let T, the set of problems and S, the set of solutions, be finite sets and

$$\begin{array}{ll} \mathsf{simple} & : T \mapsto \mathcal{P}(S) \\ \mathsf{combine} & : T \times T \times S \mapsto \mathcal{P}(S) \\ \mathsf{sub} & : T \times T \mapsto \mathsf{Bool} \\ \leq \equiv \mathsf{sub}^* & \end{array}$$

The function simple is able to solve a so called trivial problem only. On non-trivial problems, it delivers $\{\}$.

The relation sub is the sub-problem relation. A problem α is assumed to have a (possibly empty) set of sub-problems $\{\beta \mid \beta \ \mathsf{sub}\ \alpha\}$. Given a problem α, a sub-problem β of it and a solution x of β, the function combine delivers the set of solutions of α that are consistent with the pair (β, x).

We define a *recursive descent function* as an instance p of the higher order function RD.

$$\mathsf{RD} \ : \ (T \mapsto \mathcal{P}(S)) \times (T \times T \mapsto \mathsf{Bool}) \times (T \times T \times S \mapsto \mathcal{P}(S)) \mapsto (T \mapsto \mathcal{P}(S))$$

$$\begin{array}{l} \mathsf{RD\ simple\ sub\ combine} = \mathsf{p} \\ \textbf{where}\ \mathsf{p}\ \alpha = \mathsf{simple}\ \alpha + \!\!+ \ + \!\!+ /\{\mathsf{combine}\ \alpha\ \beta\ \mathsf{x} \mid \beta\ \mathsf{sub}\ \alpha,\ \mathsf{x} \in \mathsf{p}\ \beta\} \end{array}$$

We define a *recursive ascent function* as an instance p of the higher order function RA.

$$\begin{array}{l} \mathsf{RA\ simple\ sub\ combine} = \mathsf{p} \\ \textbf{where}\ \mathsf{p}\ \alpha = + \!\!+ /\{\mathsf{q}\ \beta\ \mathsf{x} \mid \beta \leq \alpha,\ \mathsf{x} \in \mathsf{simple}\ \beta\} \\ \qquad \textbf{where}\ \mathsf{q}\ \beta\ \mathsf{x} = \{\mathsf{x} \mid \beta = \alpha\} + \!\!+ \\ \qquad\qquad + \!\!+ /\{\mathsf{q}\ \gamma\ \mathsf{y} \mid \gamma \leq \alpha,\ \beta\ \mathsf{sub}\ \gamma,\ \mathsf{y} \in \mathsf{combine}\ \gamma\ \beta\ \mathsf{x}\} \end{array}$$

When the function q, which depends on α, is given a sub-problem β and a solution x of β, it returns the solutions of α that can be constructed, directly or indirectly, out of it. The RA function performs iteration from below, starting with trivial problems and constructing solutions for larger and larger problems.

In [Aug93] it is proven that

$$\mathsf{RD} = \mathsf{RA}$$

Thus, when we can find an instance of RD, we can rewrite it into the RA equivalent.

6 An iterative definition

Recall the definition of f:

$$
\begin{aligned}
&\mathsf{f\ m\ s}\ =\ \{(\langle\rangle,\mathsf{s})\} \mathbin{+\!\!+}\\
&\{((\langle \mathsf{l},\mathsf{a},\mathsf{r}\rangle,\mathsf{w}) \mid (\mathsf{l},\{\mathsf{a}\} \mathbin{+\!\!+} \mathsf{v}) \in \mathsf{f\ m\ s},\ \mathsf{l} \geq \mathsf{a},\ \mathsf{a} \geq \mathsf{m},\ (\mathsf{r},\mathsf{w}) \in \mathsf{f\ a\ v}\}
\end{aligned}
$$

This definition can be brought into the descent form by choosing:

$$
\begin{aligned}
&\mathsf{simple}\ (\mathsf{m},\mathsf{s})\ =\ \{(\langle\rangle,\mathsf{s})\}\\
&(\mathsf{n},\mathsf{t})\ \mathsf{sub}\ (\mathsf{m},\mathsf{s})\ =\ \mathsf{n}=\mathsf{m} \wedge \mathsf{t}=\mathsf{s}\\
&\mathsf{combine}\ (\mathsf{m},\mathsf{s})\ (\mathsf{m},\mathsf{s})\ (\mathsf{l},\mathsf{u})\ =\ \{((\langle \mathsf{l},\mathsf{a},\mathsf{r}\rangle,\mathsf{w}) \mid \mathsf{u} = \{\mathsf{a}\} \mathbin{+\!\!+} \mathsf{v},\ \mathsf{l} \geq \mathsf{a},\ \mathsf{a} \geq\\
&\mathsf{m},\ (\mathsf{r},\mathsf{w}) \in \mathsf{f}\ (\mathsf{a},\mathsf{v})\}\\
&\leq\ \equiv\ =
\end{aligned}
$$

Then we have:

$$
\begin{aligned}
&\quad \mathsf{f}\\
=&\quad \text{(exercise)}\\
&\quad \mathsf{RD\ simple\ sub\ combine}\\
=&\quad (\mathsf{RD} = \mathsf{RA})\\
&\quad \mathsf{RA\ simple\ sub\ combine}\\
=&\quad \text{(substitute and work out)}\\
&\quad \mathsf{p}\ \textbf{where}\\
&\quad \mathsf{p}\ (\mathsf{m},\mathsf{s})\ =\ \mathbin{+\!\!+}/\{\mathsf{q}\ (\mathsf{m},\mathsf{s})\ (\langle\rangle,\mathsf{s})\}\\
&\quad \textbf{where}\ \mathsf{q}\ (\mathsf{m},\mathsf{s})\ (\mathsf{l},\mathsf{u})\\
&\qquad\quad =\ \{(\mathsf{l},\mathsf{u})\} \mathbin{+\!\!+}\\
&\qquad\qquad \mathbin{+\!\!+}/\{\mathsf{q}\ (\mathsf{m},\mathsf{s})\ ((\langle \mathsf{l},\mathsf{a},\mathsf{r}\rangle,\mathsf{w}) \mid \mathsf{u} = \{\mathsf{a}\} \mathbin{+\!\!+} \mathsf{v},\ \mathsf{l} \geq \mathsf{a},\ \mathsf{a} \geq \mathsf{m},\ (\mathsf{r},\mathsf{w}) \in \mathsf{p}\ (\mathsf{a},
\end{aligned}
$$

The last form can be simplified by observing that all invocation of q have the same first argument (m, s), which can thus be left out.

$$
\begin{aligned}
&\mathsf{p\ m\ s}\ =\ \mathsf{q}\ \langle\rangle\ \mathsf{s}\\
&\textbf{where}\ \mathsf{q\ l\ u}\ =\ \{(\mathsf{l},\mathsf{u})\} \mathbin{+\!\!+}\\
&\qquad\qquad \mathbin{+\!\!+}/\{\mathsf{q}\ \langle \mathsf{l},\mathsf{a},\mathsf{r}\rangle\ \mathsf{w} \mid \mathsf{u} = \{\mathsf{a}\} \mathbin{+\!\!+} \mathsf{v},\ \mathsf{l} \geq \mathsf{a},\ \mathsf{a} \geq \mathsf{m},\ (\mathsf{r},\mathsf{w}) \in \mathsf{p\ a\ v}\}
\end{aligned}
$$

7 Strengthening

By the transformation from descent to ascent, one of the two recursive calls to f (i.e. p) has been removed. This was the recursive call that delivered the left-most sub-tree of a heap and we are left with a recursive call that delivers the right-most sub-tree. We can exploit this fact in strengthening the definition of p in such a way that *maximal* right-most sub-trees are delivered. The corresponding function p′ is defined by:

$$
\begin{aligned}
& \mathsf{p'\ m\ s} \\
= \ & (\text{definition of } \mathsf{p'}) \\
& \{(\mathsf{h,v}) \mid (\mathsf{h,v}) \in \mathsf{p\ m\ s},\ \mathsf{v} \leq \mathsf{m}\} \\
= \ & (\text{definition of } \mathsf{p}) \\
& \{(\mathsf{h,v}) \mid (\mathsf{h,v}) \in \mathsf{q}\ \langle\rangle\ \mathsf{s},\ \mathsf{v} \leq \mathsf{m}\} \ \textbf{where}\ \mathsf{q\ l\ u} = \textit{unchanged} \\
= \ & (\text{definition of } \mathsf{q'}) \\
& \mathsf{q'}\ \langle\rangle\ \mathsf{s} \\
& \textbf{where}\ \mathsf{q'\ l\ u} = \{(\mathsf{h,v}) \mid (\mathsf{h,v}) \in \mathsf{q\ l\ u},\ \mathsf{v} \leq \mathsf{m}\} \\
& \qquad\quad \mathsf{q\ l\ u} = \textit{unchanged}
\end{aligned}
$$

We can rewrite this definition of $\mathsf{q'}$ by substituting the definition of q into it.

$$
\begin{aligned}
\mathsf{q'\ l\ u} = \ & \{(\mathsf{l,u}) \mid \mathsf{u} \leq \mathsf{m}\} \mathbin{+\!\!\!+} \\
& \mathbin{+\!\!\!+}/\{\mathsf{q'}\ \langle \mathsf{l,a,r}\rangle\ \mathsf{w} \mid \mathsf{u} = \{\mathsf{a}\} \mathbin{+\!\!\!+} \mathsf{v},\ \mathsf{l} \geq \mathsf{a},\ \mathsf{a} \geq \mathsf{m},\ (\mathsf{r,w}) \in \mathsf{p\ a\ v}\}
\end{aligned}
$$

We can add the guard $\mathsf{w} \leq \mathsf{a}$ to the last form of q, since it can be proven that $\neg(\mathsf{w} \leq \mathsf{a}) \Rightarrow \mathsf{q'}\ \langle \mathsf{l,a,r}\rangle\ \mathsf{w} = \{\}$.

$$
\begin{aligned}
& \mathsf{q'\ l\ u} \\
= \ & (\neg(\mathsf{w} \leq \mathsf{a}) \Rightarrow \mathsf{q'}\ \langle \mathsf{l,a,r}\rangle\ \mathsf{w} = \{\}) \\
& \{(\mathsf{l,u}) \mid \mathsf{u} \leq \mathsf{m}\} \mathbin{+\!\!\!+} \\
& \mathbin{+\!\!\!+}/\{\mathsf{q'}\ \langle \mathsf{l,a,r}\rangle\ \mathsf{w} \mid \mathsf{u} = \{\mathsf{a}\} \mathbin{+\!\!\!+} \mathsf{v},\ \mathsf{l} \geq \mathsf{a},\ \mathsf{a} \geq \mathsf{m},\ (\mathsf{r,w}) \in \mathsf{p\ a\ v},\ \mathsf{w} \leq \mathsf{a}\} \\
= \ & (\text{definition of } \mathsf{p'}) \\
& \{(\mathsf{l,u}) \mid \mathsf{u} \leq \mathsf{m}\} \mathbin{+\!\!\!+} \\
& \mathbin{+\!\!\!+}/\{\mathsf{q'}\ \langle \mathsf{l,a,r}\rangle\ \mathsf{w} \mid \mathsf{u} = \{\mathsf{a}\} \mathbin{+\!\!\!+} \mathsf{v},\ \mathsf{l} \geq \mathsf{a},\ \mathsf{a} \geq \mathsf{m},\ (\mathsf{r,w}) \in \mathsf{p'\ a\ v}\}
\end{aligned}
$$

8 A non-deterministic version

We can rewrite the set-valued versions of $\mathsf{p'}$ and $\mathsf{q'}$ into non-deterministic ones by returning a single value instead of a singleton set and by replacing set union ($\mathbin{+\!\!\!+}$) by non-deterministic choice.

$$
\begin{aligned}
\mathsf{p''\ m\ s} &= \mathsf{q''\ m}\ \langle\rangle\ \mathsf{s} \\
\mathsf{q''\ m\ l\ u} &= \textbf{if}\ \ \mathsf{u} \leq \mathsf{m} \qquad\qquad\qquad\quad\ \rightarrow (\mathsf{l,u}) \\
& \quad\ [\!]\ \ \mathsf{u} = \{\mathsf{a}\} \mathbin{+\!\!\!+} \mathsf{v} \wedge \mathsf{l} \geq \mathsf{a} \wedge \mathsf{a} \geq \mathsf{m} \rightarrow \mathsf{q''\ m}\ \langle \mathsf{l,a,r}\rangle\ \mathsf{w} \\
& \qquad\qquad\qquad\qquad\qquad\qquad\qquad \textbf{where}\ (\mathsf{r,w}) = \mathsf{q''\ a}\ \langle\rangle\ \mathsf{v} \\
& \quad\ \textbf{fi}
\end{aligned}
$$

Observe that q is tail-recursive.

9 Complexity analysis

We can extend the definition of q with a couple of ghost-variables that count the number of recursive calls and storage usage. These ghost variables are:

r : the number of recursive calls
d : the recursion depth
c : the amount of heap cells allocated

We need not increase the recursion depth at the tail-recursive call.

$$
\begin{array}{ll}
\mathsf{q\ m\ l\ u\ r\ d\ c} & \\
= & \\
\textbf{if}\ \ \mathsf{u} \le \mathsf{m} & \rightarrow (\mathsf{l}, \mathsf{u}, \mathsf{r}, \mathsf{c}) \\
[\!]\ \ \mathsf{u} = \{\mathsf{a}\} +\!\!\!+ \mathsf{v} \wedge \mathsf{l} \ge \mathsf{a} \wedge \mathsf{a} \ge \mathsf{m} & \rightarrow \mathsf{q\ m}\ \langle \mathsf{l}, \mathsf{a}, \mathsf{r} \rangle\ \mathsf{w}\ (\mathsf{r}' + 1)\ \mathsf{d}\ (\mathsf{c}' + 1) \\
& \quad \textbf{where}\ (\mathsf{r}, \mathsf{w}, \mathsf{r}', \mathsf{c}') = \mathsf{q\ a}\ \langle\rangle\ \mathsf{v}\ (\mathsf{r} + 1)\ (\mathsf{d} + 1) \\
\textbf{fi} &
\end{array}
$$

With respect to these ghost variables, we can observe the following invariants (using $\#s$ to denote the length of s).

For the call $\mathsf{q\ a}\ \langle\rangle\ \mathsf{v}\ (\mathsf{r}+1)\ (\mathsf{d}+1)\ \mathsf{c}$ we need the properties that if $\mathsf{q\ m\ l\ u\ r\ d\ c} = (\mathsf{l}', \mathsf{u}'\mathsf{r}', \mathsf{c}')$ then $\#\mathsf{u} + \mathsf{c} = \#\mathsf{u}' + \mathsf{c}'$ and $\#\mathsf{u} + \mathsf{r} - \mathsf{c} = \#\mathsf{u}' + \mathsf{r}' - \mathsf{c}'$, which are proven by induction.

$$
\begin{array}{l}
\#\mathsf{v} + (\mathsf{d} + 1) + \mathsf{c} = \#\mathsf{u} + \mathsf{d} + \mathsf{c} \\
\text{and} \\
\#\mathsf{v} + (\mathsf{r} + 1) - \mathsf{c} = \#\mathsf{u} + \mathsf{r} - \mathsf{c}
\end{array}
$$

For the call $\mathsf{q\ m}\ \langle \mathsf{l}, \mathsf{a}, \mathsf{r} \rangle\ \mathsf{w}\ (\mathsf{r}' + 1)\ \mathsf{d}\ (\mathsf{c}' + 1)$:

$$
\begin{array}{lll}
\#\mathsf{w} + \mathsf{d} + (\mathsf{c}' + 1) & = \#\mathsf{w} + \mathsf{d} + 1 + \mathsf{c}' & \\
& = \#\mathsf{v} + \mathsf{d} + 1 + \mathsf{c} & \text{(by induction)} \\
& = \#\mathsf{u} + \mathsf{d} + \mathsf{c} & \\
\text{and} & & \\
\#\mathsf{w} + (\mathsf{r}' + 1) - (\mathsf{c}' + 1) & = \#\mathsf{w} + \mathsf{r}' - \mathsf{c}' & \\
& = \#\mathsf{v} + (\mathsf{r} + 1) - \mathsf{c} & \text{(by induction)} \\
& = \#\mathsf{u} + \mathsf{r} - &
\end{array}
$$

Thus, $\#\mathsf{u} + \mathsf{d} + \mathsf{c}$ and $\#\mathsf{u} + \mathsf{r} - \mathsf{c}$ are *constant* for all calls to q. The root call is given by:

$$
\mathsf{q}\ (-\infty)\ \langle\rangle\ \mathsf{input}\ 1\ 1\ 0
$$

and hence

$$
\begin{array}{l}
\#\mathsf{u} + \mathsf{d} + \mathsf{c}\ =\ \#\mathsf{input} + 1 + 0 \\
\Rightarrow \mathsf{d} + \mathsf{c} \le \#\mathsf{input} + 1
\end{array}
$$

and

$$\begin{aligned}&\#\mathsf{u}+\mathsf{r}-\mathsf{c} \;=\; \#\mathsf{input}+1-0\\ &\Rightarrow \mathsf{r}-\mathsf{c}\leq \#\mathsf{input}+1\\ &\Rightarrow \mathsf{r}\leq 2\#\mathsf{input}+2\end{aligned}$$

Since $\mathsf{d}+\mathsf{c}$ represents the storage usage of the algorithm, the algorithm needs only constant extra space, as in [Sch92]. Since each call takes constant time and the number of calls is linear, the algorithm consumes linear time.

References

[Aug93] Lex Augusteijn. *Functional Programming, Program Transformations and Compiler Construction.* Eindhoven, 1993. to appear.

[Sch92] Berry Schoenmakers. Inorder traversal of a binary heap and its inversion in optimal time and space. In *Second International Conference on the Mathematics of Program Construction, Oxford.* 1992.

A Derivation of Huffman's Algorithm

Rob R. Hoogerwoord

Eindhoven University of Technology, department of Mathematics and Computing Science,
PO Box 513, 5600 MB Eindhoven, The Netherlands

Abstract. We present a semi-formal derivation of Huffman's well-known algorithm for the construction of an *optimal encoding tree.*

0 Specification

In this paper we study *(binary) trees*, the leaves of which are labelled with *real numbers*. In the application area of source coding, these numbers constitute a probability distribution, but that is irrelevant here. Throughout this paper we use the following type conventions for our variables:

b, c	:	real number
B, B'	:	bag of real number
s, s', t, t'	:	tree
k, l	:	natural number

Trees are defined recursively, as follows:

$$\langle b \rangle \text{ is a (singleton) tree , and}$$
$$\langle s, t \rangle \text{ is a (composite) tree .}$$

Trees composed of 2 singletons we call *pairs*; for pairs we use the abbreviation $\langle b, c \rangle$, so we have:

$$\langle b, c \rangle = \langle \langle b \rangle , \langle c \rangle \rangle$$

With each tree s is associated a (real) number $w \cdot s$; informally, function w is defined as follows, where dummy x identifies the leaves of the tree:

$$w \cdot s = (\Sigma x : \text{“}x \text{ is a leaf of } s\text{”} : d \cdot x * v \cdot x) \ ,$$

where functions d and v are defined informally by:

$$d \cdot x = \text{“the distance of } x \text{ to the root of } s\text{”}$$
$$v \cdot x = \text{“the number associated with } x\text{”}$$

Each tree s also represents a bag $bag \cdot s$, defined by —where $\{b\}$ denotes the singleton bag containing b and where $+$ denotes bag summation—:

$$bag \cdot \langle b \rangle = \{b\}$$
$$bag \cdot \langle s, t \rangle = bag \cdot s + bag \cdot t$$

The remainder of this paper is devoted to a derivation of an algorithm for the computation of a solution to the following

problem 0: for given finite and nonempty bag B of real numbers, solve

$$s : bag{\cdot}s = B \ \wedge \ (\forall t : bag{\cdot}t = B : w{\cdot}s \leq w{\cdot}t)$$

□

We say that "s is optimal for B" for any solution s to this problem.

1 Solution

The key to the design of the algorithm is provided by the following observations, which lead to a recursive algorithm, based on induction on the number of elements of the bag.

For a singleton bag $\{b\}$, exactly one tree s exists for which $bag{\cdot}s = \{b\}$, namely $\langle b \rangle$; because this tree is unique, it is optimal.

For a composite —that is: non-singleton— bag all trees representing it are composite as well. For composite tree s containing a pair $\langle b, c\rangle$ as a subtree we define s' to be a tree obtained from s by replacement of one occurrence of $\langle b, c\rangle$ by the singleton tree $\langle b{+}c \rangle$. We call the act of constructing s' from s in this way ***pruning***, whereas we call the reverse operation ***grafting***. Trees s and s' thus related have the following properties:

$$\begin{aligned} bag{\cdot}s' &= bag{\cdot}s - \{b, c\} + \{b{+}c\} \\ w{\cdot}s' &= w{\cdot}s - b - c \ , \end{aligned}$$

where the latter equality follows from the following little calculation, in which $l{+}1$ denotes the distance to the root (in s) of b and c; the first formula in this calculation represents the contribution of $\langle b, c\rangle$ to $w{\cdot}s$, whereas $l*(b{+}c)$ is the contribution of $\langle b{+}c\rangle$ to $w{\cdot}s'$:

$$\begin{aligned} & (l{+}1)*b + (l{+}1)*c \\ = \quad & \{ \text{ algebra } \} \\ & l*(b{+}c) + b + c \ . \end{aligned}$$

Pruning a composite tree is always possible, as is shown by the following lemma, which can be proved by straightforward structural induction.

lemma 1: every composite tree contains a pair as subtree

□

Now we show how an optimal tree for a composite bag B can be constructed, by means of grafting, from an optimal tree for a smaller bag B'; to specify the numbers b, c to which the grafting operation pertains, we need auxiliary variables s and s'. We assume that:

s is optimal for bag B and s contains $\langle b, c\rangle$,
$B' = B - \{b, c\} + \{b{+}c\}$,
s' is obtained from s by pruning and $bag{\cdot}s' = B'$,
t' is optimal for B', so $bag{\cdot}t' = B'$,
t is obtained from t' by grafting and $bag{\cdot}t = B$;

and we calculate as follows:

$$
\begin{array}{ll}
 & w{\cdot}t \\
= & \{\ t \text{ is obtained from } t' \text{ by grafting }\} \\
 & w{\cdot}t' + b + c \\
\leq & \{\ t' \text{ is optimal and } bag{\cdot}t' = bag{\cdot}s'\ \} \\
 & w{\cdot}s' + b + c \\
= & \{\ s' \text{ is obtained from } s \text{ by pruning }\} \\
 & w{\cdot}s \\
\leq & \{\ s \text{ is optimal and } bag{\cdot}s = bag{\cdot}t\ \} \\
 & w{\cdot}t\ \ .
\end{array}
$$

Hence, $w{\cdot}t = w{\cdot}s$ which implies that t is optimal as well.

So, an optimal tree for composite bag B can be constructed as follows. Select in B elements b, c and obtain the (smaller) bag B' by replacing b, c by $b{+}c$. Construct an optimal tree for B'; grafting this tree then yields an optimal tree for B. The only difficulty here is that b and c may not be chosen arbitrarily: we have assumed that an optimal tree for B exists in which pair $\langle b, c\rangle$ occurs as subtree. To show that this assumption is realistic we need a constructive solution to our remaining

problem 2: for composite bag B, solve
$b, c\ :\ (\exists s :: \text{“}s \text{ is optimal for } B\text{”}\ \wedge\ \text{“}s \text{ contains } \langle b, c\rangle\text{”})$

□

To solve this problem we need the following stronger version of lemma 1; again, it can be proved by structural induction.

lemma 3: every composite tree contains a pair as subtree, whose two leaves have *maximal* (and equal) distances to the root.

□

The relevance of this lemma follows from the observation that a tree might contain only one pair; in that case the leaves of this one and only pair are at maximal distance from the root. A corollary of this lemma is that each composite tree has at least two leaves at maximal distance.

We now investigate what we can conclude about the numbers in leaves at maximal distance. We have:

$$
\begin{array}{ll}
 & b*k + c*l\ <\ c*k + b*l \\
\equiv & \{\ \text{algebra}\ \} \\
 & 0\ <\ (c-b)*(k-l) \\
\Leftarrow & \{\ \text{algebra}\ \}
\end{array}
$$

$b<c \wedge l<k$.

¿From this derivation it follows that a tree containing a larger number at a larger distance and a smaller number at a smaller distance is not optimal. Hence, in every optimal tree the smallest two numbers have maximal distances to the root. By lemma 3, the maximal distance occurs at least twice. Hence, in every optimal tree the smallest two numbers have ***maximal and equal*** distances to the root. Because swapping numbers at equal distances does not change the w-value of a tree and because, by lemma 3, every tree contains a pair at maximal distance, we conclude that an optimal tree exists in which the smallest two numbers occur in a pair.

So, a solution to problem 2 is the pair of the smallest two numbers in B. This concludes our derivation of the algorithm.

acknowledgement

To the Eindhoven Tuesday Afternoon Club, fŏr some useful comments on the presentation.

Lecture Notes in Computer Science

For information about Vols. 1–595
please contact your bookseller or Springer-Verlag

Vol. 596: L.-H. Eriksson, L. Hallnäs, P. Schroeder-Heister (Eds.), Extensions of Logic Programming. Proceedings, 1991. VII, 369 pages. 1992. (Subseries LNAI).

Vol. 597: H. W. Guesgen, J. Hertzberg, A Perspective of Constraint-Based Reasoning. VIII, 123 pages. 1992. (Subseries LNAI).

Vol. 598: S. Brookes, M. Main, A. Melton, M. Mislove, D. Schmidt (Eds.), Mathematical Foundations of Programming Semantics. Proceedings, 1991. VIII, 506 pages. 1992.

Vol. 599: Th. Wetter, K.-D. Althoff, J. Boose, B. R. Gaines, M. Linster, F. Schmalhofer (Eds.), Current Developments in Knowledge Acquisition - EKAW '92. Proceedings. XIII, 444 pages. 1992. (Subseries LNAI).

Vol. 600: J. W. de Bakker, C. Huizing, W. P. de Roever, G. Rozenberg (Eds.), Real-Time: Theory in Practice. Proceedings, 1991. VIII, 723 pages. 1992.

Vol. 601: D. Dolev, Z. Galil, M. Rodeh (Eds.), Theory of Computing and Systems. Proceedings, 1992. VIII, 220 pages. 1992.

Vol. 602: I. Tomek (Ed.), Computer Assisted Learning. Proceedings, 1992. X, 615 pages. 1992.

Vol. 603: J. van Katwijk (Ed.), Ada: Moving Towards 2000. Proceedings, 1992. VIII, 324 pages. 1992.

Vol. 604: F. Belli, F.-J. Radermacher (Eds.), Industrial and Engineering Applications of Artificial Intelligence and Expert Systems. Proceedings, 1992. XV, 702 pages. 1992. (Subseries LNAI).

Vol. 605: D. Etiemble, J.-C. Syre (Eds.), PARLE '92. Parallel Architectures and Languages Europe. Proceedings, 1992. XVII, 984 pages. 1992.

Vol. 606: D. E. Knuth, Axioms and Hulls. IX, 109 pages. 1992.

Vol. 607: D. Kapur (Ed.), Automated Deduction – CADE-11. Proceedings, 1992. XV, 793 pages. 1992. (Subseries LNAI).

Vol. 608: C. Frasson, G. Gauthier, G. I. McCalla (Eds.), Intelligent Tutoring Systems. Proceedings, 1992. XIV, 686 pages. 1992.

Vol. 609: G. Rozenberg (Ed.), Advances in Petri Nets 1992. VIII, 472 pages. 1992.

Vol. 610: F. von Martial, Coordinating Plans of Autonomous Agents. XII, 246 pages. 1992. (Subseries LNAI).

Vol. 611: M. P. Papazoglou, J. Zeleznikow (Eds.), The Next Generation of Information Systems: From Data to Knowledge. VIII, 310 pages. 1992. (Subseries LNAI).

Vol. 612: M. Tokoro, O. Nierstrasz, P. Wegner (Eds.), Object-Based Concurrent Computing. Proceedings, 1991. X, 265 pages. 1992.

Vol. 613: J. P. Myers, Jr., M. J. O'Donnell (Eds.), Constructivity in Computer Science. Proceedings, 1991. X, 247 pages. 1992.

Vol. 614: R. G. Herrtwich (Ed.), Network and Operating System Support for Digital Audio and Video. Proceedings, 1991. XII, 403 pages. 1992.

Vol. 615: O. Lehrmann Madsen (Ed.), ECOOP '92. European Conference on Object Oriented Programming. Proceedings. X, 426 pages. 1992.

Vol. 616: K. Jensen (Ed.), Application and Theory of Petri Nets 1992. Proceedings, 1992. VIII, 398 pages. 1992.

Vol. 617: V. Mařík, O. Štěpánková, R. Trappl (Eds.), Advanced Topics in Artificial Intelligence. Proceedings, 1992. IX, 484 pages. 1992. (Subseries LNAI).

Vol. 618: P. M. D. Gray, R. J. Lucas (Eds.), Advanced Database Systems. Proceedings, 1992. X, 260 pages. 1992.

Vol. 619: D. Pearce, H. Wansing (Eds.), Nonclassical Logics and Information Proceedings. Proceedings, 1990. VII, 171 pages. 1992. (Subseries LNAI).

Vol. 620: A. Nerode, M. Taitslin (Eds.), Logical Foundations of Computer Science – Tver '92. Proceedings. IX, 514 pages. 1992.

Vol. 621: O. Nurmi, E. Ukkonen (Eds.), Algorithm Theory – SWAT '92. Proceedings. VIII, 434 pages. 1992.

Vol. 622: F. Schmalhofer, G. Strube, Th. Wetter (Eds.), Contemporary Knowledge Engineering and Cognition. Proceedings, 1991. XII, 258 pages. 1992. (Subseries LNAI).

Vol. 623: W. Kuich (Ed.), Automata, Languages and Programming. Proceedings, 1992. XII, 721 pages. 1992.

Vol. 624: A. Voronkov (Ed.), Logic Programming and Automated Reasoning. Proceedings, 1992. XIV, 509 pages. 1992. (Subseries LNAI).

Vol. 625: W. Vogler, Modular Construction and Partial Order Semantics of Petri Nets. IX, 252 pages. 1992.

Vol. 626: E. Börger, G. Jäger, H. Kleine Büning, M. M. Richter (Eds.), Computer Science Logic. Proceedings, 1991. VIII, 428 pages. 1992.

Vol. 628: G. Vosselman, Relational Matching. IX, 190 pages. 1992.

Vol. 629: I. M. Havel, V. Koubek (Eds.), Mathematical Foundations of Computer Science 1992. Proceedings. IX, 521 pages. 1992.

Vol. 630: W. R. Cleaveland (Ed.), CONCUR '92. Proceedings. X, 580 pages. 1992.

Vol. 631: M. Bruynooghe, M. Wirsing (Eds.), Programming Language Implementation and Logic Programming. Proceedings, 1992. XI, 492 pages. 1992.

Vol. 632: H. Kirchner, G. Levi (Eds.), Algebraic and Logic Programming. Proceedings, 1992. IX, 457 pages. 1992.

Vol. 633: D. Pearce, G. Wagner (Eds.), Logics in AI. Proceedings. VIII, 410 pages. 1992. (Subseries LNAI).

Vol. 634: L. Bougé, M. Cosnard, Y. Robert, D. Trystram (Eds.), Parallel Processing: CONPAR 92 – VAPP V. Proceedings. XVII, 853 pages. 1992.

Vol. 635: J. C. Derniame (Ed.), Software Process Technology. Proceedings, 1992. VIII, 253 pages. 1992.

Vol. 636: G. Comyn, N. E. Fuchs, M. J. Ratcliffe (Eds.), Logic Programming in Action. Proceedings, 1992. X, 324 pages. 1992. (Subseries LNAI).

Vol. 637: Y. Bekkers, J. Cohen (Eds.), Memory Management. Proceedings, 1992. XI, 525 pages. 1992.

Vol. 639: A. U. Frank, I. Campari, U. Formentini (Eds.), Theories and Methods of Spatio-Temporal Reasoning in Geographic Space. Proceedings, 1992. XI, 431 pages. 1992.

Vol. 640: C. Sledge (Ed.), Software Engineering Education. Proceedings, 1992. X, 451 pages. 1992.

Vol. 641: U. Kastens, P. Pfahler (Eds.), Compiler Construction. Proceedings, 1992. VIII, 320 pages. 1992.

Vol. 642: K. P. Jantke (Ed.), Analogical and Inductive Inference. Proceedings, 1992. VIII, 319 pages. 1992. (Subseries LNAI).

Vol. 643: A. Habel, Hyperedge Replacement: Grammars and Languages. X, 214 pages. 1992.

Vol. 644: A. Apostolico, M. Crochemore, Z. Galil, U. Manber (Eds.), Combinatorial Pattern Matching. Proceedings, 1992. X, 287 pages. 1992.

Vol. 645: G. Pernul, A M. Tjoa (Eds.), Entity-Relationship Approach – ER '92. Proceedings, 1992. XI, 439 pages, 1992.

Vol. 646: J. Biskup, R. Hull (Eds.), Database Theory – ICDT '92. Proceedings, 1992. IX, 449 pages. 1992.

Vol. 647: A. Segall, S. Zaks (Eds.), Distributed Algorithms. X, 380 pages. 1992.

Vol. 648: Y. Deswarte, G. Eizenberg, J.-J. Quisquater (Eds.), Computer Security – ESORICS 92. Proceedings. XI, 451 pages. 1992.

Vol. 649: A. Pettorossi (Ed.), Meta-Programming in Logic. Proceedings, 1992. XII, 535 pages. 1992.

Vol. 650: T. Ibaraki, Y. Inagaki, K. Iwama, T. Nishizeki, M. Yamashita (Eds.), Algorithms and Computation. Proceedings, 1992. XI, 510 pages. 1992.

Vol. 651: R. Koymans, Specifying Message Passing and Time-Critical Systems with Temporal Logic. IX, 164 pages. 1992.

Vol. 652: R. Shyamasundar (Ed.), Foundations of Software Technology and Theoretical Computer Science. Proceedings, 1992. XIII, 405 pages. 1992.

Vol. 653: A. Bensoussan, J.-P. Verjus (Eds.), Future Tendencies in Computer Science, Control and Applied Mathematics. Proceedings, 1992. XV, 371 pages. 1992.

Vol. 654: A. Nakamura, M. Nivat, A. Saoudi, P. S. P. Wang, K. Inoue (Eds.), Prallel Image Analysis. Proceedings, 1992. VIII, 312 pages. 1992.

Vol. 655: M. Bidoit, C. Choppy (Eds.), Recent Trends in Data Type Specification. X, 344 pages. 1993.

Vol. 656: M. Rusinowitch, J. L. Rémy (Eds.), Conditional Term Rewriting Systems. Proceedings, 1992. XI, 501 pages. 1993.

Vol. 657: E. W. Mayr (Ed.), Graph-Theoretic Concepts in Computer Science. Proceedings, 1992. VIII, 350 pages. 1993.

Vol. 658: R. A. Rueppel (Ed.), Advances in Cryptology – EUROCRYPT '92. Proceedings, 1992. X, 493 pages. 1993.

Vol. 659: G. Brewka, K. P. Jantke, P. H. Schmitt (Eds.), Nonmonotonic and Inductive Logic. Proceedings, 1991. VIII, 332 pages. 1993. (Subseries LNAI).

Vol. 660: E. Lamma, P. Mello (Eds.), Extensions of Logic Programming. Proceedings, 1992. VIII, 417 pages. 1993. (Subseries LNAI).

Vol. 661: S. J. Hanson, W. Remmele, R. L. Rivest (Eds.), Machine Learning: From Theory to Applications. VIII, 271 pages. 1993.

Vol. 662: M. Nitzberg, D. Mumford, T. Shiota, Filtering, Segmentation and Depth. VIII, 143 pages. 1993.

Vol. 663: G. v. Bochmann, D. K. Probst (Eds.), Computer Aided Verification. Proceedings, 1992. IX, 422 pages. 1993.

Vol. 664: M. Bezem, J. F. Groote (Eds.), Typed Lambda Calculi and Applications. Proceedings, 1993. VIII, 433 pages. 1993.

Vol. 665: P. Enjalbert, A. Finkel, K. W. Wagner (Eds.), STACS 93. Proceedings, 1993. XIV, 724 pages. 1993.

Vol. 666: J. W. de Bakker, W.-P. de Roever, G. Rozenberg (Eds.), Semantics: Foundations and Applications. Proceedings, 1992. VIII, 659 pages. 1993.

Vol. 667: P. B. Brazdil (Ed.), Machine Learning: ECML – 93. Proceedings, 1993. XII, 471 pages. 1993. (Subseries LNAI).

Vol. 668: M.-C. Gaudel, J.-P. Jouannaud (Eds.), TAPSOFT '93: Theory and Practice of Software Development. Proceedings, 1993. XII, 762 pages. 1993.

Vol. 669: R. S. Bird, C. C. Morgan, J. C. P. Woodcock (Eds.), Mathematics of Program Construction. Proceedings, 1992. VIII, 378 pages. 1993.

Vol. 670: J. C. P. Woodcock, P. G. Larsen (Eds.), FME '93: Industrial-Strength Formal Methods. Proceedings, 1993. XI, 689 pages. 1993.

Vol. 671: H. J. Ohlbach (Ed.), GWAI-92: Advances in Artificial Intelligence. Proceedings, 1992. XI, 397 pages. 1993. (Subseries LNAI).

Vol. 672: A. Barak, S. Guday, R. G. Wheeler, The MOSIX Distributed Operating System. X, 221 pages. 1993.

Vol. 673: G. Cohen, T. Mora, O. Moreno (Eds.), AAECC-10: Applied, Algebra, Algebraic Algorithms and Error-Correcting Codes. Proceedings, 1993. X, 355 pages 1993.